Selected Titles in This Series

MW00845583

(Continued in the back of this publication)

Continuous Cohomology, Discrete Subgroups, and Representations of Reductive Groups

Second Edition

Mathematical
Surveys
and
Monographs

Volume 67

Continuous Cohomology, Discrete Subgroups, and Representations of Reductive Groups

Second Edition

**A. Borel
N. Wallach**

American Mathematical Society
Providence, Rhode Island

1991 *Mathematics Subject Classification.* Primary 22E41;
Secondary 22E40, 22E45, 57T15.

ABSTRACT. This is a revised and enlarged edition of the book with the same title published by the Princeton University Press in 1980 which was concerned with various types of cohomology theories pertaining to Lie groups (real or p-adic), Lie algebras, infinite dimensional representations, and to cocompact discrete subgroups of reductive groups. Apart from corrections and minor changes or amplifications, the text of the original edition has been kept. It has been augmented notably by various additions on the Zuckerman functors, the Vogan-Zuckerman theorem describing the relative Lie algebra cohomology with coefficients in an irreducible unitary representation, and sharp vanishing theorems. Furthermore, an additional chapter outlines (without proofs) how the main results on the cohomology of discrete cocompact subgroups extend to general S-arithmetic subgroups of semisimple groups over number fields. This edition can be used as a reference for research mathematicians and advanced graduate students in such diverse fields as representation theory, arithmetic groups, automorphic forms, and algebraic number theory.

Library of Congress Cataloging-in-Publication Data

Borel, Armand.
   Continuous cohomology, discrete subgroups, and representations of reductive groups / by A. Borel and N. Wallach. — 2nd ed.
      p. cm. — (Mathematical surveys and monographs, ISSN 0076-5376 ; v. 67)
   Includes bibliographical references and index.
   ISBN 0-8218-0851-6 (alk paper)
   1. Lie groups. 2. Representations of groups. 3. Homology theory. I. Wallach, Nolan R. II. Title. III. Series: Mathematical surveys and monographs; no. 67.
QA387.B64   1999
512′.55—dc21
                                                   98-44527
                                                    CIP

Contents

Introduction to the First Edition

1. This monograph is mainly concerned with two types of cohomology spaces pertaining to a reductive Lie group G (real, p-adic, or product of such groups) and a discrete cocompact subgroup Γ of G. The first one is the Eilenberg-MacLane cohomology space $H^*(\Gamma; E)$ of Γ with coefficients in a finite dimensional unitary Γ-module (or a finite dimensional G-module if G is real). The second one is attached to G, or its Lie algebra $\mathfrak{g}$ and a maximal compact subgroup K if G is real, and a representation V of G, usually infinite dimensional, and appears in various guises: continuous, smooth, or also (for G real) relative Lie algebra cohomology. Our initial interest was in the former one. However, its study may be reduced in part to the latter one (see Chapters VII and XIII), where G is the ambient group and V runs through the irreducible subspaces of $L^2(\Gamma \backslash G)$. The determination of this cohomology is then a first step towards the determination of $H^*(\Gamma; E)$. But, as this work developed, we were led to emphasize it more and more, and to treat it as our main topic rather than as an auxiliary one. In fact, ten out of thirteen chapters are devoted to it, or directly motivated by it.

The material presented here divides naturally into two parts, one devoted mainly to real Lie groups (Chapters I to IX), the other to locally compact totally disconnected groups (for short, t.d. groups), in particular reductive p-adic groups, or products of real Lie groups and t.d. groups (Chapters X to XIII). Each part in turn contains roughly three main items: general results on the cohomology used, specific ones for cohomology and representations of reductive groups, and applications to discrete cocompact subgroups.

We now give some indications on the contents of the various chapters.

2. In Chapters I to VIII, G is a real Lie group with finitely many connected components, and the underlying cohomology is the relative Lie algebra cohomology $H^*(\mathfrak{g}, \mathfrak{k}; V)$ or rather, to allow for non-connected G's, a slight modification of it denoted $H^*(\mathfrak{g}, K; V)$. Chapter I is devoted to foundational material on that cohomology. In §§1 to 4, $\mathfrak{g}$ is a finite dimensional Lie algebra over a field of characteristic zero and $\mathfrak{k}$ a subalgebra. §1 recalls the direct definition of $H^*(\mathfrak{g}, \mathfrak{k}; V)$, §2 discusses more generally the derived functors of $\mathrm{Hom}_{\mathfrak{g}}$ in the category $\mathcal{C}_{\mathfrak{g},\mathfrak{k}}$ of $(\mathfrak{g}, \mathfrak{k})$-modules, i.e., $\mathfrak{g}$-modules which are locally finite and semi-simple with respect to $\mathfrak{k}$. This approach differs only in minor details from that of G. Hochschild, in the framework of relative homological algebra. The translation in the formalism of Yoneda's long extensions is briefly recalled in §3. In §4, we give two proofs of a useful vanishing theorem of D. Wigner. From §5 on, $F = \mathbf{R}$, $\mathfrak{g}$ is the Lie algebra of G and $\mathfrak{k}$ that of a maximal compact subgroup K of G. In §5, we transpose the previous considerations to the category of $(\mathfrak{g}, K)$-modules. In §6 we introduce a slightly different category $\mathcal{C}_{\mathfrak{g},\mathfrak{k},L}$, solely as a tool to prove the existence of a Hochschild-Serre spectral sequence for $(\mathfrak{g}, K)$-modules. Also included are two results of Casselman (5.5)

and of D. Vogan (2.8) on finitely generated or admissible modules, and a Poincaré duality theorem of D. Vogan when G is semi-simple and V irreducible admissible (§7).

Chapter II is devoted to the case where $\mathfrak{g}$ is semi-simple (or reductive) and the coefficient module is the tensor product of a finite dimensional G-module E by a unitary G-module V. The cochain complex for relative Lie algebra cohomology admits then a natural scalar product. Various constructions and results of Matsushima, Matsushima-Murakami, Kuga, originating in differential geometry and Hodge theory and discussed by them in the context of discrete cocompact subgroups, are adapted to our setting in §§1 to 4, and §8; in a similar vein, §§6, 7 prove some vanishing theorems by use of spinors, suggested by results of Hotta and Parthasarathy on discrete subgroups. In §5, we consider the case where V belongs to the discrete series and show, using the characterization of the minimal K-type in V, that $H^q(\mathfrak{g}, K; E \otimes V)$ vanishes unless $2q = \dim G/K$ and V has the same infinitesimal character as the contragredient representation E^* to E.

The main topic of Chapter III is the cohomology with respect to a principal series representation. The computation uses an analogue of Shapiro's lemma (2.5), a description of K-finite vectors in induced representations (2.4), results of B. Kostant on the cohomology of nilpotent radicals of parabolic subalgebras and the Hochschild-Serre spectral sequence (§3). The results are applied in §4 to the determination of the cohomology with respect to tempered representations: in particular, it can be non-zero only in a small interval around the middle dimension and if the underlying parabolic subgroup is fundamental. These results have also been proved independently by G. Zuckerman, and those of §3 for complex semi-simple Lie algebras by P. Delorme. The last paragraph of III contains some general remarks on C^∞-vectors of induced representations, proving in particular that these are smooth functions in the cases of interest to us.

The next step is the investigation of the cohomology with respect to non-tempered representations. It is based on the Langlands classification of irreducible admissible $(\mathfrak{g}, K)$-modules and on two complements to it: some information on the Langlands parameters of the constituents of the kernel of the intertwining operators used by Langlands, and a necessary condition for unitarizability (in fact, for uniform boundedness) in terms of the Langlands parameters. The latter sharpens a result of R. Howe stating that the coefficients of a unitary representation with compact kernel vanish at infinity. These results are proved in Chapter IV (see 4.13, 5.2), which also contains a proof of the Langlands classification (4.11).

The uniform boundedness condition singles out a subset denoted $\Pi_\infty(G)$ of the set $\Pi(G)$ of infinitesimal equivalence classes of irreducible admissible $(\mathfrak{g}, K)$-modules (V, §2). It contains the unitary representations with compact kernel. Chapters V and VI are devoted to the cohomology with coefficients in $\Pi_\infty(G)$, or also in $V \otimes E$, where V represents an element of $\Pi_\infty(G)$ and E is finite dimensional, irreducible. We prove first that $H^q(\mathfrak{g}, K; V \otimes E)$ vanishes for $q < \mathrm{rk}_{\mathbf{R}} G$ (3.3), a result also obtained independently by G. Zuckerman. For E trivial, this bound is sharp in $\Pi_\infty(G)$ (but not always in the unitary dual $\widehat{G}$ of G, see (II, 8.7)): in §4, it is shown that the constituents of (an analogue of) the Steinberg representation are all in $\Pi_\infty(G)$, and that $H^q(\mathfrak{g}, K; V) \neq 0$ if $q = \mathrm{rk}_{\mathbf{R}} G$ for at least one of them. §5 reproves some results of P. Delorme on the relation between H^1 and the topology of $\widehat{G}$.

Chapter VI gives some further information on the cohomology with respect to a Langlands quotient $J_{P,\sigma,\nu}$. We need only consider the $J_{P,\sigma,\nu}$ with the same infinitesimal character as the trivial representation. The criterion IV, 5.2 gives an upper bound for ν. The general pattern which emerges is that, roughly, the bigger ν (in a suitable order relation), the lower the first non-vanishing cohomology group. Since the cohomology with respect to tempered representations is non-zero only close to the middle dimension, this suggests proceeding by increasing induction on ν. Without attempting to do this in general, we illustrate this relationship in Chapter VI by some general results when ν is minimal (§§1, 2) or $\mathrm{rk}_{\mathbf{R}}\, G = 1$ (§3), and by a complete determination of the cohomology when $G = \mathbf{SO}(n,1)$, $\mathbf{SU}(n,1)$ in §4.

Chapter VII is devoted to the cohomology of discrete subgroups. First if Γ is a discrete subgroup of the Lie group G, and E is a G-module, then we have the (well-known) formula

$$
(1) \qquad H^*(\Gamma; E) = H^*(\mathfrak{g}, K; C^\infty(\Gamma\backslash G) \otimes E)
$$

(2.7). If now Γ is cocompact, then $L^2(\Gamma\backslash G)$ admits a Hilbert discrete sum decomposition with finite multiplicities

$$
(2) \qquad L^2(\Gamma\backslash G) = \bigoplus_{\pi \in \widehat{G}} m(\pi, \Gamma) H_\pi,
$$

and (1) transforms to

$$
(3) \qquad H^*(\Gamma; E) = \bigoplus_{\pi \in \widehat{G}} m(\pi, \Gamma) H^*(\mathfrak{g}, K; H_\pi \otimes E)
$$

(5.2). There is also a counterpart to that formula when E is a unitary Γ-module, involving the decomposition of the unitarily induced representation $I^G_{\Gamma,2}(E)$ (3.2). Various consequences of the results of the previous chapters are drawn in §§4, 6.

Chapter VIII is concerned with cohomology at the $\mathbf{R}$-rank q when $G = \mathbf{SU}(p,q)$ ($p \geq q$). Let F_ℓ be the irreducible G-module whose highest weight is ℓ times the highest weight of the standard representation of $\mathbf{SU}(p,q)$ in $\mathbf{C}^{p+q}$. For each $\ell \geq q$ there is a unitary irreducible representation H_ℓ of G such that $H^q(\mathfrak{g}, K; H_\ell \otimes F_{\ell-q}) \neq 0$ (2.13). It is then shown that certain cocompact arithmetically defined subgroups of G have subgroups of finite index Γ' such that H_ℓ occurs in $L^2(\Gamma\backslash G)$, whence in particular $H^q(\Gamma'; F_{\ell-q}) \neq 0$. This extends a result of Kazhdan concerning the case where $q = 1$, which gave the first examples of discrete cocompact subgroups of $\mathbf{SU}(n,1)$ with non-vanishing first Betti number for arbitrary n. The proof uses the metaplectic representation and the duality theorem, and is quite similar to that of Kazhdan, although the context is a bit different, since Kazhdan worked with adelic groups.

3. Chapters IX to XII are devoted to continuous and smooth cohomology. §§1 to 4 of Chapter IX contain some basic material concerning derived functors in the category $\mathcal{C}_G$ of continuous G-modules (always assumed to be locally convex Hausdorff topological vector spaces over $\mathbf{C}$), when G is a locally compact group (countable at infinity). The approach is the one of Hochschild-Mostow, based on

the use of injective modules relative to G-morphisms which are strong (i.e. split for the underlying structure of topological vector spaces). After that, we are concerned with real groups (IX, §§5, 6), t.d. (totally disconnected) groups, in particular p-adic groups (X, XI), and products of such groups (XII). The formal analogies between these three cases are emphasized. In each, besides $\mathcal{C}_G$, we consider the categories $\mathcal{C}_G^\infty$ of smooth topological G-modules and $\mathcal{C}_G^f$ of non-degenerate modules over a suitable Hecke algebra. The last one (introduced in substance by Jacquet-Langlands) is abelian and the modules in it are just complex vector spaces. The Hecke algebras occurring here have no unit in general, a situation not considered in standard texts on homological algebra. However they are idempotented, and this allows one to extend some standard constructions to our case (XII, §0). In particular, $\mathcal{C}_G^f$ has enough injectives. There are natural functors

$$\mathcal{C}_G \underset{\alpha}{\overset{\gamma}{\longleftrightarrow}} \mathcal{C}_G^\infty \xrightarrow{\ \beta\ } \mathcal{C}_G^f,$$

where α (resp. β) is the passage to smooth (resp. K-finite vectors) and γ is the inclusion. γ preserves derived functors and β cohomology for quasi-complete spaces, α preserves derived functors for quasi-complete spaces in the t.d. case, and cohomology for Fréchet spaces in the other two cases.

In the real case, $\mathcal{C}_G^\infty$ consists of the usual differentiable modules, with the C^∞-topology, while, up to Chapter IX, $\mathcal{C}_G^f$ is just the category of $(\mathfrak{g}, K)$-modules. But, as is known, it may also be viewed as the category of non-degenerate modules over the Hecke algebras $\mathcal{H}(\mathfrak{g}, K)$ of bi-K-finite distributions on G with support in K. This point of view is more convenient to treat the mixed case, and is introduced later (XII, §2). The above conservation theorems for derived functors in the real case (due to Hochschild-Mostow, W. v. Est, P. Blanc) are proved in IX, §§5, 6.

If G is a t.d. group (X, §1), then a topological G-module V is smooth if every $v \in V$ is fixed under an open subgroup and V is, topologically, the inductive limit of the subspaces V^{L^*} of fixed points under compact open subgroups L of G. The Hecke algebra underlying the definition of $\mathcal{C}_G^f$ is the convolution algebra of locally constant compactly supported functions. The main case of interest is when $G = \mathcal{G}(k)$, where k is a non-archimedean local field and $\mathcal{G}$ a connected reductive k-group. If $V \in \mathcal{C}_G$, then the V-valued cochains of the Bruhat-Tits building of G provide an s-injective resolution of V (X, §2). In §4 of X we prove the results of W. Casselman which give a complete description of the cohomology of G with respect to an irreducible admissible G-module. §5 is devoted to $\mathcal{C}_G^f$, and the passage to $\mathcal{C}_G^f$ is used in §6 to prove some Künneth rules.

Chapter XI is a p-adic counterpart of IV. It discusses the analogue of the Langlands classification, and of the uniform boundedness condition. The latter is used to show that the only irreducible admissible representations with compact kernel, with respect to which G has non-vanishing cohomology in some dimension $q \neq 0$, $\mathrm{rk}_k \mathcal{G}$, are non-unitarizable (a result due to W. Casselman).

Let now $G = G_1 \times G_2$ be the product of a real Lie group G_1 and a t.d. group G_2. A topological continuous G-module V is said to be smooth if it is smooth with respect to G_1 and G_2 and if it is the topological inductive limit of the subspaces V^L, where L runs through the compact open subgroups of G_2.

There are also intermediate categories of continuous G-modules smooth with respect to one of the factors. The relations between the corresponding derived functors are discussed in §1. In §2, we fix a maximal compact subgroup K_1 of G_1 and pass to the $(K_1 \times L)$-finite vectors, where L is a compact open subgroup of G_2, which brings us to the non-degenerate modules over the Hecke algebra $\mathcal{H}(G) = \mathcal{H}(\mathfrak{g}_1, K_1) \otimes \mathcal{H}(G_2)$. §3 is devoted to some Künneth rules and to applications to the cohomology of products of reductive groups or of adelic groups.

In Chapter XIII, we consider the cohomology space $H^*(\Gamma; E)$, where Γ is a discrete cocompact subgroup of G and E a finite dimensional unitary Γ-module, first in general (§1), then when G is a product of reductive groups G_s $(s \in S)$. In the latter case, we have a formula quite similar to (3), except that $L^2(\Gamma \backslash G)$ is replaced by the unitarily induced representation from E. Furthermore, since the G_s's are of type I, each $\pi \in \widehat{G}$ is a Hilbert tensor product $\pi = \widehat{\bigotimes}_s \pi_s$ $(\pi_s \in \widehat{G}_s)$, and the Künneth rule gives

$$(4) \qquad H^*_{ct}(G; H_\pi) = \bigotimes_s H_{ct}(G_s; H_{\pi_s}).$$

This allows us to apply the earlier results on continuous cohomology of real or p-adic groups. We then pass to some applications. We prove the Casselman vanishing theorem (2.6) and extend it to the case where Γ is irreducible (3.1) in a product of semi-simple groups over non-archimedean fields (3.6). Following a suggestion of G. Prasad, we also show it to be valid when E is a finite dimensional vector space over an arbitrary field of characteristic zero, and G has rank ≥ 2, using a theorem of Margulis (3.7). Finally, we prove that if $G = \mathcal{G}(A)$ is the adèle group of a semi-simple anisotropic group $\mathcal{G}$ over a global field, then $H^*(\mathcal{G}(k); \mathbf{R})$ reduces to the continuous cohomology of the archimedean factor of $\mathcal{G}(A)$ (3.9).

A survey of some of the main results on vanishing and non-vanishing cohomology is given at the end of the book.

4. This monograph is an outgrowth of a seminar on the "Cohomology of discrete subgroups of semi-simple Lie groups" held at The Institute for Advanced Study in 1976–77. A first set of notes was written and distributed at that time. Most of the material of these notes is incorporated in Chapters I to IX, except for some results which were rendered somewhat obsolete by others found in the course of the seminar. There was also some discussion of the p-adic case in the seminar, but it was not written up then. In the first version, we kept track of who did what and each chapter was accordingly authored or coauthored. It would have been quite awkward to do so in the present version, which represents a considerable reorganization and expansion of the first one. Rather, we prefer to take joint responsibility for the results and mistakes in this book, except however that the first (resp. second) named author wishes to leave credit for Chapters IV, VIII, XI (resp. VII, IX, XII, XIII) to the second (resp. first) named author.

The transition from the first to the final version was a rather painful process, involving a long series of changes, additions, amplifications, corrections upon corrections, reshuffling and renumbering. We are very grateful to the secretaries of the School of Mathematics, and in particular to Peggy Murray, who had by far the greatest load, for having taken care so skillfully and so speedily of this endless series of changes upon changes, which required expertise not only in typing but in cutting, pasting and collage as well.

A reference such as 3.4 (resp. 3.4(1)) refers to section 3.4 (resp. relation 3.4(1)) of the same chapter; if preceded by a capitalized Roman numeral it refers to the corresponding section or relation of the chapter denoted by that numeral.

A. BOREL, N. WALLACH[*]
July 1978

THE INSTITUTE FOR ADVANCED STUDY, Princeton, N.J. 08540
RUTGERS UNIVERSITY, New Brunswick, N.J. 08903

[*]The second named author did part of this work while enjoying the hospitality of Brandeis University. He also wishes to acknowledge partial support from NSF grant number MCS 77-04278 AO1.

Introduction to the Second Edition

This second edition includes a number of corrections, minor changes or amplifications to the original text, as well as some further material that reports on later relevant developments.

The numbering in the first edition has been maintained. The new additions have been inserted either at the beginning or the end of a paragraph, or a chapter. This explains some numbering that is a bit unusual: In section 3 of Chapter 0, in particular, there is a subsection 3.0 (which has subsections). The main new topics are:

I, §8, which gives a construction, in the framework of this book, of the Zuckerman functors and describes their main properties.

II, §10 provides sharp bounds, case by case, for the vanishing theorems, due to Enright, Kumaresan, Parthasarathy, Vogan-Zuckerman, which in many cases are improvements of the ones given originally.

VI, §0 introduces the translation functors and their relationship with relative Lie algebra cohomology.

VI, §5 is devoted to the Vogan-Zuckerman theorem, which describes $\mathrm{Ext}^*_{\mathfrak{g},K}(F,V)$, where V runs through the irreducible unitary $(\mathfrak{g},K)$-modules and F through the finite dimensional irreducible $(\mathfrak{g},K)$-modules.

XIII, §4 studies the cohomology of an S-arithmetic subgroup of G with coefficients in a rational G-module.

Moreover, a new Chapter XIV has been added. It outlines how the main results proved in Chapters VII, VIII and XIII for the cohomology of discrete cocompact subgroups extend to general S-arithmetic subgroups of semisimple algebraic groups over number fields.

It has been almost 20 years since the publication of the original version of this book. During that time the methods of homological algebra have become increasingly important in the construction of admissible representations and in the study of arithmetic groups. Although some of the original material in this book has been superseded, it is still a useful reference. We thank the American Mathematical Society, in particular S. Gelfand, for having encouraged us to publish this second edition. The authors would also like to thank the editorial staff for an extremely helpful and thorough reading of the manuscript.

A. BOREL, N. WALLACH
1999

THE INSTITUTE FOR ADVANCED STUDY, Princeton, NJ 08540
UNIVERSITY OF CALIFORNIA, San Diego, La Jolla, CA 92014

CHAPTER 0

Notation and Preliminaries

§1 contains some general notation, §2 some definitions and facts on representations of Lie groups, and §3 fixes a number of conventions on reductive groups. The notation introduced here will often be used without reference.

1. Notation

1.1. As usual, $\mathbf{Z}$ is the ring of integers, $\mathbf{N} = \{z \in \mathbf{Z} \mid z \geq 0\}$ the set of natural integers, $\mathbf{Q}$ (resp. $\mathbf{R}$, resp. $\mathbf{C}$) the field of rational (resp. real, resp. complex) numbers, $\mathbf{R}_+^*$ the multiplicative group of strictly positive real numbers.

If A is an algebra with identity, then A^* is the group of units of A.

1.1.1. If $V = \bigoplus_{i \in \mathbf{Z}} V^i$ is a vector space graded by $\mathbf{Z}$ and if $m \in \mathbf{Z}$, then $V[m]$ denotes the graded vector space defined by

$$V[m]^i = V^{i+m} \ (i \in \mathbf{Z}).$$

1.1.2. Let V be a complex vector space. If V has the structure of a module over a group or a Lie algebra and if $m \in \mathbf{N}$, then we have consistently written mV for the direct sum of m copies of V, with the corresponding diagonal action, thus committing an abuse of notation. To adopt a correct one would entail an amount of changes that we found too daunting. We thereby, regretfully, announce that we shall maintain our original convention.

1.2. If G is a group, and M a subset of G, then $\mathcal{Z}_G(M)$ or $\mathcal{Z}(M)$ is the centralizer of M and $\mathcal{N}_G(M)$ or $\mathcal{N}(M)$ the normalizer of M:

$$\mathcal{Z}_G(M) = \{g \in G \mid g \cdot m = m \cdot g \ (m \in M)\},$$
$$\mathcal{N}_G(M) = \{g \in G \mid g \cdot M \cdot g^{-1} \subset M\}.$$

Int g is the inner automorphism $x \mapsto g \cdot x \cdot g^{-1}$. We also write ${}^g x$ for Int $g(x)$, and ${}^g M = \text{Int } g(M)$. The center of G is denoted $\mathcal{Z}(G)$ or $\mathcal{C}(G)$, and $\mathcal{D}G$ is the derived group of G.

1.3. If $g \in G$, then ℓ_g (resp. r_g) denotes the left (resp. right) translation by g on G, or on functions f on G. In particular

(1) $$\ell_g f(x) = f(g^{-1} \cdot x), \qquad r_g f(x) = f(x \cdot g) \qquad (x \in G).$$

Thus $\ell_{g \cdot h} = \ell_g \cdot \ell_h$, $r_{gh} = r_g \cdot r_h$ $(g, h \in G)$.

1.4. If G is a topological group, then G^0 is the connected component of the identity in G.

1.5. The Lie algebra of a real Lie group $G, H, \cdots$ will be denoted by the corresponding German lower case letter $\mathfrak{g}, \mathfrak{h}, \cdots$, and the exponential map $\mathfrak{g} \to G$ is denoted exp. We also write e^x for $\exp x (x \in \mathfrak{g})$. If $\mathfrak{m}$ is a subspace of $\mathfrak{g}$, then $\mathfrak{m}_c$ stands for the complexification $m \otimes_{\mathbf{R}} \mathbf{C}$ of $\mathfrak{m}$.

The universal enveloping algebra over $\mathbf{C}$ of $\mathfrak{g}$ is denoted $U(\mathfrak{g})$. Its center is denoted $Z(\mathfrak{g})$.

The centralizer (resp. normalizer) of $\mathfrak{m}$ in $\mathfrak{g}$ is denoted $\mathfrak{z}(\mathfrak{m})$ or $\mathfrak{z}_{\mathfrak{g}}(\mathfrak{m})$, resp. $\mathfrak{n}(\mathfrak{m})$ or $\mathfrak{n}_{\mathfrak{g}}(\mathfrak{m})$:

$$\mathfrak{z}_{\mathfrak{g}}(\mathfrak{m}) = \{x \in \mathfrak{g}/[\mathfrak{m}, x] = 0\}, \qquad \mathfrak{n}_{\mathfrak{g}}(\mathfrak{m}) = \{x \in \mathfrak{g}/[x, m] \in \mathfrak{m} \; (m \in \mathfrak{M})\}.$$

As usual the differential of $\operatorname{Int} x$ $(x \in G)$ at 1 is denoted $\operatorname{Ad} x$, and, for $x \in \mathfrak{g}$, $\operatorname{ad} x \colon \mathfrak{g} \mapsto \mathfrak{g}$ is defined by $\operatorname{ad} x(y) = [x, y]$. For $\mathfrak{m} \subset \mathfrak{g}$, we let

$$\mathcal{Z}_G(\mathfrak{m}) = \{x \in G \mid \operatorname{Ad} x(m) = m \; (m \in \mathfrak{m})\},$$
$$\mathcal{N}_G(\mathfrak{m}) = \{x \in G \mid \operatorname{Ad} x(\mathfrak{m}) = \mathfrak{m}\}.$$

1.6. If G is a Lie group, then $X(G)$ is the group of continuous homomorphisms of G into $\mathbf{R}^*$ and

$$^0G = \bigcap_{\chi \in X(G)} \ker |\chi|.$$

It is a normal subgroup which contains the derived group and all compact subgroups of G.

1.7. Unless otherwise stated, topological vector spaces are assumed to be over $\mathbf{R}$ or $\mathbf{C}$, Hausdorff locally convex and quasi-complete, and manifolds to be C^∞ and countable at infinity. If M is a manifold and V a topological vector space, then $C^\infty(M; V)$ is the space of C^∞-functions of M, with values in V, endowed with the C^∞-topology. The space of V-valued smooth differential p-forms $(p \in \mathbf{N})$ on M is denoted $A^p(M; V)$, and $A^*(M; V)$ is the direct sum of the spaces $A^p(M; V)$. Thus $A^0(M; V) = C^\infty(M; V)$. If V is a Fréchet space, then so is $A^p(M; V)$ $(p \in \mathbf{N})$.

If M, N are manifolds, then $C^\infty(A, B)$ is the space of smooth maps $A \to B$, endowed with the C^∞-topology.

2. Representations of Lie groups

2.1. Let G be a Lie group with finite component group. By a topological G-module (or simply a G-module) V, where V is assumed to be a locally convex and locally complete Hausdorff topological vector space over $\mathbf{C}$, we mean a homomorphism $G \to \operatorname{Aut} V$ defined by a *continuous* map $G \times V \to V$. It will be denoted (π, V), or V or π. The action of g on v is often denoted $g.v$ or gv rather than $\pi(g)v$. We shall denote by $\mathcal{C}_G$ the category of topological G-modules and equivariant continuous linear maps.

V is said to be finitely generated if there is a finite subset S of V such that the span of the vectors $g.c$ $(g \in G, c \in S)$ is dense in V.

2.2. Let $(\pi, V) \in \mathcal{C}_G$. For $v \in V$ we let $c_v \colon G \to V$ denote the orbit map $c_v(g) = \pi(g)v$. It is continuous. If $\widetilde{v}$ is a continuous functional on V, then the function $c_{v,\widetilde{v}}$ on G defined by

(1) $$c_{v,\widetilde{v}}(g) = \langle \pi(g)v, \widetilde{v} \rangle = \langle c_v(g), \widetilde{v} \rangle \qquad (g \in G)$$

is called a *coefficient* of π.

An elementary calculation shows that we have

$$(2) \qquad\qquad c_{x \cdot v, \tilde{v}} = r_x c_{v, \tilde{v}} \qquad (x \in G).$$

If V is a Hilbert space, then the coefficients may also be defined to be the functions $c_{v,w} \colon g \mapsto (\pi(g)v, w)$, where $v, w \in V$ and $(\ ,\ )$ is the scalar product on V.

2.3. Let $(\pi, V) \in \mathcal{C}_G$. The vector $v \in V$ is said to be differentiable (resp. analytic) if c_v is C^∞ (resp. analytic). The space of differentiable (resp. analytic) vectors is denoted V^∞ (resp. V^ω). It is stable under G. The representation π defines a representation of $\mathfrak{g}$ or $U(\mathfrak{g})$ on V^∞ (resp. V^ω) which is denoted π_∞ (resp. π_ω) or simply π.

A continuous representation (π, V) is differentiable if $V = V^\infty$ and if the map $v \mapsto \{g \mapsto g \cdot v\}$ is a topological isomorphism of V onto its image in $C^\infty(G; V)$, endowed with the topology induced from that of $C^\infty(G; V)$, to be called the C^∞-topology. We let $\mathcal{C}_G^\infty$ be the category of differentiable G-modules and continuous G-morphisms.

Let (π, V) be a continuous G-module. Then V^∞, endowed with the C^∞-topology and the given action of G, is a differentiable G-module. We denote it (π_∞, V^∞). If V is a Fréchet space, then so is V^∞. The map $(\pi, V) \mapsto (\pi_\infty, V^\infty)$ is a functor. If V is a Hilbert space, then, by the principle of uniform boundedness, the topology on V^∞ is defined by the semi-norms $v \mapsto \|Xv\|$ ($X \in U(\mathfrak{g})$).

2.4. A vector $v \in V$ is *G-finite* if it is contained in a finite dimensional subspace stable under G. A G-module is *locally finite* if every element is G-finite.

Let K be a compact subgroup of G. We let $V_{(K)}$ denote the space of K-finite vectors. It is the union of the images $V_{(W)}$ of the maps

$$\mathrm{Hom}_K(W, V) \otimes W \to V$$

defined by

$$\tau \otimes w \mapsto \tau(w) \qquad (\tau \in \mathrm{Hom}_K(W, V), w \in W),$$

where W runs through all finite dimensional K-modules. If W is irreducible, then $V_{(W)}$ is the *isotypic subspace* of type W. We say that V is *admissible* if all isotypic subspaces are finite dimensional (or equivalently all of the $V_{(W)}$ are finite dimensional for all finite dimensional W).

Assume a maximal compact subgroup K of G has been fixed. Then we set $V_0 = V^\infty \cap V_{(K)}$. This space is stable under $\mathfrak{g}$. Note that if an isotypic subspace of K in V is finite dimensional, then it is contained in V^∞. We say that π is *admissible* if the isotypic subspaces in V are all finite dimensional. In this case $V_0 = V_{(K)}$.

2.5. A $(\mathfrak{g}, K)$-module is a real or complex vector space which is a $\mathfrak{g}$-module, a locally finite and semi-simple K-module and such that the operations of $\mathfrak{g}$ and K satisfy the following compatibility conditions:

1) $\pi(k) \cdot (\pi(X)) \cdot v = \pi(\mathrm{Ad}\, k(X)) \cdot \pi(k) \cdot v$ ($k \in K; X \in U(\mathfrak{g}); v \in V$);

2) if F is a K-stable finite dimensional subspace of V, then the representation of K on F is differentiable, and has $\pi|_{\mathfrak{k}}$ as its differential.

A $(\mathfrak{g}, K)$-module is *admissible* if it is admissible as K-module.

Let V be a vector space on which $\mathfrak{g}$ and K operate so as to satisfy 1) and 2) and in which every K-stable finite dimensional subspace is K-semi-simple. Then the subspace $V_{(K)}$ of K-finite vectors in V is K-semi-simple and stable under $\mathfrak{g}$, hence is a $(\mathfrak{g}, K)$-module.

If (π, V) is a $(\mathfrak{g}, K)$-module, then $\mathfrak{g}$ and K operate as usual on the dual space V' of V. The above conditions are met. The space of K-finite vectors in V' is then a $(\mathfrak{g}, K)$-module, to be called the *contragredient* (g, K)-*module to* V, and to be denoted $(\widetilde{\pi}, \widetilde{V})$. It is admissible if and only if V is. In that case, V is contragredient to $\widetilde{V}$.

A $(\mathfrak{g}, V)$-module (π, V) is *unitary* if V is endowed with a positive non-degenerate scalar product $(\ ,\ )$ which is invariant under K and (infinitesimally) invariant under $\mathfrak{g}$:

$$(\pi(k) \cdot v, \pi(k) \cdot w) = (v, w),$$
$$(\pi(x)v, w) + (v, \pi(x) \cdot w) = 0 \quad (v, w \in V, k \in K, x \in \mathfrak{g}).$$

We let $\mathcal{C}_{\mathfrak{g},K}$ be the category of $(\mathfrak{g}, K)$-modules and $(\mathfrak{g}, K)$-morphisms, and $\Pi(G)$ the set of isomorphisms classes of irreducible admissible $(\mathfrak{g}, K)$-modules.

A $(\mathfrak{g}, K)$-module (π, V) (or a differentiable G-module) is said to have an *infinitesimal character* χ if there is a homomorphism $Z(\mathfrak{g}) \to \mathbf{C}$ such that $\pi(z) = \chi(z) \cdot \mathrm{Id}$ for all $z \in Z(\mathfrak{g})$. This is in particular the case if (π, V) is irreducible and admissible.

2.6. Let $(\pi, V) \in \mathcal{C}_G$. Then V_0 is a $(\mathfrak{g}, K)$-module. We denote it sometimes (π_0, V_0). It is admissible (resp. unitary) if (π, V) is so, and it is finitely generated as a $\mathfrak{g}$-module if (π, V) is finitely generated as a G-module.

It is known that every irreducible admissible $(\mathfrak{g}, K)$-module can be realized as the space of K-finite vectors in an irreducible admissible differentiable G-module [**77**]. In fact, this statement is true more generally for finitely generated admissible $(\mathfrak{g}, K)$-modules, but we shall not need this fact.

Two smooth representations are *infinitesimally equivalent* if the two associated $(\mathfrak{g}, K)$-modules of K-finite vectors are isomorphic.

2.7. We let $Z(\mathfrak{g}, K)$ denote the subgroup of elements of the center of K which act trivially on $\mathfrak{g}$. If G is connected, with compact center, then $Z(\mathfrak{g}, K)$ is just the center of G. We say that a $(\mathfrak{g}, K)$-module (π, V) has a central character ω_π if there exists a character $\omega_\pi \colon Z(\mathfrak{g}, K) \to \mathbf{C}^*$ such that $\pi(z) = \omega_\pi(z) \cdot \mathrm{Id}$ for all $z \in Z(\mathfrak{g}, K)$. If (π, V) is admissible and irreducible, then it has both an infinitesimal character and a central character.

2.8. The set of equivalence classes of irreducible unitary representations of G is denoted $\mathcal{E}(G)$ or $\widehat{G}$.

Let (π, V) be unitary, irreducible. There exists then a unitary character ω_π of $\mathcal{C}(G)$ such that $\pi(z) = \omega_\pi(z) \mathrm{Id}$ for $z \in \mathcal{C}(G)$. Therefore $|c_{u,v}|$ $(u, v \in V)$ is a function on $G/\mathcal{C}(G)$. The representation π is said to be in the discrete series if it is unitary, irreducible and if its coefficients are square integrable modulo the center, i.e. on $G/\mathcal{C}(G)$. We let $\mathcal{E}_d(G)$ be the set of equivalence classes of discrete series representations of G.

If G is compact, then $\mathcal{E}(G) = \mathcal{E}_d(G)$.

3. Linear algebraic and reductive groups

3.0. In this book, up to Chapter XII, we are mainly concerned with real or complex Lie groups. The point of view of algebraic groups becomes more prominent in XIII, XIV. Our general reference for linear algebraic groups is [**124**]. We review some basic concepts in characteristic 0.

k is a field of characteristic 0, and K an algebraically closed extension of k.

3.0.1.   A subgroup $\mathcal{G} \subset \mathbf{GL}_n(K)$ is *linear algebraic* if there exist polynomials $P_\alpha \in K[X_{11}, X_{12}, \ldots, X_{nn}]$, $\alpha \in I$, such that

$$\mathcal{G} = \{g = (g_{ij}) \in \mathbf{GL}_n(K) \mid P_\alpha(g_{11}, \ldots, g_{nn}) = 0 \ (\alpha \in I)\}.$$

It is defined over k if the ideal of polynomials vanishing on $\mathcal{G}$ is generated by elements of $k[X_{11}, X_{12}, \ldots, X_{nn}]$. Then we set $\mathcal{G}(k) = \mathcal{G} \cap \mathbf{GL}_n(k)$.

The group $\mathcal{G}$ is connected (in the Zariski topology) if and only if it is irreducible as an algebraic variety. If $K = \mathbf{C}$, $\mathcal{G}$ is also a complex Lie group, and it is connected if and only if it is connected as a manifold. Moreover, if it is defined over $\mathbf{R}$, then $\mathcal{G}(\mathbf{R})$ is a Lie group which may have several (but at most finitely many) connected components in the ordinary topology.

3.0.2.   The group $\mathbf{GL}_1$ may be identified with the group K^*. The linear algebraic group $\mathcal{G}$ is an (algebraic) *torus* if it is isomorphic to a product of a finite number of $\mathbf{GL}_1$'s. This is equivalent with the requirement that it is diagonalizable. If $\mathcal{G}$ is moreover defined over k and the isomorphism can be defined over k, then it is said to be split over k or *k-split*. (This condition is equivalent with the requirement that there exist $g \in \mathbf{GL}_n(k)$ such that $g\mathcal{G}g^{-1}$ is diagonal.)

3.0.3.   Let $\mathcal{G}$ be a linear algebraic group defined over k. The maximal k-split tori of $\mathcal{G}$ are all conjugate under $\mathcal{G}(k)$. Their common dimension is the *k-rank*, $\mathrm{rk}_k(\mathcal{G})$ of $\mathcal{G}$ [18] (see also [124], 2.0.9, 19.2).

3.0.4.   The group $\mathcal{G}$ is reductive if its Lie algebra is reductive.

Assume that $\mathcal{G}$ is connected. Then a closed subgroup $\mathcal{P}$ of $\mathcal{G}$ is parabolic if $\mathcal{G}/\mathcal{P}$ is a projective variety.

3.1.   In this book, a real Lie group G is said to be *reductive* if there exists a linear algebraic group $\mathcal{G}$ defined over $\mathbf{R}$, whose identity component (in the Zariski topology) is reductive and a morphism $\nu \colon G \to \mathcal{G}(\mathbf{R})$ with finite kernel, whose image is an open subgroup of finite index of $\mathcal{G}(\mathbf{R})$. Unless otherwise stated, we also assume that G is of "connected type", i.e. that $\mathrm{Ad}\, G$ is contained in $\mathrm{Ad}(\mathfrak{g}_c)$.

This implies in particular that the identity component $\mathcal{Z}(G^0)^0$ of the center of G^0 is also central in G.

3.2.   The usual terminology of algebraic groups will be extended to such groups. In particular, a subgroup T of G is a torus (resp. $\mathbf{R}$-split torus) if it is the inverse image of the group of real points $\mathcal{S}(\mathbf{R})$ of an $\mathbf{R}$-torus (resp. $\mathbf{R}$-split torus) $\mathcal{S}$ of $\mathbf{G}$. The *split component of a torus* T is the identity component of its greatest $\mathbf{R}$-split subtorus. The maximal $\mathbf{R}$-split tori of G are conjugate under G^0. Their common dimension is the $\mathbf{R}$-rank or split rank $\mathrm{rk}_{\mathbf{R}}(G)$ of G. The *split component* of G is the identity component of the greatest split torus in the center of G (or, equivalently, of G^0, cf. 3.1). The group G is the direct product of its split component by ^0G.

3.3.   A *Cartan involution* θ of G is an involutive automorphism of G whose fixed point set is a maximal compact subgroup and which is the inversion on the split component of G. Given K, there is exactly one Cartan involution with fixed point set K. If $\mathfrak{s}$ is the (-1)-eigenspace of $d\theta$, then $(k, x) \mapsto k \cdot \exp x$ $(k \in K, x \in \mathfrak{s})$ is an isomorphism of analytic manifolds of $K \times \mathfrak{s}$ onto G. In particular $S = \exp \mathfrak{s}$ is a closed subspace isomorphic to $\mathfrak{s}$ under the exponential mapping, on which θ acts by inversion. The Cartan involutions are conjugate under automorphisms of G.

3.4. A *parabolic subgroup* P is the normalizer of a parabolic subalgebra $\mathfrak{p}$ of $\mathfrak{g}$. It is the inverse image of the group of real points $\mathcal{P}(\mathbf{R})$ of a parabolic subgroup $\mathcal{P}$ defined over $\mathbf{R}$ of $\mathcal{G}$. The unipotent radical N or N_P of P is the analytic subgroup generated by the nilradical of $\mathfrak{p}$. A Levi subgroup M of P is the inverse image of a Levi $\mathbf{R}$-subgroup $\mathcal{M}$ of $\mathcal{P}$. A *split component* A of P is the split component of a maximal torus in the radical of P. If A_P or A is one, then it is a split component of $\mathcal{Z}_G(A)$, and $\mathcal{Z}_G(A)$ is a Levi subgroup of P. We have

$$(1) \qquad P = M \ltimes N, \quad M = A \times {}^0M, \quad \text{hence} \quad P = MN = A \cdot {}^0M \cdot N.$$

In particular, $P \cap \theta(P)$ is the unique θ-stable Levi subgroup of P. Its split component is $P \cap S$. We always have $G = P \cdot K$, and $K \cap P$ is a maximal compact subgroup of $P \cap \theta(P)$. The dimension of A is the *parabolic rank* $\mathrm{prk}(P)$ of P.

A p-pair is a pair (P, A) consisting of a parabolic subgroup and a split component A of P. The standard Levi decomposition of P is $P = M \cdot N$ with $M = \mathcal{Z}_G(A)$. A p-pair (P', A') dominates (P, A) (written $(P', A') \succ (P, A)$) if $P' \supset P$, $A' \subset A$. The minimal p-pairs are conjugate under inner automorphisms of G^0, or even K^0. If a minimal parabolic subgroup P_0 (resp. a minimal p-pair (P_0, A_0)) is chosen, the *standard parabolic subgroups* (resp. p-pairs) are the parabolic subgroups containing P_0 (resp. the p-pairs dominating (P_0, A_0)). A p-pair (P, A) is *semi-standard* if $A \subset A_0$.

The p-pair $(\overline{P}, A)$ opposite to (P, A) consists of A and of the parabolic subgroup $\overline{P}$ opposite to P and containing $M = \mathcal{Z}_G(A)$. Thus $\overline{P} = M \cdot \overline{N}$, $\overline{N} = N_{\overline{P}}$. If M is θ-stable, then $\overline{P} = \theta(P)$.

3.5. Let (P, A) be a p-pair. Then $\Phi(P, A)$ is the set of roots of P with respect to A and $\Delta(P, A)$ the set of simple roots in $\Phi(P, A)$. We shall indifferently view it also as the set $\Phi(\mathfrak{p}, \mathfrak{a})$ of roots of $\mathfrak{p}$ with respect to $\mathfrak{a}$, i.e. we make no distinction between a character α of A and its differential. The value of a character α on $a \in A$ is denoted $\alpha(a)$ or a^α. Moreover we let

$$(1) \qquad\qquad \rho_P(a) = \left(\det \mathrm{Ad}\, a\big|_{\mathfrak{n}}\right)^{1/2} \quad (a \in A);$$

more generally

$$(2) \qquad\qquad \rho_P(m) = \left|\left(\det \mathrm{Ad}\, m\big|_{\mathfrak{n}}\right)\right|^{1/2} \quad (m \in M),$$

where $P = M \cdot N$ is the standard Levi decomposition of P. Thus, in the Lie algebra language, ρ_P is half the sum of the elements of $\Phi(\mathfrak{p}, \mathfrak{a})$, each counted with its multiplicity. Every element of $\Phi(P, A)$ is a linear combination with coefficients in $\mathbf{N}$ of elements in $\Delta(P, A)$. The latter are linearly independent, and their number is equal to $\dim A \cap \mathcal{D}G$. We have $\Phi(P, A) = -\Phi(\overline{P}, A)$.

3.6. If $\mathfrak{h}$ is a Cartan subalgebra of $\mathfrak{g}$, then $\Phi = \Phi(\mathfrak{g}_c, \mathfrak{h}_c)$ is the set of roots of $\mathfrak{g}_c$ with respect to $\mathfrak{h}_c$. If $\mathfrak{a}_0$ is the Lie algebra of a maximal $\mathbf{R}$-split torus, then ${}_{\mathbf{R}}\Phi = {}_{\mathbf{R}}\Phi(\mathfrak{g}, \mathfrak{a}_0)$ is the set of $\mathbf{R}$-roots, i.e. of roots of $\mathfrak{g}$ with respect to $\mathfrak{a}_0$. The algebras $\mathfrak{a}_0$ are the Lie algebras of the split components of the minimal parabolic subgroups of G. If (P_0, A_0) is a minimal p-pair, then

$$_{\mathbf{R}}\Phi(\mathfrak{g}, \mathfrak{a}) = \Phi(P_0, A_0) \cup (-\Phi(P_0, A_0)) = \Phi(P_0, A_0) \cup \Phi(\overline{P}_0, A_0),$$

and $\Phi(P_0, A_0)$ is the set of positive elements in ${}_{\mathbf{R}}\Phi(\mathfrak{g}, \mathfrak{a})$ for some ordering.

CHAPTER I

Relative Lie Algebra Cohomology

In this chapter, F is a commutative field, $\mathfrak{g}$ a finite dimensional Lie algebra over F, $\mathfrak{k}$ a subalgebra of $\mathfrak{g}$, $U(\mathfrak{g})$ (resp. $U(\mathfrak{k})$) the universal enveloping algebra of $\mathfrak{g}$ (resp. $\mathfrak{k}$). We let $R = U(\mathfrak{g})$ and $S = U(\mathfrak{k})$, except in §3, where S denotes Yoneda extensions. From 2.4 on, F is of characteristic zero and $\mathfrak{k}$ is reductive in $\mathfrak{g}$.

1. Lie algebra cohomology

1.1. We review here the standard definitions in the cohomology of Lie algebras (see [**31**, **74**]). A $\mathfrak{g}$-module is a vector space V over F on which $\mathfrak{g}$ acts via a homomorphism $\pi \colon \mathfrak{g} \to \mathfrak{gl}(V)$. It will be denoted by V, or by the pair (π, V). It will often be infinite dimensional. If V is a $\mathfrak{g}$-module, and $q \in \mathbf{N}$, then

$$(1) \qquad C^q = C^q(\mathfrak{g}; V) = \operatorname{Hom}_F(\Lambda^q \mathfrak{g}, V),$$

and $d \colon C^q \to C^{q+1}$ is defined by

$$
\begin{aligned}
(2) \qquad df(x_0, \ldots, x_q) &= \sum_i (-1)^i x_i \cdot f(x_0, \ldots, \widehat{x}_i, \ldots, x_q) \\
&\quad + \sum_{i<j} (-1)^{i+j} f([x_i, x_j], x_0, \ldots, \widehat{x}_i, \ldots, \widehat{x}_j, \ldots, x_q),
\end{aligned}
$$

where, as usual, $\widehat{}$ over an argument means that the argument should be omitted. Then $d^2 = 0$ and $H^*(\mathfrak{g}; V)$ is the cohomology of the complex $\{C^q\}$.

To $x \in \mathfrak{g}$ there is associated an endomorphism θ_x of C^q and a linear map $i_x \colon C^q \to C^{q-1}$ (the interior product) defined by

$$(3) \qquad (\theta_x f)(x_1, \ldots, x_q) = \sum_i f(x_1, \ldots, [x_i, x], \ldots, x_q) + x \cdot f(x_1, \ldots, x_q),$$

$$(4) \qquad (i_x f)(x_1, \ldots, x_{q-1}) = f(x, x_1, \ldots, x_{q-1}).$$

The maps d, i_x, θ_x are related by

$$(5) \qquad \theta_x = d \cdot i_x + i_x \cdot d.$$

Write $C^q(\mathfrak{g}; V)$ as $\Lambda^q \mathfrak{g}^* \otimes V$. Let $\{x_i\}$ be a basis of $\mathfrak{g}$ and $\{x^i\}$ the dual basis of $\mathfrak{g}^*$. Denote by $\varepsilon(x)$ the left exterior product by x in $\Lambda \mathfrak{g}^*$ and by d_0 the differential of $C^*(\mathfrak{g})$. Then (2) translates to

$$(6) \qquad d = d_0 \otimes 1 + \sum_i \varepsilon(x^i) \otimes \pi(x_i),$$

$$(7) \qquad 2 \cdot d_0 = \sum_i \varepsilon(x^i) \cdot \theta_{x_i}$$

(cf. [**74**, 3.4] for (7), and [**44**, 5.26] for the general case).

1.2. Let $C^q(\mathfrak{g}, \mathfrak{k}, V)$ be the subspace of $C^q(\mathfrak{g}, V)$ consisting of the elements annihilated by the maps i_x and θ_x for all $x \in \mathfrak{k}$. Then $C^q(\mathfrak{g}, \mathfrak{k}; V)$ is stable under d and its cohomology groups are the *relative cohomology* groups $H^q(\mathfrak{g}, \mathfrak{k}; V)$ of $\mathfrak{g}$ mod $\mathfrak{k}$, with coefficients in V. Note that we have

$$(1) \qquad\qquad C^q(\mathfrak{g}, \mathfrak{k}; V) = \mathrm{Hom}_{\mathfrak{k}}(\Lambda^q(\mathfrak{g}/\mathfrak{k}), V),$$

where the action of $\mathfrak{k}$ on $\Lambda^q(\mathfrak{g}/\mathfrak{k})$ is induced by the adjoint representation, i.e., $C^q(\mathfrak{g}, \mathfrak{k}, V)$ may be identified with the subspace of elements $f \in \mathrm{Hom}_F(\Lambda^q(\mathfrak{g}/\mathfrak{k}), V)$ which satisfy the relation

$$(2) \quad \sum_i f(x_1, \ldots, [x, x_i], \ldots, x_q) = x \cdot f(x_1, \ldots, x_q) \quad (x \in \mathfrak{k}; x_i \in \mathfrak{g}/\mathfrak{k}, i = 1, \ldots, q).$$

We have in particular

$$(3) \qquad H^0(\mathfrak{g}, V) = H^0(\mathfrak{g}, \mathfrak{k}; V) = V^{\mathfrak{g}} = \{v \in V \mid x \cdot v = 0 \text{ for all } x \in \mathfrak{g}\}.$$

Since $\Lambda^* \mathfrak{g}/\mathfrak{k}$ is finite dimensional, it is clear that

$$(4) \qquad \textit{The functor } V \mapsto H^*(\mathfrak{g}, \mathfrak{k}; V) \textit{ commutes with inductive limits.}$$

1.3. These cohomology groups obey the *Künneth rule*. To simplify notation, we just consider the case of two factors. Assume then

$$(1) \qquad \begin{aligned} \mathfrak{g} = \mathfrak{g}_1 \oplus \mathfrak{g}_2, \qquad \mathfrak{k} = \mathfrak{k}_1 \oplus \mathfrak{k}_2, \qquad V = V_1 \otimes V_2 \\ (\mathfrak{k}_i \subset \mathfrak{g}_i, \quad V_i \text{ a } \mathfrak{g}_i\text{-module}, \quad i = 1, 2). \end{aligned}$$

Then, for all q's,

$$(2) \qquad H^q(\mathfrak{g}, \mathfrak{k}; V) = \bigoplus_{a+b=q} H^a(\mathfrak{g}_1, \mathfrak{k}_1; V_1) \otimes H^b(\mathfrak{g}_2, \mathfrak{k}_2; V_2).$$

To see this, note that we can write

$$C^q(\mathfrak{g}, \mathfrak{k}; V) = (\Lambda^q(\mathfrak{g}/\mathfrak{k})^* \otimes V)^{\mathfrak{k}}, \qquad \Lambda(\mathfrak{g}/\mathfrak{k})^* = \Lambda(\mathfrak{g}_1/\mathfrak{k}_1)^* \otimes \Lambda(\mathfrak{g}_2/\mathfrak{k}_2)^*.$$

However, if A_i, U_i are $\mathfrak{k}_i$-modules $(i = 1, 2)$ and $A_1 \otimes A_2$, $U_1 \otimes U_2$ are viewed as $\mathfrak{k}$-modules in the obvious way, then

$$(3) \qquad (A_1 \otimes A_2 \otimes U_1 \otimes U_2)^{\mathfrak{k}} = (A_1 \otimes U_1)^{\mathfrak{k}_1} \otimes (A_2 \otimes U_2)^{\mathfrak{k}_2}.$$

Therefore

$$(4) \qquad C^*(\mathfrak{g}, \mathfrak{k}; V) = C^*(\mathfrak{g}_1, \mathfrak{k}_1; V) \otimes C^*(\mathfrak{g}_2, \mathfrak{k}_2; V)$$

(graded tensor product), whence our assertion.

1.4. In this subsection and the next one, F is of characteristic zero, $\mathfrak{g}$ unimodular, $\mathfrak{k}$ reductive in $\mathfrak{g}$ and $n = \dim \mathfrak{g}$, $m = \dim \mathfrak{g}/\mathfrak{k}$. The algebra $\mathfrak{k}$ is then also unimodular, hence acts trivially on $\Lambda^m(\mathfrak{g}/\mathfrak{k})$.

It is known that $H^*(\mathfrak{g}, \mathfrak{k}; F)$ satisfies Poincaré duality ([**74**, §12], [**44**, 10.27, 10.28]). In particular, $H^q(\mathfrak{g}, \mathfrak{k}; F)$ is canonically isomorphic to the dual of $H^{m-q}(\mathfrak{g}, \mathfrak{k}; F)$ for all $q \in \mathbf{Z}$. This implies in particular (since $C^m(\mathfrak{g}, \mathfrak{k}; F)$ is one-dimensional)

$$(1) \qquad C^m(\mathfrak{g}, \mathfrak{k}; F) = H^m(\mathfrak{g}, \mathfrak{k}; F) = F, \ dC^{m-1}(\mathfrak{g}, \mathfrak{k}; F) = 0.$$

1.5. PROPOSITION. *Assume V and W are two $\mathfrak{g}$-modules in perfect duality with respect to a $\mathfrak{g}$-invariant pairing $\langle \, , \, \rangle$, and that $H^*(\mathfrak{g}, \mathfrak{k}; V)$ and $H^*(\mathfrak{g}, \mathfrak{k}; W)$ are finite dimensional. Then $H^q(\mathfrak{g}, \mathfrak{k}; V)$ is canonically isomorphic to the dual of $H^{m-q}(\mathfrak{g}, \mathfrak{k}; W)$ for all $q \in \mathbf{Z}$.*

We can view $C^q(V) = C^q(\mathfrak{g}, \mathfrak{k}; V)$ as the space $(\Lambda(\mathfrak{g}/\mathfrak{k})^* \otimes V)^{\mathfrak{k}}$ of $\mathfrak{k}$-invariants in $\Lambda(\mathfrak{g}/\mathfrak{k})^* \otimes V$, and similarly for W. Then the map

$$C^q(V) \times C^{m-q}(W) \to \Lambda^m(\mathfrak{g}/\mathfrak{k})^*,$$

defined by

$$\langle y \otimes v, y' \otimes w \rangle = \langle v, w \rangle y \wedge y'$$
$$(v \in V; \quad w \in W; \quad y \in \Lambda^q(\mathfrak{g}/\mathfrak{k})^*, \quad y' \in \Lambda^{m-q}(\mathfrak{g}/\mathfrak{k})^*)$$

defines a perfect pairing between $C^q(V)$ and $C^{m-q}(W)$, once a basis element of $\Lambda^m(\mathfrak{g}/\mathfrak{k})^*$ is chosen. It is easily checked, using 1.1(6) and 1.4(1), that we have

(1) $\langle du, b \rangle = (-1)^{q+1} \langle u, db \rangle$ $(u \in C^q(V), \; b \in C^{m-q-1}(W)).$

From this 1.5 follows immediately.

1.6. The real case. Let $F = \mathbf{R}$ and let G be a Lie group with Lie algebra $\mathfrak{g}$, K a closed connected subgroup of G with Lie algebra $\mathfrak{k}$.

If V is a smooth G-module, then we let G operate on the space $A(G/K; V)$ of V-valued differential forms on G/K by the rule

(1) $(g \circ \omega)(x, Y) = g(\omega(g^{-1} \cdot x, g^{-1} \cdot Y))$

where $g \in G$, $x \in G/K$, and Y is a q-vector at x. It is then readily seen that the evaluation map at the origin, which assigns to $\omega \in A(G/K; V)$ its value at e, defines an isomorphism of the space $A(G/K; V)^G$ of G-invariant differential forms onto $C(\mathfrak{g}, \mathfrak{k}; V)$, which carries the exterior differential to the differential of 1.1. Thus, $H^*(\mathfrak{g}, \mathfrak{k}; V)$ is the cohomology of the space of G-invariant V-forms on G/K.

Assume G to be compact connected, V to be finite dimensional and acted upon trivially by G. Then a standard averaging argument shows that $H^*(A(G/K; V)^G) = H^*(A(G/K; V))$; hence, by the de Rham theorem

(2) $H^*(\mathfrak{g}, \mathfrak{k}; V) = H^*(G/K; V).$

This is a result of E. Cartan which is in fact at the origin of the notion of Lie algebra cohomology. A bit more precisely, E. Cartan conjectured two theorems, which were proved later by de Rham, and stated that, modulo those results, the cohomology of G/K could be computed using invariant differential forms. In fact, he was mainly concerned with compact symmetric spaces, for which all invariant forms are closed and even harmonic (see II, 3.2).

2. The Ext functors for $(\mathfrak{g}, \mathfrak{k})$-modules

2.1. It is well known that the groups $H^q(\mathfrak{g}; V)$ may be viewed as the derived functors of $V \mapsto V^{\mathfrak{g}}$ in the category of R-modules. More generally, one may define the derived functors $\mathrm{Ext}^q_R(U, V)$ of $(U, V) \mapsto \mathrm{Hom}_{\mathfrak{g}}(U, V)$, and we have

(1) $\mathrm{Ext}^q_R(F, V) = H^q(\mathfrak{g}; V), \quad \mathrm{Ext}^q_R(U, V) = H^q(\mathfrak{g}, \mathrm{Hom}_F(U, V))$ $(q \in \mathbf{Z}),$

where F is viewed as the trivial $\mathfrak{g}$-module (see XIII and IX, 4.3 in [**31**]).

We shall need similar facts in the relative case. A general theory was developed by G. Hochschild [**59**] in the context of relative homological algebra with respect to the pair (R, S). However, in order to prove the equality

(2) $\mathrm{Ext}^q_{R,S}(F, V) = H^q(\mathfrak{g}, \mathfrak{k}; V),$

he had to assume F to be of characteristic zero and $\mathfrak{k}$ to be reductive in $\mathfrak{g}$. This is at any rate the only case of interest in this book (with in fact F either $\mathbf{R}$ or $\mathbf{C}$). In the relative theory, one accepts only exact sequences of R-modules which split over S. We shall adopt here a slightly different point of view, using the usual absolute theory, but in a more restricted category, that of $(\mathfrak{g}, \mathfrak{k})$-modules, defined below. In principle, this is a bit less general than Hochschild's approach, but sufficient for our purposes.

2.2. Let V be a $\mathfrak{k}$-module. An element $v \in V$ is $\mathfrak{k}$-*finite* if $U(\mathfrak{k}) \cdot v$ is a finite dimensional subspace. The $\mathfrak{k}$-module V is *locally $\mathfrak{k}$-finite* if every element is $\mathfrak{k}$-finite. Thus V is locally $\mathfrak{k}$-finite if and only if every finite dimensional subspace is contained in a finite dimensional subspace stable under $\mathfrak{k}$.

A vector space V over F is a $(\mathfrak{g}, \mathfrak{k})$-module if it is a $\mathfrak{g}$-module which is locally $\mathfrak{k}$-finite and is semi-simple as a $\mathfrak{k}$-module. In particular, every $\mathfrak{k}$-simple submodule is finite dimensional. It suffices to require that V be locally $\mathfrak{k}$-finite and that every finite dimensional $\mathfrak{k}$-stable subspace be semi-simple [**21**, §3, n° 3].

A $(\mathfrak{g}, \mathfrak{k})$-module V is *admissible* if the isotypic subspaces for $\mathfrak{k}$ are all finite dimensional. If V is admissible, it is clearly a direct sum of simple $\mathfrak{k}$-modules.

Let $\mathcal{C}$ or $\mathcal{C}_{\mathfrak{g},\mathfrak{k}}$ be the category of $(\mathfrak{g}, \mathfrak{k})$-modules. It is closed under direct sums. If $V \in \mathcal{C}$, then every $\mathfrak{g}$-submodule of V and every $\mathfrak{g}$-module quotient of V belong to $\mathcal{C}$.

Since $\mathfrak{g}$-modules are canonically R-modules and vice versa, we get equivalent notions if we replace above $\mathfrak{g}$ and $\mathfrak{k}$ by R and S. We shall use both interchangeably. Since all our modules are semi-simple for S, it is clear that all exact sequences in $\mathcal{C}$ split over S. If F *is of characteristic zero*, then the tensor product over F of two elements of $\mathcal{C}$ also belongs to $\mathcal{C}$. This follows from the fact that in characteristic zero, the tensor product of two finite dimensional semi-simple modules for a Lie algebra $\mathfrak{m}$ is also semi-simple [**25**, §6, n° 5, Cor. 1].

Let (π, V) be a $\mathfrak{g}$-module. Then the subspace $V_{(\mathfrak{k})}$ spanned by the finite dimensional $\mathfrak{k}$-stable subspaces of V is stable under $\mathfrak{g}$. Therefore, if these subspaces are semi-simple $\mathfrak{k}$-modules, the space $V_{(\mathfrak{k})}$ is a $(\mathfrak{g}, \mathfrak{k})$-module. We note that the image of $\mathrm{Hom}_{\mathfrak{k}}(\Lambda(\mathfrak{g}/\mathfrak{k}), V)$ is necessarily contained in $V_{(\mathfrak{k})}$. Therefore the inclusion $V_{(\mathfrak{k})} \subset V$ induces isomorphisms

$$(1) \qquad C^*(\mathfrak{g}, \mathfrak{k}; V_{(\mathfrak{k})}) \overset{\sim}{\to} C^*(\mathfrak{g}, \mathfrak{k}; V), \qquad H^*(\mathfrak{g}, \mathfrak{k}; V_{(\mathfrak{k})}) \overset{\sim}{\to} H^*(\mathfrak{g}, \mathfrak{k}; V),$$

i.e., in computing cohomology, we can always replace a $\mathfrak{g}$-module V by the subspace of $\mathfrak{k}$-finite vectors in V.

If the $\mathfrak{k}$-types occurring in $\Lambda(\mathfrak{g}/\mathfrak{k})$ have finite multiplicities in $V_{(\mathfrak{k})}$ (in particular, if V is admissible), then $C^*(\mathfrak{g}, \mathfrak{k}; V)$ is finite dimensional, and hence $H^*(\mathfrak{g}, \mathfrak{k}; V)$ is finite dimensional.

Now let (π, V) be a $(\mathfrak{g}, \mathfrak{k})$-module. By analogy with 0, 2.5, the *contragredient module* $(\widetilde{\pi}, \widetilde{V})$ is by definition the space $V'_{(\mathfrak{k})}$ spanned by the $\mathfrak{k}$-stable finite dimensional subspaces in the dual space V' to V, acted upon by the usual contragredient representation, i.e., $\widetilde{\pi}(x) = {}^t\pi(-x)$ $(x \in \mathfrak{g})$, where ${}^t\pi$ is the transpose of π. If U is a finite dimensional $\mathfrak{k}$-stable subspace of V', then U is the dual space to the quotient of V by the annihilator of U in V, hence is a semi-simple $\mathfrak{k}$-module. Therefore $(\widetilde{\pi}, \widetilde{V}) \in \mathcal{C}$.

As usual, the center of the universal enveloping algebra of a Lie algebra $\mathfrak{m}$ over F will be denoted $Z(\mathfrak{m})$.

A $\mathfrak{g}$-module (π, V) is said to have an *infinitesimal character* if there exists a character of $Z(\mathfrak{g})$, i.e., a unital F-algebra homomorphism: $Z(\mathfrak{g}) \to F$, to be denoted

χ or χ_π or χ_V, such that

$$\pi(z) = \chi_\pi(z) \cdot \mathrm{Id} \qquad (z \in Z(\mathfrak{g})).$$

This is the case in particular if (π, V) is an absolutely irreducible and admissible $(\mathfrak{g}, \mathfrak{k})$-module.

2.3. EXAMPLE. Let $F = \mathbf{R}$. Let G be a connected Lie group, $\mathfrak{g}$ its Lie algebra, and $\mathfrak{k}$ the Lie algebra of a compact subgroup of G. Then $\mathfrak{k}$ is reductive in $\mathfrak{g}$, and every $(\mathfrak{g}, K)$-module is a $(\mathfrak{g}, \mathfrak{k})$-module.

This example is the one which has motivated the above definition, and in fact, later, besides finite dimensional modules, we shall mainly consider $(\mathfrak{g}, \mathfrak{k})$-modules associated in this way to unitary representations.

2.4. Projective modules. We recall that from now on F is of characteristic zero and $\mathfrak{k}$ is *reductive in* $\mathfrak{g}$, i.e., $\mathfrak{g}$ is a semi-simple $\mathfrak{k}$-module with respect to the adjoint representation. The algebra $\mathfrak{k}$ operates on $\mathfrak{g}$ by the adjoint representation, whence a representation on the tensor algebra $T(\mathfrak{g})$ of $\mathfrak{g}$ and on $U(\mathfrak{g})$. Under this representation, both $T(\mathfrak{g})$ and $U(\mathfrak{g})$ are locally $\mathfrak{k}$-finite and semi-simple (see [**25**, §6, n° 5, Cor. 2]).

LEMMA. *Let U be a locally $\mathfrak{k}$-finite semi-simple $\mathfrak{k}$-module. Then the induced module $I(U) = I_{S,R}(U) = R \otimes_S U$ is a projective $(\mathfrak{g}, \mathfrak{k})$-module.*

Although not stated in this way, this is in effect proved in [**59**]. We sketch the argument. First, by standard "Frobenius reciprocity" we have for every $(\mathfrak{g}, \mathfrak{k})$-module V a canonical isomorphism

$$(1) \qquad\qquad m \colon \mathrm{Hom}_R(I(U), V) \overset{\sim}{\to} \mathrm{Hom}_S(U, V),$$

defined by the restriction to $1 \otimes U$. Now let $A, B \in \mathcal{C}$, $f \colon B \to A$ a surjective morphism, and $s \colon I(U) \to A$ a morphism. We have to show the existence of $t \colon I(U) \to B$ such that $f \circ t = s$. Since A is a direct S-summand of B, we can find an S-module homomorphism $t' \colon U \to B$ such that $m(s) = f \circ t'$. We then put $t = m^{-1}(t')$. It remains to see that $I(U)$ belongs to $\mathcal{C}$.

The R-module structure on $I(U)$ understood here comes from left translations on R. It gives by restriction an action of S; call it the ordinary action. On the other hand, S acts on R via the adjoint representation on $\mathfrak{g}$. With respect to this action, R is locally S-finite and semi-simple, as remarked above. Then $R \otimes_F U$, with the tensor product of these actions of S, is also S-semi-simple and locally finite. The operation of $s \in S$ is given by

$$(2) \qquad s \circ (r \otimes u) = (s \cdot r - r \cdot s) \otimes u + r \otimes s \cdot u \qquad (s \in \mathfrak{k}; \ r \in R, \ u \in U).$$

It is readily seen to leave stable the kernel M of the canonical map $R \otimes_F U \to I(U)$, and thereby induces an action on $I(U)$, with respect to which $I(U)$ is locally finite and semi-simple. However, the sum of the last two terms on the right hand side of (2) belongs to M. Hence this new S-action coincides with the ordinary one on $I(U)$, which proves our contention.

2.5. The functors Ext. The map $(r, u) \mapsto r \cdot u$ induces a surjective morphism $I(U) \to U$. Thus every element of $\mathcal{C}$ is a quotient of a projective one, and we can construct projective resolutions in the usual way. If

$$(1) \qquad \longrightarrow X_q \overset{\partial_q}{\longrightarrow} X_{q-1} \overset{\partial_{q-1}}{\longrightarrow} \cdots \longrightarrow X_0 \overset{\varepsilon}{\longrightarrow} U \longrightarrow 0$$

is one for U, and $V \in \mathcal{C}$, then the groups $\operatorname{Ext}^q(U, V)$ are by definition the cohomology groups of the complex

(2)
$$\operatorname{Hom}_R(X_0, V) \xrightarrow{d_0} \operatorname{Hom}_R(X_1, V) \xrightarrow{d_1} \cdots \xrightarrow{d_{q-1}} \operatorname{Hom}_R(X_q, V) \longrightarrow \cdots .$$

As usual, they do not depend on the choice of the projective resolution. Moreover, it follows from [**31**, IX, 4.3] that

(3)
$$\operatorname{Ext}^q(F, \operatorname{Hom}_F(U, V)) = \operatorname{Ext}^q(U, V).$$

We should check that 2.1(2) is satisfied. Let $X_q = R \otimes_S \Lambda^q(\mathfrak{g}/\mathfrak{k})$. Define $\partial_q \colon X_q \to X_{q-1}$ by

(4)
$$\partial_q(r \otimes x_1 \wedge \cdots \wedge x_q) = \sum (-1)^{i-1} x_i \cdot r \otimes x_1 \wedge \cdots \wedge \widehat{x}_i \wedge \cdots \wedge x_q$$
$$+ \sum_{i<j} (-1)^{i+j} r \otimes [x_i, x_j] \wedge x_1 \wedge \cdots \wedge \widehat{x}_i \wedge \cdots \wedge \widehat{x}_j \wedge \cdots \wedge x_q$$

and let $\varepsilon \colon X_0 = R \xrightarrow{\varepsilon} F$ be the augmentation. Then the X_i are projective (2.4) and

(5)
$$\longrightarrow X_q \xrightarrow{\partial_q} \cdots \xrightarrow{\partial_1} X_0 \xrightarrow{\varepsilon} F \longrightarrow 0$$

is easily seen to be exact [**59**]. Hence (5) is a projective resolution of F. In view of 2.4(1), we have $\operatorname{Hom}_R(X_q, V) = \operatorname{Hom}_S(\Lambda^q(\mathfrak{g}/\mathfrak{k}), V)$, and it follows immediately from (4) that the complex $\{\operatorname{Hom}_R(X_q, V)\}$ may be identified with the one used in 1.2 to define relative Lie algebra cohomology, whence 2.1(2).

2.6. Injective modules. We have used projective resolutions in $\mathcal{C}$, which will suffice for our purposes. But our category also contains enough injectives. We briefly outline their construction.

Let V be a locally finite semi-simple $\mathfrak{k}$-module and
$$P^0(V) = P^0_{R,S}(V) = \operatorname{Hom}_S(R, V)$$
the usual coinduced or "produced" module from S to R. Let
$$P(V) = P_{R,S}(V) = \operatorname{Hom}_S(R, V)_{(S)}$$
be the subspace of $P^0(V)$ spanned by the S-finite elements. We claim that $P(V)$ is an injective module in $\mathcal{C}$.

We view $P^0(V)$ as a subspace of $\operatorname{Hom}_F(R, V)$. On the latter, S acts first by left translations on R, the "ordinary action", and second, as above, it acts via the given action on V and the operation on R stemming from the adjoint representation of $\mathfrak{k}$ on $\mathfrak{g}$. As in 2.4, it is first checked that these two actions coincide on $P(V)$. Let us prove now that every finite dimensional subspace M of $P^0(V)$ stable under S is a semi-simple S-module. Let $\{R_j\}_{j=0,1,\cdots}$ be the usual increasing filtration of R [**25**]. There exists j such that the restriction map $\operatorname{Hom}_F(R, V) \to \operatorname{Hom}_F(R_j, V)$ is injective on M; hence it identifies M to a subspace of $\operatorname{Hom}_F(R_j, V)$. Since R_j is finite dimensional, $\operatorname{Hom}_F(R_j, V) = R_j^* \otimes_F V$ is $\mathfrak{k}$-semi-simple, hence so is M. It follows that $P(V)$ can also be defined as the subspace of $P^0(V)$ generated by the S-invariant finite dimensional subspaces of $P^0(V)$. It is then clearly an R-module, and then an (R, S)-module. Furthermore, if N is an (R, S)-module and $N \to P^0(V)$ an R-morphism, then $\operatorname{Im} N \subset P(V)$. Since $P^0(V)$ is injective with respect to (R, S)-modules (the argument is the dual to that of 2.4, see [59]), it follows that $P(V)$ is injective in $\mathcal{C}$. Now let V be a $(\mathfrak{g}, \mathfrak{k})$-module. As usual, the map which associates to $v \in V$ the homomorphism $r \mapsto r \cdot v$ of R into V yields an injective morphism of V into $P(V)$. Hence every element of $\mathcal{C}$ is contained in an injective module in $\mathcal{C}$, and we can construct injective resolutions in the usual way.

Let X_q be as in 2.5(5). Since it is projective, $A^q = \mathrm{Hom}_S(X_q, V)_{(S)}$ is injective; therefore $\{A^q\}$ provides an injective resolution such that

$$(6) \qquad\qquad A^{q,\mathfrak{g}} = C^q(\mathfrak{g}, \mathfrak{k}; V) \qquad (q \in \mathbf{N}).$$

In particular, if V is admissible, then V admits an injective resolution $0 \to V \to A^*$ such that $A^{*\mathfrak{g}}$ is finite dimensional.

2.7. Finitely generated $(\mathfrak{g}, \mathfrak{k})$-modules. Let U be a finitely generated $(\mathfrak{g}, \mathfrak{k})$-module. Then there exists a $\mathfrak{k}$-stable finite dimensional subspace E of U such that $U = R.E$. Then U is a quotient of the $(\mathfrak{g}, \mathfrak{k})$-module $R \otimes_S E$, which is projective and finitely generated. It follows that U admits a projective resolution by finitely generated $(\mathfrak{g}, \mathfrak{k})$-modules. Therefore, if U and V are finitely generated, $\mathrm{Ext}^*_{\mathfrak{g}, \mathfrak{k}}(U, V)$ can be computed within the category of finitely generated $(\mathfrak{g}, \mathfrak{k})$-modules.

The following proposition was communicated to one of us by D. Vogan in the case where U and V are admissible and irreducible. The proof is a mild simplification of his.

2.8. PROPOSITION (D. Vogan). *Let U be a finitely generated and V an admissible $(\mathfrak{g}, \mathfrak{k})$-module. Then $\mathrm{Ext}^*_{\mathfrak{g}, \mathfrak{k}}(U, V)$ may be computed as the cohomology of a finite-dimensional complex. In particular, $\mathrm{Ext}^*_{\mathfrak{g}, \mathfrak{k}}(U, V)$ is finite-dimensional.*

We note first that if A is a finitely generated and B an admissible $(\mathfrak{g}, \mathfrak{k})$-module, then $\mathrm{Hom}_{\mathfrak{g}}(A, B)$ is finite dimensional. In fact, there exists a $\mathfrak{k}$-stable finite dimensional subspace E of A such that $A = R.E$; hence $\mathrm{Hom}_{\mathfrak{g}}(A, B) \subset \mathrm{Hom}_{\mathfrak{k}}(E, B)$. But this last space is finite dimensional if B is admissible.

$\mathrm{Ext}^*_{\mathfrak{g}, \mathfrak{k}}(U, V)$ is the cohomology of the complex

$$\{C^q = \mathrm{Hom}_{\mathfrak{g}}(X_q, V)\},$$

where X_q is any projective resolution of U in $C_{\mathfrak{g}, \mathfrak{k}}$. By 2.7, we may assume the X_q's to be finitely generated. Then C^q is finite dimensional for every q by our initial remark. Since moreover $\mathrm{Ext}^q_{\mathfrak{g}, \mathfrak{k}} = 0$ for $q > m = \dim(\mathfrak{g}, \mathfrak{k})$, we may replace C^{m+1} by dC^m for $m = \dim \mathfrak{g}/\mathfrak{k}$, and C^q by 0 for $q > m + 1$, whence the proposition.

2.9. PROPOSITION. *Let U and V be admissible, and assume one of them to be finite dimensional. Let $m = \dim(\mathfrak{g}/\mathfrak{k})$. Then $\mathrm{Ext}^q(U, V)$ is canonically isomorphic to the dual of $\mathrm{Ext}^{m-q}(\widetilde{U}, \widetilde{V})$.*

Via 2.5(3), this follows from 1.5 and 2.8.

3. Long exact sequences and Ext

3.1. Long exact sequences. We recall here the interpretation of $\mathrm{Ext}^q(U, V)$ in terms of long exact sequences. For more details, see [78, Chap. III]. Given $q \geq 1$ and $U, V \in \mathcal{C}_{\mathfrak{g}, \mathfrak{k}}$, let $S_q(U, V)$ be the set of exact sequences in $\mathcal{C}$ of the form

$$S : 0 \to V \to E_{q-1} \to \cdots \to E_0 \to U \to 0.$$

If $U', V' \in \mathcal{C}$ and $S' \in S_q(U', V')$, then a homomorphism $\gamma \colon S \to S'$ is given by morphisms $E_i \to E'_i$, $\gamma_U \colon U \to U'$, $\gamma_V \colon V \to V'$, which yield a commutative diagram

(1)
$$\begin{array}{ccccccccccc}
0 & \longrightarrow & V & \longrightarrow & E_{q-1} & \longrightarrow & \cdots & \longrightarrow & E_0 & \longrightarrow & U & \longrightarrow & 0 \\
& & \downarrow{\scriptstyle \gamma_V} & & \downarrow & & & & \downarrow & & \downarrow{\scriptstyle \gamma_U} & & \\
0 & \longrightarrow & V' & \longrightarrow & E'_{q-1} & \longrightarrow & \cdots & \longrightarrow & E'_0 & \longrightarrow & U' & \longrightarrow & 0.
\end{array}$$

In $S_q(U, V)$ we consider the smallest equivalence relation $\equiv$ such that $S \equiv S'$ if there exists either a morphism $S \to S'$ or a morphism $S' \to S$ which is the identity at both ends. Let $\mathrm{Ext}'^q(U, V)$ be the set of such equivalence classes. Then it is well-known that:

(i) *There is an addition on* $\mathrm{Ext}'^q(U, V)$, *defined by the Baer sum* (*cf.* 3.2.13), *with respect to which* $\mathrm{Ext}'^q(U, V)$ *is a commutative group, whose zero element is represented by split exact sequences* (*at each stage, the kernel is a direct R-summand*).

(ii) *The group* $\mathrm{Ext}'^q(U, V)$ *is canonically isomorphic to* $\mathrm{Ext}^q(U, V)$.

This is all proved in [**78**, III]. We just recall some of the relevant constructions and facts in the next subsection.

3.2. (1) Given $S \in S_q(U, V)$ and $\gamma \colon V \to V'$, there is associated an element

$$\gamma S \in S_q(U, V') \colon 0 \to V' \to E'_{q-1} \to \cdots \to E'_0 \to U \to 0$$

endowed with a morphism $\alpha \colon S \to S'$ such that $\alpha_U = \mathrm{id}$, $\alpha_V = \gamma$, called the *push-out* of S. The module E'_{q-1} is by definition the quotient of $E_{q-1} \oplus V'$ by the subgroup generated by the elements $(\mu v, -\gamma v)$ ($v \in V$, where $\mu \colon V \to E_{q-1}$ is given by S). The other modules E'_i are constructed similarly by induction.

(2) Given $S \in S_q(U, V)$, $U' \in \mathcal{C}$ and $\delta \colon U' \to U$, there exists $S\delta = S' \in S_q(U', V)$, the *pull-back* of S, endowed with a morphism $\beta \colon S' \to S$ such that $\beta_V = \mathrm{id}$, $\beta_{U'} = \delta$. The module E'_0 is the pull-back of U', and E_0 and the E'_i are constructed similarly by induction.

(3) Let $S, S' \in S_q(U, V)$. Then $S \oplus S' \in S_q(U \oplus U, V \oplus V)$. Let $S_1 \in S_q(U, V \oplus V)$ be the pull-back of $S \oplus S'$ via the diagonal map $U \to U \oplus U$. Then the Baer sum $S + S' \in S_q(U, V)$ is the push-out of S_1 by the map $V \oplus V \to V$ defined by the addition in V.

(4) It is elementary, and follows from [**78**, III, 5.3], that we have

(1) $$1.S \equiv S, \qquad S.0 \equiv 0.$$

Furthermore, if $U', V' \in \mathcal{C}$, $S' \in S_q(U', V')$ and $\gamma \colon S \to S'$ is a morphism, then $\gamma_V . S \equiv S . \gamma_U$ [**78**, III, 5.1]. As a consequence, we see that if $S \in S_q(U, V)$ admits an endomorphism γ such that $\gamma_V = 1$, $\gamma_U = 0$, then $S \equiv 0$. Indeed, we have then $0 \equiv S.0 \equiv 1.S \equiv S$.

(5) We now define the maps which yield the isomorphisms of 3.1(ii). Fix a projective resolution (X_i) of U. Let $S \in S_q(U, V)$. The resolution (X_i), being projective, can be mapped into S; then we get a commutative diagram:

$$\begin{array}{ccccccccccc}
X_{q+1} & \longrightarrow & X_q & \longrightarrow & X_{q-1} & \longrightarrow & \cdots & \longrightarrow & X_0 & \longrightarrow & U & \longrightarrow & 0 \\
\downarrow & & \downarrow{\scriptstyle a_q} & & \downarrow{\scriptstyle a_{q-1}} & & & & \downarrow{\scriptstyle a_0} & & \| & & \\
0 & \longrightarrow & V & \longrightarrow & E_{q-1} & \longrightarrow & \cdots & \longrightarrow & E_0 & \longrightarrow & U & \longrightarrow & 0.
\end{array}$$

Then $a_q \in \mathrm{Hom}_{U(\mathfrak{g})}(X_q, V)$ is zero on ∂X_{q+1}, hence is a cocycle. The assignment $S \mapsto a_q$ then yields a map from $S_q(U, V)$ to the space of q-cocycles, which can be proved to induce an isomorphism μ of Ext'^q onto Ext^q [**78**, III, 6.4].

Conversely, a q-cocycle z_q can be viewed as a $U(\mathfrak{g})$-morphism δ of ∂X_q into V. To z_q we associate the push-out $\delta S'$ of

$$S': 0 \to \partial X_q \to X_{q-1} \to \cdots \to X_0 \to U \to 0.$$

This yields the inverse isomorphism to μ (cf. [**78**, III, 6.4]).

4. A vanishing theorem

4.1. THEOREM. *Let U, V be two $(\mathfrak{g}, \mathfrak{k})$-modules with infinitesimal characters χ_U, χ_V. If $\chi_U \neq \chi_V$, then $\mathrm{Ext}^q(U, V) = 0$ for all q's.*

To prove this theorem, we use the interpretation of Ext^q in terms of long exact sequences. If $\chi_U \neq \chi_V$, then we can find $z \in Z(\mathfrak{g})$ such that $\chi_V(z) = 1$, $\chi_U(z) = 0$. Let $S \in S_q(U, V)$. Then z operates on each term of S and defines an endomorphism $\gamma(z)$ of S. By construction $\gamma(z)_V = 1$, $\gamma(z)_U = 0$, hence $S \equiv 0$ by 3.2(4).

REMARK. This theorem is an analogue of a result of D. Wigner about the continuous cohomology of real Lie groups (see [**12**, 2.4]). The proof is exactly the same.

4.2. COROLLARY. *Let U be finite dimensional. If $\chi_{\widetilde{U}} \neq \chi_V$, then $H^q(\mathfrak{g}, \mathfrak{k}; U \otimes V) = 0$ for all q's.*

We have $U \otimes_F V = \mathrm{Hom}_F(U', V)$. Since

$$H^q(\mathfrak{g}, \mathfrak{k}; \mathrm{Hom}_F(U', V)) = \mathrm{Ext}^q(F, \mathrm{Hom}_F(U', V)) = \mathrm{Ext}^q(U', V),$$

we are reduced to 4.1.

4.3. The proof of 4.1 was based on the use of Yoneda extensions. It was pointed out recently by T. A. Springer to the first named author that it could in fact be proved directly in the context of section 2. We sketch his argument.

We note first that $\mathrm{Ext}^q(U, V)$ is a functor in each variable, contravariant in the first one, covariant in the second one. In particular, if $f\colon V \to V'$ (resp. $g\colon U \to U'$) is a morphism in $\mathcal{C}_{\mathfrak{g}, \mathfrak{k}}$, there is associated to it canonically a homomorphism

$$f_2\colon \mathrm{Ext}^q(U, V) \to \mathrm{Ext}^q(U, V')$$
$$(\text{resp. } g_1\colon \mathrm{Ext}^q(U', V) \to \mathrm{Ext}^q(U, V)) \qquad (q \in \mathbf{N}).$$

For instance, if C^* (resp. C'^*) defines an injective resolution of V (resp. V'), then f extends, uniquely up to homotopy, to a morphism of C^* into C'^*, and hence it extends to a morphism of the complexes defining $\mathrm{Ext}^*(U, V)$ and $\mathrm{Ext}^*(U, V')$, and g extends obviously to a homomorphism of $\mathrm{Hom}_{\mathfrak{g}}(U', C^*)$ into $\mathrm{Hom}_{\mathfrak{g}}(U, C^*)$.

Consider in particular the case where $U = U'$, $V = V'$, $f(v) = z \cdot v$ (resp. $g(u) = z \cdot u$) for some fixed $z \in Z(\mathfrak{g})$.

4.4. LEMMA. *Let $z \in Z(\mathfrak{g})$, $U, V \in \mathcal{C}_{\mathfrak{g}, \mathfrak{k}}$, and $q \in \mathbf{N}$. Then the homomorphisms $z_1, z_2\colon \mathrm{Ext}^q(U, V) \to \mathrm{Ext}^q(U, V)$ associated to $z\colon U \to U$ and $z\colon V \to V$ defined by $x \mapsto z \cdot x$ are identical.*

First let $q = 0$. If $f \in \mathrm{Hom}_{\mathfrak{g}}(U, V)$, then $z_1 f(u) = z \cdot f(u)$ and $z_2 \cdot f(u) = f(z \cdot u)$ ($u \in U$); hence $z_1 = z_2$ in this case. Let $q \geq 1$ and assume our assertion proved up to $q - 1$. Let

(1)
$$0 \to V \to C \to V' \to 0$$

be a short exact sequence in $\mathcal{C}_{\mathfrak{g}, \mathfrak{k}}$, where C is injective (2.6). Since $\mathrm{Ext}^j(U, C) = 0$ for $j \geq 1$, the exact cohomology sequence associated to the exact sequence

(2)
$$0 \to \mathrm{Hom}_F(U, V) \to \mathrm{Hom}_F(U, C) \to \mathrm{Hom}_F(U, V') \to 0$$

gives rise to a homomorphism

(3)
$$\mathrm{Ext}^{j-1}(U, V') \to \mathrm{Ext}^j(U, V) \qquad (j = 1, 2, \cdots),$$

which is surjective for $j = 1$, an isomorphism for $j \geq 2$. Since this homomorphism commutes with z_1 and z_2, the passage from $q - 1$ to q follows.

4.5. Second proof of 4.1. Let z be as in the above proof of 4.1. Then $z : U \to U$ is the zero map; hence $z_1 \cdot \mathrm{Ext}^q(U, V) = 0$, while $z : V \to V$ is the identity, hence z_2 is the identity of $\mathrm{Ext}^q(U, V)$ ($q \in \mathbf{N}$). Since $z_1 = z_2$, this implies $\mathrm{Ext}^q(U, V) = 0$.

5. Extension to $(\mathfrak{g}, K)$-modules

In this section $F = \mathbf{R}$, G is a Lie group with finite component group, and K a maximal compact subgroup of G.

5.1. Cohomology. Let K^0 be the identity component of K. Let (π, V) be a $(\mathfrak{g}, K)$-module (0, 2.5). We put

(1)
$$C^q(\mathfrak{g}, K; V) = \mathrm{Hom}_K(\Lambda^q(\mathfrak{g}/\mathfrak{k}), V),$$

where K acts on $\mathfrak{g}/\mathfrak{k}$ via the adjoint representation. Clearly

(2)
$$C^q(\mathfrak{g}, K; V) \subset C^q(\mathfrak{g}, K^0; V) = C^q(\mathfrak{g}, \mathfrak{k}, V).$$

Moreover, K/K^0 acts naturally on $C^q(\mathfrak{g}, \mathfrak{k}; V)$ and we have

(3)
$$C^q(\mathfrak{g}, K; V) = C^q(\mathfrak{g}, \mathfrak{k}; V)^{K/K^0}.$$

Obviously, the $C^q(\mathfrak{g}, K; V)$ form a subcomplex of $C(\mathfrak{g}, \mathfrak{k}; V)$. The resulting cohomology groups are denoted $H^p(\mathfrak{g}, K; V)$. It follows immediately from (3) that we have

(4)
$$H^q(\mathfrak{g}, K; V) = H^q(\mathfrak{g}, \mathfrak{k}; V)^{K/K^0}.$$

5.2. The functors $\mathrm{Ext}^q(U, V)$. The group G also acts on the tensor algebra of $\mathfrak{g}$ and on $R = U(\mathfrak{g})$ by extension of the adjoint representation. As in 2.4, it is seen that R thus becomes a $(\mathfrak{g}, K)$-module. It follows then that if U is a locally finite semi-simple K-module, then $I(U) = R \otimes_S U$ ($S = U(\mathfrak{k})$), endowed with the K-action stemming from the tensor product of its actions on R and U, and with the $\mathfrak{g}$-action given by left translations on R, is a $(\mathfrak{g}, K)$-module, which is projective in $\mathcal{C}$. If U and V are $(\mathfrak{g}, K)$-modules, then $\mathrm{Ext}^q(U, V)$ is defined as the q-th cohomology group of the complex $\{\mathrm{Hom}_{\mathfrak{g}, K}(X_i, V)\}$, where (X_i) is a projective resolution of U. There is a natural action of K/K^0 on $\mathrm{Hom}_{\mathfrak{g}, K^0}(U, V)$ and on the complex $\mathrm{Hom}_{\mathfrak{g}, K^0}(X_i, V)$, and we have

(1)
$$\mathrm{Hom}_{\mathfrak{g}, K}(X_i, V) = (\mathrm{Hom}_{\mathfrak{g}, K^0}(X_i, V))^{K/K^0}.$$

Hence

(2)
$$\operatorname{Ext}^q_{\mathfrak{g},K}(U, V) = (\operatorname{Ext}^q_{\mathfrak{g},K^0}(U, V))^{K/K^0}.$$

In particular,

(3)
$$\operatorname{Ext}^q_{\mathfrak{g},K}(U, V) = 0 \qquad (q \neq 0),$$
$$\operatorname{Ext}^0_{\mathfrak{g},K}(U, V) = \operatorname{Hom}_K(U, V) \quad \text{if } \mathfrak{g} = \mathfrak{k}.$$

Moreover, it follows from the definitions that we also have

(4) $\operatorname{Ext}^q_{R,S}(U, V) = \operatorname{Ext}^q_{\mathfrak{g},K^0}(U, V) \qquad (q \in \mathbf{N}; U, V \in \mathcal{C}_{\mathfrak{g},K^0}).$

The identification with classes of long exact sequences proceeds as in §3.

REMARK. If (π, V) is a smooth G-module, then the evaluation map of 1.4 induces an isomorphism

$$A(G/K; V) \overset{\sim}{\to} C^*(\mathfrak{g}, K; V) \cong C^*(\mathfrak{g}, K; V_{(K)});$$

hence

$$H^*(A(G/K; V)) = H^*(\mathfrak{g}, K; V) = H^*(\mathfrak{g}, K; V_{(K)}).$$

5.3. THEOREM. *Let U, V be $(\mathfrak{g}, K)$-modules. Assume that they have infinitesimal characters χ_U, χ_V (resp. central characters ω_U, ω_V).*

(i) *If $\chi_U \neq \chi_V$ (resp. $\omega_U \neq \omega_V$), then $\operatorname{Ext}^q_{\mathfrak{g},K}(U, V) = 0$ for all q's.*

(ii) *Let U be finite dimensional. If $\chi_{\widetilde{U}} \neq \chi_V$ (resp. $\omega_{\widetilde{U}} \neq \omega_V$), then $H^q(\mathfrak{g}, K; U \otimes V) = 0$ for all q's.*

The reduction of (ii) to (i) is as in 4.2. The assertion (i) for the infinitesimal characters can be proved as in 4.1, or reduced to 4.1 using 5.2(2),(4).

Given a $(\mathfrak{g}, K)$-module M, any element $z \in Z$ defines an automorphism of M commuting with $\mathfrak{g}$ and K; hence the group algebra $\mathbf{R}[Z]$ of Z operates on M as an algebra of endomorphisms of $(\mathfrak{g}, K)$-module. Now if $\omega_U \neq \omega_V$, there exists $z \in \mathbf{R}[Z]$ such that $\omega_U(z) = 0$, $\omega_V(z) = 1$. The vanishing of $\operatorname{Ext}^q(U, V)$ then follows as in 4.1. More directly, one can let Z operate on resolutions, using the fact that every $(\mathfrak{g}, K)$-module is a direct sum of eigenspaces for Z.

5.4. COROLLARY. *Assume G to be connected, reductive $(0, §3)$. Let H be the derived group of G and S the connected center of G. Let $G_0 = S \times H_0$, where H_0 is the analytic subgroup with Lie algebra $\mathfrak{h}$ in the simply connected complex Lie group H_c with Lie algebra $\mathfrak{h}_c$, and K_0 the maximal compact subgroup of G_0 with Lie algebra $\mathfrak{k}$. Assume U to be finite dimensional and $\operatorname{Ext}^q_{\mathfrak{g},K}(U, V) \neq 0$ for some $q \in \mathbf{N}$. Then V is also a $(\mathfrak{g}, K_0)$-module, and*

(1) $\operatorname{Ext}^i_{\mathfrak{g},K}(U, V) = \operatorname{Ext}^i_{\mathfrak{g},K_0}(U, V) \quad \text{for all } i \in \mathbf{N}.$

The groups H and H_0 have a common finite covering $\widetilde{H}$; hence $\widetilde{G} = S \times \widetilde{H}$ is a common finite covering of G and G_0. Let $\widetilde{K}$ be the inverse image of K in $\widetilde{G}$. It is the maximal compact subgroup of $\widetilde{G}$ with Lie algebra $\mathfrak{k}$. Let $\widetilde{Z}$ be the centralizer of $\mathfrak{g}$ in K and $\alpha \colon \widetilde{G} \to G_0$, $\beta \colon \widetilde{G} \to G$ be the canonical projections. U and V can be viewed as $(\mathfrak{g}, \widetilde{K})$-modules, with central characters $\omega_U \circ \alpha$ and $\omega_V \circ \beta$. Since U is finite dimensional, it is also a G_0-module; therefore $\omega_U \circ \beta$ factors through α, and we have $\ker(\alpha|_{\widetilde{Z}}) \subset \ker(\omega_U \circ \beta)$. But $\omega_V = \omega_U$ by 5.3; hence this inclusion is also true with ω_V instead of ω_U, and the $\widetilde{K}$-module structure of V goes down to one of

K_0-module. This proves our first assertion. Then (1) follows from the fact that the Ext^i in question are both equal to $\mathrm{Ext}^i_{\mathfrak{g},\mathfrak{k}}$.

5.5. PROPOSITION (W. Casselman). *Let $\mathfrak{g}$ be reductive and U, V be finitely generated admissible $(\mathfrak{g}, K)$-modules. Then $\mathrm{Ext}^q_{\mathfrak{g},K}(U, V)$ can be computed using long exact sequences in the category of finitely generated admissible modules.*

(a) It follows from 2.7 that we may compute $\mathrm{Ext}_{\mathfrak{g},K}(U, V)$ using long exact sequences of finitely generated $(\mathfrak{g}, K)$-modules.

(b) Let $A \in \mathcal{C}_{\mathfrak{g},K}$ be finitely generated, and J an ideal of finite codimension of $Z(\mathfrak{g})$. Then $A/J \cdot A$ is admissible.

By 2.7, A is a quotient of a finite sum of R-modules of the form $R/R \cdot I$, where I is an ideal of finite codimension of S such that S/I is a simple $\mathfrak{k}$-module. It suffices to prove (b) for A of this form, but in this case it follows from a theorem of Harish-Chandra asserting that each isotypic subspace of A is a finite $Z(\mathfrak{g})$-module (cf. [**113**, 2.2.1.1]).

(c) Now let A be finitely generated and admissible. Then it is of finite length. This follows from the fact, proved by Harish-Chandra, that there are, up to infinitesimal equivalence, only finitely many admissible irreducible $(\mathfrak{g}, K)$-modules with a given infinitesimal character (cf. [**151**, 8.4.1]). It follows in particular that A contains a finite dimensional subspace C, sum of isotypic subspaces for K, such that if B is a $\mathfrak{g}$-submodule of A and $B \cap C = \{0\}$, then $B = \{0\}$.

(d) Let $0 \to V \xrightarrow{j} A$ be exact, with A, V finitely generated, and V admissible. Let W be a subspace of V playing for V the same role as C for A in (c). Let J be an ideal of $Z(\mathfrak{g})$ which annihilates $j(W)$. By the Artin-Rees lemma, there exists r such that

$$J^n A \cap j(W) \subset J(J^{n-r}A \cap j(W)) = 0,$$

for n big enough. It follows that, for such an n, the mapping $V \to A/J^n A$ is injective. Moreover, by (b), $A/J^n A$ is admissible. Thus V maps injectively into an admissible quotient of A.

The proposition now follows by standard homological algebra. We sketch the argument. Let

$$0 \longrightarrow V \xrightarrow{j} A_1 \xrightarrow{u} A_2 \longrightarrow \cdots \longrightarrow A_n \longrightarrow U \longrightarrow 0$$

be an exact sequence of $(\mathfrak{g}, K)$-modules, where the A_i's are finitely generated. We want to see that it is equivalent to a sequence in which all the terms are admissible. If $n = 1$, then A_1 is already admissible. Assume our assertion proved for $n-1$. By (d) there exists an ideal $J \subset \mathfrak{z}$ of finite codimension such that $V \cap J \cdot A_1 = 0$. Then (1) is equivalent to

(2) $0 \to V \to A_1/JA_1 \to A_2/u(JA_1) \to A_2 \to \cdots \to A_n \to U \to 0,$

where now A_1/JA_1 is admissible; then we can pass to an equivalent sequence of admissible modules using the induction assumption.

The following result was pointed out to us by the referee.

5.6. COROLLARY. *Let V, W be finitely generated admissible $(\mathfrak{g}, K)$-modules. Let $\widetilde{V}$, $\widetilde{W}$ denote (as usual) the K-finite duals of V and W respectively. Then $\mathrm{Ext}^q_{\mathfrak{g},K}(V, W)$ is isomorphic with $\mathrm{Ext}^q_{\mathfrak{g},K}(\widetilde{W}, \widetilde{V})$.*

It is a result of Harish-Chandra that $\widetilde{V}$, $\widetilde{W}$ are admissible and finitely generated. If

$$E: 0 \longrightarrow W \xrightarrow{\ i\ } E_q \xrightarrow{\ \alpha_q\ } \cdots \longrightarrow E_1 \xrightarrow{\ \alpha_1\ } V \longrightarrow 0$$

represents $\xi \in \mathrm{Ext}_{\mathfrak{g},K}^q(V,W)$, define

$$\widetilde{E}: 0 \longrightarrow \widetilde{V} \xrightarrow{\ \alpha_1^*\ } \widetilde{E}_1 \longrightarrow \cdots \xrightarrow{\ \alpha_q^*\ } \widetilde{E}_q \xrightarrow{\ i^*\ } \widetilde{W} \longrightarrow 0$$

to be the K-finite dual sequence (α_i^* and i^* are the transpose mappings). If $E \xrightarrow{\phi} E'$ or $E' \xrightarrow{\psi} E$ is a morphism in $S_q(V,W)$ (see 3.1), then $\widetilde{E}' \xrightarrow{\phi^*} \widetilde{E}$ or $\widetilde{E} \xrightarrow{\psi^*} \widetilde{E}'$ is a morphism in $S_q(\widetilde{W}, \widetilde{V})$. This implies that if we set $\widetilde{\xi}$ equal to the class of $\widetilde{E}$ relative to $\equiv$ (see 3.1), then $\xi \mapsto \widetilde{\xi}$ is well defined. It is clear (see 3.2) that $\xi \mapsto \widetilde{\xi}$ defines a linear map of $\mathrm{Ext}_{\mathfrak{g},K}^q(V,W)$ into $\mathrm{Ext}_{\mathfrak{g},K}^q(\widetilde{W}, \widetilde{V})$. It is also clear by construction that $(\widetilde{\xi})^\sim = \xi$. Hence $\xi \mapsto \widetilde{\xi}$ is bijective.

6. $(\mathfrak{g}, \mathfrak{k}, L)$-modules.
A Hochschild-Serre spectral sequence in the relative case

In this section we prove the existence of a Hochschild-Serre spectral sequence in relative Lie algebra cohomology. We shall limit ourselves to our main case of interest, that of $(\mathfrak{g}, K)$-modules over $\mathbf{C}$, but there is an obvious variation for $(\mathfrak{g}, \mathfrak{k})$-modules (see 6.7).

6.1. The category of $(\mathfrak{g}, \mathfrak{k}, L)$-modules. Let L be a compact Lie group, whose Lie algebra $\mathfrak{l}$ contains an ideal isomorphic to $\mathfrak{k}$ (also to be denoted $\mathfrak{k}$), $\alpha \colon L \to \mathrm{Aut}\,\mathfrak{g}$ a continuous representation of L in $\mathrm{Aut}\,\mathfrak{g}$ by automorphisms which leave $\mathfrak{k}$ stable. Let K be the analytic subgroup of L with Lie algebra $\mathfrak{k}$.

A real vector space V is a $(\mathfrak{g}, \mathfrak{k}, L)$-module if the following conditions are fulfilled:

(i) $\mathfrak{g}$, hence $U(\mathfrak{g})$, and L operate on V. With respect to L, the space V is locally finite and semi-simple. The representation of L on any finite dimensional L-stable subspace is differentiable.

(ii) L is a group of operators for the $U(\mathfrak{g})$-module structure, i.e.,

$$x(u \cdot v) = x(u) \cdot x(v) \qquad (x \in L;\ u \in U(\mathfrak{g});\ v \in V).$$

(iii) Any finite dimensional K-stable subspace M of V is stable under $\mathfrak{k}$, and the differential of the representation of K in M is the representation of $\mathfrak{k}$ obtained by restriction of the representation of $\mathfrak{g}$.

Thus, V is a $(\mathfrak{g}, K)$-module with an additional group of operators L. We let $\mathcal{C}_{\mathfrak{g},\mathfrak{k},L}$ be the category of $(\mathfrak{g}, \mathfrak{k}, L)$-modules, the morphisms being the linear maps commuting with both $\mathfrak{g}$ and L. It is a subcategory of $\mathcal{C}_{\mathfrak{g},K}$.

6.2. Cohomology spaces. The complex $C^*(\mathfrak{g}, \mathfrak{k}; V) = \mathrm{Hom}_{\mathfrak{k}}(\Lambda(\mathfrak{g}/\mathfrak{k}), V)$ has a natural L-module structure, stemming from the actions on $\mathfrak{g}/\mathfrak{k}$ and V, which commutes with the differentials. Hence there is an L-module structure on $H^*(\mathfrak{g}, \mathfrak{k}; V)$, with respect to which this space is locally finite and semi-simple. Furthermore, we define $H^q(\mathfrak{g}, L; V)$ to be the q-th cohomology space of the complex $\mathrm{Hom}_L(\Lambda(\mathfrak{g}/\mathfrak{k}), V)$ ($q = 0, 1, 2, \cdots$). Since the L action is semi-simple, taking fixed points is an exact functor. Hence

$$(1) \qquad\qquad H^q(\mathfrak{g}, L; V) = H^q(\mathfrak{g}, \mathfrak{k}; V)^{L/K}.$$

The case considered in §5 is the one where K is open in L. However, our main reason for introducing this greater generality is to be able to consider also the case where $\mathfrak{k} = (0)$.

6.3. **Ext functors.** In $\mathcal{C}_{\mathfrak{g},\mathfrak{k},L}$ we may consider projective and injective modules, and derived functors of $\mathrm{Hom}_{\mathfrak{g},\mathfrak{k}}$ and of $\mathrm{Hom}_{\mathfrak{g},L}$. The actions of L on $\mathfrak{g}$ and $\mathfrak{k}$ extend to representations of L in R and S, with respect to which these are locally finite and semi-simple L-modules. The argument of 2.4 shows that if $V \in \mathcal{C}_{\mathfrak{g},\mathfrak{k},L}$, then $I(V) = R \otimes_S V$, endowed with the L-module structure given by the tensor product of the actions on the two factors, is a projective $(\mathfrak{g}, \mathfrak{k}, L)$-module. It follows that there is at least one projective resolution of V in $\mathcal{C}_{\mathfrak{g},\mathfrak{k},L}$ which is at the same time a projective resolution in $\mathcal{C}_{\mathfrak{g},\mathfrak{k}}$. In fact, the projective resolution $\{X_q\}$ of the groundfield given in 2.5 is one in $\mathcal{C}_{\mathfrak{g},\mathfrak{k},L}$. Consequently, the derived functors of $\mathrm{Hom}_{\mathfrak{g}}$ in $\mathcal{C}_{\mathfrak{g},\mathfrak{k},L}$ are the same as in $\mathcal{C}_{\mathfrak{g},\mathfrak{k}}$; but they are endowed moreover with a canonical structure of locally finite and semi-simple L-module, which may be defined from the action of L on any projective resolution in $\mathcal{C}_{\mathfrak{g},\mathfrak{k},L}$; standard arguments show it to be independent of the resolution.

As in 2.6, let $P^0(V) = \mathrm{Hom}_S(R, V)$, where $V \in \mathcal{C}_{\mathfrak{g},\mathfrak{k},L}$. It is an L-module in the obvious way. Let $P(V) = \mathrm{Hom}_S(R, V)_{(L)}$ be the space of L-finite vectors. By an argument similar to the one of 2.6, one sees that the representation of L on any finite dimensional L-stable subspace is differentiable, and therefore semi-simple. Thus $P(V) \in \mathcal{C}_{\mathfrak{g},\mathfrak{k},L}$ and is again injective. Hence there are injective resolutions of V in $\mathcal{C}_{\mathfrak{g},\mathfrak{k},L}$ which are injective resolutions in $\mathcal{C}_{\mathfrak{g},\mathfrak{k}}$.

We denote again by $\mathrm{Ext}_{\mathfrak{g},\mathfrak{k}}$ the derived functors of $\mathrm{Hom}_{\mathfrak{g}}$ in $\mathcal{C}_{\mathfrak{g},\mathfrak{k},L}$, and moreover let $\mathrm{Ext}_{\mathfrak{g},L}$ be the derived functors of $\mathrm{Hom}_{\mathfrak{g},L}$ in that category. If $U, V \in \mathcal{C}_{\mathfrak{g},\mathfrak{k},L}$, and if $0 \to V \to C^0 \to \cdots$ is an injective resolution of V in $\mathcal{C}_{\mathfrak{g},\mathfrak{k},L}$, then $\mathrm{Ext}_{\mathfrak{g},\mathfrak{k}}(U, V)$ is the q-th cohomology of $\mathrm{Hom}_{\mathfrak{g}}(U, C^q)$, while $\mathrm{Ext}^q_{\mathfrak{g},L}(U, V)$ is the q-th cohomology space of the complex $\{\mathrm{Hom}_{\mathfrak{g},L}(U, C^i)\}$.

6.4. LEMMA. *Let $\mathfrak{n}$ be an ideal of $\mathfrak{g}$ which is stable under L, and H a closed subgroup of L. Let V be an injective $(\mathfrak{g}, \mathfrak{k}, L)$-module. Then V is also injective as an $(\mathfrak{n}, \mathfrak{k} \cap \mathfrak{n}, H)$-module.*

Since $[\mathfrak{n}, \mathfrak{k}] \subset \mathfrak{n}$, the algebra $\mathfrak{k}$ is the direct sum of two ideals $\mathfrak{k}_1 = \mathfrak{k} \cap \mathfrak{n}$ and $\mathfrak{k}_2$. Let $S_i = U(\mathfrak{k}_i)$ $(i = 1, 2)$.

There are L-invariant subspaces $\mathfrak{m}$, $\mathfrak{m}'$ of $\mathfrak{g}$ such that $\mathfrak{g} = \mathfrak{n} \oplus \mathfrak{m}$ and $\mathfrak{m} = \mathfrak{m}' \oplus \mathfrak{k}_2$. Using the Poincaré-Birkhoff-Witt theorem [**25**], we see that we can write

(1) $$R = U(\mathfrak{n}) \otimes M, \qquad M = M' \otimes U(\mathfrak{k}_2) = M' \otimes S_2,$$

with M and M' stable under L. Also, M is invariant under right translations by S_2. By the so-called adjoint associativity between Hom and $\otimes$ (see e.g. [**78**, VI, (8.7)]), we have

(2) $$\mathrm{Hom}_S(R, U) = \mathrm{Hom}_{S_1}(U(\mathfrak{n}), \mathrm{Hom}_{S_2}(M, U)),$$

where S_1 acts on $U(\mathfrak{n})$ by left translations, S_2 acts on M by right translations, and S_1 acts on $\mathrm{Hom}_{S_2}(M, U)$ by the given action on U (this is compatible with the S_2-action, since S_1 and S_2 commute). Moreover, this isomorphism is compatible with the natural operations of L, whence an isomorphism

(3) $$\mathrm{Hom}_S(R, U)_{(H)} = \mathrm{Hom}_{S_1}(U(\mathfrak{n}), \mathrm{Hom}_{S_2}(M, U))_{(H)}.$$

Thus $\operatorname{Hom}_S(R, U)_{(H)}$ can be written in $\mathcal{C}_{\mathfrak{n}, \mathfrak{k}_1, H}$ in the form $P(U')$, for some $U' \in \mathcal{C}_{\mathfrak{n}, \mathfrak{k}_1, H}$. Hence it is injective in that category.

6.5. THEOREM. *Let $\mathfrak{n}$ be an ideal in $\mathfrak{g}$ stable under L, $\mathfrak{k}_1 = \mathfrak{k} \cap \mathfrak{n}$, K_1 a closed normal subgroup of L with Lie algebra $\mathfrak{k}_1$, and $V \in \mathcal{C}_{\mathfrak{g}, \mathfrak{k}, L}$. Then there exist a spectral sequence which abuts to $H^*(\mathfrak{g}, \mathfrak{k}; V)$, in which $E_2^{p,q} = H^p(\mathfrak{g}/\mathfrak{n}, \mathfrak{k}/\mathfrak{k}_1; H^q(\mathfrak{n}, \mathfrak{k}_1; V))$, and a spectral sequence which abuts to $H^*(\mathfrak{g}, L; V)$ and in which $E_2^{p,q} = H^p(\mathfrak{g}/\mathfrak{n}, L/K_1, H^q(\mathfrak{n}, K_1; V))$ $(p, q \in \mathbf{Z})$.*

The argument is the standard one. Start from an injective resolution

(1)
$$0 \to V \to C^0 \to C^1 \to \cdots$$

of V in $\mathcal{C}_{\mathfrak{g}, \mathfrak{k}, L}$ and consider the subcomplex $D^* = \{C^{i, \mathfrak{n}, K_1}\}$ of elements in C^* fixed under $\mathfrak{n}$ and K_1. By 6.4, (1) is also an injective resolution of V in $\mathcal{C}_{\mathfrak{n}, \mathfrak{k}_1, K_1}$; therefore,

(2)
$$H^q(D^*) = H^q(\mathfrak{n}, K_1; V) \qquad (q \in \mathbf{N}).$$

The complex D^* is a complex of $(\mathfrak{g}/\mathfrak{n}, \mathfrak{k}/\mathfrak{k}_1, L/K_1)$-modules, whence a natural structure of $(\mathfrak{g}/\mathfrak{n}, \mathfrak{k}/\mathfrak{k}_1, L/K_1)$-module on the right hand side of (2). It follows immediately from the definitions that if M is injective in $\mathcal{C}_{\mathfrak{g}, \mathfrak{k}, L}$, then $M^{\mathfrak{n}, K_1}$ is injective in $\mathcal{C}_{\mathfrak{g}/\mathfrak{n}, \mathfrak{k}/\mathfrak{k}_1, L/K_1}$. Thus the D^i's are injective in the latter category. In particular they are acyclic.

Let F^{**} be the direct sum of the complexes

(3)
$$F^{*q} = C^*(\mathfrak{g}/\mathfrak{n}, \mathfrak{k}/\mathfrak{k}_1; D^{*q}) \qquad (q \in \mathbf{N}).$$

It is a "first quadrant" double complex in the usual way. We consider the two spectral sequences $('E_r)$ and $(''E_r)$ associated to the two filtrations of F^{**} defined by the partial degrees. If the degree in D^{*n} is used, giving rise to the "second filtration", then

(4)
$$''E_0^{*,q} = C^*(\mathfrak{g}/\mathfrak{n}, \mathfrak{k}/\mathfrak{k}_1; D^q),$$

and the differential d_0'' is that of §1, hence

(5)
$$''E_1^{p,q} = H^p(\mathfrak{g}/\mathfrak{n}, \mathfrak{k}/\mathfrak{k}_1; D^q).$$

Since the D^q are injective, hence acyclic, in the category of $(\mathfrak{g}/\mathfrak{n}, \mathfrak{k}/\mathfrak{k}_1)$-modules, we have

(6)
$$''E_1^{p,q} = 0 \quad \text{if } p \neq 0,$$
$$''E_1^{0,q} = (D^q)^{\mathfrak{g}/\mathfrak{n}} = C^{q, \mathfrak{g}, K_1} \qquad (q \in \mathbf{N}).$$

Then the differential d_1'' is induced by that of C^*, and hence

(7)
$$''E_2^{0,p} = H^q(C^{*, \mathfrak{g}, K_1}) = ''E_\infty^{0,p} = H^q(\mathfrak{g}, \mathfrak{k}; V)^{K_1},$$
$$''E_2^{p,q} = ''E_\infty^{p,q} = 0 \quad \text{if } p \neq 0,$$
$$H^q(F^{**}) = H^q(\mathfrak{g}, \mathfrak{k}; V)^{K_1} \qquad (q \in \mathbf{N}).$$

We now consider the spectral sequence $('E_r)$ associated to the "first filtration" (by the degree in C^*). We have then

(8)
$$'E_0^{p,*} = C^p(\mathfrak{g}/\mathfrak{n}, \mathfrak{k}/\mathfrak{k}_1; D^*) \qquad (p \in \mathbf{N}),$$

the differential d'_0 being induced by the differential of D^*. It is clear from the definition that the formation of the relative Lie algebra cochain complex is an exact functor, therefore

$$(9) \qquad 'E_1^{p,q} = C^q(\mathfrak{g}/\mathfrak{n}, \mathfrak{k}/\mathfrak{k}_1; H^q(D^*)) \qquad (p, q \in \mathbf{N}).$$

It follows then from (2) and (7) that, if K_1 is connected, $('E_r)$ gives the first spectral sequence of the theorem. Furthermore, it is clear that L acts as a group of operators on the whole situation, and that the E_r's are locally finite semi-simple L-modules. The second spectral sequence is then obtained by taking L-invariants in the first one.

6.6. COROLLARY. *Let V be a $(\mathfrak{g}, K)$-module. Let H be a closed normal subgroup of K whose Lie algebra $\mathfrak{h}$ is normal in $\mathfrak{g}$. Then*

$$H^*(\mathfrak{g}, K; V) = H^*(\mathfrak{g}/\mathfrak{h}, K/H; V^H).$$

In fact, we have (see 5.2(3))

$$(1) \qquad H^q(\mathfrak{h}, H; V) = 0 \qquad (q \neq 0), \qquad H^0(\mathfrak{h}, H; V) = V^H.$$

We then consider the second spectral sequence of 6.5 in the case where $L = K$, $\mathfrak{n} = \mathfrak{h}$. The equalities (1) show that $E_2 = E_\infty$, $E_2^{p,q} = 0$ for $q \neq 0$, whence the result.

6.7. REMARK. There is also a Hochschild-Serre spectral sequence for $(\mathfrak{g}, \mathfrak{k})$-modules, which can be discussed in the framework of §2. The proof of its existence is analogous to the one above, but simpler. Let $\mathfrak{l}$ be a Lie algebra of derivations of $\mathfrak{g}$ leaving $\mathfrak{k}$ stable, under which $\mathfrak{g}$ is fully reducible. Let us define a $(\mathfrak{g}, \mathfrak{k}, \mathfrak{l})$-module V to be a $(\mathfrak{g}, \mathfrak{k})$-module and a $\mathfrak{l}$-module, which is locally finite and semi-simple with respect to $\mathfrak{l}$, such that

$$x(y \cdot v) = (x \cdot y) \cdot z + y \cdot (x \cdot z) \qquad (x \in \mathfrak{l}, \ y \in \mathfrak{g}, \ v \in V).$$

Again the derived functors of $\mathrm{Hom}_\mathfrak{g}$ in the category of $(\mathfrak{g}, \mathfrak{k}, \mathfrak{l})$-modules are the same as in $\mathcal{C}_{\mathfrak{g},\mathfrak{k}}$, but are $\mathfrak{l}$-modules in a natural way. Of course, $\mathcal{C}_{\mathfrak{g},\mathfrak{k}}$ may be identified with $\mathcal{C}_{\mathfrak{g},\mathfrak{k},\mathfrak{k}}$. Using this, one deduces, exactly as in 6.5, the existence of a Hochschild-Serre spectral sequence in $\mathcal{C}_{\mathfrak{g},\mathfrak{k}}$.

7. Poincaré duality

The main results of this section, (7.3) to (7.6), are due to D. Vogan (unpublished).

7.1. Let G be a connected reductive group $(0, §3)$. We recall that a Cartan subalgebra of $\mathfrak{g}$ is *fundamental* if it contains a Cartan subalgebra of a maximal compact subgroup of G. The fundamental Cartan subalgebras form one conjugacy class under inner automorphisms [**113**, 1.3.3.3], and the corresponding Cartan subgroups (also said to be fundamental) are connected [**113**, 1.4.1.4]. If H is a normal connected subgroup of G, and $\mathfrak{c}$ (resp. C) a fundamental Cartan subalgebra (resp. subgroup) of $\mathfrak{g}$ (resp. G), then $\mathfrak{c} \cap \mathfrak{h}$ is a fundamental Cartan subalgebra (resp. subgroup) of $\mathfrak{h}$ (resp. H).

7.2. LEMMA. *Let $\mathfrak{c}$ be a fundamental Cartan subalgebra of $\mathfrak{g}$, $C = \mathcal{Z}_G(\mathfrak{c})$, and μ an automorphism of $\mathfrak{g}$ which fixes $\mathfrak{c}$ pointwise. Then $\mu = \mathrm{Ad}\, c$, for some $c \in C$. In particular, μ is inner.*

By the remark at the end of 7.1, we may replace G by its derived group, hence take G to be semi-simple. Also, since C is connected, we may replace G by an isogeneous group, and assume it is linear.

We may assume $\mathfrak{c}$ to be stable under the Cartan involution θ associated to a given maximal compact subgroup K of G. Then $\mathfrak{t} = \mathfrak{k} \cap \mathfrak{c}$ is a Cartan subalgebra of $\mathfrak{k}$, contains elements which are regular in $\mathfrak{g}$, and $C = \mathcal{Z}_G(\mathfrak{t})^0$. The automorphism μ extends to an automorphism of $\mathfrak{g}_c$ which is the identity on the Cartan subalgebra $\mathfrak{c}_c$, hence of the form $b = \exp x$, with $x \in \mathfrak{c}_c$.

It is known that any two maximal compact subgroups of G are conjugate by an element which centralizes their intersection. Hence, after having composed μ with an automorphism $\operatorname{Ad} a$ $(a \in C)$, we may assume that μ stabilizes K. Its restriction to K is then of the form $\operatorname{Ad} t^{-1}$, with $t \in T = \mathcal{Z}_K(\mathfrak{t}) = \exp \mathfrak{t}$. Replacing μ by $\operatorname{Ad} t \circ \mu$, we are reduced to the case where μ fixes both $\mathfrak{k}$ and $\mathfrak{c}$ pointwise. Then μ commutes with θ and the complex conjugation τ of $\mathfrak{g}_c$ with respect to $\mathfrak{g}$. The automorphisms θ and τ commute and generate a finite group of automorphisms of $\mathfrak{g}_c$ (viewed as a real Lie algebra) leaving $\mathfrak{c}_c$ stable. Replacing x by its average y over that group, we see that $b = \exp y$, with $y \in \mathfrak{c}_c$ fixed by θ and τ; hence $y \in \mathfrak{t}$. But then $b \in T \subset C$.

7.3. PROPOSITION. *Let $\mathfrak{g}$ be a real semi-simple Lie algebra. There exists one and only one connected component of $\operatorname{Aut} \mathfrak{g}$ with the following property: given a Cartan subalgebra $\mathfrak{c}$ of $\mathfrak{g}$, there exists $\mu \in Q$ which is equal to $-\operatorname{Id}$ on $\mathfrak{c}$.*

We let $G = \operatorname{Ad} \mathfrak{g}$. The uniqueness of Q follows from 7.2. It remains to prove its existence. We fix a Cartan involution θ of $\mathfrak{g}$ and let $\mathfrak{k}$ be its fixed point set.

a) Let $\mathfrak{g} = \mathfrak{sl}_2(\mathbf{R})$. Then $\operatorname{Aut} \mathfrak{g}$ has two connected components. We claim that the component $Q \neq \operatorname{Ad} \mathfrak{g}$ satisfies our conditions. In fact, up to inner automorphisms, $\mathfrak{g}$ has two Cartan subalgebras, the Lie algebras $\mathfrak{c}_1$ of skew symmetric matrices and $\mathfrak{c}_2$ of diagonal matrices with trace zero. Then $\operatorname{Ad} x_i$, where

$$x_1 = \begin{pmatrix} 1 & 0 \\ 0 & -1 \end{pmatrix}, \qquad x_2 = \begin{pmatrix} 0 & 1 \\ 1 & 0 \end{pmatrix},$$

induces $-\operatorname{Id}$ on $\mathfrak{c}_i$ $(i = 1, 2)$ and belongs to Q.

b) Assume now $\mathfrak{c}$ to be a fundamental Cartan subalgebra of $\mathfrak{g}$. The existence of an automorphism μ of $\mathfrak{g}$ which induces $-\operatorname{Id}$ on $\mathfrak{c}$ follows from 3.5 in [**6**] (which in turn is based on results of F. Gantmacher). We explain briefly how to reduce the proof to that lemma, using the notation introduced there. We take $\mathfrak{h}_c = \mathfrak{c}_c$, and fix a compact form $\mathfrak{g}_u$ of $\mathfrak{g}$ and a Chevalley basis of $\mathfrak{g}_c$ as in loc. cit. Let θ be a Cartan involution of $\mathfrak{g}$ leaving $\mathfrak{c}$ stable. Then, since $\mathfrak{c}$ is fundamental, it contains an element fixed under θ and regular in $\mathfrak{g}$. Then we may take this θ for the θ of [**6**, 3.5]. It follows then from line 3 on p. 115 of [**6**] that

$$\mu \colon h \mapsto -h \quad (h \in \mathfrak{c}_c), \qquad y_b \mapsto y_{-b} \quad (b \in \Phi),$$

is an automorphism of $\mathfrak{g}_c$ leaving $\mathfrak{g}_u$ stable and commuting with θ. But, then, it also leaves $\mathfrak{g}$ stable, and induces $-\operatorname{Id}$ on $\mathfrak{c}$. [To be complete, we should remark that in the equality $\theta(y_b) = \pm y_b$ of [**6**, p. 115], for $b = \theta(b)$, the signs for b and $-b$ have to be the same; this follows from the fact that θ must fix h_b.]

c) By 7.2 and b) there exists a unique connected component Q of $\operatorname{Aut} \mathfrak{g}$ which satisfies our condition for fundamental Cartan subalgebras. We now prove that it satisfies it for all Cartan subalgebras $\mathfrak{c}$, by induction on the dimension of the split part of $\mathfrak{c}$. Let $\mathfrak{c}$ be non-fundamental. Then it has a real root [**113**, 1.3.3.4], call it α. Let $\mathfrak{u}$ be the kernel of α. Then $\mathfrak{z}(\mathfrak{u}) = \mathfrak{u} \oplus \mathfrak{m}$, with $\mathfrak{m}$ isomorphic to $\mathfrak{sl}_2(\mathbf{R})$. We have $\mathfrak{c} = \mathfrak{u} \oplus \mathfrak{h}$, where $\mathfrak{h} = \mathfrak{m} \cap \mathfrak{c}$ is a split Cartan subalgebra of $\mathfrak{m}$. Let $\mathfrak{t}$ be a compact Cartan subalgebra of $\mathfrak{m}$. Then $\mathfrak{u} \oplus \mathfrak{t}$ is a Cartan subalgebra of $\mathfrak{g}$, whose split part has strictly smaller dimension than the split part of $\mathfrak{c}$. By our induction assumption, there exists $\mu \in Q$ which is equal to $-\operatorname{Id}$ on $\mathfrak{u} \oplus \mathfrak{t}$. Then μ leaves $\mathfrak{u}$ and therefore $\mathfrak{m}$ stable. By (a) and 7.2, there exists an inner automorphism ν of $\mathfrak{m}$ such that $\nu \circ \mu\big|_{\mathfrak{m}}$ is $-\operatorname{Id}$ on $\mathfrak{h}$. But $\operatorname{Ad}\mathfrak{m}$ imbeds into $\operatorname{Ad}\mathfrak{g}$, and so $\nu \circ \mu$ may be viewed as an automorphism of $\mathfrak{g}$. Then, clearly, $\nu \circ \mu$ is in Q and is equal to $-\operatorname{Id}$ on $\mathfrak{c}$.

7.4. COROLLARY. *Let G be a connected linear semi-simple group whose complexification G_c is simply connected. Then there exists a connected component Q of $\operatorname{Aut} G$ with the following property: given a Cartan subgroup C of G, there exists $\mu \in Q$ which induces the inversion $x \mapsto x^{-1}$ on C.*

The group G_c may be viewed as an algebraic group, $\mathcal{G}$, defined over $\mathbf{R}$, whose group of real points is G, and $\operatorname{Aut}\mathfrak{g}$ may be identified with the group of automorphisms of G_c which are defined over $\mathbf{R}$, or also with the group $\operatorname{Aut} G$ of Lie group automorphisms of G. Therefore, if C is a Cartan subgroup of G, any $\mu \in \operatorname{Aut}\mathfrak{g}$ which leaves $\mathfrak{c}$ stable and induces $-\operatorname{Id}$ on $\mathfrak{c}$, viewed as an automorphism of G, automatically leaves C stable and induces $x \mapsto x^{-1}$ on C. Thus 7.4 is just a translation of 7.3.

7.5. COROLLARY. *Let G and Q be as in 7.4, K a maximal compact subgroup of G, and μ an element of Q which leaves K stable. Let (π, V) be an irreducible admissible $(\mathfrak{g}, K)$-module. Then $({}^{\mu}\pi, V)$ is isomorphic to the contragredient $(\mathfrak{g}, K)$-module $(\widetilde{\pi}, \widetilde{V})$ to (π, V).*

[Here ${}^{\mu}\pi$ is defined as usual by ${}^{\mu}\pi(x) = \pi(\mu^{-1}(x))$ (for $x \in \mathfrak{g} \cup K$).]

Let (σ, E) be a differentiable irreducible admissible G-module such that $(\pi, V) = (\sigma_0, E_0)$ is the space of K-finite vectors of E [**77**], and let $\tau = {}^{\mu}\sigma$. Let θ_{σ} and θ_{τ} be the characters of σ and τ. They are locally summable functions which are analytic on the set of regular elements. To prove that τ is isomorphic to the contragredient representation $\widetilde{\sigma}$ of σ, it suffices to prove that if C is a Cartan subgroup, then $\theta_{\sigma}(h) = \theta_{\tau}(h^{-1})$ for every $h \in C$ which is regular in G. By 7.4, there exists $\nu \in Q$ which induces the inversion on C. Since μ and ν are in the same connected component of $\operatorname{Aut} G$, the representations τ and ${}^{\nu}\sigma$ are equivalent, hence have the same character. Therefore

$$\theta_{\tau}(h) = \theta_{\sigma}(\nu^{-1}(h)) = \theta_{\sigma}(h^{-1}),$$

for all $h \in C$ which are regular in G. The corollary follows.

7.6. PROPOSITION. *Let G be a connected reductive group, K a maximal compact subgroup of G and $m = \dim G/K$. Let (σ, E) be a finite dimensional representation of G and (π, V) an irreducible admissible $(\mathfrak{g}, K)$-module. Then $H^q(\mathfrak{g}, K; E \otimes V)$ is canonically isomorphic to the dual of $H^{m-q}(\mathfrak{g}, K; E \otimes V)$ for all $q \in \mathbf{Z}$.*

We may assume σ to be irreducible. Moreover, by 5.4, it suffices to consider the case where $G = S \times H$, with S commutative and H linear and semi-simple with simply connected complexification. The representations σ and π are then tensor products of representations of S and H of the same type, so that, by the Künneth rule (1.3), we are reduced to the cases where G is commutative, or is linear and semi-simple with simply connected complexification.

If G is commutative, then $E \otimes V$ is one-dimensional. If it is a non-trivial module, then $H^*(\mathfrak{g}, K; E \otimes V) = 0$, say by 5.3; otherwise

$$H^*(\mathfrak{g}, K; E \otimes V) = H^*(\mathfrak{g}, \mathfrak{k}; \mathbf{C}) = H^*(\mathfrak{g}/\mathfrak{k}; \mathbf{C}) = \Lambda(\mathfrak{g}/\mathfrak{k})^*,$$

whence our assertion follows in this case. In the other case, the automorphism μ of 7.5 transforms σ and π into the contragredient representations $\widetilde{\sigma}$, $\widetilde{\pi}$; therefore it induces isomorphisms

$$H^q(\mathfrak{g}, K; E \otimes V) \xrightarrow{\sim} H^q(\mathfrak{g}, K; \check{E} \otimes \check{V}) \qquad (q \in \mathbf{N}).$$

Our assertion then follows from 2.9.

8. The Zuckerman functors

Let $\mathfrak{g}$ be a Lie algebra over $\mathbf{C}$ and let K be a compact Lie group with Lie algebra $\mathfrak{k}$ that is a subalgebra of $\mathfrak{g}$. We assume that there is a representation, Ad, of K on $\mathfrak{g}$ such that relative to that action $\mathfrak{g}$ is a $(\mathfrak{k}, K)$-module. Let M be a closed subgroup of K. G. Zuckerman has assigned to a $(\mathfrak{g}, M)$-module V a finite sequence of $(\mathfrak{g}, K)$-modules $R^i\Gamma^{\mathfrak{g},K}_{\mathfrak{g},M} V$ $(0 \le i \le \dim K/M)$. Our purpose here is to provide an equivalent construction in the framework of this chapter which makes the so-called "duality theorem" (8.11) more transparent (this method is also described in [**152**].) The results of this section will be applied in VI to the discussion of the Vogan-Zuckerman theorem [**149**].

8.1. We look upon $C^\infty(K)$ as a right and left K-module under right and left translations. Let $\mathcal{H}(K)$ denote the space of all right (hence left) K-finite functions on K. Since it is an invariant subspace under both structures, it is endowed with two commuting $(\mathfrak{k}, K)$-module structures.

Let $M \subset K$ be a closed subgroup and let V be a $(\mathfrak{g}, M)$-module with action π. We view $V \otimes \mathcal{H}(K)$ as a $(\mathfrak{k}, M)$-module under $\pi \otimes l$ and as a $(\mathfrak{k}, K)$ module under $I \otimes r$. We now define an action of $\mathfrak{g}$ that commutes with the first action. For this we need a bit of notation. If $\lambda \in \mathfrak{g}^*$ and if $X \in \mathfrak{g}$, $k \in K$, then we set

$$c_{\lambda, X}(k) = \lambda(\mathrm{Ad}(k)X).$$

Then $c_{\lambda, X} \in \mathcal{H}(K)$. Let $X_1, \ldots, X_n$ be a basis of $\mathfrak{g}$ and let $\lambda_1, \ldots, \lambda_n$ be the dual basis of $\mathfrak{g}^*$. If $X \in \mathfrak{g}$, then we define

$$\mu(X) \colon V \otimes \mathcal{H}(K) \to V \otimes \mathcal{H}(K)$$

by

$$\mu(X)(v \otimes f) = \sum_i X_i v \otimes c_{\lambda_i, X} f \qquad (v \in V, f \in \mathcal{H}(K)).$$

The following result is proved by doing the obvious calculations.

LEMMA. 1) *If $X, Y \in \mathfrak{g}$, then $\mu([X, Y]) = \mu(X)\mu(Y) - \mu(Y)\mu(X)$.*
2) *If $Y \in \mathfrak{k}$ and $X \in \mathfrak{g}$, then $\mu(X) \circ (\pi \otimes l)(Y) = (\pi \otimes l)(Y) \circ \mu(X)$.*
3) *If $m \in M$ and $X \in \mathfrak{g}$, then $\mu(X)(\pi(m) \otimes l(m)) = (\pi(m) \otimes l(m))\mu(X)$.*

We will use the notation

(1) $$(\Gamma^K_M)^i(V) = H^i(\mathfrak{k}, M; V \otimes \mathcal{H}(K))$$

for the relative Lie algebra cohomology with coefficients in $V \otimes \mathcal{H}(K)$ viewed as a $(\mathfrak{k}, M)$-module under $\pi \otimes l$. If M and K are understood, then we will use the notation $\Gamma^i(V)$. Then the lemma implies that μ induces a $\mathfrak{g}$-module structure on $(\Gamma_M^K)^i(V)$ and, moreover, $I \otimes r$ defines a $(\mathfrak{k}, K)$-module structure on $(\Gamma_M^K)^i(V)$.

8.2. PROPOSITION. *Under the two actions described above, $(\Gamma_M^K)^i(V)$ is a $(\mathfrak{g}, K)$-module.*

The proof of this result is in subsection 8.4. We will first study the special case $i = 0$, which is simpler and contains the basic idea (to which we will refer) of the proof of the full result.

8.3. Clearly $(\Gamma_M^K)^0(V) = (V \otimes \mathcal{H}(K))^K$ relative to the action $\pi \otimes l$. The assertion that

$$\mu(X)(I \otimes r(k)) = (I \otimes r(k))\mu(\mathrm{Ad}(k)^{-1}X) \quad \text{for } X \in \mathfrak{g}, k \in K$$

follows from the following calculation

$$\mu(X)(I \otimes r(k))(v \otimes f) = \sum_i X_i v \otimes c_{\lambda_i, X} r(k) f$$

$$= \sum_i X_i v \otimes r(k)(r(k)^{-1}(c_{\lambda_i, X})f) \quad (v \in V, \ f \in \mathcal{H}(K)).$$

We note that

$$(r(k)^{-1}(c_{\lambda_i, X})f)(x) = c_{\lambda_i, X}(\mathrm{Ad}(x)\,\mathrm{Ad}(k)^{-1}X)f(x) = c_{\lambda_i, \mathrm{Ad}(k)^{-1}X}f.$$

In light of the first part of the computation the assertion now follows. We also must see that if $Y \in \mathfrak{k}$, then the action $\mu(Y)$ on $(\Gamma_M^K)^0(V)$ coincides with the action induced by $I \otimes r(Y)$. We choose our basis so that $X_1, \ldots, X_k$ give a basis of $\mathfrak{k}$ and $X_{k+1}, \ldots, X_n$ give a basis for an $\mathrm{Ad}(k)$-invariant complement to $\mathfrak{k}$. If $\sum_j v_j \otimes f_j \in (\Gamma_M^K)^0(V)$ $(v_j \in V, \ f_j \in \mathcal{H}(K))$, then

$$\mu(Y)(v \otimes f) = \sum_j \sum_{i=1}^k X_i v_j \otimes c_{\lambda_i, Y} f_j = -\sum_j \sum_{i=1}^k v_j \otimes l(X_i)(c_{\lambda_i, Y} f_j).$$

We now note that, since K is compact, $\mathrm{tr}(\mathrm{ad}(Y)_{|\mathfrak{k}}) = 0$ for $Y \in \mathfrak{k}$. Also if $x \in K$, then

$$\sum_{i=1}^k l(X_i)(c_{\lambda_i, Y} f_j)(x) = -\sum_i \lambda_i(\mathrm{ad}(X_i)\,\mathrm{Ad}(x)Y)f_j(x)$$

$$- \sum_i \lambda_i(\mathrm{Ad}(x)Y)l(X_i)f(x).$$

The above observation implies that the first term in the expression is 0, and a little thought tells us that the second is $r(Y)f$. The 0-th case is now completely proved. In the literature the $(\mathfrak{g}, K)$-module with this structure is denoted $(\Gamma_{\mathfrak{g}, M}^{\mathfrak{g}, K})(V)$.

8.4. We will now prove 8.2 by downward induction on i. If $i > p = \dim K/M$, then $\Gamma^i V = 0$ for all V. We consider $i = p$. Since $\mathrm{tr}(\mathrm{ad}(Y)_{|\mathfrak{k}}) = 0$ for $Y \in \mathfrak{k}$, we see that

$$\Gamma^p(V) = ((V \otimes \mathcal{H}(K))/((\pi \otimes l)(\mathfrak{k})(V \otimes \mathcal{H})))^M$$

(the M-invariants with respect to the action $\pi \otimes l$) with the action of K induced by $I \otimes r$ and the action of $\mathfrak{g}$ induced by μ. The same argument as in 8.3 proves the assertion for $i = p$. Assume the result for all V and all $p \geq j > i$. To establish it for i we consider $I(V) = U(\mathfrak{g}) \otimes_{U(\mathfrak{m})} V$ (as in 2.6). This module is projective in the category $\mathcal{C}(\mathfrak{g}, M)$. Let $\Phi \colon I(V) \to V$ be given by $\Phi(g \otimes v) = gv$. Set $W = \ker \Phi$. Note that Φ is surjective, so we have the short exact sequence

(1) $$0 \to W \to I(V) \to V \to 0.$$

We now claim

i) *As a* $(\mathfrak{k}, M)$*-module,* $I(V) \cong U(\mathfrak{k}) \otimes_{U(\mathfrak{m})} (Z \otimes V)$ *with* Z *an* $\mathrm{ad}(\mathfrak{k}), M$ *invariant subspace of* $U(\mathfrak{g})$ *such that* $U(\mathfrak{g})$ *is isomorphic with* $U(\mathfrak{k}) \otimes Z$ *as a* $(\mathfrak{k}, M)$*-module.*

This is deduced from the following observation of Lepowsky, which can be found in [**137**] (which we will also use later in this proof).

ii) *Let* $\mathfrak{g}$ *be a Lie algebra and let* $\mathfrak{m}$ *be a Lie subalgebra. If* W *is an* $\mathfrak{m}$*-module and if* V *is a* $\mathfrak{g}$ *module, then*

$$U(\mathfrak{g}) \otimes_{U(\mathfrak{m})} (W \otimes V) \cong (U(\mathfrak{g}) \otimes_{U(\mathfrak{m})} W) \otimes V$$

under the map $g \otimes (w \otimes v) \mapsto g \cdot ((1 \otimes w) \otimes v)$.

(For a proof, cf. [**151**, Lemma 6.A.1.2].)

Now

$$I(V) \cong (Z \otimes U(\mathfrak{k})) \otimes_{U(\mathfrak{m})} V \cong Z \otimes (U(\mathfrak{k}) \otimes_{U(\mathfrak{m})} V)$$
$$\cong (U(\mathfrak{k}) \otimes_{U(\mathfrak{m})} V) \otimes Z \cong U(\mathfrak{k}) \otimes_{U(\mathfrak{m})} (Z \otimes V).$$

Here the $\cong$ indicates either the obvious natural isomorphism or the one indicated in ii.

Thus, $I(V)$ is projective in $\mathcal{C}(\mathfrak{k}, M)$. We now apply Lepowsky's observation again and see that since

$$(U(\mathfrak{g}) \otimes_{U(\mathfrak{m})} V) \otimes \mathcal{H}(K) \cong U(\mathfrak{k}) \otimes_{U(\mathfrak{m})} (Z \otimes V) \otimes \mathcal{H}(K)$$
$$\cong U(\mathfrak{k}) \otimes_{U(\mathfrak{m})} (Z \otimes V \otimes \mathcal{H}(K))$$

(the action on $\mathcal{H}(K)$ is via l), $I(V) \otimes \mathcal{H}(K)$ is projective in $\mathcal{C}(\mathfrak{k}, M)$. If we tensor (1) with $\mathcal{H}(K)$, then we have the short exact sequence

(2) $$0 \to W \otimes \mathcal{H}(K) \to I(V) \otimes \mathcal{H}(K) \to V \otimes \mathcal{H}(K) \to 0$$

with the maps given by the given maps tensored with the identity mapping. This exact sequence is also compatible with the action $I \otimes r$ of K and μ of $\mathfrak{g}$.

The long exact sequence of cohomology now yields the shorter sequences

$$\to \Gamma^i(I(V)) \to \Gamma^i(V) \overset{\delta}{\to} \Gamma^{i+1}(W) \to \Gamma^{i+1}(I(V)).$$

Since $\Gamma^i(I(V)) = H^i(\mathfrak{k}, M; I(V) \otimes \mathcal{H}(K))$, $i < p$, and 1) implies that $I(V) \otimes \mathcal{H}(K)$ is projective in $\mathcal{C}(\mathfrak{k}, M)$, we see that $\Gamma^i(I(V)) = 0$. Thus δ is injective. Since δ is constructed by chasing diagrams, we see that δ intertwines the actions of K induced by $I \otimes r$ and the actions of $\mathfrak{g}$ given by μ. Since the compatibility is true for $i + 1$ and δ is injective, the compatibility is true for i. This completes the proof.

8.5. We now note two facts about this construction. The first is that

1) $\Gamma^i(V) \cong L_{p-i} \Gamma^p(V)$.

Here L_i indicates the i-th left derived functor from the category $\mathcal{C}(\mathfrak{g}, M)$ to $\mathcal{C}(\mathfrak{g}, K)$. The functor $\Gamma^p V$ is denoted by $\Pi_{\mathfrak{g}, M}^{\mathfrak{g}, K} V$ in [**140**]. This follows directly from 8.4.

The second, which will be established in 8.6, is

2) $\Gamma^i(V) \cong R^i \Gamma_{\mathfrak{g},M}^{\mathfrak{g},K}(V)$.

Here R^i stands for the right derived functor of $V \mapsto \Gamma_{\mathfrak{g},M}^{\mathfrak{g},K}V$ from the category $\mathcal{C}(\mathfrak{g}, M)$ to the category $\mathcal{C}(\mathfrak{g}, K)$.

8.6. The basic difference in the proof of 2) is that if $U \in \mathcal{C}(\mathfrak{m}, M)$, then as a $(\mathfrak{k}, M)$-module the injective module

$$P(U) = \operatorname{Hom}_{U(\mathfrak{m})}(U(\mathfrak{g}), U)_{(M)}$$

in $\mathcal{C}(\mathfrak{g}, M)$ is not obviously injective in $\mathcal{C}(\mathfrak{k}, M)$. To prove this we will need some additional notation.

For $V \in \mathcal{C}(\mathfrak{m}, M)$, we look upon V^* as an $\mathfrak{m}$ (resp. M)-module via $X \cdot \lambda(v) = -\lambda(Xv)$ (resp. $m \cdot \lambda(v) = \lambda(m^{-1}v)$) for $\lambda \in V^*$, $v \in V$, $X \in \mathfrak{m}$ (resp. $m \in M$). $\operatorname{Hom}_{U(\mathfrak{m})}(U(\mathfrak{k}), V^*)$ is a $\mathfrak{k}$- and an M-module under $X \cdot f(k) = f(kX)$ and $m \cdot f(k) = m(f(\operatorname{Ad}(m)^{-1}k))$ for $f \in \operatorname{Hom}_{U(\mathfrak{m})}(U(\mathfrak{k}), V^*)$, $k \in U(\mathfrak{k})$, $X \in \mathfrak{k}$ and $m \in M$.

LEMMA. $\operatorname{Hom}_{U(\mathfrak{m})}(U(\mathfrak{k}), V^*)_{(M)} = \operatorname{Hom}_{U(\mathfrak{m})}(U(\mathfrak{k}), V_{(M)}^*)_{(M)}$.

For $f \in \operatorname{Hom}_{U(\mathfrak{m})}(U(\mathfrak{k}), V^*)_{(M)}$, we denote by W_f the complex linear span of $M \cdot f$. Then W_f is an $(\mathfrak{m}, M)$ submodule. Let $T: W_f \to V^*$ be defined by $T(u) = u(1)$. Then $T(m \cdot u) = mT(u)$. Thus $T(W_f) \subset V_{(M)}^*$. This implies that if $f \in \operatorname{Hom}_{U(\mathfrak{m})}(U(\mathfrak{k}), V^*)_{(M)}$, then $f(1) \in V_{(M)}^*$. If $k \in K$, then $k \cdot f \in \operatorname{Hom}_{U(\mathfrak{m})}(U(\mathfrak{k}), V^*)_{(M)}$, and thus $f(k) = (k \cdot f)(1) \in V_{(M)}^*$. Consequently

$$\operatorname{Hom}_{U(\mathfrak{m})}(U(\mathfrak{k}), V^*)_{(M)} \subset \operatorname{Hom}_{U(\mathfrak{m})}(U(\mathfrak{k}), V_{(M)}^*)_{(M)}.$$

Since the reverse inclusion is obvious, the lemma follows.

We will now use the lemma to show that, if $U \in \mathcal{C}(\mathfrak{m}, M)$, then $P(U)$ is injective as an element of $\mathcal{C}(\mathfrak{k}, M)$. Clearly, we may assume that U is finite dimensional. Let Z be as in 8.4. Then, as an element of $\mathcal{C}(\mathfrak{k}, M)$,

$$P(U) = \operatorname{Hom}_{U(\mathfrak{m})}(U(\mathfrak{k}) \otimes Z, U)_{(M)} = \operatorname{Hom}_{U(\mathfrak{m})}(U(\mathfrak{k}), Z^* \otimes U)_{(M)}.$$

If we set $V = Z \otimes U^*$, then the lemma implies that $P(U)$ is isomorphic with $\operatorname{Hom}_{U(\mathfrak{m})}(U(\mathfrak{k}), (Z^* \otimes U)_{(M)})_{(M)}$, which is injective in the category $\mathcal{C}(\mathfrak{k}, M)$. This implies that $\Gamma^i P(U) = 0$ for $i > 0$. We can now argue as in 8.4 to prove the assertion 2) in 8.5 above.

8.7. For the record, we mention that the same arguments as in 8.4 imply

THEOREM. *If $V \in \mathcal{C}(\mathfrak{g}, K)$ and $W \in \mathcal{C}(\mathfrak{g}, M)$, then $\Gamma^i(W \otimes V) \cong \Gamma^i(W) \otimes V$ in $\mathcal{C}(\mathfrak{g}, K)$.*

8.8. We note the following consequence of the definitions.

THEOREM. *Let $V \in \mathcal{C}(\mathfrak{g}, M)$, and let $W \in \mathcal{C}(\mathfrak{k}, K)$ be irreducible. Then*

$$\dim \operatorname{Hom}_{\mathfrak{k},K}(W, \Gamma^i V) = \dim H^i(\mathfrak{k}, M; V \otimes W^*).$$

Indeed, $\mathcal{H}(K) = \bigoplus_{\gamma \in \widehat{K}} V_\gamma^* \otimes V_\gamma$ as a (K, K) bi-module (the left factor corresponds to the left regular representation). Thus we have the equivalence

$$\Gamma^i V = \bigoplus_{\gamma \in \widehat{K}} H^i(\mathfrak{k}, M; V \otimes V_\gamma^*) \otimes V_\gamma$$

as $(\mathfrak{k}, K)$-modules. The theorem now follows.

8.9. The functors $\Gamma_{\mathfrak{g},M}^{\mathfrak{g},K}$ and $\Pi_{\mathfrak{g},M}^{\mathfrak{g},K}$ have more direct definitions. If K is connected and if $V \in \mathcal{C}(\mathfrak{g}, K)$, then $\Gamma_{\mathfrak{g},M}^{\mathfrak{g},K} V$ is the space of all $v \in V$ such that the $\mathfrak{k}$-module action $U(\mathfrak{k})v$ is the differential of a finite dimensional representation of K (see below). In the general case we note that $\mathfrak{g}$ acts by its action on V:

$$\Gamma_{\mathfrak{g},M}^{\mathfrak{g},K} V = H^0(\mathfrak{k}, M; V \otimes \mathcal{H}(K)) = ((V \otimes \mathcal{H}(K))^{\mathfrak{k}})^M,$$

all relative to the left regular action of K on $\mathcal{H}(K)$. That is,

$$\Gamma_{\mathfrak{g},M}^{\mathfrak{g},K} V = \bigoplus_{\gamma \in \widehat{K}} \mathrm{Hom}_{\mathfrak{k}}(V_\gamma, V)^M \otimes V_\gamma.$$

Here M acts on $\mathrm{Hom}_{\mathfrak{k}}(U, V)$ by $(m \cdot T)(u) = mT(m^{-1}u)$.

Thus if K is connected we have the desired interpretation. If not, then set $M_1 = K^0 \cap M$ (K^0 is, as usual, the identity component of K). Then we have

$$\bigoplus_{\gamma \in \widehat{K}} \mathrm{Hom}_{\mathfrak{k}}(V_\gamma, V)^M \otimes V_\gamma = \mathrm{Ind}_{MK^0}^K (\Gamma_{\mathfrak{g},M_1}^{\mathfrak{g},K^0} V).$$

Here M acts on $\Gamma_{\mathfrak{g},M_1}^{\mathfrak{g},K^0} V$ via the action of M on V (K^0 being connected, we are allowed to use the above interpretation). Since $K/(MK^0)$ is finite, the induced representation is defined as the set of all functions $f \colon K \to \Gamma_{\mathfrak{g},M_1}^{\mathfrak{g},K^0} V$ such that $f(uk) = uf(k)$ for $u \in MK^0$ and $k \in K$. If $X \in \mathfrak{g}$, then $(Xf)(u) = (\mathrm{Ad}(u)X) \cdot f(u)$ and $(kf)(u) = f(uk)$ for $k, u \in K$.

We will only be dealing with the functors $\Gamma_{\mathfrak{g},M}^{\mathfrak{g},K}$ in this book. For a more direct definition of the functors $\Pi_{\mathfrak{g},M}^{\mathfrak{g},K}$ we refer the reader to [**140**].

8.10. In this subsection we give a proof of the duality theorem of [**135**] since it is the basis of all of the proofs of unitarizability of Zuckerman functors and since it is a simple calculation in the context of this section. As usual, if $V \in \mathcal{C}(\mathfrak{g}, M)$, then $\widetilde{V}$ will denote the contragredient $(\mathfrak{g}, M)$-module and $\check{V}$ will denote the conjugate dual $(\mathfrak{g}, M)$-module. Let dk denote the normalized Haar measure on K. On $\mathcal{H}(K)$ we define a symmetric form $(\ ,\ )$ and a sesquilinear form $\langle\ ,\ \rangle$ by

$$(f, g) = \int_K f(k)g(k)\, dk \qquad (f, g \in \mathcal{H}(K))$$

and

$$\langle f, g \rangle = \int_K f(k)\overline{g(k)}\, dk \qquad (f, g \in \mathcal{H}(K)).$$

If $V \in \mathcal{C}(\mathfrak{k}, M)$, then we use the symmetric (resp. sesquilinear) pairing of $V \otimes \mathcal{H}(K)$ with $\widetilde{V} \otimes \mathcal{H}(K)$ (resp. $\check{V} \otimes \mathcal{H}(K)$) given by the natural pairing between V and $\widetilde{V}$ (resp. $\check{V}$) tensored with $(\ ,\ )$ (resp. $\langle\ ,\ \rangle$). In light of section 2.9, Poincaré duality for $(\mathfrak{k}, M)$-cohomology implies that if $d = \dim K/M$, then we have a non-degenerate bilinear (resp. sesquilinear) pairing between $\Gamma^i(V)$ and $\Gamma^{d-i}(\widetilde{V})$ (resp. $\Gamma^{d-i}(\check{V})$). Here we use the obvious sesquilinear pairing of $\Lambda^i(\mathfrak{k}/\mathfrak{m})_\mathbf{C}^*$ with $\Lambda^{d-i}(\mathfrak{k}/\mathfrak{m})_\mathbf{C}^*$. We take $\mathfrak{g}$ to be a Lie algebra over $\mathbf{R}$ and the action of K to be real.

8.11. THEOREM. *Let $V \in \mathcal{C}(\mathfrak{g}, M)$. For each $0 \leq i \leq d$ the natural bilinear (resp. sesquilinear) pairing between $\Gamma^i(V)$ and $\Gamma^{d-i}(\widetilde{V})$ (resp. $\Gamma^{d-i}(\check{V})$) is $(\mathfrak{g}, K)$-invariant.*

We will prove the result in the sesquilinear case and leave the other (slightly simpler) case to the reader. We take $X_1, \ldots, X_n$ to be a basis of $\mathfrak{g}$ over $\mathbf{R}$. The functions $c_{\lambda_i, X}$ are real valued for all i and all $X \in \mathfrak{g}$. We show that if $v \in V$, $u \in \check{V}$ and $f, g \in \mathcal{H}(K)$, then

$$\langle \mu(X)(v \otimes f), u \otimes g \rangle = -\langle v \otimes f, \mu(X)(u \otimes g) \rangle$$

by doing the obvious calculation:

$$\langle \mu(X)(v \otimes f), u \otimes g \rangle = \sum_i \langle X_i v \otimes c_{\lambda_i, X} f, u \otimes g \rangle$$

$$= \sum_i \langle X_i v, u \rangle \langle c_{\lambda_i, X} f, g \rangle = \sum_i (-\langle v, X_i u \rangle) \langle f, c_{\lambda_i, X} g \rangle$$

$$= -\langle v \otimes f, \mu(X)(u \otimes g) \rangle.$$

Since the natural sesquilinear pairing between the i-th and $d-i$-th cohomology spaces is K-invariant, this proves that it is $(\mathfrak{g}, K)$ invariant.

CHAPTER II

Scalar Product, Laplacian and Casimir Element

1. Notation and general remarks

1.1. In this chapter G is a connected reductive Lie group $(0, \S3)$, K a maximal compact subgroup of G, and θ the Cartan involution associated to K. Our general reference for properties of maximal compact subgroups and Cartan involutions is [**125**]. We have the Cartan decomposition

$$(3) \qquad \mathfrak{g} = \mathfrak{k} \oplus \mathfrak{p}, \quad \text{where } \mathfrak{p} = \{x \in \mathfrak{g} \mid \theta(x) = -x\}.$$

We let B be a G- and θ-invariant non-degenerate symmetric bilinear form on $\mathfrak{g}$, whose restriction to $\mathfrak{k}$ (resp. $\mathfrak{p}$) is negative (resp. positive) non-degenerate. If $\mathfrak{g}$ is semi-simple, B will be the Killing form of $\mathfrak{g}$. We have

$$(4) \qquad B(\mathfrak{k}, \mathfrak{p}) = 0, \quad [\mathfrak{k}, \mathfrak{k}] \subset \mathfrak{k}, \quad [\mathfrak{k}, \mathfrak{p}] \subset \mathfrak{p}, \quad [\mathfrak{p}, \mathfrak{p}] \subset \mathfrak{k}.$$

It is also well known that if $\mathfrak{k}$ does not contain any non-zero ideal of $\mathfrak{g}$, then $[\mathfrak{p}, \mathfrak{p}] = \mathfrak{k}$.

In the sequel $m = \dim \mathfrak{p}$, $n = \dim \mathfrak{g}$, $(x_i)_{1 \le i \le m}$ is an orthonormal basis of $\mathfrak{p}$ and $(x_a)_{m < a \le n}$ a pseudo-orthonormal basis of $\mathfrak{k}$ with respect to B, i.e.,

$$(5) \qquad \begin{aligned} B(x_i, x_j) &= \delta_{ij} \quad (1 \le i, j \le m), \\ B(x_a, x_b) &= -\delta_{ab} \quad (m < a, b \le n). \end{aligned}$$

In general, we make the convention that indices i, j, k, l run from 1 to m, and indices a, b, c, d from $m+1$ to n. In view of (2), we have, with this convention

$$(6) \qquad [x_i, x_j] = \sum_a c_{i,j}^a x_a, \qquad [x_a, x_i] = \sum_j c_{a,i}^j x_j.$$

As usual, the structure constants are antisymmetric in the two lower indices. Moreover,

1.2. LEMMA. *We have $c_{i,j}^a = c_{aj}^i$ $(1 \le i, j \le m; m < a \le n)$.*

In fact, since B is invariant, we have

$$(1) \qquad B([x_i, x_j], x_a) + B(x_i, [x_a, x_j]) = 0.$$

By construction, the first term is equal to $-c_{ij}^a$ and the second one to c_{aj}^i.

1.3. We recall that if (y_s) is a basis of $\mathfrak{g}$ and (y_s') the dual basis of $\mathfrak{g}$ with respect to B, then

$$(1) \qquad C = \sum_{1 \le s \le n} y_s \cdot y_s'$$

represents an element in the center of the universal enveloping algebra $U(\mathfrak{g})$ of $\mathfrak{g}$, which is independent of the choice of the basis, and is called the *Casimir element* of $U(\mathfrak{g})$. With the notation of 1.1, we have in particular

$$(2) \qquad C = \sum x_j^2 - \sum x_a^2.$$

1.4. Relative Lie algebra cohomology. Let (π, V) be a $(\mathfrak{g}, \mathfrak{k})$-module. We may write

$$(1) \qquad C^q(V) = C^q(\mathfrak{g}, \mathfrak{k}; V) = \mathrm{Hom}_{\mathfrak{k}}(\Lambda^q \mathfrak{p}, V) = (\Lambda^q \mathfrak{p}^* \otimes V)^{\mathfrak{k}}.$$

Moreover, in view of the relation $[\mathfrak{p}, \mathfrak{p}] \subset \mathfrak{k}$, there are no bracket terms in the formula for the coboundary operator (I, 1.1(2)); therefore

$$(2) \qquad d\eta(y_0, \ldots, y_q) = \sum_i (-1)^i y_i \cdot \eta(y_0, \ldots, \widehat{y_i}, \ldots, y_q) \qquad (\eta \in C^q(V)).$$

Let

$$(3) \qquad D^q(V) = \mathrm{Hom}_{\mathbf{R}}(\Lambda^q \mathfrak{p}, V).$$

Evidently, $D^q(V)$ contains $C^q(V)$. We note that (2) also makes sense on D^q, hence defines a linear operator $D^q(V) \to D^{q+1}(V)$, also to be denoted d. The space $D^q(V)$ may be identified with the subspace of $C^q(\mathfrak{g}, V)$ whose elements are annihilated by the interior products i_x $(x \in \mathfrak{k})$. Let d_0 be the coboundary operator in $C^q(\mathfrak{g}; V)$. Then, of course, $d_0 f = df$ if $f \in C^q(\mathfrak{g}, k; V)$. However, if $f \in D^q(V)$, then df is the restriction of $d_0 f$ to $\Lambda^{q+1}\mathfrak{p}$, but is not equal to df in general. In particular, we do not necessarily have $d^2 = 0$ on $D^q(V)$. Since d_0 commutes with the θ_x $(x \in \mathfrak{g})$ and $D^q(V)$ is stable under θ_x for $x \in \mathfrak{k}$, we have

$$(4) \qquad d \circ \theta_x = \theta_x \circ d \quad \text{on } D^q(V) \text{ for all } x \in \mathfrak{k}.$$

1.5. Notation for cochains. If A is a finite set, then $|A|$ denotes its cardinality. For a positive integer s, let $I_s = \{1, 2, \ldots, s\}$.

We shall denote by (ω^a, ω^i) the basis of $\mathfrak{g}^*$ dual to (x_a, x_i). The elements ω^i will also be viewed as forming a basis of $\mathfrak{p}^*$ dual to (x_i). For $I \subset I_m$ with $|I| = q$, we put

$$(1) \qquad \omega^I = \omega^{j_i} \wedge \cdots \wedge \omega^{j_q}, \quad \text{if } I = \{j_1, \ldots, j_q\}.$$

If $\eta \in D^q(V)$, let

$$(2) \qquad \eta_I = \eta_{j_1, \ldots, j_q} = \eta(x_{j_1}, \ldots, x_{j_q}) \qquad (x_{j_i} \in \mathfrak{p}, \ 1 \le i \le q).$$

Then η can be written

$$(3) \qquad \eta = \sum_{I \subset I_m, |I|=q} \eta_I \cdot \omega^I,$$

or also

$$(4) \qquad \eta = (q!)^{-1} \sum_{j_1, \ldots, j_q \in I_q} \eta_{j_1, \ldots, j_q} \omega^{j_1} \wedge \cdots \wedge \omega^{j_q}.$$

If $I = (j_1, \ldots, j_q)$ and $u \in I_q$, then $I(u)$ denotes I with the u-th entry removed. The equality 1.4(2) can then also be written

$$(5) \quad (d\eta)_I = \sum_{1 \le u \le q+1} (-1)^{u-1} \pi(X_u) \cdot \eta_{I(u)} \qquad (\eta \in D^q(V); \ I \subset I_m, \ |I| = q+1).$$

Note also that we have, for $1 \leq u \leq q$,

$$(6) \qquad \eta_I = \eta_{j_1,\ldots,j_q} = (-1)^{u-1}\eta_{j_u,j_1,\ldots,\widehat{j_u},\ldots,j_q} = (-1)^{u-1}\eta_{j_u \cup I(u)}.$$

2. Scalar product

2.1. We shall be interested in the case $V = H \otimes E$, where (ρ, E) is a finite dimensional complex continuous representation of G and (σ, H) is a unitary $(\mathfrak{g}, \mathfrak{k})$-module. The latter condition means that H is a complex vector space endowed with a positive non-degenerate scalar product $(\ ,\ )_H$, such that $(Xu,v)_H + (u, Xv)_H = 0$ for all $u, v \in H$ and $X \in \mathfrak{g}$. It is not required that H be complete. We let $\tau = \rho \otimes \sigma$. For $x \in \mathfrak{g}$ we shall often write $\tau(x) = \sigma(x) + \rho(x)$ as a shorthand for $\tau(x) = \sigma(x) \otimes 1 + 1 \otimes \rho(x)$.

2.2. On E there is always a so-called *admissible scalar product*, i.e. one which is invariant under $\mathfrak{k}$, and such that $\rho(x)$ is self-adjoint for $x \in \mathfrak{p}$. We assume that E is endowed with one, to be denoted $(\ ,\ )_E$, and then introduce on

$$(1) \qquad D^q(V) = \Lambda^q \mathfrak{p}^* \otimes H \otimes E$$

the scalar product which is the tensor product of $(\ ,\ )_V = (\ ,\ )_H \otimes (\ ,\ )_E$ with the scalar product on $\Lambda^q \mathfrak{p}^*$ defined by the form B (1.1). In particular, if

$$(2) \qquad \mu = \sum_{I \subset I_m} \mu_I \cdot \omega^I, \qquad \eta = \sum_I \nu_I \cdot \omega^I$$

(notation of 1.5), then

$$(3) \qquad (\mu, \nu) = \sum_I (\mu_I, \eta_I)_V.$$

Since these scalar products are invariant under $\mathfrak{k}$, we have

$$(4) \qquad (\theta_x \mu, \nu) + (\mu, \theta_x \nu) = 0 \qquad (\mu, \nu \in D^q(V),\ x \in \mathfrak{k}).$$

For $x \in \mathfrak{g}$, we let $\tau(x)^*$ be the adjoint of $\tau(x)$ with respect to $(\ ,\ )_V$. Thus

$$(5) \qquad \begin{aligned} \tau(x)^* &= -\tau(x) & \text{if } x \in \mathfrak{k}, \\ \tau(x)^* &= \rho(x) - \sigma(x) & \text{if } x \in \mathfrak{p}. \end{aligned}$$

Note that we may replace $\mathfrak{p}$ by its complexification $\mathfrak{p}_c$. In this case, the scalar production $\Lambda^q \mathfrak{p}_c$ is the positive Hermitian product which extends the scalar product defined by the invariant form B on $\Lambda^q \mathfrak{p}$, i.e. $(x, y) = B(x, \overline{y})$, $(x, y \in \mathfrak{p}_c)$ where $\overline{}$ is the complex conjugation with respect to $\mathfrak{p}$.

2.3. PROPOSITION. *Let* $\partial \colon D^q(V) \to D^{q-1}(V)$ *be defined by*

$$(1) \qquad (\partial \eta)_J = \sum_{1 \leq j \leq m} \tau(x_j)^* \eta_{\{j\} \cup J} \qquad (J \subset I_m,\ |J| = q - 1).$$

Then ∂ *commutes with the* θ_x $(x \in \mathfrak{k})$, *maps* $C^q(V)$ *into* $C^{q-1}(V)$ *and is adjoint to* d, *i.e.*

$$(2) \qquad (\partial \eta, \mu) = (\eta, d\mu) \qquad (\eta \in D^q(V),\ \mu \in D^{q-1}(V)).$$

Using 1.5(5) and 2.2(3), we have

$$(\eta, d\mu) = \sum_{|I|=q} (\eta_I, (d\mu)_I)_V = \sum_I \left(\eta_I, \sum_u (-1)^{u-1} \tau(x_{j_u}) \cdot \mu_{I(u)} \right),$$

where $I = \{j_1, \ldots, j_q\}$. Hence

$$(\eta, d\mu) = \sum_{u,I} ((-1)^{u-1} \tau(x_{j_u})^* \eta_I, \mu_{I(u)}).$$

This combined with 1.5(6) proves (2). (2) together with 1.4(4) and 2.2(4) implies that ∂ commutes with θ_x ($x \in \mathfrak{k}$). Since $C^q(V)$ is the subspace of $D^q(V)$ annihilated by the θ_x ($x \in \mathfrak{k}$), it follows that $\partial C^q(V) \subset C^{q-1}(V)$.

2.4. We let $\Delta = d\partial + \partial d$ be the *Laplacian*. For each q, Δ is an endomorphism of $D^q(V)$ which leaves $C^q(V)$ stable. For $\eta \in D^q(V)$, we have by 2.3

(1) $$(\Delta\eta, \eta) = (d\eta, d\eta) + (\partial\eta, \partial\eta).$$

Since the scalar product is positive non-degenerate, this implies

(2) $$\Delta\eta = 0 \Leftrightarrow d\eta = \partial\eta = 0 \Leftrightarrow (\Delta\eta, \eta) = 0.$$

The element η is *harmonic* if it satisfies the conditions of (2). The space of harmonic forms in $C^q(V)$ will be denoted $\mathcal{H}^q(V)$. As usual, η is said to be *closed* if $d\eta = 0$, *coclosed* if $\partial\eta = 0$.

2.5. THEOREM. *Let $\pi = \sigma, \rho$ or $\tau = \sigma \otimes \rho$, and view V as a $(\mathfrak{g}, \mathfrak{k})$-module under π. Let Δ_π be the corresponding Laplacian. Then*
(i)

$$(\Delta_\pi \cdot \eta)_I = \sum_{1 \le j \le m} \pi(x_j) \cdot \pi(x_j)^* \cdot \eta_I$$
$$+ \sum_{\substack{1 \le j \le m \\ 1 \le u \le q}} (-1)^{u-1} [\pi(x_{j_u}), \pi(x_j)^*] \eta_{j \cup I(u)}$$

$$(\eta \in D^q(V), \ I \subset I_m, \ |I| = q).$$

(ii) *We have $\Delta_\tau = \Delta_\sigma + \Delta_\rho$ on $D^q(V)$.*
(iii) (Kuga) *If $\eta \in C^q(V)$, then*

$$(\Delta_\tau \eta)_I = (\rho(C) - \sigma(C)) \cdot \eta_I \qquad (I \subset I_m, \ |I| = q),$$

where C is the Casimir element (1.3).

(i) We view V as a $(\mathfrak{g}, \mathfrak{k})$-module under π, but still denote by d the coboundary operator and by ∂ its adjoint. In this proof, $I \subset I_m$, $|I| = q$, the index a (resp. j, resp. u) runs from $m+1$ to n (resp. 1 to m, resp. 1 to q). Then

$$(\partial d\eta)_I = \sum \pi(x_j)^* (d\eta)_{j \cup I}$$
$$= \sum_j \pi(x_j)^* \left(\pi(x_j) \cdot \eta_I + \sum_u (-1)^u \pi(x_{j_u}) \eta_{j \cup I(u)} \right),$$

(1) $$(\partial d\eta)_I = \sum_j \pi(x_j)^* \pi(x_j) \eta_I + \sum_{j,u} (-1)^u \pi(x_j)^* \pi(x_{j_u}) \eta_{j \cup I(u)}.$$

On the other hand,

$$(d\partial\eta)_I = \sum_{1 \le u \le q} (-1)^{u-1}\pi(x_{j_u}) \cdot (\partial\eta)_{I(u)},$$

$$(d\partial\eta)_I = \sum_{u,j}(-1)^{u-1}\pi(x_{j_u})\pi(x_j)^*\eta_{j \cup I(u)},$$

and hence

(2) $$\qquad (\Delta\eta)_I = \sum_j \pi(x_j)^*\pi(x_j)\eta_I + \sum_{u,j}(-1)^{u-1}[\pi(x_{j_u}), \pi(x_j)^*]\eta_{j \cup I(u)}.$$

This proves (i).

(ii) Now let $\pi = \tau$. Since $\sigma \otimes 1$ and $1 \otimes \rho$ commute, we have

(3)
$$[\pi(x_{j_u}), \pi(x_j)^*] = [\sigma(x_{j_u}) + \rho(x_{j_u}), \sigma(x_j)^* + \rho(x_j)^*],$$

$$[\pi(x_{j_u}), \pi(x_j)^*] = [\sigma(x_{j_u}), \sigma(x_j)^*] + [\rho(x_{j_u}), \rho(x_j)^*].$$

Moreover, the equalities $\sigma(x_j)^* = -\sigma(x_j)$ and $\rho(x_j) = \rho(x_j)^*$ yield

(4) $$\qquad \pi(x_j)\pi(x_j)^* = \rho(x_j)^2 - \sigma(x_j)^2 = \rho(x_j)^*\rho(x_j) + \sigma(x_j)\sigma(x_j)^*.$$

The assertion (ii) follows from (i), (3), (4).

(iii) The first sum on the right hand side of (i) is equal to

$$\sum_j (\rho(x_j)^2 - \sigma(x_j)^2) \cdot \eta_I.$$

To prove (iii), it remains to show that

(5) $$\qquad \sum_a (\sigma(x_a)^2 - \rho(x_a)^2)\eta_I = \sum(-1)^{u-1}[\tau(x_{j_u}), \tau(x_j)^*]\eta_{j \cup I(u)}.$$

Call the right hand side Q. By (3) and 2.2(5),

$$[\tau(x_{j_u}), \tau(x_j)^*] = [\sigma(x_j), \sigma(x_{j_u})] - [\rho(x_j), \rho(x_{j_u})],$$

$$[\tau(x_{j_u}), \tau(x_j)^*] = \sum c^a_{j,j_u}(\sigma(x_a) - \rho(x_a)).$$

Therefore

(6) $$\qquad Q = \sum_a (\sigma(x_a) - \rho(x_a))\left(\sum_{j,u}(-1)^{u-1}c^a_{j,j_u}\eta_{j \cup I(u)}\right).$$

But $c^a_{j,j_u} = c^j_{a,j_u}$ by 1.2. Hence

$$L_a = \sum_{j,u}(-1)^{u-1}c^a_{j,j_u}\eta_{j \cup I(u)} = \sum c^j_{a,j_u}\eta(x_{j_1}, \ldots, x_j, \ldots, x_{j_q}),$$

where x_j is at the u-th place; this can be written

$$L_a = \sum_u \eta(x_{j_1}, \ldots, [x_a, x_{j_u}], \ldots, x_{j_q}).$$

Since $\eta \in C^q(V)$, it is annihilated by the θ_x ($x \in \mathfrak{k}$). Hence

(7) $$\qquad L_a = \tau(x_a) \cdot \eta(x_{j_1}, \ldots, x_{j_q}) = \tau(x_a) \cdot \eta_I,$$

and (5) follows from (6) and (7).

2.6. COROLLARY. *Let $\eta \in D^q(V)$. Then $\Delta_\tau \eta = 0$ if and only if $\Delta_\rho \eta = \Delta_\sigma \eta = 0$.*

This follows from 2.4(2) and 2.5(ii).

2.7. The results of this section (and the next one) have been known for some time, but do not seem to have been formulated in this way in the literature. They have their origin in the work of Matsushima and Murakami [**82**, **83**] on the cohomology of discrete cocompact subgroups of G, where they are proved when H is the space of K-finite smooth functions on the quotient $\Gamma \backslash G$ on G by one such subgroup. More precisely, it is shown in [**83**] that

$$(1) \qquad\qquad H^*(\Gamma; E) = H^*(\mathfrak{g}, \mathfrak{k}; H \otimes E)$$

(see also Chapter VII). This being granted, the computations made here are substantially those of [**82**]. In particular, see [**82**, §6] for 2.3 and 2.5(iii). In that special case, 2.5(i),(ii) are also implicit in [**82**, §7] and are made explicit in [**93**, §1].

3. Special cases

3.1. PROPOSITION. *Assume that $\sigma(C) = s \cdot \mathrm{Id}$, $\rho(C) = r \cdot \mathrm{Id}$.*
(a) *If $r \neq s$, then $H^q(\mathfrak{g}, \mathfrak{k}; H \otimes E) = 0$ for all q's.*
(b) *If $r = s$, then all cochains are closed, harmonic, and we have*

$$H^q(\mathfrak{g}, \mathfrak{k}; H \otimes E) = C^q(\mathfrak{g}, \mathfrak{k}; H \otimes E) = \mathrm{Hom}_\mathfrak{k}(\Lambda^q \mathfrak{p}, H \otimes E) \quad \textit{for all } q\text{'s.}$$

By 2.5(iii), $\Delta = (r - s) \cdot \mathrm{Id}$ on $C^q(H \otimes E)$ for all q's.
Assume that $r \neq s$. Let η be a q-cocycle. Then $\Delta\eta = d\partial\eta$; hence

$$\eta = (r - s)^{-1}\Delta\eta = (r - s)^{-1} \cdot d\partial\eta$$

is a coboundary, whence (a).

Now let $r = s$. Then $\Delta = 0$, and all cochains are harmonic, hence closed and coclosed by 2.4. This yields (b).

3.2. COROLLARY. *Let (ρ, E) be irreducible. If ρ is non-trivial, then $H^q(\mathfrak{g}, \mathfrak{k}; E) = 0$ for all q's. If ρ is the trivial representation, then $H^q(\mathfrak{g}, \mathfrak{k}; E) = C^q(\mathfrak{g}, \mathfrak{k}; E) = (\Lambda^q \mathfrak{p}^*)^\mathfrak{k}$ for all q's.*

If ρ is irreducible, then $\rho(C) = r \cdot \mathrm{Id}$, and it is well known that $r = 0$ if and only if ρ is the trivial representation. 3.2 then follows from 3.1 applied to the case where (σ, H) is the trivial representation.

REMARK. If we identify $(\Lambda^q \mathfrak{p}^*)^\mathfrak{k}$ with the G-invariant differential forms on G/K (cf. I, 1.6), the corollary in the case of the trivial representation asserts that on G/K all invariant forms are harmonic, closed and coclosed. This is a well-known result of E. Cartan.

3.3. COROLLARY. *Let H be the space of K-finite vectors in the space of an irreducible unitary representation of G. If $\sigma(C) = 0$, then $H^q(\mathfrak{g}, \mathfrak{k}; H) = \mathrm{Hom}_\mathfrak{k}(\Lambda^q \mathfrak{p}, H)$, and if $\sigma(C) \neq 0$, then $H^q(\mathfrak{g}, \mathfrak{k}; H) = 0$, for all q's.*

Under our assumption, $\sigma(C)$ is a multiple of the identity. 3.3 then follows from 3.1, applied to the case where ρ is the trivial representation.

3.4. Now assume H to be an admissible $(\mathfrak{g}, \mathfrak{k})$-module. Since $C^q(V)$ may be written as

$$(1) \qquad\qquad C^q(V) = \mathrm{Hom}_{\mathfrak{k}}(\Lambda^q\mathfrak{p} \otimes E^*, H),$$

it is finite dimensional. Our complex $C^*(V)$ is then finite dimensional and the elementary "Hodge theory" in finite dimensional vector spaces obtains: we have an orthogonal decomposition

$$(2) \qquad\qquad C^q(V) = H^q(V) \oplus dC^{q-1}(V) \oplus \partial C^{q+1}(V),$$

and Δ is an invertible operator on $dC^{q-1}(V) \oplus \partial C^{q+1}(V)$. As usual, this implies

$$(3) \qquad\qquad H^q(\mathfrak{g}, \mathfrak{k}; V) \cong \mathcal{H}^q(V),$$

i.e. every cohomology class is represented by a unique harmonic form.

Let A_q be the sum of the isotypic subspaces of H corresponding to the $\mathfrak{k}$-types occurring in $\Lambda^q\mathfrak{p} \otimes E^*$. It is finite dimensional. Let $B_q = E \otimes A_q$, and C_q the subspace of B_q annihilated by $\rho(C) - \sigma(C)$. Then

$$(4) \qquad\qquad \mathcal{H}^q(V) \cong \mathrm{Hom}_{\mathfrak{k}}(\Lambda^q\mathfrak{p}, C_q).$$

Note that $\Lambda^q\mathfrak{p}$ and $\Lambda^{m-q}\mathfrak{p}$ are isomorphic $\mathfrak{k}$-modules. Therefore C_q and C_{m-q} are isomorphic $\mathfrak{k}$-modules, and we have

$$(5) \qquad H^q(\mathfrak{g}, \mathfrak{k}; V) = \mathcal{H}^q(V) = H^{m-q}(\mathfrak{g}, \mathfrak{k}; V) \qquad (q \in \mathbf{N}).$$

REMARK. As in 2.7, let Γ be a cocompact discrete subgroup of G and H the space of K-finite smooth functions on $\Gamma \backslash G$. Then (5) is also valid, although H is not admissible. Modulo 2.7(1), this is proved in [**82**, 6.2] using the Hodge theory of harmonic forms on $\Gamma \backslash X$. (For this [**82**] assumes Γ to be torsion free so that $\Gamma \backslash X$ is a smooth manifold, but the reduction to that case is easy; one could also use Hodge theory on V-manifolds.) Another proof of (5) in this case will be given in Chapter VII.

4. The bigrading in the bounded symmetric domain case

4.1. In this section, we assume that G has compact center and $X = G/K$ carries an invariant complex structure. As is known, X is then equivalent to a bounded symmetric domain. The complexification $\mathfrak{p}_c$ of $\mathfrak{p}$ decomposes into the sum $\mathfrak{p} = \mathfrak{p}^+ \oplus \mathfrak{p}^-$ of two commutative subalgebras of $\mathfrak{g}_c$, normalized by $\mathfrak{k}_c$, consisting of nilpotent elements, and there is an element z_0 in the center of $\mathfrak{k}$ such that $\mathrm{ad}\, z_0$ is the multiplication by $\pm i$ on $\mathfrak{p}^{\pm}$. Conversely, if there is such an z_0 in $\mathfrak{k}$, then $\mathrm{ad}\, z_0|_{\mathfrak{p}} = J$ defines a complex structure on $\mathfrak{p}$ which is invariant under $\mathfrak{k}$, and hence an invariant almost complex structure on X. It is well known to be integrable, hence to give rise to an invariant complex structure on X.

The cochain complex $C^*(\mathfrak{g}, \mathfrak{k}; V)$ is then bigraded. In the situation of §2, this bigrading induces one in cohomology (cf. 4.5). This was shown in [**82**, II, §§2, 3]; at any rate in the context of that paper (cf. 2.7), but the proofs are the same, and we shall omit some details. This reflects the fact that G/K carries an invariant Kaehler metric. Similarly, the familiar results on *primitive cohomology* of Kaehler manifolds extend to our situation (4.8).

4.2. We fix notation so that the projection $\mathfrak{p} \to \mathfrak{p}^+$ is a **C**-isomorphism. We let $^-$ denote complex conjugation of $\mathfrak{g}_c$ with respect to $\mathfrak{g}$. Then $\mathfrak{p}^-$ and $\mathfrak{p}^+$ are complex conjugates of each other. In the present case, the dimension m of $\mathfrak{p}$ is even. Let $m' = m/2$. Let $\{x_i\}_{1 \leq i \leq m'}$ be a basis of $\mathfrak{p}^+$. Then $\{\overline{x}_i\}$ is a basis of $\mathfrak{p}^-$. The invariant form B is non-degenerate on $\mathfrak{p}$ and identically zero on $\mathfrak{p}^+$ and $\mathfrak{p}^-$. Furthermore, we may assume that

$$(1) \qquad\qquad B(x_i, \overline{x}_j) = \delta_{ij} \qquad (1 \leq i, j \leq m').$$

We let $\{\omega^i, \overline{\omega}^j\}$ be the corresponding dual basis in $\mathfrak{p}_c^*$, and

$$(2) \qquad \omega^I = \omega^{i_1} \wedge \cdots \wedge \omega^{i_q}, \overline{\omega}^I = \overline{\omega}^{i_1} \wedge \cdots \wedge \overline{\omega}^{i_q}$$

$$(I = \{i_1, \ldots, i_q\} \subset I_{m'}).$$

On the relative Lie algebra cochains, we now consider the bigrading defined by

$$(3) \qquad C^{p,q} = C^{p,q}(\mathfrak{g}, \mathfrak{k}; V) = \mathrm{Hom}_{\mathfrak{k}}(\Lambda^p \mathfrak{p}^+ \otimes \Lambda^q \mathfrak{p}^-, V) \qquad (p, q \in \mathbf{N}).$$

In the isomorphism of I, 1.6 with the space $A(X; V)^G$ of G-invariant differential forms on X, they correspond to the forms of type p, q. An element $\eta \in C^{p,q}$ can be written in the form

$$(4) \qquad \eta = \sum_{I,J} \eta_{I,J} \omega^I \wedge \overline{\omega}^J \qquad (I, J \subset I_{m'}; |I| = p, |J| = q).$$

The operator d is now a sum $d = d' + d''$, with d' (resp. d'') of bidegree $(1, 0)$, (resp. $(0, 1)$). It follows readily from 1.5(5) that we have

$$(5) \qquad (d'\eta)_{I,J} = \sum_{1 \leq u \leq p+1} (-1)^{u-1} \pi(x_u) \cdot \eta_{I(u),J}$$

$$(|I| = p + 1, \ |J| = q),$$

$$(6) \qquad (d''\eta)_{I,J} = \sum_{1 \leq u \leq q+1} (-1)^{p+u-1} \pi(\overline{x}_u) \eta_{I,J(u)}$$

$$(|I| = p, \ |J| = q + 1).$$

4.3. We now revert to the assumptions of §2. The representation τ is extended to $\mathfrak{g}_c$ in the obvious way. We have then

$$(1) \qquad \tau(x)^* = -\tau(\overline{x}), (x \in \mathfrak{k}_c), \tau(x)^* = \rho(\overline{x}) - \sigma(\overline{x}) \qquad (x \in \mathfrak{p}_c).$$

The boundary operator ∂ adjoint to d decomposes into $\partial = \partial' + \partial''$, with ∂' (resp. ∂'') of bidegree $(-1, 0)$, (resp. $(0, -1)$) adjoint to d' (resp. d''). It follows from (1) and 2.3 that we have, for $\eta \in C^{p,q}$,

$$(2) \qquad (\partial'\eta)_{I,J} = \sum_{1 \leq u \leq m'} \tau(\overline{x}_\gamma)^* \eta_{\{u\} \cup I, J},$$

where $I, J \subset I_{m'}, |I| = p - 1, |J| = q$, and

$$(3) \qquad (\partial''\eta)_{I,J} = (-1)^p \sum_{1 \leq u \leq m'} \tau(x_\gamma) \eta_{I, \{u\} \cup J},$$

where $I, J \subset I_{m'}, |I| = p, |J| = q - 1$.

4.4. PROPOSITION. *Let* $\Delta' = d'\partial' + \partial'd'$, $\Delta'' = d''\partial'' + \partial''d''$. *Then* $\Delta = \Delta' + \Delta''$.

By definition

$$\Delta = (d' + d'')(\partial' + \partial'') + (\partial' + \partial'')(d' + d''),$$

$$\Delta = \Delta' + \Delta'' + (d'\partial'' + d''\partial') + (\partial''d' + \partial'd'').$$

It suffices to show that the two last sums vanish. This computation is made in [**82**, p. 404]. As in [**82**], this implies

4.5. COROLLARY. *The Laplacian preserves the bidegree. If* $\eta \in C^*(\mathfrak{g}, \mathfrak{k}; V)$ *is harmonic, then its bihomogeneous components are also harmonic. We have*

$$H^*(\mathfrak{g}, \mathfrak{k}; V) = \bigoplus_{p,q} H^{p,q}(\mathfrak{g}, \mathfrak{k}; V),$$

where $H^{p,q}$ *is the space of cohomology classes represented by harmonic forms of bidegree* (p, q).

4.6. With J the complex structure on $\mathfrak{p}$ defined in 4.1, let

$$\omega(x, y) = B(x, Jy), h(x, y) = B(x, y) + i \cdot \omega(x, y) \qquad (x, y \in \mathfrak{p}).$$

Then ω is alternating, non-degenerate of type $(1, 1)$, invariant under $\operatorname{Ad} K$, and h is a positive non-degenerate K-invariant Hermitian scalar product. h allows one to define a G-invariant Hermitian metric on G/K, which is Kaehlerian, because the 2-form on G/K defined by ω is G-invariant, hence closed. We now transcribe in our framework some results of Kaehlerian geometry [**115**].

Let $L: \Lambda\mathfrak{p}_c^* \to \Lambda\mathfrak{p}_c^*$ be the left multiplication by ω. It is a linear transformation of bidegree $(1, 1)$ which preserves $\Lambda\mathfrak{p}$ and commutes with K. By [**115**, Cor. p. 28], an element $x \in \Lambda^p\mathfrak{p}_c$ is primitive if and only if $p \leq m'$, and $L^{m'-p+1}x = 0$. We shall take this as the definition of a primitive element. Let Pr^p be the space of primitive elements of degree $p \leq m'$. If $p > m'$, we let $\operatorname{Pr}^p = 0$ by definition. Then L^a is injective on Pr^p for $0 \leq a \leq m' - p$, and we have a direct sum decomposition

(1) $$\Lambda^p\mathfrak{p}_c^* = \bigoplus_{s \geq 0} L^s \cdot \operatorname{Pr}^{p-2s}$$

[**115**, p. 28]. Here, s need only run through the s's in $[0, p - m']$. Note that, since L is bihomogeneous, Pr^p is the direct sum of its intersections $\operatorname{Pr}^{a,b}$ with the spaces $\Lambda^a\mathfrak{p}^+ \otimes \Lambda^b\mathfrak{p}^-$ $(a + b = p)$, and (1) is also compatible with the bidegree.

4.7. Now let (π, V) be a $(\mathfrak{g}, \mathfrak{k})$-module. We extend the definition of L to $C^p(\mathfrak{g}, \mathfrak{k}; V) = (\Lambda^p\mathfrak{p}_c^* \otimes V)^{\mathfrak{k}}$ by making it operate on the first factor as above. This defines an endomorphism of $C^*(\mathfrak{g}, \mathfrak{k}; V)$ of bidegree $(1, 1)$, which commutes with d since $d\omega = 0$. Therefore it goes over to cohomology. Let

$$H_{\operatorname{pr}}^{a,b} = H_{\operatorname{pr}}^{a,b}(\mathfrak{g}, \mathfrak{k}; V) \qquad (a, b \in \mathbf{N}, \ a + b \leq m')$$

be the space of *primitive elements* in $H^{a,b}$, i.e. of elements annihilated by L^s for $s > m' - a - b$, and put $H_{\operatorname{pr}}^{a,b} = 0$ if $a + b > m'$.

4.8. THEOREM. *Let $V = H \otimes E$ as in §2, and assume that the Casimir operator operates by scalars on H and E. Then L^s is injective on $H^{a,b}_{\mathrm{pr}}$ for $s \leq m' - a - b$, and we have a direct sum decomposition*

$$H^{a,b}(\mathfrak{g}, \mathfrak{k}; H \otimes E) = \bigoplus_s L^s \cdot H^{a-2s, b-2s}_{\mathrm{pr}} \qquad (a, b \in I_{m'}).$$

There is something to prove only if the eigenvalues of C on H and E are equal (3.1). In that case, all cochains are closed, the cohomology identifies to the cochain complex, and we are immediately reduced to 4.6.

5. Cohomology with respect to square integrable representations

5.1. In this section, G is semi-simple. Let T be a maximal torus of K. By a well-known theorem of Harish-Chandra [**52**], G has a discrete series if and only if T is a Cartan subgroup of G. We assume the latter condition in this section. We are interested in $\mathrm{Ext}^*_{\mathfrak{g}, \mathfrak{k}}(F, H)$ when F is a finite dimensional $\mathfrak{g}$-module and $H = V_K$ is the space of K-finite vectors in the space of a square integrable representation (π, V) of G. By I 5.4, if this group is non-zero, then V is also a representation space for the real form G_0 with Lie algebra $\mathfrak{g}$ of the simply connected group with Lie algebra $\mathfrak{g}_c$. Therefore we may (and do) assume G to be equal to G_0.

Let Φ (resp. Φ_k) be the root system of $\mathfrak{g}_c$ (resp. $\mathfrak{k}_c$) with respect to the complexification $\mathfrak{t}_c$ of the Lie algebra $\mathfrak{t}$ of T, W (resp. W_k) the Weyl group and $P(\Phi)$ (resp. $P(\Phi_k)$) the set of weights of Φ (resp. Φ_k). In particular, $P(\Phi)$ is a lattice in $i \cdot \mathfrak{t}^*$. The equivalence classes of irreducible square integrable representations of G correspond canonically and bijectively to the orbits of W_k in the set of regular elements in $P(\Phi)$. We let ω_Λ be the class of representations associated to a regular element $\Lambda \in P(\Phi)$. Its elements have the infinitesimal character χ_Λ, in the usual parametrization. More precisely, let

(1) $P^+ = P(\Phi)^+ = \{\alpha \in P(\Phi) \mid \langle \Lambda, \alpha \rangle > 0\},$

(2) $P^+_k = P(\Phi_k) \cap P^+, \quad \Phi^+ = \Phi \cap P^+, \quad \Phi^+_k = \Phi_k \cap P^+,$

where $\langle \, , \, \rangle$ is a W-invariant scalar product on $i \cdot \mathfrak{t}^*$, say the one defined via the Killing form. Let

(3) $$2\rho = \sum_{\alpha \in \Phi^+} \alpha, \qquad 2 \cdot \rho_k = \sum_{\alpha \in \Phi^+_k} \alpha.$$

Then χ_Λ is the infinitesimal character of the finite dimensional irreducible representation with highest weight $\Lambda - \rho$.

If $\mu \in P(\Phi_k)$, we let F_μ be a representation space of an irreducible representation of $\mathfrak{k}$ with extremal weight μ.

5.2. PROPOSITION. *Let $\Lambda \in P(\Phi)$ be regular and $(\pi, V) \in \omega_\Lambda$. Then*
1) $\dim \mathrm{Hom}_K(F_{\Lambda + \rho - 2\rho_k}, V) = 1$,
2) if $\mathrm{Hom}(F_\mu, V) \neq 0$ with $\mu \in P^+_k$, then $\mu = \Lambda + \rho - 2\rho_k + Q$, where Q is a sum of elements in Φ^+.

This proposition states that $\Lambda + \rho - 2\rho_k$ is the lowest K-weight of ω_Λ, and that it has multiplicity-one. It can be viewed as a consequence of the truth of Blattner's conjecture [**57**]. However, it admits a simpler proof, and was in fact proved before Blattner's conjecture [**97, 109**].

5.3. THEOREM. *Let* Λ, Φ^+, Φ_k^+ *be as in 5.1. Let* $(\pi, V) \in \omega_\Lambda$. *Let* H *be the* $(\mathfrak{g}, K)$-*module of* K-*finite vectors in* V. *Let* (σ, F) *be an irreducible finite dimensional representation of* G.

a) *If the highest weight of* (σ, F) *relative to* Φ^+ *is not* $\Lambda - \rho$, *then* $\mathrm{Ext}_{\mathfrak{g},\mathfrak{k}}^i(F, H) = 0$ *for all* i.

b) *If the highest weight of* (σ, F) *is* $\Lambda - \rho$, *then* $\dim \mathrm{Ext}_{\mathfrak{g},\mathfrak{k}}^i(F, H) = \delta_{i,q}$, *where* $q = (\dim G/K)/2$.

PROOF. a) follows from (I, 4.2) and 5.1.

b) By 3.1, we have

(1) $$\mathrm{Ext}_{\mathfrak{g},\mathfrak{k}}^i(F, H) = H^i(\mathfrak{g}, \mathfrak{k}; F^* \otimes H) = \mathrm{Hom}_{\mathfrak{k}}(\Lambda^i \mathfrak{p} \otimes F, H) \qquad (i \in \mathbf{N}).$$

Thus we must compute $\mathrm{Hom}_K(\Lambda^i \mathfrak{p}_c \otimes F, H)$.

Let $\Phi_n = \Phi - \Phi_k$ and $\Phi_n^+ = \Phi_n \cap \Phi^+$. Then the weights of T on $\Lambda^i \mathfrak{p}_c$ are of the form $\alpha_1 + \cdots + \alpha_i$ with $\alpha_1, \ldots, \alpha_i$ distinct elements of Φ_n. Set $\rho_n = \rho - \rho_k$. Then the weights of $\Lambda^i \mathfrak{p}_c$ are of the form $\alpha_1 + \cdots + \alpha_j - (\alpha_{j+1} + \cdots + \alpha_i)$, where $(\alpha_1 \cdots \alpha_j)$ (resp. $\alpha_{j+1} \cdots \alpha_i$) are distinct elements of Φ_n^+. Now $\alpha_1 + \cdots + \alpha_j = 2\rho_n - \gamma_1 - \cdots - \gamma_t$, with $\{\gamma_1, \ldots, \gamma_t\} \cup \{\alpha_1, \ldots, \alpha_j\} \subset \Phi_n^+$ and $\gamma_1, \ldots, \gamma_t$ distinct. Hence the weights in $\Lambda^i \mathfrak{p}_c$ are of the form $2\rho_n - Q$ with Q a sum of elements of Φ_n^+. Furthermore, if $2\rho_n$ is a weight in $\Lambda^i \mathfrak{p}_c$, then $i = q$ and $2\rho_n$ is a weight in $\Lambda^q \mathfrak{p}_c$ of multiplicity 1. The weights of (σ, F) relative to T are of the form $\Lambda - \rho - Q$ with Q a sum of elements of Φ^+, and $\Lambda - \rho$ is a weight of multiplicity 1 (this is the theorem of the highest weight). This implies:

(i) *The weights of* T *on* $\Lambda^i \mathfrak{p}_c \otimes F$ *are of the form* $2\rho_n + \Lambda - \rho - Q$, *with* Q *a sum of elements of* Φ^+. *If* $2\rho_n + \Lambda - \rho + Q'$ *is a weight in* $\Lambda^i \mathfrak{p}_c \otimes F$, *then* $Q' = 0$, $i = q$ *and the weight* $2\rho_n + \Lambda - \rho = \Lambda + \rho - 2\rho_k$ *has multiplicity 1 in* $\Lambda^q \mathfrak{p}_c \otimes F$.

This implies in turn

(ii) *If* λ *is* Φ_k^+-*dominant integral and* $\mathrm{Hom}_K(F_\lambda, \Lambda^i \mathfrak{p}_c \otimes F) \neq 0$, *then* $\lambda = \Lambda + \rho - 2\rho_k - Q$ *with* Q *a sum of elements of* Φ^+. *If* $\lambda = \Lambda + \rho - 2\rho_k + Q$ *with* Q *a sum of elements of* Φ^+, *then* $\mathrm{Hom}_K(F_\lambda, \Lambda^i \mathfrak{p}_c \otimes F) \neq 0$ *implies* $Q = 0$ *and* $i = q$. *Furthermore,* $\dim \mathrm{Hom}_K(F_{\Lambda+\rho-2\rho_k}, \Lambda^q \mathfrak{p}_c \otimes F) = 1$.

The assertion b) now follows from (1), (ii) and 5.2.

5.4. THEOREM. *Let* M *be a reductive group (see 0, §3) whose identity component has a compact center. Let* (π, V) *be a discrete series representation of* M *and* (σ, F) *a finite dimensional irreducible representation of* M. *Let* $q = (\dim M/K)/2$. *Then* $\mathrm{Ext}_{g,K}^i(F, V) = 0$ *for* $i \neq q$.

The restriction of (π, V) to M^0 is the direct sum of finitely many irreducible representations (see below), which are then clearly square integrable. In view of I, 5.1(4), this reduces us to the case where M is connected. We have then $M = M' \cdot S$, where M' is semi-simple and S is a central torus. We may view V and F as $M' \times S$ modules. V is then the tensor product of a one-dimensional representation of S by an irreducible representation of M', which is then also square integrable. Since F is finite dimensional, it is a direct sum of irreducible representations, each of which is a tensor product of a one-dimensional representation of S by an irreducible representation of M'. Using the Künneth rule (I, 1.3), and the fact that q is also equal to $(\dim M'/(M' \cap K))/2$, we see that we may assume M' semi-simple and F irreducible. This reduces us to 5.3.

In this proof, we have used the following fact.

5.5. LEMMA. *Let L be a reductive group, L' an open normal subgroup of L and (π, V) an irreducible admissible L-module. Then V is the direct sum of finitely many irreducible admissible L'-modules.*

This is well known. Not knowing a good reference for it, we include a proof for the sake of completeness. Let Q be a maximal compact subgroup of L and $Q' = Q \cap L$. Then Q' is a maximal compact subgroup of L' and is a normal subgroup of finite index of Q. It follows from Frobenius reciprocity that if $\sigma \in \widehat{L'}$, then there exist only finitely many $\tau \in \widehat{L}$ whose restriction to L' contains σ. Therefore V is an admissible L'-module. It suffices then to show the existence of one irreducible L'-submodule $U \subset V$, because then V is the sum of the transforms $x(U)$ $(x \in L/L')$, hence the direct sum of finitely many of them. To prove the existence of U, one may use the fact (proved by Harish-Chandra) that, up to infinitesimal equivalence, there are only finitely many representations with a given infinitesimal character. A simpler argument, suggested by H. Jacquet, is the following: Let $(\widetilde{\pi}, \widetilde{V})$ be the contragredient representation to (π, V). It is a simple L-module, hence a finitely generated L'-module. Consequently, it has a proper simple quotient. But, since we deal with admissible representations, (π, V) is infinitesimally equivalent to the contragredient of $(\widetilde{\pi}, \widetilde{V})$. As a consequence, it has a proper simple L'-submodule.

5.6. LEMMA. *Let L, L' be as in 5.5, K a maximal compact subgroup of L and $K' = K \cap L'$. Let (π, E) be an irreducible admissible $(\mathfrak{l}, K)$-module. Then E is the direct sum of finitely many irreducible admissible $(\mathfrak{l}, K')$-modules.*

By [**77**], E may be viewed as the module of K-finite vectors of an irreducible admissible smooth L-module E. The lemma then follows from 5.5 and from the fact that the module of K-finite vectors of an irreducible admissible smooth L'-module is algebraically irreducible.

5.7. PROPOSITION. *We keep the assumptions of 5.4 and assume moreover that F is irreducible with respect to M^0. Then*

$$(1) \qquad \dim H^q(\mathfrak{m}, K; V \otimes F) \leq 1, \quad \text{for } q = q(G).$$

Let V_0 be an irreducible $(\mathfrak{m}, K^0)$-submodule of V (see 5.6). Let U be the $(\mathfrak{m}, K)$-module induced from the $(\mathfrak{m}, K^0)$-module V_0. As a vector space, $U = I_{K^0}^K(V_0)$. Then $U \otimes F$ may be viewed as the $(\mathfrak{m}, K)$-module induced from $V_0 \otimes F$. By Frobenius reciprocity, we have an exact sequence of $(\mathfrak{m}, K)$-modules

$$(2) \qquad 0 \to V \otimes F \to U \otimes F \to V' \otimes F \to 0,$$

and moreover

$$(3) \qquad \operatorname{Hom}_K(\Lambda(\mathfrak{m}/\mathfrak{k}), U \otimes F) = \operatorname{Hom}_{K^0}(\Lambda(\mathfrak{m}/\mathfrak{k}), V_0 \otimes F);$$

hence

$$(4) \qquad H^i(\mathfrak{m}, K; U \otimes F) = H^i(\mathfrak{m}, K^0; V_0 \otimes F) \qquad (i \in \mathbf{Z}).$$

Let $W = V, V'$. We have

$$H^i(\mathfrak{m}, K; W \otimes F) = (H^i(\mathfrak{m}, K^0; W \otimes F))^{K/K^0} \qquad (i \in \mathbf{Z}).$$

Since W is a direct sum of finitely many discrete series representations of M^0, 5.4 shows that

$$H^i(\mathfrak{m}, K; W \otimes F) = 0, \quad \text{for } i \neq q(G).$$

Hence, by the cohomology sequence associated to (2), we have an embedding

$$0 \to H^q(\mathfrak{m}, K; V \otimes F) \to H^q(\mathfrak{m}, K; U \otimes F), \quad \text{for } q = q(G).$$

This reduces us to the case where $V = U$. But then, (4) brings us back to the case where M is connected. As in the proof of 5.4, write $M = M' \cdot S$, where M' is semi-simple connected and S is compact commutative. Then V_0 and F are the tensor products of one-dimensional representations of S by irreducible representations of M', and our assertion follows from 5.3 and the Künneth rule.

6. Spinors and the spin Laplacian

6.1.   Let $(V, \langle \ , \ \rangle)$ be a pair consisting of a finite dimensional vector space V over $\mathbf{R}$ and an inner product (strictly positive) on V. Let $n = \dim V$. A *space of spinors* for $(V, \langle \ , \ \rangle)$ is a pair (γ, S), where S is a finite dimensional vector space over $\mathbf{C}$ and $\gamma \colon V \to \operatorname{End}(S)$ an $\mathbf{R}$ linear map satisfying:
  1) $\gamma(v)^2 = -\langle v, v \rangle I,\ v \in V$,
  2) If $W \subset S$ is a subspace such that $\gamma(v)W \subset W$ for all $v \in V$, then $W = (0)$ or $W = S$.
  If (γ, S) and (γ', S') are spaces of spinors for $(V, \langle \ , \ \rangle)$, we say that they are *equivalent* if there exists a linear bijection $A \colon S \to S'$ such that $A \circ \gamma(v) = \gamma'(v) \circ A$ for all $v \in V$.

6.2. LEMMA. *Set $n = \dim V$. If n is even, then up to automorphism there exists exactly one space of spinors for $(V, \langle \ , \ \rangle)$. If n is odd, there are exactly 2. Fix a space of spinors for $(V, \langle \ , \ \rangle)$. Then there exists a unitary structure $\langle \ , \ \rangle$ on $S(V)$ such that $\langle \gamma(v)u, w \rangle = -\langle u, \gamma(v)w \rangle$ for $v \in V$, $u, w \in S(V)$.*

This lemma is due to C. Chevalley [**133**]. A proof can also be found in [**151**, Lemma 9.2.1, p. 359].

6.3.   Let $(\gamma, S(V))$ be a space of spinors for V. Let $\mathfrak{so}(V) = \{X \in \operatorname{End}(V) \mid \langle Xv, w \rangle = -\langle v, Xw \rangle\}$. Let $v_1, \dots, v_n$ be an orthonormal basis of V. Let $E_{ij} \in \operatorname{End}(V)$ be defined by $E_{ij}v_k = \delta_{jk}v_i$. If $i \neq j$ set

$$\sigma(E_{ij} - E_{ji}) = -\tfrac{1}{2}\gamma(v_i)\gamma(v_j) \in \operatorname{End}(S(V)).$$

Then it is an easy exercise to show that $(\sigma, S(V))$ is a representation of $\mathfrak{so}(V)$ on $S(V)$ which up to equivalence is independent of the choice of space of spinors. Set

$$h_j = E_{2j-1,2j} - E_{2j,2j-1}, \qquad j = 1, \dots, [n/2] = r,$$

$$\mathfrak{h} = \sum \mathbf{R} h_j.$$

Let $\{\lambda_j\}$ be the basis of $\mathfrak{h}^*$ dual to $\{h_j\}$. Then $\mathfrak{h}$ is a maximal abelian subalgebra of $\mathfrak{so}(V)$, the weights of σ on $\mathfrak{h}$ are precisely the linear forms

$$i\left(\tfrac{1}{2}(\lambda_1 + \cdots + \lambda_r) - \lambda_{j_1} - \cdots - \lambda_{j_k}\right),$$

with $1 \leq j_1 < \cdots < j_k \leq r$, and each occurs with multiplicity 1.

6.4. LEMMA. *Let $1 \leq i, j, k, l \leq n$, and let $R_{ijkl} \in \mathbf{C}$ satisfy*
  1) $R_{ijkl} = R_{klij}$,
  2) $R_{ijkl} = -R_{jikl}$,

3) $R_{ijkl} + R_{kijl} + R_{jkil} = 0$. *Then*

$$\sum_{ijkl} R_{ijkl} \gamma(v_i) \gamma(v_j) \gamma(v_k) \gamma(v_l) = 2 \left(\sum_{ij} R_{ijji} \right) I.$$

This lemma is proved by the obvious computation.

6.5. LEMMA. *Let μ be the natural representation of $\mathbf{SO}(V)$ on ΛV_c.*
1) *If n is even, then μ is equivalent with $\sigma \otimes \sigma$.*
2) *If n is odd, then μ is equivalent with $\sigma \otimes \sigma \oplus \sigma \otimes \sigma$.*

This lemma is an easy consequence of the results in 6.3.

6.6.   We now specialize to the case where $\mathfrak{g}$ is a semi-simple Lie algebra over $\mathbf{R}$, $\mathfrak{g} = \mathfrak{k} \oplus \mathfrak{p}$ is a Cartan decomposition and $V = \mathfrak{p}$, $\langle \, , \, \rangle = B|_{\mathfrak{p} \times \mathfrak{p}}$. We let C_k be the Casimir operator of $\mathfrak{k}$ associated with $B|_{\mathfrak{k}}$. Let $\tau_0(Y) = \operatorname{ad} Y|_{\mathfrak{p}}$ for $Y \in \mathfrak{k}$. Then $\tau_0 \colon \mathfrak{k} \to \mathfrak{so}(\mathfrak{p})$ is a Lie algebra homomorphism. Set $s(Y) = \sigma \circ \tau_0(Y)$, $Y \in \mathfrak{k}$. Then $(s, S(\mathfrak{p}))$ is a unitary representation of $\mathfrak{k}$.

Let $\mathfrak{h}^+ \subset \mathfrak{k}$ be a maximal abelian subalgebra of $\mathfrak{k}$. Let $\mathfrak{h}$ be the centralizer in $\mathfrak{g}$ of $\mathfrak{h}^+$. Then $\mathfrak{h}$ is a Cartan subalgebra of $\mathfrak{k}$. Let Φ be the root system of $(\mathfrak{g}_c, \mathfrak{h}_c)$ and let Φ_k be the root system of $(\mathfrak{k}_c, \mathfrak{h}_c^+)$. Fix a set Φ_k^+ of positive roots for Φ_k. A set of positive roots Φ^+ of Φ will be said to be *compatible* with Φ_k^+ if the following two conditions are satisfied:
1) if $\alpha \in \Phi_k^+$, then $\alpha = \beta|_{\mathfrak{h}_c^+}$ for some $\beta \in \Phi^+$;
2) if $\alpha \in \Phi^+$, then $\theta \alpha \in \Phi^+$ (here θ is the Cartan involution of $(\mathfrak{g}, \mathfrak{k})$).

6.7.   Fix compatible sets Φ_k^+ and Φ^+ as in 6.6. Set $\mathfrak{h}_c^- = \{ h \in \mathfrak{h}_c \mid \theta h = -h \}$. We identify $(\mathfrak{h}_c^+)^*$ with $\{ \lambda \in \mathfrak{h}_c^* \mid \theta \lambda = \lambda \}$ and $(\mathfrak{h}_c^-)^*$ with $\{ \lambda \in \mathfrak{h}_c^* \mid \theta \lambda = -\lambda \}$. For $\lambda \in \mathfrak{h}_c^*$ we write $\lambda = \lambda^+ + \lambda^-$, $\lambda^\pm \in (\mathfrak{h}_c^\pm)^*$. Set

$$2\rho = \sum_{\alpha \in \Phi^+} \alpha, \qquad 2\rho_k = \sum_{\alpha \in \Phi_k^+} \alpha, \qquad \rho_n = \rho - \rho_k.$$

For $\lambda \in (\mathfrak{h}_c^+)^*$, let $\mathfrak{p}_\lambda = \{ x \in \mathfrak{p}_c \mid \operatorname{ad} h \cdot x = \lambda(h) x, \; h \in \mathfrak{h}_c^+ \}$. Set $\mathfrak{p}^\pm = \sum_{\pm \lambda \in \Phi^+|_{(\mathfrak{h}_c^+)}} \mathfrak{p}_\lambda$. Then 6.3 implies that the weights of $(s, S(\mathfrak{p}))$ are of the form

$$\rho_n - \mu_{i_1} - \cdots - \mu_{i_r} \quad \text{with } 0 \neq \mathfrak{p}_{\mu_i} \subset \mathfrak{p}^+ \mu = \mu_{i_j} \; (1 \leq j \leq r).$$

6.8. SCHOLIUM (Kostant [**72**]). *Let $\Lambda_1, \ldots, \Lambda_r$ be Φ^+ dominant integral forms on $\mathfrak{h}_c$.*

Let F_i be the irreducible $\mathfrak{g}_c$ module with highest weight Λ_i, $i = 1, \ldots, r$. If λ is a weight of $F_1 \otimes \cdots \otimes F_r$, then

$$|\lambda| \leq |\Lambda_1 + \cdots + \Lambda_r|,$$

and equality occurs if and only if there is $s \in W(\Phi)$ such that $\lambda = s(\Lambda_1 + \cdots + \Lambda_r)$.

Let $s \in W(\Phi)$ be such that $s\lambda$ is Φ^+ dominant integral. Then $|s\lambda| = |\lambda|$. We have $s\lambda = \lambda_1 + \cdots + \lambda_r$, with λ_i a weight of F_i. This implies that $\lambda_i = \Lambda_i - \xi_i$, with ξ_i a sum of elements of Φ^+. Hence $s\lambda = \Lambda_1 + \cdots + \Lambda_r - \xi_1 - \cdots - \xi_r$. Set $\Lambda = \Lambda_1 + \cdots + \Lambda_r$, $\xi = \xi_1 + \cdots + \xi_r$. Then

$$\langle \lambda, \lambda \rangle = \langle s\lambda, \Lambda - \xi \rangle = \langle s\lambda, \Lambda \rangle - \langle s\lambda, \xi \rangle \leq \langle s\lambda, \Lambda \rangle$$
$$= \langle \Lambda - \xi, \Lambda \rangle \leq \langle \Lambda, \Lambda \rangle$$

with equality if and only if $0 = \langle s\lambda, \xi \rangle = \langle \Lambda, \xi \rangle$. Hence equality occurs if and only if $\langle \xi, \xi \rangle = 0$.

6.9. LEMMA. *Set $W^1 = \{t \in W \mid t\Phi^+ \text{ is compatible with } \Phi_k^+\}$. Let τ_λ denote the irreducible representation of $\mathfrak{k}_c$ with highest weight λ relative to Φ_k^+. Then*

$$s = \bigoplus_{t \in W^1} 2^{[l_0/2]} \tau_{t\rho - \rho_k},$$

where $l_0 = \dim \mathfrak{h}_c^-$ and $m\tau_\lambda$ means a direct sum of m copies of τ_λ.

Every weight of s is of the form $\rho_n - \xi$, where ξ is a weight of $\mathfrak{h}_c^+$ on $\Lambda\mathfrak{p}^+$, and every such weight occurs. If $Q \subset \Phi^+$ set $\langle Q \rangle = \sum_{\alpha \in Q} \alpha$. Then every weight of s is of the form $\rho_n - \langle Q \rangle\big|_{\mathfrak{h}^+}$, $Q \subset \Phi^+$. In particular, we see that ρ_n is a weight of s. But then ρ_n must be an extreme weight. This implies that τ_{ρ_n} occurs in s.

Let $(x_i)_{1 \leq i \leq m}$, $(x_a)_{m < a \leq n}$ be as in 1.1. Set $R_{ijkl} = B([x_i, x_j], [x_k, x_l])$.

$$(1) \qquad\qquad s(C_k) = cI \quad \text{with} \quad 8c = \sum_{ij} R_{ijji}.$$

Indeed, if $x \in \mathfrak{k}$, then using 6.3 we see that

$$4s(x) = \sum_{ij} \langle [x, x_i], x_j \rangle \gamma(x_i) \gamma(x_j).$$

This implies that

$$\begin{aligned}
16s(C_k) &= -\sum_{i,j,k,l,a} \langle [x_a, x_i], x_j \rangle \langle [x_a, x_k], x_l \rangle \gamma(x_i)\gamma(x_j)\gamma(x_k)\gamma(x_l) \\
&= \sum_{i,j,k,l} R_{ijkl} \gamma(x_i)\gamma(x_j)\gamma(x_k)\gamma(x_l).
\end{aligned}$$

Now apply Lemma 6.4.

Formula (1), combined with the fact that τ_{ρ_n} occurs in s, implies

$$(2) \qquad\qquad s(C_k) = (|\rho|^2 - |\rho_k|^2)I.$$

Suppose now that τ_λ occurs in s. Then $\lambda = \rho_n - \langle Q \rangle\big|_{\mathfrak{h}^+}$ and $Q \subset \Phi^+$. Hence $\lambda + \rho_k = \rho - \langle Q \rangle\big|_{\mathfrak{h}^+}$. (2) implies that $|\lambda + \rho_k|^2 = |\rho|^2$. Hence $|\rho - \langle Q \rangle\big|_{\mathfrak{h}^+}|^2 = |\rho|^2$. Since $(\rho - \langle Q \rangle)^+ = \rho - \langle Q \rangle\big|_{\mathfrak{h}^+}$ we see that

$$(3) \qquad\qquad |\rho - \langle Q \rangle|^2 \geq |\rho|^2.$$

As is well known, the weights of the finite dimensional representation of $\mathfrak{g}_c$, with highest weight ρ are precisely the forms $\rho - \langle Q \rangle$, with $Q \subset \Phi^+$. The relation (3), combined with 6.7, implies that $\lambda + \rho_k = t\rho$ with $t \in W$. But then $t\Phi^+$ is compatible with Φ_k^+.

This implies that if τ_λ occurs in s, then $\lambda = t\rho - \rho_k$ with $t \in W^1$. Replacing Φ^+ by $t\Phi^+$, we see that each $t\rho - \rho_k$, $t \in W^1$, is an extreme weight of s and the multiplicity of $\tau_{t\rho - \rho_k}$ in s is precisely the dimension of the $t\rho - \rho_k$ weight space of s. But this multiplicity is easily seen to be $2^{[l_0/2]}$.

6.10.   Let (π, H) be a unitary $(\mathfrak{g}, \mathfrak{k})$-module. We give $H \otimes S$ $(S = S(\mathfrak{p}))$ the tensor product inner product. Define $D \colon H \otimes S \to H \otimes S$ by

$$D = \sum \pi(x_i) \otimes \gamma(x_i).$$

6.11.   LEMMA (compare Schmid [97]). 1) *If* $x, y \in H \otimes S$, *then* $\langle Dx, y \rangle = \langle x, Dy \rangle$.
   2) $D^2 = -\pi(C) \otimes I - (|\rho|^2 - |\rho_k|^2)I + (\pi \otimes s)(C_k)$.

1) is obvious.
To prove 2), observe that

$$\begin{aligned}
D^2 &= \sum_{ij} \pi(x_i)\pi(x_j) \otimes \gamma(x_i)\gamma(x_j) \\
&= -\sum_i \pi(x_i)^2 \otimes I + \sum_{i \neq j} \pi(x_i)\pi(x_j) \otimes \gamma(x_i)\gamma(x_j) \\
&= -\pi(C) \otimes I + \pi(C_k) \otimes I + \sum_{i \neq j} \pi(x_i)\pi(x_j) \otimes \gamma(x_i)\gamma(x_j).
\end{aligned}$$

Since $\gamma(x_i)\gamma(x_j) = -\gamma(x_j)\gamma(x_i)$ for $i \neq j$, we find that

$$\begin{aligned}
D^2 &= -\pi(C) \otimes I + \pi(C_k) \otimes I + \frac{1}{2}\sum_{i,j} \pi([x_i, x_j]) \otimes \gamma(x_i)\gamma(x_j) \\
&= -\pi(C) \otimes I + \pi(C_k) \otimes I - \frac{1}{2}\sum_{i,j,a} B([x_i, x_j], x_a)\pi(x_a) \otimes \gamma(x_i)\gamma(x_j) \\
&= -\pi(C) \otimes I + \pi(C_k) \otimes I - 2\sum_a \pi(x_a) \otimes s(x_a) \\
&= -\pi(C) \otimes I + \pi(C_k) \otimes I + (\pi \otimes s)(C_k) - \pi(C_k) \otimes I - I \otimes s(C_k) \\
&= -\pi(C) \otimes I - I \otimes s(C_k) + (\pi \otimes s)(C_k).
\end{aligned}$$

Lemma 6.11 now follows from 6.9 (2).

6.12.   PROPOSITION. *Let* F *be the irreducible finite dimensional representation with highest weight* $\Lambda - \rho$ *relative to* Φ^+. *Let* (π, H) *be a unitary* $(\mathfrak{g}, \mathfrak{k})$-module *with* $\pi(C)$ *a scalar operator.*
   1) *If* $\theta\Lambda \neq \Lambda$, *then* $\operatorname{Ext}^*_{\mathfrak{g}, \mathfrak{k}}(F, H) = 0$.
   2) *Suppose that* (π, H) *is admissible, and that* $\pi(C) = (|\Lambda|^2 - |\rho|^2)I$. *Then*

$$\dim \operatorname{Hom}_{\mathfrak{k}}(\Lambda\mathfrak{p}, H \otimes F^*) = 2^{[l_0/2]+\varepsilon} \sum_{t \in W^1} \dim \operatorname{Hom}_{\mathfrak{k}}(F_{t\Lambda - \rho_k}, H \otimes S),$$

where F_λ *is as in 5.1 and* $\varepsilon = 0$ *or* 1, $\varepsilon \equiv (\dim \mathfrak{p}) \bmod 2$.

Let $\pi(C) = \lambda I$. Assume that $H^r(\mathfrak{g}, \mathfrak{k}; H \otimes F^*) \neq (0)$ for some r. Then $\lambda = |\Lambda|^2 - |\rho|^2$ (see 3.1) and $\operatorname{Hom}_{\mathfrak{k}}(\Lambda^r\mathfrak{p} \otimes F, H) \neq (0)$. Now, as a $\mathfrak{k}$-module, $\Lambda\mathfrak{p} = S \otimes S$ or $S \otimes S \oplus S \otimes S$ depending on whether $\dim \mathfrak{p}$ is even or odd. Thus, since $S \equiv S^*$ as a $\mathfrak{k}$-module, $\operatorname{Hom}_{\mathfrak{k}}(\Lambda\mathfrak{p} \otimes F, H)$ is equal to $\operatorname{Hom}_{\mathfrak{k}}(F \otimes S, H \otimes S)$ (resp. $\operatorname{Hom}_{\mathfrak{k}}(F \otimes S, H \otimes S) \oplus \operatorname{Hom}_{\mathfrak{k}}(F \otimes S, H \otimes S)$) if $\dim \mathfrak{p}$ is even (resp. odd).

Let $F \otimes S = \bigoplus m_v F_v$ as a $\mathfrak{k}$-module. Suppose that $\operatorname{Hom}_{\mathfrak{k}}(F_v, H \otimes S) \neq (0)$ for some v with $m_v \neq (0)$. 6.11 implies that if $\xi \in H \otimes S$, then

$$\langle D^2\xi, \xi \rangle = (-\lambda - |\rho|^2 + |\rho_k|^2)\langle \xi, \xi \rangle + \langle (\pi \otimes s)(C_k)\xi, \xi \rangle.$$

Since $\langle D^2\xi,\xi\rangle = \langle D\xi,D\xi\rangle \geq 0$, we see that $\langle(\pi\otimes s)(C_k)\xi,\xi\rangle \geq (\lambda + |\rho|^2 - |\rho_k|^2)\langle\xi,\xi\rangle$. But $\lambda = |\Lambda|^2 - |\rho|^2$, and hence we have

$$\langle(\pi\otimes s)(C_k)\xi,\xi\rangle \geq (|\Lambda|^2 - |\rho_k|^2)|\xi|^2.$$

This in turn implies that if $\mathrm{Hom}_{\mathfrak{k}}(F_v, H\otimes S) \neq (0)$, then

$$|v + \rho_k|^2 \geq |\Lambda|^2.$$

If $m_v \neq 0$, then $v = \mu^+ + t\rho - \rho_k$, with μ a weight of F and $t \in W^1$. Hence $|\Lambda|^2 \leq |v + \rho_k|^2 = |\mu^+ + t\rho|^2 \leq |\mu + t\rho|^2$. But μ is a weight of F and $t\rho$ is a weight of the finite dimensional representation with highest weight ρ. Thus $|\mu + t\rho|^2 \leq |\Lambda - \rho + \rho|^2 = |\Lambda|^2$, and equality occurs if and only if $\mu + t\rho = u(\Lambda - \rho) + u\rho$, $u \in W$. That is, $u^{-1}\mu + u^{-1}t\rho = \Lambda$. But $u^{-1}\mu = \Lambda - \rho - \xi$ and $u^{-1}t\rho = \rho - \langle Q\rangle$, where $Q \subset \Phi^+$ and ξ is a sum of elements of Φ^+. Hence $\Lambda - \xi - \langle Q\rangle = \Lambda$. Thus $\xi = \langle Q\rangle = (0)$. This implies $t = u$ and $\mu = t(\Lambda - \rho)$. But then we have

$$|\Lambda|^2 \leq |t(\Lambda - \rho)^+ + t\rho|^2 \leq |t\Lambda|^2 = |\Lambda|^2.$$

Since $(t\rho)^+ = t\rho$, this implies that $|\Lambda|^2 = |(t\Lambda^+)|^2 = |\Lambda^+|^2$. Thus $|\Lambda^-|^2 = 0$. This proves 1). Since we have shown that $v + \rho_k = t\Lambda$, $t \in W^1$ if $m_v \neq 0$ and $\mathrm{Hom}_{\mathfrak{k}}(F_v, H\otimes S) \neq 0$, Assertion 2) follows by the argument at the end of 6.9.

6.13. It should be observed that 6.12 1) can be proved for (π, H) admitting an infinitesimal character χ_π as follows: Let $x \mapsto {}^tx$ be the canonical anti-involution on $U(\mathfrak{g})$. Let $x \mapsto \overline{x}$ be the conjugation in $\mathfrak{g}_c$ relative to $\mathfrak{g}$, extended canonically to $U(\mathfrak{g})$. Then $\chi_\pi(z) = \overline{\chi_\pi({}^t\overline{z})}$, since π is unitary. It is not hard to show that if Λ is as in 6.12, then $\overline{\chi_\Lambda({}^t\overline{z})} = \chi_{\theta\Lambda}(z)$. If $\mathrm{Ext}^*_{\mathfrak{g},\mathfrak{k}}(F, H) \neq 0$, then $\chi_\Lambda = \chi_\pi$. Hence $\chi_{\theta\Lambda} = \chi_\Lambda$. But then $\theta\Lambda = t\Lambda$, $t \in W$. Since $\theta\Lambda$ and Λ are both Φ^+-dominant integral, this implies $\theta\Lambda = \Lambda$.

7. Vanishing theorems using spinors

7.1. If $P \subset \Phi$ is a system of positive roots compatible with Φ_k^+, we let $\mathfrak{p}^+(P) = \sum \mathfrak{p}_\lambda$, where the sum is over all λ such that $\mathfrak{p}_\lambda \neq 0$ and $\lambda \in P|_{\mathfrak{h}^+}$.

The following vanishing theorem uses ideas in Hotta-Parthasarathy [**61**].

7.2. THEOREM. *Let F, Λ and (π, H) be as in 6.12. Suppose that $\pi(C) = (|\Lambda|^2 - |\rho|^2)I$ and that $\theta\Lambda = \Lambda$. Suppose that whenever $t \in W^1$ and ξ is a weight of $\Lambda\mathfrak{p}^+(t\Phi^+)$ so that $t\Lambda + t\rho - 2\rho_k - \xi$ is Φ_k^+-dominant integral, then $t\Lambda - \rho_k - \xi$ is Φ_k^+-dominant integral. Then $H^j(\mathfrak{g},\mathfrak{k}; H\otimes F^*) = 0$ for $0 \leq j < \dim\mathfrak{p}^+$.*

(Note that if Λ satisfies our condition, then $\Lambda + \mu$ satisfies this condition for μ Φ_k^+-dominant integral. Also if Λ is as in 6.12, then $k\Lambda$ satisfies this condition for k large.)

PROOF OF 7.2. Proposition 3.1 implies that

$$H^j(\mathfrak{g},\mathfrak{k}; H\otimes F^*) = \mathrm{Hom}_{\mathfrak{k}}(\Lambda^j\mathfrak{p}, H\otimes F^*) = \mathrm{Hom}_{\mathfrak{k}}(F\otimes\Lambda^j\mathfrak{p}, H) \qquad (j \in \mathbf{N}).$$

We compute $\mathrm{Hom}_{\mathfrak{k}}(F \otimes \Lambda\mathfrak{p}, H)$. Now $\Lambda\mathfrak{p} = S \otimes S$ or $S \otimes S \oplus S \otimes S$. Thus we really must compute

$$
\begin{aligned}
\mathrm{Hom}_{\mathfrak{k}}(F \otimes S \otimes S, H) &= \mathrm{Hom}_{\mathfrak{k}}(F \otimes S, H \otimes S) \\
&= \sum_{t \in W^1} \mathrm{Hom}_{\mathfrak{k}}(2^{[l_0/2]} F_{t\Lambda - \rho_k}, H \otimes S) \\
&= \sum_{t \in W^1} \mathrm{Hom}_{\mathfrak{k}}(2^{[l_0/2]} F_{t\Lambda - \rho_k} \otimes S, H) \ (6.12(2)).
\end{aligned}
$$

We look at $\mathrm{Hom}_{\mathfrak{k}}(F_{t\Lambda - \rho_k} \otimes S, H)$. We have $\tau_{t\Lambda - \rho_k} \otimes S = \sum m_\lambda \tau_\lambda$, and if $m_\lambda \neq 0$, then $\lambda = t\Lambda - 2\rho_k + t\rho - \xi$, where ξ is a weight of $\Lambda\mathfrak{p}^+(t\Phi^+)$. Now the hypothesis of this theorem implies that $t\Lambda - \rho_k - \xi$ is Φ_k^+-dominant integral if $m_\lambda \neq 0$. It follows that $\tau_\lambda \otimes S$ contains $\tau_{\lambda - \rho_n}$ since S contains $\tau_{\rho_n}^*$. If $\mathrm{Hom}_{\mathfrak{k}}(F_\lambda, H) \neq 0$, then, arguing as in the proof of 6.12, we find that the lowest eigenvalue of $(\tau_\lambda \otimes S)(C_k)$ is greater than or equal to $|\Lambda|^2 - |\rho_k|^2$. This implies that $|t\Lambda - \xi|^2 \geq |\Lambda|^2$. Now ξ is a weight of $\Lambda\mathfrak{p}^+(t\Phi^+)$. Hence $\xi = t\langle Q \rangle|_{h^+}$ with $Q \subset \Phi^+$. Hence $|t\Lambda - t\langle Q \rangle|_{h^+}|^2 \geq |\Lambda|^2$. But then $|\Lambda - \langle Q \rangle|^2 \geq |\Lambda - \langle Q \rangle^+|^2 \geq |\Lambda|^2$. $\Lambda = \Lambda - \rho + \rho$ and $\Lambda - \rho$ is the highest weight of F. Hence $|\Lambda - \rho + \rho - \langle Q \rangle|^2 \geq |\Lambda|^2$. Thus $t(\Lambda - \rho) + t\rho - t\langle Q \rangle = u(\Lambda - \rho) + u\rho$ for some $u \in W$. Arguing as in the proof of 6.12, we see that $u = t$ and $Q = \varnothing$.

We have shown that if $F \otimes \Lambda\mathfrak{p} = \sum n_\lambda \tau_\lambda$, then

$$
\mathrm{Hom}_{\mathfrak{k}}(F \otimes \Lambda\mathfrak{p}, H) = \sum_{t \in W^1} \mathrm{Hom}_{\mathfrak{k}}(n_{t(\Lambda + \rho) - 2\rho_k} F_{t(\Lambda + \rho) - 2\rho_k}, H).
$$

To complete the proof of the theorem we must show that

$$
\mathrm{Hom}_{\mathfrak{k}}(F_{t(\Lambda + \rho) - 2\rho_k}, F \otimes \Lambda^j \mathfrak{p}) = 0
$$

for $0 \leq j < \dim \mathfrak{p}^+$ and for $j > \dim \mathfrak{p}^+ + l_0$. By replacing Φ^+ by $t\Phi^+$, we can assume $t = 1$. The weights of F relative to $\mathfrak{h}_c^+$ are of the form $\Lambda - \rho - \xi^+$, where ξ is a sum of elements of Φ^+. The weights of $\Lambda^j \mathfrak{p}_c$ are of the form $\lambda_{i_1} + \cdots + \lambda_{i_j}$, where the λ_{i_j} are weights of F in $\mathfrak{p}_c$.

Hence the weights of $F \otimes \Lambda^j \mathfrak{p}_c$ are of the form $\Lambda - \rho - \xi^+ + 2\rho_n - \xi_1^+$, with ξ_1 a sum of elements of Φ^+. Thus the highest possible weight is $\Lambda + \rho - 2\rho_k$, and this occurs only if $2\rho_n$ is a weight of $\Lambda^j \mathfrak{p}$. But $2\rho_n$ is a weight of $\Lambda^j \mathfrak{p}_c$ only if $\dim \mathfrak{p}_c^+ \leq j \leq \dim \mathfrak{p}_c^+ + l_0$.   Q.E.D.

7.3.   We now assume that (π, H) is irreducible and admissible, and that Λ satisfies the conditions of Theorem 7.2.

THEOREM. *If $H^*(\mathfrak{g}, \mathfrak{k}; H \otimes F^*) \neq 0$, then:*
1) (π, H) *is in the fundamental series for $\mathfrak{g}$ relative to $t\Phi^+$ for some $t \in W^1$ (see [38] for the definitions) and (π, H) has lowest $\mathfrak{k}$-type $\tau_{t\Lambda + t\rho - 2\rho_k}$.*
2) $\dim H^j(\mathfrak{g}, \mathfrak{k}; H \otimes F^*) = \binom{l_0}{j - q}$, *where $q = \dim \mathfrak{p}_c^+$ $(j \in \mathbf{N})$.*

PROOF. By the proof of Theorem 7.2, (π, H) must contain $\tau_{t\Lambda + t\rho - 2\rho_k}$, and cannot contain any $\mathfrak{k}$-type of the form $\tau_{t\Lambda + t\rho - 2\rho_k - \xi}$ with ξ a weight of $\Lambda\mathfrak{p}^+(t\Phi^+)$. Theorem 6.3 of [38] now implies 1) and $\dim \mathrm{Hom}_{\mathfrak{k}}(F_{t(\Lambda + \rho) - 2\rho_k}, H) = 1$, whence 2). For a more general statement see III, 5.1.

7.4.   Let $\Lambda^j \mathfrak{p} = n_{0,j} F_0 + \sum n_{\lambda, j} F_\lambda$ as a $\mathfrak{k}$-module.
Let $B_j^+ = \{\lambda \mid n_{\lambda, j} \neq 0$ and $(\tau_\lambda \otimes s)(C_k)$ has lowest eigenvalue at least $|\rho|^2 - |\rho_k|^2\}$.

7.5. LEMMA. *Let* (π, H) *be a unitary* $(\mathfrak{g}, \mathfrak{k})$-*module with* $\pi(C)$ *a scalar and* $H^{\mathfrak{g}} = (0)$. *If* $B_j^+ = \varnothing$, *then* $H^j(\mathfrak{g}, \mathfrak{k}; H) = 0$.

PROOF. If $H^j(\mathfrak{g}, \mathfrak{k}; H) \neq 0$, then $\pi(C) = 0$ (3.1) and $H^j(\mathfrak{g}, \mathfrak{k}; H) = \mathrm{Hom}_{\mathfrak{k}}(\Lambda^j \mathfrak{p}_c, H)$. If $H^{\mathfrak{g}} = 0$, then $H^{\mathfrak{k}} = 0$. Hence $\mathrm{Hom}_{\mathfrak{k}}(F_\lambda, H) \neq 0$ for some $\lambda \neq 0$, so that $n_{\lambda, j} \neq 0$. But then 6.10 implies that $(\tau_\lambda \otimes s)(C_k)$ has lowest eigenvalue at least $|\rho|^2 - |\rho_k|^2$.   Q.E.D.

7.6.   We assume that $\mathfrak{g}$ is simple as a real Lie algebra. Then there are two possibilities for $\mathfrak{p}_c$ as a $\mathfrak{k}$-module.
1) $\mathfrak{p}_c$ is an irreducible $\mathfrak{k}$-module.
2) $\mathfrak{p}_c = V_1 \oplus V_2$ with V_1, V_2 irreducible $\mathfrak{k}$ submodules.

7.7. LEMMA. *Let* Φ_k^+ *be a system of positive roots for* Φ_k. *Let* Φ^+ *be a compatible system of positive roots for* Φ. *If* 7.6, 1) *holds, let* λ *be the highest weight relative to* Φ_k^+ *of* $\mathfrak{p}_c$ *as a* $\mathfrak{k}$-*module. If* 7.6, 2) *holds, let* λ_1, λ_2 *be the highest weights relative to* Φ_k^+ *for* V_1 *and* V_2 *respectively. Assume*
1) *if* 7.6, 1) *holds, there is* $t \in W^1$ *so that* $t\rho - \rho_k - \lambda$ *is* Φ_k^+-*dominant and* λ *is not a simple root in* $t\Phi^+$;
2) *if* 7.6, 2) *holds, then for* $i = 1, 2$ *there exists* $t_i \in W^1$ *so that* $t_i \rho - \rho_k - \lambda_i$ *is* Φ_k^+-*dominant and* $\lambda_i = \alpha$ *for some* $\alpha \in t_i \Phi^+$, *but* λ_i *is not a simple root in* $t_i \Phi^+$, $i = 1, 2$.
If (π, H) *is a unitary* $(\mathfrak{g}, \mathfrak{k})$-*module with* $\pi(C)$ *a scalar, then* $H^1(\mathfrak{g}, \mathfrak{k}; H) = 0$.

PROOF. Assume 7.6, 1), holds. Then $\mathfrak{p} = F_\lambda$. We have

$$F_\lambda \otimes S = \sum_{t \in W^1} 2^{[l_0/2]} F_\lambda \otimes F_{t\rho - \rho_k}.$$

The lowest weight of F_λ is $-\lambda$. Hence for t as in 1), $F_\lambda \otimes S \supset F_{t\rho - \rho_k - \lambda}$. Now $|t\rho - \rho_k - \lambda + \rho_k|^2 = |t\rho - \lambda|^2$. We have $\lambda = \alpha^+$ for some $\alpha \in t\Phi^+$ (λ is the highest weight). Hence $|t\rho - \lambda|^2 = |t\rho - \alpha^+|^2 \leq |t\rho - \alpha|^2 \leq |\rho|^2$. If $|t\rho - \lambda|^2 \geq |\rho|^2$, then we must have $|t\rho - \alpha|^2 = |\rho|^2$; hence α is $t\Phi^+$ simple and $|t\rho - \alpha^+|^2 = |t\rho - \alpha|^2$. Hence $\alpha = \alpha^+$. This contradicts 1); hence $B_1^+ = \varnothing$.
2) Use the same proof for $i = 1, 2$.

7.8. PROPOSITION. *Suppose that* $\mathfrak{g}$ *is isomorphic with* $\mathfrak{sp}(n, 1)$, $n \geq 2$, *or with the* **R**-*rank* 1 *real form of* **F**$_4$. *Let* (π, H) *be a unitary* $(\mathfrak{g}, \mathfrak{k})$-*module with* $\pi(C)$ *a scalar. Then* $H^1(\mathfrak{g}, \mathfrak{k}; H) = 0$.

We use the notation of Bourbaki [27].
1) $\mathfrak{g} = \mathfrak{sp}(n, 1)$. Then $\mathfrak{g}_c = \mathbf{C}_{n+1}$. We label the roots as in [27], p. 254. Then Φ_k^+ has simple roots

$$2\varepsilon_1 = 2 \sum_{i=1}^{n} \alpha_i + \alpha_{n+1}, \alpha_2, \ldots, \alpha_{n+1}.$$

We note that $S_{\alpha_1} \in W^1$. Also

$$\rho = \sum_{j=1}^{n+1} (n + 2 - j)\alpha_j, \qquad \rho_k = \varepsilon_1 + \sum_{j=2}^{n+1} (n + 2 - j)\varepsilon_j.$$

Hence

$$\rho - \rho_k = n\varepsilon_1, \qquad s_{\alpha_1}\rho - \rho_k = n\varepsilon_1 - \alpha_1 = (n - 1)\varepsilon_1 + \varepsilon_2.$$

If λ is as in 7.7, then

$$\lambda = \alpha_1 + 2 \sum_{i=2}^{n} \alpha_i + \alpha_{n+1} = \varepsilon_1 + \varepsilon_2.$$

Thus

$$s_{\alpha_1} \rho - \rho_k - \lambda = (n-2)\varepsilon_1.$$

We therefore see that if $n \geq 2$, then $s_{\alpha_1} \rho - \rho_k - \lambda$ is Φ_k^+-dominant integral. The simple roots of $s_{\alpha_1} \Phi^+$ are

$$-\alpha_1, \alpha_1 + \alpha_2, \alpha_3, \ldots, \alpha_{n+1}.$$

Thus if $n \geq 2$, then λ is not $s_{\alpha_1} \Phi^+$ simple. The result in this case now follows from 7.7.

2) $\mathfrak{g}$ *is the* **R**-*rank* 1 *real form of* $\mathbf{F}_4$.
Here we use pp. 272, 273 of [**27**]. The simple roots of Φ_k^+ are

$$\varepsilon_1 - \varepsilon_2, \quad \varepsilon_2 - \varepsilon_3, \quad \varepsilon_3 - \varepsilon_4, \quad \varepsilon_4.$$

The simple roots of Φ^+ are

$$\alpha_1 = \varepsilon_2 - \varepsilon_3, \quad \alpha_2 = \varepsilon_3 - \varepsilon_4, \quad \alpha_3 = \varepsilon_4, \quad \alpha_4 = \tfrac{1}{2}(\varepsilon_1 - \varepsilon_2 - \varepsilon_3 - \varepsilon_4).$$

It is clear that $s_{\alpha_4} \in W^1$. Also

$$2\rho = 11\varepsilon_1 + 5\varepsilon_2 + 3\varepsilon_3 + \varepsilon_4, \qquad 2\rho_k = 7\varepsilon_1 + 5\varepsilon_2 + 3\varepsilon_3 + \varepsilon_4.$$

(See [**27**], p. 253.) Hence

$$\rho - \rho_k = 2\varepsilon_1, \qquad s_{\alpha_4} \rho - \rho_k = \tfrac{1}{2}(3\varepsilon_1 + \varepsilon_2 + \varepsilon_3 + \varepsilon_4).$$

If λ is as in 7.7, then

$$\lambda = \tfrac{1}{2}(\varepsilon_1 + \varepsilon_2 + \varepsilon_3 + \varepsilon_4).$$

Thus $s_{\alpha_4} \rho - \rho_k = -\lambda = \varepsilon_1$, which is Φ_k^+-dominant integral. The simple roots of $s_{\alpha_4} \Phi^+$ are $\alpha_1, \alpha_2, \alpha_3 + \alpha_4, -\alpha_4$. Since

$$\lambda = \alpha_1 + 2\alpha_2 + 3\alpha_3 + \alpha_4,$$

λ is not $s_{\alpha_4} \Phi^+$ simple. The result, in this case, now follows from 7.7.

8. Matsushima's vanishing theorem

In this section, we assume that $\mathfrak{g}$ *is semi-simple and has no compact factor.*

8.1. Let $L(\ ,\ )$ be the symmetric bilinear form on $\mathfrak{k}$ defined by

(1) $$L(x,y) = \operatorname{tr}(\operatorname{ad}_{\mathfrak{p}} x \circ \operatorname{ad}_{\mathfrak{p}} y) \qquad (x, y \in \mathfrak{k}).$$

We have

(2) $$B(x,y) = B_{\mathfrak{k}}(x,y) + L(x,y) \qquad (x, y \in \mathfrak{k}),$$

where $B_{\mathfrak{k}}$ (resp. B) is the Killing form of $\mathfrak{k}$ (resp. $\mathfrak{g}$). The eigenvalues of the endomorphisms $\operatorname{ad} x$ $(x \in \mathfrak{k})$ are purely imaginary, and our assumption on $\mathfrak{g}$ insures that $\mathfrak{k}$ acts faithfully on $\mathfrak{p}$ via the adjoint representation. Hence $L(\ ,\ )$ is negative and non-degenerate. We let

(3) $$A = \min_{x \in \mathfrak{k}, B(x)=-1} -L(x,x).$$

Then $0 < A \leq 1$.

We use the notation and conventions of §1. We have, taking (1.2) into account,

(4)
$$L(x_a, x_b) = \sum_{i,j} c_{aj}^i \cdot c_{bi}^j = \sum_{i,j} c_{ij}^a \cdot c_{ji}^b,$$

(5)
$$L(x_a, x_b) = -\sum_{ij} c_{ij}^a \cdot c_{ij}^b \qquad (m < a, b \le n).$$

In the sequel, we assume moreover that the X_a's form an orthogonal basis with respect to $L(\, , \,)$. Set

(6)
$$R(x, y) = -\operatorname{ad}[x, y]_{\mathfrak{p}} \qquad (x, y \in \mathfrak{p}),$$

hence

(7)
$$R(x, y) \cdot z = [[y, x], z] \qquad (x, y, z \in \mathfrak{p}),$$

and put

(8)
$$R_{ijkl} = B([[x_l, x_k], x_j], x_i) = B([x_l, x_k], [x_j, x_i]).$$

Therefore

(9)
$$R_{ijkl} = -\sum_a c_{kl}^a \cdot c_{ij}^a.$$

As is well known, $R(\, , \,)$ is the curvature tensor on G/K, for the invariant Riemannian metric which, on $\mathfrak{p} = T(G/K)_e$, is equal to the restriction of the Killing form. However, this interpretation will not be needed here.

8.2. The form $F_{\mathfrak{g}}^q$. We denote by η_{ij} the coordinates of an element $\eta \in \mathfrak{p} \otimes \mathfrak{p}$ with respect to the basis $x_i \otimes x_j$ ($1 \le i, j \le m$), and put, for $q = 1, 2, \cdots$,

(1)
$$F_{\mathfrak{g}}^q(\xi, \eta) = (A/2q) \cdot \sum_{ij} \xi_{ij} \cdot \eta_{ij} + \sum_{ijkl} R_{ijkl} \xi_{il} \eta_{jk}$$

with A given by 8.1(3). Let

(2)
$$m(\mathfrak{g}) = \max(\{0\} \cup \{q \mid F_{\mathfrak{g}}^q > 0\}).$$

8.3. THEOREM. *Let (π, V) be a unitary $(\mathfrak{g}, \mathfrak{k})$-module on which the Casimir element acts by a scalar multiple of the identity and such that $V^{\mathfrak{g}} = 0$. Then $H^q(\mathfrak{g}, \mathfrak{k}; V) = 0$ for $q \le m(\mathfrak{g})$.*

The assumptions on V are satisfied if (π, V) is irreducible, admissible and non-trivial. Therefore

8.4. COROLLARY. *If (σ, H) is a non-trivial irreducible admissible unitary $(\mathfrak{g}, \mathfrak{k})$-module, then $H^q(\mathfrak{g}, \mathfrak{k}; H) = 0$ for $q \le m(\mathfrak{g})$.*

8.5. This theorem is the representation theoretic analogue of a theorem of Matsushima on the cohomology of cocompact discrete subgroups [**80**], to be discussed in VII. The proof given here is essentially the same as Matsushima's.

Theorem 8.3 also applies to any admissible unitary $(\mathfrak{g}, K)$-module (π, V) for which $V^{\mathfrak{g}} = 0$. In fact, V is then a direct sum of primary subspaces with respect to the Casimir element C. Moreover, using unitarity and admissibility, one sees that C acts by scalars on each of those; this reduces us to the theorem.

8.6. Proof of 8.3. If C acts non-trivially on V, then $H^q(\mathfrak{g}, \mathfrak{k}; V) = 0$ for all q's by (3.1). From now on, we assume that $\pi(C) = 0$. We shall prove that if there exists $q \leq m(\mathfrak{g})$ such that $H^q(\mathfrak{g}, \mathfrak{k}; V) \neq 0$, then $V^{\mathfrak{g}} \neq 0$. If $q = 0$, this is clear. So let $q \geq 1$. Since $\pi(C) = 0$, all cochains are closed, harmonic and $H^q(\mathfrak{g}, \mathfrak{k}; V) = C^q(\mathfrak{g}, \mathfrak{k}; V)$ (3.1). We have then to show that if

$$
(1) \qquad\qquad \eta = \sum_I \eta_I \cdot \omega^I,
$$

is a q-cochain, then $\eta_I \in V^{\mathfrak{g}}$, i.e.

$$
(2) \qquad\qquad x_i \eta_I = x_a \eta_I = 0,
$$

for all i, a, I subject to our conditions. Since $[\mathfrak{p}, \mathfrak{p}] = \mathfrak{k}$, it suffices in fact to prove that

$$
(3) \qquad\qquad x_i \eta_I = 0 \qquad (1 \leq i \leq m;\ I \subset I_m,\ |I| = q).
$$

That (3) implies (2) also follows from

$$
(4) \qquad\qquad 0 = (C\eta_I, \eta_I) = \sum_a \|x_a \eta_I\|^2 - \sum_i \|x_i \eta_I\|^2,
$$

which, incidentally, also shows that if $v \in V^{\mathfrak{k}}$, then $v \in V^{\mathfrak{g}}$. In the sequel, u runs from 1 to q, i, j, k, l, j_u from 1 to m, a, b, c from $m + 1$ to n, and I through the subsets of q elements of $I_m = \{1, 2, \ldots, m\}$. Let

$$
(5) \qquad \Phi(\eta) = \frac{(q-1)!}{2} \sum_{i,j,I} \|[x_i, x_j]\eta_I\|^2 = (2q)^{-1} \cdot \sum_{\substack{i,j \\ j_1,\ldots,j_q}} \|[x_i, x_j]\eta_{j_1 \cdots j_q}\|^2.
$$

We shall transform $\Phi(\eta)$ in two ways. First, using $[x_i, x_j] = \sum c_{ij}^a x_a$, we can write

$$
(6) \qquad\qquad \Phi(\eta) = \frac{(q-1)!}{2} \sum_{a,b,i,j,I} c_{ij}^a \cdot c_{ij}^b (x_a \eta_I, x_b \eta_I).
$$

In view of 8.1(5), this gives

$$
(7) \qquad\qquad \Phi(\eta) = -\frac{(q-1)!}{2} \sum_{a,b,I} L(x_a, x_b) \cdot (x_a \eta_I, x_b \eta_I).
$$

Since the x_a's are assumed to be orthogonal with respect to L, the sum is in fact over $a = b$ and, by the definition of A (8.1(3)), we have

$$
(8) \qquad\qquad \Phi(\eta) \geq \frac{A \cdot (q-1)!}{2} \sum_{a,I} \|x_a \eta_I\|^2.
$$

If we use the formula for $[x_i, x_j]$ on one term of each of the scalar products in (5), we get

$$
(9) \qquad\qquad \Phi(\eta) = (2q)^{-1} \sum_{\substack{i,j,a \\ j_1,\cdots,j_q}} c_{ij}^a (x_a \cdot \eta_{j_1 \cdots j_q}, [x_i, x_j] \cdot \eta_{j_1 \cdots j_q}).
$$

Since c_{ij}^a and $[x_i, x_j]$ are antisymmetric in i, j, this gives

$$
(10) \qquad\qquad \Phi(\eta) = q^{-1} \sum_{\substack{i,j,a \\ j_1,\ldots,j_q}} c_{ij}^a (x_a \cdot \eta_{j_1 \cdots j_q}, x_i x_j \cdot \eta_{j_1 \cdots j_q}).
$$

By assumption, $\eta \in C^q(\mathfrak{g}, \mathfrak{k}; V)$. Therefore

$$x_a \cdot \eta_{j_1 \cdots j_q} = \sum_u \eta(x_{j_1}, \ldots, [x_a, x_{j_u}], \ldots, x_{j_q})$$

$$= \sum_{a,k,u} c_{a,k_u}^k \eta(x_{j_1}, \ldots, x_k, \ldots, x_{j_q})$$

$$= \sum_{a,k,u} (-1)^{u-1} \cdot c_{a,j_u}^k \cdot \eta(x_{j_1}, x_{j_1}, \ldots, \widehat{x}_{j_u}, \ldots, x_{j_q}).$$

Then we have, using (1.2),

$$q \cdot \Phi(\eta) = \sum_{\substack{i,j,k,u \\ j_1,\ldots,j_q}} (-1)^{u-1} \left(\sum_a c_{ij}^a \cdot c_{kj_u}^a \right) (\eta_{k,j_1,\ldots,j_u,\ldots,j_q}, x_i \cdot x_j \cdot \eta_{j_1 \cdots j_q}).$$

Since (π, V) is unitary, we have

$$(\eta_{k,j_1,\ldots,\widehat{j}_u,\ldots,j_q}, x_i \cdot x_j \cdot \eta_{j_1 \cdots j_q}) = -(x_i \eta_{k,j_1,\ldots,\widehat{j}_u,\ldots,j_q}, x_j \cdot \eta_{j_1,\ldots,j_q}).$$

Taking 8.1(8) into account, we get

$$q\Phi(\eta) = \sum_{\substack{i,j,k,u \\ j_1,\ldots,j_q}} (-1)^{u-1} R_{ijkj_u} (x_i \cdot \eta_{k,j_1,\ldots,\widehat{j}_u,\ldots,k_q}, x_j \cdot \eta_{j_1,\ldots,j_q}),$$

$$q\Phi(\eta) = \sum_{\substack{i,j,k,u \\ j_1,\ldots,j_q}} R_{ijkj_u} (x_i \cdot \eta_{k,j_1,\ldots,\widehat{j}_u,\ldots,j_q}, x_j \cdot \eta_{j_u,j_1,\ldots,\widehat{j}_u,\ldots,j_q}).$$

This can be written

$$q\Phi(\eta) = q \sum_{\substack{i,j,k,l \\ j_2,\ldots,j_q}} R_{ijkl} (x_i \eta_{k,j_2,\ldots,j_q}, x_j \eta_{l,j_2,\ldots,j_q}).$$

Since R_{ijkl} is antisymmetric in the last two indices (see 8.1(8)), we get finally

$$\Phi(\eta) = - \sum_{\substack{i,j,k,l \\ j_2,\ldots,j_q}} R_{ijkl} (x_i \eta_{k,j_2,\ldots,j_q}, x_j \eta_{l,j_2,\ldots,j_q}). \tag{11}$$

Together with (8), this yields

$$\sum_{j_2,\ldots,j_q} \left\{ \frac{A}{2q} \sum_{i,j} \|x_j \eta_{i,j_2,\ldots,j_q}\|^2 \right.$$

$$\left. + \sum_{i,j,k,l} R_{ijkl} (x_i \eta_{k,j_2,\ldots,j_q}, x_j \eta_{l,j_2,\ldots,j_q}) \right\} \leq 0. \tag{12}$$

On $\mathfrak{p} \otimes \mathfrak{p} \otimes V$, we consider the tensor product $F_{\mathfrak{g},V}^q$ of $F_{\mathfrak{g}}^q$ and of the given scalar product on V. It is positive non-degenerate since $q \leq m(\mathfrak{g})$. The inequality (12) can now be written

$$\sum_j F_{\mathfrak{g},V}^q (\{x_j \cdot \eta_{i \cup J}\}) \leq 0, \tag{13}$$

where J runs through the subsets of $I_m = \{1, \ldots, m\}$ having $q-1$ elements. Since $F_{\mathfrak{g},V}^q$ is positive non-degenerate, we get

(14) $x_j \cdot \eta_{i \cup J} = 0 \qquad (1 \le i, j \le m; J \subset I_m, |J| = q-1)$,

which is just (3).

8.7.   The value of $m(\mathfrak{g})$ for $\mathfrak{g}$ simple non-compact has been determined case by case [**68, 80**]. In view of the vanishing theorem proved in Chapter V, we need be concerned only with the cases where $m(\mathfrak{g}) \ge \mathrm{rk}_{\mathbf{R}}\,\mathfrak{g}$. This occurs in the following cases only:

$\mathfrak{g}$ is the complex form of $\mathbf{F}_4$ (resp. $\mathbf{E}_7$, resp. $\mathbf{E}_8$); $m(\mathfrak{g})$ is equal to 4 (resp. 8, resp. 14).

$\mathfrak{g}$ is the real form of $\mathbf{E}_8$ with real rank 4 and maximal compact subalgebra isomorphic to $\mathbf{E}_7 + \mathfrak{su}(2)$. Then $m(\mathfrak{g}) = 5$.

9. Direct products

9.1.   Let (ρ, E) be a finite dimensional $\mathfrak{g}$-module. Then we let $M(\mathfrak{g}, \rho)$ be the greatest integer such that $H^q(\mathfrak{g}, \mathfrak{k}; H \otimes E) = 0$ for all irreducible admissible non-trivial unitary $(\mathfrak{g}, \mathfrak{k})$-modules and all $q \le M(\mathfrak{g}, \rho)$. If ρ is the trivial one-dimensional representation of $\mathfrak{g}$, we write $M(\mathfrak{g})$ for $M(\mathfrak{g}, \rho)$. In particular, $m(\mathfrak{g}) \le M(\mathfrak{g})$.

9.2.   Let $\mathfrak{g} = \mathfrak{g}' \oplus \mathfrak{g}''$ be a direct product. First, assume $\mathfrak{g}''$ to be *compact*. Then, if (π, V) is any $(\mathfrak{g}, \mathfrak{k})$-module, we have

(1) $H^q(\mathfrak{g}, \mathfrak{k}; V) = H^q(\mathfrak{g}', \mathfrak{k}'; V^{\mathfrak{g}''}) \qquad (q \ge 0)$,

where $\mathfrak{k}' = \mathfrak{k} \cap \mathfrak{g}'$, and hence $\mathfrak{k} = \mathfrak{k}' \oplus \mathfrak{g}''$. In fact, $\mathfrak{g}''$ operates trivially on $\mathfrak{g}/\mathfrak{k} = \mathfrak{p} = \mathfrak{g}'/\mathfrak{k}'$; therefore

(2) $\mathrm{Hom}_{\mathfrak{k}}(\Lambda^q \mathfrak{p}, V) \xrightarrow{\sim} \mathrm{Hom}_{\mathfrak{k}'}(\Lambda^q \mathfrak{p}, V^{\mathfrak{g}''})$,

i.e. we have canonical isomorphisms

(3) $C^q(\mathfrak{g}, \mathfrak{k}; V) \xrightarrow{\sim} C^q(\mathfrak{g}', \mathfrak{k}', V^{\mathfrak{g}''}) \qquad (q \ge 0)$.

This yields (1). Note also that, since $\mathfrak{g}'' \subset \mathfrak{k}$, the module V is locally finite and semi-simple with respect to $\mathfrak{g}''$, so that $V = V^{\mathfrak{g}''} \oplus V'$, where $\mathfrak{g}'' \cdot V' = 0$, and $V^{\mathfrak{g}''}$, V' are both stable under $\mathfrak{g}$. This reduces us to the case where $\mathfrak{g}$ has no compact factor.

9.3.   Assume now that $\mathfrak{g} = \mathfrak{g}_1 \oplus \cdots \oplus \mathfrak{g}_s$, with $\mathfrak{g}_i$ simple non-compact for $i = 1, \ldots, s$. Write accordingly

(1)

$\mathfrak{k} = \mathfrak{k}_1 + \cdots + \mathfrak{k}_s, \mathfrak{p} = \mathfrak{p}_1 \oplus \cdots \oplus \mathfrak{p}_s$,   where $\mathfrak{k}_i = \mathfrak{g}_i \cap \mathfrak{k}$, $\mathfrak{p}_i = \mathfrak{g}_i \cap \mathfrak{p}$ $(1 \le i \le s)$.

Let (ρ, E) be irreducible. Then

(2) $E = E_1 \otimes \cdots \otimes E_s, \qquad \rho = \rho_1 \otimes \cdots \otimes \rho_s$,

where (ρ_i, E_i) is an irreducible $\mathfrak{g}_i$-module. If H is also a tensor product of $(\mathfrak{g}_i, \mathfrak{k}_i)$-modules, then we can apply the Künneth rule (I, 1.3).

9.4. We keep the notation of 9.3 and assume moreover that (σ, H) is an irreducible admissible unitary $(\mathfrak{g}, K)$-module. If $H^q(\mathfrak{g}, \mathfrak{k}; H \otimes E) \neq 0$ for some q, then, by I, 5.4, we may assume that K is the direct product of the analytic subgroups K_i, where K_i has Lie algebra $\mathfrak{k}_i$ (cf. 9.3(1)). It follows that we have a tensor product decomposition

$$(1) \qquad\qquad H = H_1 \otimes \cdots \otimes H_s, \qquad \sigma = \sigma_1 \otimes \cdots \otimes \sigma_s,$$

where (σ_i, H_i) is an irreducible unitary admissible $(\mathfrak{g}_i, K_i)$-module. If I is the set of indices for which σ_i is not trivial, we have then

$$(2) \qquad\qquad H^q(\mathfrak{g}, \mathfrak{k}; H \otimes E) = 0, \quad \text{for } q < \sum_{i \in I} (M(\mathfrak{g}_i, \rho_i) + 1).$$

10. Sharp vanishing theorems

In this section we will discuss a vanishing theorem due to Enright [**134**], Parthasarathy [**145**], Kumaresan [**141**] and Vogan-Zuckerman [**149**] which is based on the ideas in §6. The critical technique is due to Kumaresan (extending methods of Parthasarathy). The most general version of the theorem is due to Vogan and Zuckerman, who also laid the groundwork for proving that the result was best possible (see VI, §5 for a discussion).

10.1. If $\mathfrak{q}$ is a θ-stable parabolic subalgebra and if $\mathfrak{u}(\mathfrak{q})$ is the nil-radical of $\mathfrak{q}$, then clearly $\theta\mathfrak{u}(\mathfrak{q}) = \mathfrak{u}(\mathfrak{q})$. Let $\mathfrak{u}_n(\mathfrak{q}) = \mathfrak{u}(\mathfrak{q}) \cap \mathfrak{p}$. Let F be an irreducible finite dimensional $(\mathfrak{g}, K)$-module. Let $\mathcal{P}(F)$ denote the set of all proper θ-stable parabolic subalgebras such that $\dim F^{\mathfrak{u}(\mathfrak{q})} = 1$. Obviously, $\mathcal{P}(\mathbf{C})$ is the set of all proper θ-stable parabolic subalgebras. Set

$$c(F) = \min\{\dim \mathfrak{u}_n(\mathfrak{q}) \mid \mathfrak{q} \in \mathcal{P}(F)\}.$$

We can now state the vanishing theorem.

THEOREM. *We assume that $\mathfrak{g}$ is semisimple. Let V be an irreducible unitary $(\mathfrak{g}, K)$-module with kernel contained in $\mathfrak{k}$ and let F be an irreducible finite dimensional $(\mathfrak{g}, K)$-module. Then*

$$H^i(\mathfrak{g}, K; V \otimes F^*) = 0 \quad \text{for } i < c(F).$$

NOTE. $c(F) \geq c(\mathbf{C})$ for all irreducible finite dimensional $(\mathfrak{g}, K)$-modules. One can show that $c(\mathbf{C}) \geq \mathrm{rk}_{\mathbf{R}}(G)$ (see VI, 5.4 and V, 3.4). In 10.3 we will tabulate the cases when $c(\mathbf{C}) > \mathrm{rk}_{\mathbf{R}}(G)$.

10.2. We now give a discussion of how the proof of the theorem of [**149**] (a full exposition can be found in [**151**, 9.4, 9.5]) relates to the material of this chapter. The argument begins as in the proof of 6.12 (we will use the notation therein). That is, if $\mathrm{Ext}^*_{\mathfrak{g}, K}(F, V) \neq 0$, then $\mathrm{Hom}_K(V \otimes S, F \otimes S) \neq 0$ and F has the same infinitesimal character as V.

Let $\gamma \in \widehat{K}$ and $V_\gamma \in \gamma$ (as usual). We may now assume that there exists $\gamma \in \widehat{K}$ such that $\mathrm{Hom}_K(V_\gamma, V \otimes S) \neq 0$ and $\mathrm{Hom}_K(V_\gamma, F \otimes S) \neq 0$. 6.11 implies that if $T \in \mathrm{Hom}_K(V_\gamma, V \otimes S)$, $u \in V_\gamma$ and $v = Tu$ is such that $\langle v, v \rangle = 1$, then

$$(1) \qquad\qquad \langle Dv, Dv \rangle = -\nu - (|\rho|^2 + |\rho_k|^2) + \mu,$$

where ρ (resp. ρ_k) is the half sum of a system of positive roots for $\mathfrak{g}_c$ (resp. $\mathfrak{k}_c$), λ is the eigenvalue of the Casimir operator C of $\mathfrak{g}$ on V (hence on F) and ν is the eigenvalue of the Casimir operator C_k of $\mathfrak{k}$ on V_γ. We will now use the notation in 6.6. We denote by Λ the highest weight of F with respect to Φ^+, and by λ_γ the highest weight of V_γ with respect to Φ_k^+. Then $\nu = |\Lambda + \rho|^2 - |\rho|^2$ and $\mu = |\lambda_\gamma + \rho_k|^2 - |\rho_k|^2$. If we now make the obvious substitution in (1), we have

$$(2) \qquad\qquad \langle Dv, Dv \rangle = |\lambda_\gamma + \rho_k|^2 - |\Lambda + \rho|^2.$$

Since $\langle Dv, Dv \rangle \geq 0$, this implies

$$(3) \qquad\qquad |\lambda_\gamma + \rho_k|^2 \geq |\Lambda + \rho|^2.$$

This is the simplest form of the *Dirac inequality*. We note that F has played no role as yet. The relationship of γ with F is then used by Vogan and Zuckerman to highly constrain the possibilities for γ (cf. [**151**, 9.5.2]). The rest of the argument (for the most part due to Kumaresan) is extremely delicate and would take us too far afield (cf. [**151**, 9.5.3–7]).

10.3. The purpose of this subsection is to give a tabulation of the cases when G is connected and simple over $\mathbf{R}$ and $c(\mathbf{C})$ is larger than the real rank. We first consider the case when G is a simple Lie group over $\mathbf{C}$ looked upon as a Lie group over $\mathbf{R}$. The labeling is the usual A–G classification of Killing-Cartan.

Cartan Type	$c(\mathbf{C})$
B_n	$2n - 1$
C_n	$2n - 1$
D_n	$2n - 2$
E_6	16
E_7	27
E_8	57
F_4	15
G_2	5

The next list consists of those cases when $\mathfrak{g}_c$ is simple. If G is locally isomorphic with a classical group we give its classical name; otherwise we use the Cartan label of the corresponding symmetric space (cf. [**58**, p. 354]).

Classical name	Cartan Type	$c(\mathbf{C})$
$\mathbf{SU}^*(2n)$, $n \geq 3$	AII	$2(n - 1)$
$\mathbf{SU}^*(6)$	AII	3
$\mathbf{SO}^*(2n)$, $n \geq 4$	DIII	$n - 1$
$\mathbf{Sp}(p, q)$, $1 \leq p \leq q$	CII	$2p$
	EI	13
	EII	8
	EIII	8
	EIV	6
	EV	15
	EVI	12
	EVII	11
	EVIII	29
	EIX	24
	FI	8
	FII	4
	G	3

10.4. REMARK. As was pointed out earlier, 8.4 is a representation theoretic version of Matsushima's theorem on the cohomology of cocompact discrete subgroups. The realization (in the Spring of 1976) that Matsushima's argument had a representation theoretic transcription was in fact the starting point for the 1976–1977 seminar, of which the first edition of this book was an outgrowth. His results have been featured in view of their importance to the genesis of this book and since, in some cases, they implied better bounds than (V, 3.4). But those have been replaced by the sharper ones in VI, 10.3 and could therefore be omitted from this second edition. However, Matsushima's idea has resurfaced in a different context: that of the so-called "geometric superrigidity" in [**143**]. In Matsushima's work the point was to show that certain Γ-invariant harmonic forms were G-invariant (Γ a discrete cocompact subgroup of G). In [**143**] the goal is to prove that certain harmonic maps have totally geodesic images. This is achieved by using a non-linear version of Matsushima's approach (see section 13 in [**143**] for this discussion).

CHAPTER III

Cohomology with Respect to an Induced Representation

This chapter is mainly devoted to the computation of $H^*(\mathfrak{g}, K; F \otimes V)$, where G is connected and reductive, F is a finite dimensional representation of G and V is the space of K-finite vectors in a representation induced from a representation, W, of a parabolic subgroup P of G. It is expressed essentially in terms of groups $H^*(\mathfrak{m}, K \cap P; W \otimes F)$, where M is a Levi subgroup of P, with respect to the tensor product of the representation of M giving rise to V and of a suitable finite dimensional representation of M (3.3). Together with the results of II, §5, this yields an essentially complete description of $H^*(\mathfrak{g}, K; F \otimes V)$ when V is tempered (5.2). In particular it is concentrated in an interval of length $l_0(G) = \operatorname{rk} G - \operatorname{rk} K$ (K a maximal compact subgroup) around $(\dim G/K)/2$, and is zero if P is not fundamental (in the sense of 4.1). If V is induced from a tempered irreducible representation, then the cohomology is zero outside an interval of length at most the **R**-rank of G (6.1).

After this work was done, we were informed that G. Zuckerman had obtained independently similar results (since then published in [**119**]). Our own starting point was a formula proved by P. Delorme and describing the cohomology of complex semi-simple groups with coefficients in certain degenerate principal series. We thank him very much for communicating it to us.

By (I, 5.4), there is no loss in generality in assuming that the derived group of G is linear, and has a simply connected complexification. We shall do so.

1. Notation and conventions

1.1. In this chapter, G is a connected, reductive Lie group, K a maximal compact subgroup of G, A_0 a maximal connected commutative **R**-split subgroup whose Lie algebra is orthogonal to that of K and P_0 is a minimal parabolic subgroup with split component A_0. A parabolic pair (P, A) is *standard* (resp. *semi-standard*) if $P \supset P_0$, $A \subset A_0$ (resp. $A \subset A_0$).

1.2. We fix a Cartan subalgebra $\mathfrak{h}$ of $\mathfrak{g}$ containing $\mathfrak{a}_0$, and let $H = Z_G(\mathfrak{h})$ be the corresponding Cartan subgroup. If (P, A) is semi-standard, then

$$(1) \qquad \mathfrak{h} = \mathfrak{b} \oplus \mathfrak{a}, \quad \text{where } \mathfrak{b} = \mathfrak{b}_P = \mathfrak{h} \cap {}^0\mathfrak{m},$$

and $\mathfrak{b}$ is a Cartan subalgebra of ${}^0\mathfrak{m}$. We also have

$$(2) \qquad H = B \times A, \quad \text{where } B = {}^0M \cap H \text{ is a Cartan subgroup of } {}^0M.$$

We have then a canonical isomorphism

$$(3) \qquad \mathfrak{h}_c^* = \mathfrak{b}_c^* + \mathfrak{a}_c^*,$$

where $\mathfrak{b}_c^*$ (resp. $\mathfrak{a}_c^*$) is identified to the space of linear forms on $\mathfrak{h}_0$ which are zero on $\mathfrak{a}$ (resp. $\mathfrak{b}$).

1.3. Let $\Phi = \Phi(\mathfrak{g}_c, \mathfrak{h}_c)$ (resp. $_{\mathbf{R}}\Phi = \Phi(\mathfrak{g}_c, \mathfrak{a}_{0c})$) be the set of roots of $\mathfrak{g}_c$ with respect to $\mathfrak{h}_c$ (resp. $\mathfrak{a}_{0c}$). Its elements will also be viewed as roots of G_c with respect to H (resp. A), i.e. we make no distinction between a "global" root and its differential at the origin. The elements of $_{\mathbf{R}}\Phi$ are the **R**-roots. The value of a root α on an element a is denoted $\alpha(a)$ or a^α. If (P, A) is a p-pair, then $\Phi(P, A)$ is the set of roots of P with respect to A, i.e. the characters of A in $\mathfrak{n}$ with respect to the adjoint action, and $\Delta(P, A)$ the set of "simple" elements in $\Phi(P, A)$. We recall that $\Delta(P, A)$ is a basis of $(\mathfrak{a} \cap \mathcal{D}\mathfrak{g})^*$ and that every element in $\Phi(P, A)$ is linear combination with coefficients in $\mathbf{N}$ of elements in $\Delta(P, A)$. The dimension of A is the parabolic rank $\operatorname{prk} P$ of P. As usual we let $\rho_P \in \mathfrak{a}^*$ be defined by

$$(1) \qquad \rho_P(a) = (\det \operatorname{Ad} a\big|_{\mathfrak{n}_P})^{1/2} \qquad (a \in A).$$

If an ordering on Φ (resp. $_{\mathbf{R}}\Phi$) is fixed, then Δ (resp. $_{\mathbf{R}}\Delta$) denotes the set of simple roots (resp. **R**-roots) and Φ^+ (resp. $_{\mathbf{R}}\Phi^+$) the set of positive roots (resp. **R**-roots). Orderings on Φ and $_{\mathbf{R}}\Phi$ are compatible if the restriction of a positive element is positive. The choice of an ordering on $_{\mathbf{R}}\Phi$ is equivalent to that of a minimal parabolic subgroup $P_0 \supset A_0$, and then $_{\mathbf{R}}\Phi^+ = \Phi(P_0, A_0)$.

Fix an ordering on Φ. The fundamental highest weights ϖ_α ($\alpha \in \Delta$) are then defined by

$$(2) \qquad (\varpi_\alpha, \beta) = \delta_{\alpha\beta}(\beta, \beta)/2 \qquad (\alpha, \beta \in \Delta),$$

where $(\ ,\ )$ is a scalar product invariant under the Weyl group. We recall that

$$(3) \qquad \varpi_\alpha = \sum_{\gamma \in \Delta} d_{\alpha\gamma}\gamma, \quad \text{with } d_{\alpha\gamma} \geq 0, \ d_{\alpha\alpha} > 0$$

(and more precisely $d_{\alpha\gamma} > 0$ if and only if α, γ belong to the same simple factor of $\mathfrak{g}_c$).

If (P, A) is a semi-standard p-pair, then $\Phi(\mathfrak{m}_c, \mathfrak{h}_c) = \Phi(^0\mathfrak{m}_c, \mathfrak{b}_c)$ may be identified to the set of roots which are zero on $\mathfrak{a}$, and $\Delta_m = \Delta \cap \Phi(\mathfrak{m}_c, \mathfrak{h}_c)$ is the set of simple roots for the ordering induced from the given one on Φ. Moreover, if we let

$$(4) \qquad 2\rho = \sum_{\alpha \in \Phi^+} \alpha, \qquad 2\rho_{0M} = \sum_{\alpha \in \Phi(^0\mathfrak{m}_c, \mathfrak{b}_c)^+} \alpha,$$

then

$$(5) \qquad \rho\big|_{\mathfrak{b}} = \rho_{0M}.$$

If (P, A) is a standard p-pair, then

$$(6) \qquad \rho_P(a) = \rho(a) = \frac{1}{2} \sum_{\alpha \in \Phi(\mathfrak{p}_c, \mathfrak{h}_c)} \alpha(a) \qquad (a \in \mathfrak{a}_P).$$

1.4. Weyl groups. Let $W = W(\mathfrak{g}_c, \mathfrak{h}_c)$ be the Weyl group of $\mathfrak{g}_c$ with respect to $\mathfrak{h}_c$, and similarly $W_M = W(\mathfrak{m}_c, \mathfrak{h}_c) = W(^0\mathfrak{m}_c, \mathfrak{b}_c)$. We put

$$(1) \qquad W^P = \{w \in W \mid w^{-1}(\alpha) > 0 \ (\alpha \in \Delta_M)\}.$$

Then W^P is a set of representatives for the right cosets $W_M \cdot w$ in W. As usual, the length $l(w)$ of $w \in W$ is meant with respect to the set S of reflections $s_\alpha \in W$

$(\alpha \in \Delta)$. We recall that if $t \in W$, the minimum of $l(w)$ on $W_M \cdot w$ is attained on $W^P \cap \{W_M \cdot w\}$, and only on that element [**72**, 5.13].

1.5. Infinitesimal characters. If (π, V) is an irreducible admissible representation of a linear reductive group L of connected type $(0, 3.1)$, then χ_π or χ_V denotes its infinitesimal character. We shall use the *standard parametrization* of the infinitesimal characters by $\mathfrak{q}_c^*$ modulo the Weyl group, where $\mathfrak{q}_c$ is a Cartan subalgebra of $\mathfrak{l}_c$: if V is finite dimensional, with highest weight μ, then $\chi_\pi = \chi_{\mu+\rho}$.

2. Induced representations and their K-finite vectors

2.1. If R is a closed subgroup of a Lie group Q and (π, V_π) a differentiable R-module, then the representation of Q *induced* from π is the representation defined by right translations on

(1)
$$\mathrm{Ind}_R^Q(\pi) = \mathrm{Ind}_R^Q(V_\pi) = \{f \in C^\infty(Q; V_\pi) \mid f(r \cdot q) = \pi(r) \cdot f(q) \; (q \in Q, \; r \in R)\}.$$

It is differentiable. If (τ, U_τ) is a finite dimensional continuous representation of Q, then there is a canonical isomorphism

(2)
$$\zeta \colon \mathrm{Ind}_R^Q(V_\pi \otimes U_\tau) \xrightarrow{\sim} (\mathrm{Ind}_R^Q(V_\pi)) \otimes U_\tau,$$

given by

(3)
$$\zeta(f)(q) = \tau(q)^{-1} \cdot f(q) \qquad (f \in \mathrm{Ind}_R^Q(V_\pi \otimes U_\tau); \; q \in Q).$$

2.2. Let (P, A) be a semi-standard parabolic p-pair in G, $P = M \cdot N$ the standard Levi decomposition of P, and $K_P = K \cap P$. Thus K_P is a maximal compact subgroup of P contained in M, or even in ^0M $(0, 1.6)$. Let (σ, H) be a differentiable admissible representation of M into a Hilbert space H, and H_0 the $(\mathfrak{m}, K_P)$-module of K_P-finite vectors in H. As usual, H is also viewed as a P-module on which N acts trivially. We then let

$$\mathrm{Ind}_P^G(\sigma) = \mathrm{Ind}_P^G(H) = \{f \in C^\infty(G; H) \mid f(p \cdot g) = \sigma(p) \cdot f(g) \; (p \in P; \; g \in G)\}.$$

We shall also write $I(\sigma)$ for $\mathrm{Ind}_P^G(\sigma)$. We assume that σ possesses central character. Our purpose in this section is to give an algebraic description of the space I_0 of K-finite vectors in $I(\sigma)$ (2.4) and to use it to give a form of "Shapiro's lemma" in the context of these $(\mathfrak{g}, K)$-modules (2.5).

2.3. Let V be an $(\mathfrak{m}, K_P)$-module. We set

$$U_0 = \mathrm{Hom}_{U(\mathfrak{p})}(U(\mathfrak{g}), V) = \{f \colon U(\mathfrak{g}) \to V \mid f(pg) = p \cdot f(g), \; g \in U(\mathfrak{g}), \; p \in U(\mathfrak{p})\}.$$

We look upon U_0 as a $U(\mathfrak{g})$-module under right multiplication. That is, $xf(y) = f(yx)$ for $f \in U_0$, $x, y \in U(\mathfrak{g})$. Let U_1 denote the set of all $\mathfrak{k}$-finite vectors f in U_0 such that the cyclic space $U(\mathfrak{k})f$ is completely reducible as a $\mathfrak{k}$-module. Let $\widetilde{K}$ denote the simply connected covering group of K. Then since U_1 is a direct sum of irreducible representations of $\mathfrak{k}$, there is a $\widetilde{K}$-module structure on U_1 such that the differential of the representation agrees with the given action of $\mathfrak{k}$. Let $p \colon \widetilde{K} \to K$ denote the covering homomorphism of $\widetilde{K}$ onto K. Let $Z = \ker p$. We note that $Z \subset \widetilde{K}_P = \widetilde{K} \cap p^{-1}(P)$. We set U equal to the set of all $f \in U_1$ such if $m \in \widetilde{K}_P$, then

$$(mf)(x) = p(m) \cdot f(\mathrm{Ad}(m)^{-1}x).$$

We now show that U is a $(\mathfrak{g}, K)$-module. If $y \in U(\mathfrak{g})$, $m \in \widetilde{K}_P$ and $f \in U$, then

$$
\begin{aligned}
m(kf)(x) = (\mathrm{Ad}(m)y)(mf)(x) &= mf(x \, \mathrm{Ad}(m)y) \\
&= p(m) \cdot f(\mathrm{Ad}(m)^{-1}(x \, \mathrm{Ad}(m)y)) \\
&= p(m) \cdot f((\mathrm{Ad}(m)^{-1}x)y) = p(m) \cdot yf(\mathrm{Ad}(m)^{-1}x).
\end{aligned}
$$

Thus $yf \in U$. Hence U is a $\mathfrak{g}$- and $\widetilde{K}$-invariant subspace of U_1. Since Z acts by the identity on U, we see that the action of $\widetilde{K}$ on U pushes down to K. Hence U is indeed a $(\mathfrak{g}, K)$-module.

We use the notation $\mathrm{Ind}_{(\mathfrak{p}, K_P)}^{(\mathfrak{g}, K)}(V)$ for U and call the above-constructed representation the $(\mathfrak{g}, K)$-module parabolically induced from V. If $\mathfrak{g}$, K, P are understood, we will use the notation $I(V)$.

2.4. PROPOSITION. *Let (σ, H) be a differentiable admissible representation of M and let H_0 be the underlying $(\mathfrak{m}, K_P)$-module. Let $T \colon I(\sigma) \to \mathrm{Hom}_{U(\mathfrak{p})}(U(\mathfrak{g}), H_0)$ be defined by $Tf(x) = xf(1)$ for $f \in I(\sigma)$, $x \in U(\mathfrak{g})$. If I_0 is the space of all K-finite vectors in $I(\sigma)$, then $T(I_0) = I(H_0)$ and T defines an isomorphism of $(\mathfrak{g}, K)$-modules.*

Let $x, y \in U(\mathfrak{g})$. Then

$$
y \cdot Tf(x) = Tf(xy) = xyf(1) = yf(x) = T(yf)(1);
$$

hence T commutes with $U(\mathfrak{g})$. Let $p \in \mathfrak{p}$, $x \in U(\mathfrak{g})$. Then

$$
\begin{aligned}
(Tf)(px) = pxf(1) &= \frac{d}{dt}xf(e^{tp})\big|_{t=0} \\
&= \frac{d}{dt}\sigma(e^{tp}) \cdot xf(1)\big|_{t=0} = \sigma(p) \cdot Tf.
\end{aligned}
$$

This then extends to $p \in U(\mathfrak{p})$ and shows that $\mathrm{Im}\, T \subset U_2$. Since it is a $\mathfrak{g}$-morphism, it follows that T is a $(\mathfrak{g}, K)$-morphism of I_0 into U_1. We want to show that $\mathrm{Im}\, T$ is actually in U. Let $f \in I_0$, $x \in \mathfrak{g}$, $m \in K_P$. Then

$$
(m \cdot Tf)(x) = T(mf)(x) = x \cdot (mf)(1) = \frac{d}{dt}(mf)(e^{tx})\big|_{t=0}.
$$

For $y \in G$, we have

$$
(mf)(y) = f(ym) = f(m \cdot m^{-1} \cdot y \cdot m) = \sigma(m) \cdot f(m^{-1} \cdot y \cdot m),
$$

whence, for $x \in \mathfrak{g}$,

$$
\begin{aligned}
\frac{d}{dt}(mf)(e^{tx})\big|_{t=0} &= \frac{d}{dt}\sigma(m) \cdot f(\mathrm{Ad}\, m^{-1}(e^{tx}))\big|_{t=0} \\
&= (\sigma(m) \cdot (\mathrm{Ad}\, m^{-1})(x) \cdot f)(1).
\end{aligned}
$$

Hence

$$
mTf(x) = \sigma(m) \cdot f(\mathrm{Ad}\, m^{-1}(x)) \qquad (m \in K_M, \ x \in \mathfrak{g}, \ f \in I_0).
$$

This then extends to $x \in U(\mathfrak{g})$, and shows that $Tf \in U$. If $Tf = 0$, then $xf(1) = 0$ for all $x \in U(\mathfrak{g})$. But the elements of I_0 are K-finite and $\mathfrak{z}(\mathfrak{g})$-finite, hence analytic; therefore $f = 0$, and T is injective.

We want to construct an inverse S to T. Let $f \in U$. We define Sf on K by

(1) $$ Sf(k) = (kf)(1). $$

We have

$$Sf(mk) = (mkf)(1) = \sigma(m)(kf)(\operatorname{Ad} m^{-1}(1))$$
$$= \sigma(m)(kf)(1) = \sigma(m) \cdot Sf(k).$$

Therefore, we may extend Sf to G by the rule

$$(2) \qquad\qquad Sf(p \cdot k) = \sigma(p) \cdot Sf(k) \qquad (p \in P; \ k \in K).$$

It is immediate that Sf is K-finite, and hence $Sf \in I_0$; moreover, (1) implies

$$(3) \qquad\qquad y \cdot Sf(1) = f(y) \qquad (y \in U(\mathfrak{k}), \ f \in U),$$

since the K-action on U is obtained by integrating right translations. To show that T is an isomorphism, it suffices to prove that $T \cdot S = \operatorname{Id}$. We have $U(\mathfrak{g}) = U(\mathfrak{p}) \cdot U(\mathfrak{k})$. It suffices therefore to prove

$$(4) \qquad\qquad TS(f)(xy) = f(xy) \qquad (f \in U, \ x \in U(\mathfrak{p}), \ y \in U(\mathfrak{k})).$$

We have, using (3),

$$TSf(xy) = \sigma(x)TSf(y) = \sigma(x) \cdot y(Sf)(1) = \sigma(x)f(y) = f(xy),$$

whence (4).

2.5. PROPOSITION. *Let H_0 be as above. Let V be a $(\mathfrak{g}, K)$-module. There are canonical isomorphisms*

(i) $\operatorname{Hom}_{\mathfrak{p}, K_P}(V, H_0) \xrightarrow{\sim} \operatorname{Hom}_{\mathfrak{g}, K}(V, I(H_0))$,

(ii) $H^*(\mathfrak{p}, K_P; H_0) \xrightarrow{\sim} H^*(\mathfrak{g}, K; I(H_0))$.

Let $U = I(H_0)$.

(i) Let $f \in \operatorname{Hom}_{\mathfrak{g}, K}(V, U)$. We let $Tf: V \to H_0$ be defined by $Tf(v) = f(v)(1)$. Given $g \in \operatorname{Hom}_{\mathfrak{p}, K_P}(V, H_0)$, define $Sg: V \to \operatorname{Hom}(U(\mathfrak{g}), H_0)$ by

$$Sg(v)(r) = g(r \cdot v) \qquad (v \in V; \ r \in U(\mathfrak{g})).$$

Routine checking shows that

$$TS = \operatorname{Id}, \qquad \operatorname{Im} S \subset \operatorname{Hom}_{\mathfrak{g}, K}(V, U), \qquad \operatorname{Im} T \subset \operatorname{Hom}_{\mathfrak{p}, K_P}(V, H_0),$$

and that T is injective, whence our first assertion.

(ii) The left-hand side is the cohomology of the complex C^*, where

$$(1) \qquad\qquad C^i = \operatorname{Hom}_{\mathfrak{g}, K}(U(\mathfrak{g}) \otimes_{U(\mathfrak{k})} \Lambda^i(\mathfrak{g}/\mathfrak{k}), U) \qquad (i \in \mathbf{N}).$$

Similarly, the right-hand side is the cohomology of the complex D^*, where

$$(2) \qquad\qquad D^i = \operatorname{Hom}_{\mathfrak{p}, K_P}(U(\mathfrak{p}) \otimes_{U(\mathfrak{k}_P)} \Lambda^i(\mathfrak{p}/\mathfrak{k}_P), H_0) \qquad (i \in \mathbf{N}).$$

By (i), we have

$$(3) \qquad\qquad C^i = \operatorname{Hom}_{\mathfrak{p}, K_P}(U(\mathfrak{g}) \otimes_{U(\mathfrak{k})} \Lambda^i(\mathfrak{g}/\mathfrak{k}), H_0) \qquad (i \in \mathbf{N}).$$

We have $\mathfrak{g} = \mathfrak{k} + \mathfrak{p}$; hence $\mathfrak{p}/\mathfrak{k}_P = \mathfrak{g}/\mathfrak{k}$, $U(\mathfrak{g}) = U(\mathfrak{p}) \cdot U(\mathfrak{k})$. More precisely, there exists a subspace Q of $U(\mathfrak{p})$, stable under the adjoint representation restricted to K_P, such that $U(\mathfrak{p}) = Q \otimes U(\mathfrak{k}_P)$. We have vector space isomorphisms

$$U(\mathfrak{g}) = Q \otimes U(\mathfrak{k}), \qquad U(\mathfrak{p}) = Q \otimes U(\mathfrak{k}_P).$$

It follows that the natural map

$$U(\mathfrak{p}) \otimes_{U(\mathfrak{k}_P)} \Lambda^i(\mathfrak{p}/\mathfrak{k}_P) \to U(\mathfrak{g}) \otimes_{U(\mathfrak{k})} \Lambda^i(\mathfrak{g}/\mathfrak{k})$$

defined by inclusions is an isomorphism of $(\mathfrak{p}, K_P)$-modules. In view of (2) and (3), this yields an isomorphism of C^* onto D^*, whence (ii).

3. Cohomology with respect to principal series representations

We first state, in the form needed below, a special case of a theorem of Kostant [**72**, Thm. 5.14].

3.1. THEOREM (B. Kostant). *Let λ be a dominant weight of G_c and F_λ a finite dimensional G-module with highest weight λ. Let (P, A) be a standard p-pair, $P = M \cdot N$ its standard Levi decomposition. For $\mu \in \mathfrak{b}_c^*$ ($\mathfrak{b} = {}^0\mathfrak{m} \cap \mathfrak{h}$, cf. 1.2), let E_μ denote an irreducible M_c-module with extreme weight μ. Let $j \in \mathbf{N}$. Then, there is an isomorphism of ${}^0M$-modules*

$$H^j(\mathfrak{n}_P, F_\lambda) = \bigoplus_{s \in W^P, l(s)=j} E_{s(\rho+\lambda)-\rho}.$$

Note that the weights $s(\rho + \lambda) - \rho$ are all dominant and distinct, as s ranges through W^P (loc. cit.); hence the decomposition of $H^*(\mathfrak{n}; F_\lambda)$ as an M_c-module is multiplicity free.

3.2.   We fix a standard p-pair (P, A) of G and let $P = M \cdot N$ be the standard Levi decomposition of P. Let (σ, H_σ) be a differentiable admissible Fréchet ${}^0M$-module with an infinitesimal character χ_σ, and let $\nu \in \mathfrak{a}_c^*$. Then the induced representation $(\pi_{P,\sigma,\nu}, I_{P,\sigma,\nu})$ is the representation defined by right translations on

$$(1) \qquad I_{P,\sigma,\nu} = \{f \in C^\infty(G; H_\sigma) \mid f(man \cdot g) = a^{(\rho_P+\nu)} \cdot \sigma(m) \cdot f(g)\},$$

($g \in G$, $m \in {}^0M$, $a \in A$, $n \in N$). Thus, in the notation of 2.2,

$$(2) \qquad\qquad I_{P,\sigma,\nu} = \operatorname{Ind}_P^G(H_\sigma \otimes \mathbf{C}_{\rho_P+\nu}),$$

where, for $\mu \in \mathfrak{a}_c^*$, we let $\mathbf{C}_\mu$ denote $\mathbf{C}$ acted upon via μ by A.

It is an admissible finitely generated Fréchet G-module whose infinitesimal character is $\chi_{\lambda_\sigma+\nu}$ (cf. 1.5) if $\lambda_\sigma \in \mathfrak{b}_c^*$ is such that $\chi_\sigma = \chi_{\lambda_\sigma}$.

3.3. THEOREM. *Let P, A, M, N, σ, ν be as in 3.2. Write I for $I_{P,\sigma,\nu}$. Let $\lambda \in \mathfrak{h}_c^*$ be a dominant weight and F_λ a simple G_c-module with highest weight λ.*
   (i) *If $H^*(\mathfrak{g}, K; I \otimes F_\lambda) \neq 0$, then there exists $s \in W^P$ such that*
      (1) $s(\rho + \lambda)\big|_A + \nu = 0$,
      (2) $\chi_\sigma = \chi_{-s(\rho+\lambda)}\big|_{\mathfrak{b}_c}$.
Such an s is unique.
   (ii) *If $s \in W^P$ satisfies (1) and (2), then, for every $q \in \mathbf{N}$, we have*

$$(3) \qquad H^{q+l(s)}(\mathfrak{g}, K; I \otimes F_\lambda) = (H^*({}^0\mathfrak{m}, K_P; H_\sigma \otimes E_{(s(\rho+\lambda)-\rho)}) \otimes \Lambda\mathfrak{a}_c^*)^q,$$

where ${}^0\mathfrak{m}$ is the Lie algebra of ${}^0M$.

REMARKS. 1) The conditions (1) and (2) are equivalent to

$$-(\rho + \lambda) \in W(\lambda\sigma + \nu).$$

Condition (1) implies that ν is real valued.

2) In (3), $E_{s(\rho+\lambda)-\rho}$ is viewed as an ${}^0M$-module by restriction. Since M is the direct product of ${}^0M$ by a commutative group, $E_{s(\rho+\lambda)-\rho}$ is an irreducible ${}^0M$-module. Its restriction to ${}^0M^0$ is a multiple of the irreducible representation with highest weight $(s(\rho + \lambda) - \rho) \mid \mathfrak{b}_c = s(\rho + \lambda) \mid \mathfrak{b}_c - \rho_{0_M}$.

3.4. Proof of the theorem. By 3.2(2) and 2.1(2) we have

(1) $$I \otimes F_\lambda = \mathrm{Ind}_P^G(F_\lambda \otimes H_\sigma \otimes \mathbf{C}_{\nu+\rho}).$$

Since we can replace a differentiable module by the space of K-finite vectors to compute cohomology (I, 2.2), 2.5 implies

(2) $$H^*(\mathfrak{g}, K; I \otimes F_\lambda) = H^*(\mathfrak{p}, K_P; F_\lambda \otimes H_\sigma \otimes \mathbf{C}_{\nu+\rho}).$$

By definition, $\mathfrak{n}$ acts trivially on $H_\sigma \otimes \mathbf{C}_{\nu+\rho}$. By the Künneth rule (I, 1.3) we have then

$$H^*(\mathfrak{n}; F_\lambda \otimes H_\sigma \otimes \mathbf{C}_{\rho+\nu}) = H^*(\mathfrak{n}; F_\lambda) \otimes H_\sigma \otimes \mathbf{C}_{\rho+\nu}.$$

We apply (I, 6.5) to the case where $\mathfrak{g} = \mathfrak{p}$, $L = K_P$, $K_1 = \{1\}$, $V = F_\lambda \otimes H_\sigma \otimes \mathbf{C}_{\rho+\nu}$. There exists a spectral sequence (E_r) abutting on $H^*(\mathfrak{p}, K_P; F_\lambda \otimes H_\sigma \otimes \mathbf{C}_{\rho+\nu})$ and in which

(3) $$E_2^{p,q} = H^p(\mathfrak{m}, K_P; H^q(\mathfrak{n}; F_\lambda) \otimes H_\sigma \otimes \mathbf{C}_{p+\nu}).$$

Kostant's theorem (3.1) then yields

(4) $$H^q(\mathfrak{n}; F_\lambda) = \bigoplus_{s \in W^P, l(s) = q} L_s, \quad \text{where } L_s = E_{s(\lambda+\rho)-\rho}.$$

Therefore

(5) $$E_2^{p,q} = \bigoplus_{s \in W^P, l(s) = q} H^p(\mathfrak{m}, K_P; H_\sigma \otimes \mathbf{C}_{\rho+\nu} \otimes L_s).$$

Since $M = {}^0M \times A$, the M-module L_s may be viewed as the tensor product of an irreducible ${}^0M$-module by the one-dimensional A-module $\mathbf{C}_{(s(\rho+\lambda)-\rho)|A}$. Let

(6) $$\nu_s = s(\rho + \lambda)\big|_A - \rho\big|_A + \rho_P + \nu.$$

Since (P, A) is assumed to be standard, we have $\rho\big|_A = \rho_P$; hence

(7) $$\nu_s = s(\rho + \lambda)\big|_A + \nu.$$

Using I, 1.3 and I, 5.1(4), we can apply the Künneth formula and get

(8) $$H^*(\mathfrak{m}, K_P; L_s \otimes H_\sigma \otimes \mathbf{C}_{\nu_s}) = H^*({}^0\mathfrak{m}, K_P; L_s \otimes H_\sigma) \otimes H^*(\mathfrak{a}, \mathbf{C}_{\nu_s}).$$

If $\nu_s \neq 0$, then $H^*(\mathfrak{a}; \mathbf{C}_{\nu_s}) = 0$ by I, 4.1, and then $E_2 = 0$ in view of (8) and (5), which proves the necessity of (1).

If now $\nu_s = 0$, then

(9) $$H^*(\mathfrak{a}; \mathbf{C}) = \Lambda \mathfrak{a}_c^*,$$

and we have

(10) $$H^*(\mathfrak{m}, K_P; L_s \otimes H_\sigma \otimes \mathbf{C}_{\nu_s}) = H^*({}^0\mathfrak{m}, K_P; L_s \otimes H_\sigma) \otimes \Lambda \mathfrak{a}_c^*.$$

By I, 5.3, the space $H^*({}^0\mathfrak{m}, K_P; L_s \otimes H_\sigma)$ is zero if χ_σ is not equal to the infinitesimal character of the representation $\widetilde{L}_s$ contragredient to L_s. Since the highest weight of L_s is $(s(\rho + \lambda) - \rho)\big|_{\mathfrak{b}_c}$ and $\rho\big|_{\mathfrak{b}_c} = \rho_{0_M}$, the infinitesimal character of $\widetilde{L}_s$ is $\chi_{-(s(\lambda+\rho))|\mathfrak{b}_c}$. This proves the necessity of (2) in (i).

These two conditions determine $s(\rho + \lambda)$ uniquely; but $\rho + \lambda$ is regular, so they fix $s \in W$ as well, and the uniqueness assertion of (i) follows.

Now let $s \in W^P$ satisfy those conditions. By the previous argument, we have

(11) $$H^*(\mathfrak{m}, K_P; L_t \otimes H_\sigma \otimes \mathbf{C}_{\rho+\lambda}) = 0, \quad \text{if } t \in W^P, \ t \neq s.$$

Then (5) and (11) imply

(12) $$E_2^{p,q} = 0, \quad \text{if } q \neq l(s),$$

and (5) and (10) yield

(13) $$E_2^{p,l(s)} = (H^*(^0\mathfrak{m}, K_P; L_s \otimes H_\sigma) \otimes \Lambda\mathfrak{a}_c^*)^p \quad (p \in \mathbf{N}).$$

(12) and (13) show that the spectral sequence (E_r) degenerates and that we have

(14) $$H^j(\mathfrak{g}, K; I \otimes F_\lambda) = E_2^{j-l(s),l(s)} \quad (j \in \mathbf{N});$$

(3) now follows from (13) and (14).

3.5. Let (τ, U_τ) be a continuous finite dimensional representation of A which is *quasi-unipotent*, i.e., there exists $\nu \in \mathfrak{a}_c^*$, called the weight of ν, such that $(\tau(a) - a^\nu \cdot \mathrm{Id})$ is nilpotent for every $a \in A$. Any continuous finite dimensional representation of A is a direct sum of quasi-unipotent ones. Theorem 3.4 holds true if $\mathbf{C}_\nu$ is replaced by U_τ, provided that in (3), the factor $\Lambda\mathfrak{a}_c^*$ is replaced by $H^*(\mathfrak{a}_c; U_\tau)$. The proof of (i) is reduced to the case considered above by using a Jordan-Hölder decomposition of (τ, U_τ). The proof of (ii) is then the same as above.

4. Fundamental parabolic subgroups

4.1. Let L be a reductive group of connected type $(0, 3.1)$ and L_1 the greatest connected normal semi-simple group of L^0. A Cartan subgroup C of L is *fundamental* if and only if it contains a maximal torus of L. This condition is equivalent to $C \cap L_1$ being fundamental in L_1. The fundamental Cartan subgroups of L form one conjugacy class [**113**, 1.4.1.4, p. 110]. A parabolic subgroup P of L is *fundamental* if it is minimal among those which contain a fundamental Cartan subgroup. P is fundamental if and only if $P \cap L_1$ is fundamental in L_1. Those parabolic subgroups form one class of associated parabolic subgroups: if C is a fundamental Cartan subgroup of L^0 and C_d^0 its greatest connected $\mathbf{R}$-split subgroup, then $\mathcal{Z}_{L^0}(C_d^0)$ is a Levi subgroup of P for all fundamental parabolic subgroups of L^0 containing C. In particular, $\mathrm{prk}\, P$ is equal to the difference $\mathrm{rk}\, L - \mathrm{rk}\, Q$, where Q is a maximal compact subgroup of L. If $\mathrm{rk}\, L = \mathrm{rk}\, Q$, i.e., if L has a discrete series, then L is its own fundamental parabolic subgroup.

Recall that a parabolic pair (P, A) is *cuspidal* if ^0M_P has a compact Cartan subgroup. If so, the center of ^0M_P is compact. A fundamental parabolic subgroup is cuspidal.

4.2. LEMMA. *Let (P, A) be a cuspidal p-pair in G, $M = \mathcal{Z}(A)$, $N = R_uP$.*

(i) *If P is fundamental, then all root spaces in $\mathfrak{n}$ are even dimensional. In particular, $\dim \mathfrak{n}$ is even. Moreover,*

$$\dim \mathfrak{n} \geq \max(2 \cdot \dim A, 2 \cdot \mathrm{rk}\, K).$$

(ii) *If P is not fundamental, then the Cartan subalgebras of $^0\mathfrak{m}_c$ are singular in $\mathfrak{g}_c$.*

(i) Assume P to be fundamental. Let S be a maximal torus of ^0M_P. Then S is also a Cartan subgroup in a maximal compact subgroup of G; hence it contains elements which are regular in $\mathfrak{g}_c$ [**113**, 1.3.3.2], and the Cartan subgroup $S \cdot A$ is the centralizer of some element in S. In particular, the representation of S in $\mathfrak{n}$ given by the adjoint representation does not contain any trivial representation. It is therefore a sum of two-dimensional real irreducible representations. Since S

leaves all root spaces stable, this proves the first assertion of (i), and also shows that $\dim \mathfrak{n} \geq 2 \cdot \dim S = 2 \cdot \operatorname{rk} K$. Since A acts faithfully on $\mathfrak{n}$, there are at least $\dim A$ linearly independent roots; hence $\dim \mathfrak{n} \geq 2 \dim A$.

(ii) Assume now P is not fundamental. Let T be a maximal torus of G containing S. Then $T \subset \mathcal{Z}(S)$ and $T \neq S$. The group $R = \mathcal{Z}(S)/S$ is reductive. The group A maps isomorphically onto the identity component of a Cartan subgroup of R. It is $\mathbf{R}$-split. But R contains a non-trivial torus, namely T/S; hence its Cartan subgroups are not all conjugate to each other. As a consequence, R is not commutative; therefore $Z(S)$ has a non-trivial semi-simple subgroup. But then $\mathfrak{s}$ is singular. Since $\mathfrak{s}_c$ is a Cartan subalgebra of $^0\mathfrak{m}_c$, this proves (ii).

4.3. Let L be a Lie group with finitely many connected components and Q a maximal compact subgroup of L. We put

(1) $$2 \cdot q(L) = \dim L - \dim Q.$$

Assume the Lie algebra of L to be reductive. Then we let

(2) $$l_0(L) = \operatorname{rk} L - \operatorname{rk} Q, \qquad 2 \cdot q_0(L) = 2q(L) - l_0(L).$$

Since the rank and the dimension of a reductive Lie algebra are congruent mod 2, $q_0(L)$ is an integer.

4.4. LEMMA. *Let L be a reductive group with compact center. Then $q_0(L) \geq \operatorname{rk}_{\mathbf{R}} L$ and $q_0(L) + l_0(L) \leq 2 \cdot q(L) - \operatorname{rk}_{\mathbf{R}} L$.*

We may assume L to be connected. Then $L = L' \cdot S$, with S central compact, L' semi-simple, and $L' \cap S$ finite. $q_0(\ ), l_0(\ ), \operatorname{rk}_{\mathbf{R}}$, and $q(\ )$ are the same for L and L'; this reduces us to the case where L is connected semi-simple. Passing to a finite covering does not change these constants, so we may assume $L = G$ and use our standard notation.

The set $_{\mathbf{R}}\Delta$ has $\dim A_0$ elements; hence $\dim N_0 \geq \dim A_0$. By the Iwasawa decomposition $G = K \cdot A_0 \cdot N_0$, we have then

(1) $$2q(G) = \dim A_0 + \dim N_0 \geq 2 \dim A_0 = 2 \operatorname{rk}_{\mathbf{R}} G.$$

This proves the lemma when $l_0(G) = 0$. Now let (P, A) be a standard fundamental p-pair of G, $P = MN$ the standard Levi decomposition of P and S a maximal torus of ^0M. The group ^0M has compact center; hence (1) also yields

(2) $$q(^0M) \geq \operatorname{rk}_{\mathbf{R}}(^0M).$$

We have

(3) $$\operatorname{rk}_{\mathbf{R}} G = \operatorname{rk}_{\mathbf{R}} {}^0M + \dim A, \qquad \dim A = l_0(G).$$

Since P is standard, the Iwasawa decomposition $G = K \cdot A_0 \cdot N_0$ induces one on ^0M, whence

(4) $$2q(G) = 2q(^0M) + \dim N + \dim A = 2q(^0M) + \dim N + l_0(G),$$

(5) $$2q_0(G) = 2q(^0M) + \dim N.$$

Using (2), 4.2 and (3), we get

(6) $$q_0(G) \geq \operatorname{rk}_{\mathbf{R}} {}^0M + (\dim N)/2 \geq \operatorname{rk}_{\mathbf{R}} {}^0M + \dim A = \operatorname{rk}_{\mathbf{R}} G.$$

On the other hand, by (4), (5)

(7) $$2q(G) - \operatorname{rk}_{\mathbf{R}} G = 2 \cdot q(^0M) + \dim N - \operatorname{rk}_{\mathbf{R}} {}^0M = 2 \cdot q_0(G) - \operatorname{rk}_{\mathbf{R}} {}^0M;$$

(6) then yields

(8) $\qquad 2q(G) - \mathrm{rk}_{\mathbf{R}}\, G \geq q_0(G) + \mathrm{rk}_{\mathbf{R}}\, G - \mathrm{rk}_{\mathbf{R}}\, {}^0M = q_0(G) + l_0(G).$

4.5. LEMMA. *Let L be a reductive group whose identity component has a compact center.*

(i) *We have $q(L) = \mathrm{rk}_{\mathbf{R}}\, L$ if and only if the non-compact normal subgroups of L^0 are of type $\mathbf{SL}_2(\mathbf{R})$.*

(ii) *We have $q_0(L) = \mathrm{rk}_{\mathbf{R}}\, L$ if and only if every non-compact simple factor of L^0 is of type $\mathbf{SL}_2(\mathbf{R})$, $\mathbf{SL}_2(\mathbf{C})$ or $\mathbf{SL}_3(\mathbf{R})$.*

PROOF. The reduction to the case where L is connected, simple and non-compact is immediate, and is left to the reader. So assume L to be so. Fix a minimal p-pair (P_0, A_0) and let $P_0 = M_0 \cdot N_0$ be the standard Levi decomposition of P_0.

(i) By 4.4(1), the condition $q(L) = \mathrm{rk}_{\mathbf{R}}\, L$ is equivalent to

(1) $\qquad\qquad\qquad \dim N_0 = \mathrm{rk}_{\mathbf{R}}\, L = \dim A_0.$

Since L is simple, and $\dim N_0 \geq (\mathrm{Card}_{\mathbf{R}}\, \Phi^+)$, this is possible only if $\dim A_0 = 1$. Also, $\mathrm{rk}\, L = 1$, because any maximal torus of M_0 acts necessarily trivially on the one-dimensional space N_0, hence is reduced to $\{1\}$. Then L is locally isomorphic to $\mathbf{SL}_2(\mathbf{R})$. The converse is clear.

(ii) Now let $q(L) \neq q_0(L)$ and $q_0(L) = \mathrm{rk}_{\mathbf{R}}\, G$. Let (P, A) be a standard fundamental p-pair, $P = M \cdot N$ the standard Levi decomposition of P. The group P is cuspidal, hence $q({}^0M) = q_0({}^0M)$. By 4.4(5), 4.4 and 4.2,

(2) $\qquad q_0(L) = q({}^0M) + (\dim N)/2, \quad q({}^0M) \geq \mathrm{rk}_{\mathbf{R}}\, {}^0M, \quad \dim N \geq 2 \dim A.$

In view of 4.4(6),

(3) $\qquad q_0(L) = \mathrm{rk}_{\mathbf{R}}\, L \Leftrightarrow q({}^0M) = \mathrm{rk}_{\mathbf{R}}\, {}^0M, \ \dim N = 2 \cdot \dim A.$

By (i) the first equality on the right hand side is equivalent to ${}^0M$ having all its non-compact simple factors of type $\mathbf{SL}_2(\mathbf{R})$. In view of 4.2, the second one yields

(4) $\qquad\qquad\qquad \Phi(P, A) = \Delta(P, A).$

Assume now L to be absolutely simple. Then (4) implies, by standard facts on roots, that $\dim A = 1$; hence, by 4.2, $\mathrm{rk}({}^0M) \leq 1$. If $\mathrm{rk}({}^0M) = 0$, then L is of type $\mathbf{SL}_2(\mathbf{R})$, and $q(L) = q_0(L)$, in contradiction with our present assumption. Hence $\mathrm{rk}({}^0M) = 1$, and therefore $\mathrm{rk}\, L = 2$. The representation of ${}^0M$ in $\mathfrak{n}$ given by the adjoint representation has finite kernel; hence ${}^0M^0$ is either a circle group or locally isomorphic to $\mathbf{SL}_2(\mathbf{R})$. In the former case, no root of L would restrict to zero on $\mathfrak{a}$, and $\Phi(P, A)$ would have at least two elements. Therefore ${}^0M^0$ is of type $\mathbf{SL}_2(\mathbf{R})$. We have a semi-direct product decomposition $N_0 = N \cdot ({}^0M \cap N_0)$, where ${}^0M \cap N$ is one-dimensional; hence $\dim N_0 = 3$. From this it follows readily that L is locally isomorphic to $\mathbf{SL}_3(\mathbf{R})$.

Finally, assume L not to be absolutely simple. Then there exists an absolutely simple complex group R such that L is R, viewed as a real Lie group. In this case, $\Phi(P, A)$ may be viewed as the set of positive roots in the root system $\Phi(R)$ of R, for some ordering. Then (4) shows that R has rank 1, i.e., R is locally isomorphic to $\mathbf{SL}_2(\mathbf{C})$.

5. Tempered representations

5.1. THEOREM. *Let (P, A) be a standard cuspidal p-pair of G, (σ, H_0) a discrete series representation of ^0M and $\nu \in \mathfrak{a}_c^*$ purely imaginary. Let $I = I_{P,\sigma,\nu}$ (3.2), and let F_λ be a finite dimensional irreducible G-module with highest weight λ. Assume $H^*(\mathfrak{g}, K; I \otimes F_\lambda) \neq 0$. Then $\nu = 0$, P is fundamental (4.1), the length $l(s)$ of the element $s \in W^P$ satisfying 3.3(1), (2) is equal to $(\dim N)/2$, and we have*

$$(1) \quad \dim H^q(\mathfrak{g}, K; I \otimes F) = \binom{l_0}{q - q_0} \qquad (q \in \mathbf{N}; \; q_0 = q_0(G), \; l_0 = l_0(G)).$$

In particular, $H^q(\mathfrak{g}, K; I \otimes F_\lambda) = 0$ if $q \notin [q_0, q_0 + l_0]$.

The non-vanishing of the cohomology implies that ν is real (3.3); hence $\nu = 0$. We must then have, by 3.3(1),

$$(2) \qquad\qquad s(\rho + \lambda)\big|_A = 0,$$

which means that $s(\rho + \lambda) \in \mathfrak{b}_c^*$ (notation of §2). Since $s(\rho + \lambda)$ is regular, it follows that $\mathfrak{b}_c^*$ is not orthogonal to any root, or, equivalently, that, $\mathfrak{b}$ contains regular elements of $\mathfrak{g}_c$. Then 4.2(ii) shows that P is fundamental. Consequently,

$$(3) \qquad\qquad \dim A = l_0(G).$$

We now use 3.3(3), writing L_s for $E_{s(\rho+\lambda)-\rho}$. By assumption, σ belongs to the discrete series of ^0M. By II, 5.4 and 5.7, $H^*(^0\mathfrak{m}, K_P; H_\sigma \otimes L_s)$ is concentrated in dimension $q(^0M)$ and has dimension one (since it is $\neq 0$). We have then

$$(4) \qquad H^{q+l(s)}(\mathfrak{g}, K; I \otimes F_\lambda) = \Lambda^j \mathfrak{a}_c^* \qquad (q \in \mathbf{N}; \; j = q - q(^0M)).$$

In particular, the lowest and highest dimensions in which the left-hand group is not zero are $q(^0M) + l(s)$ and $q(^0M) + l(s) + l_0(G)$. The representation $\pi_{P,\sigma,\gamma}$ is unitary since σ is, and ν is purely imaginary. Therefore $H^*(\mathfrak{g}, K; I \otimes F_\lambda)$ satisfies Poincaré duality (II, 3.4), and we have

$$2 \cdot q(^0M) + 2 \cdot l(s) + l_0(G) = 2 \cdot q(G).$$

Then, (3) and 4.4(4) show that $2 \cdot l(s) = \dim N$, and the theorem follows.

5.2. COROLLARY. (i) $H^q(\mathfrak{g}, K; I \otimes F_\lambda) = 0$ *if $q < \mathrm{rk}_{\mathbf{R}} G$ or $q > 2q(G) - \mathrm{rk}_{\mathbf{R}} G$.*
(ii) *If $H^q(\mathfrak{g}, K; I \otimes F_\lambda) \neq 0$ for $q = \mathrm{rk}_{\mathbf{R}} G$, then each non-compact simple factor of G is isomorphic to $\mathbf{SL}_2(\mathbf{R})$, $\mathbf{SL}_2(\mathbf{C})$ or $\mathbf{SL}_3(\mathbf{R})$.*
(iii) *Let (π, V) be an irreducible tempered $(\mathfrak{g}, K)$-module. Then*

$$H^q(\mathfrak{g}, K; V \otimes F_\lambda) = 0 \quad \text{if } q \notin [q_0(G), q_0(G) + l_0(G)].$$

If V is not a fundamental principal series representation, then $H^*(\mathfrak{g}, K; V \otimes F_\lambda) = 0$.

(i) follows from 5.1 and 4.4, (ii) from 5.1 and 4.5. If V is as in (iii), then it is a direct G-summand of a representation $I = I_{P,\sigma,\nu}$ with σ and ν as in 5.1. Hence $H^*(\mathfrak{g}, K; V \otimes F_\lambda)$ is a direct summand of $H^*(\mathfrak{g}, K; I \otimes F_\lambda)$, and (iii) is a consequence of 5.1.

5.3. PROPOSITION. *Let L be a reductive group of connected type $(0, 3.1)$ with compact center, Q a maximal compact subgroup of L. Let (π, V) be an irreducible* *tempered $(\mathfrak{l}, Q)$-module and (σ, F) a finite dimensional rational representation of L. Then*

(1)
$$H^q(\mathfrak{l}, Q; V \otimes F) = 0 \quad \text{for } q \notin [q_0(L), q_0(L) + l_0(L)],$$
$$q < \mathrm{rk}_\mathbf{R} L, \ q > 2q(L) - \mathrm{rk}_\mathbf{R} L.$$

We have $H^*(\mathfrak{l}, Q; V \otimes F) = H^*(\mathfrak{l}, Q^0; V \otimes F)^{Q/Q^0}$. Moreover, the restriction (π, V) to L^0 is a direct sum of finitely many tempered irreducible representations (cf. II, 5.5). This reduces us to the case where L is connected. There is a finite covering $L' \to L$, where L' is reductive, $L' = L_1 \times L_2$, with L_1 compact, L_2 semi-simple. In view of (I, 6.6), we may assume that $L = L'$. We may write (π, V) as a tensor product of irreducible $(\mathfrak{l}_i, Q \cap L_i)$-modules. Since the rational representations of L are fully reducible, we may assume F to be irreducible, and then write it as a tensor product. $F = F_1 \otimes F_2$, where F_i is an irreducible representation of L_i $(i = 1, 2)$. We have then, by the Künneth rule (I, 1.3)

$$H^*(\mathfrak{l}, Q; V \otimes F) = H^*(\mathfrak{l}_1, Q_1; V_1 \otimes F_1) \otimes H^*(\mathfrak{l}_2, Q_2; V_2 \otimes F_2),$$

where $Q_i = L_i \cap Q$ $(i = 1, 2)$. Since L_1 is compact, the first factor is trivial (I, 5.2(3)); Corollary 5.2 applies to the second factor. The proposition follows immediately from this and 4.4.

5.4. COROLLARY. *If $H^q(\mathfrak{l}, Q; V \otimes F) \neq 0$ for $q = \mathrm{rk}_\mathbf{R} L$, then each non-compact simple factor of L^0 is locally isomorphic to $\mathbf{SL}_2(\mathbf{R})$, $\mathbf{SL}_3(\mathbf{R})$ or $\mathbf{SL}_2(\mathbf{C})$.*

This follows from 4.5 and 5.3.

6. Representations induced from tempered ones

For later reference, we formulate a consequence of the previous results.

6.1. THEOREM. *Let (P, A) be a standard p-pair in G, $M = \mathcal{Z}_G(A)$, (σ, H_σ) an irreducible admissible tempered $(^0\mathfrak{m}, K \cap M)$-module and $\nu \in \mathfrak{a}_c^*$. Let F_λ be a finite dimensional irreducible representation of G with highest weight λ, and $I = I_{P,\sigma,\nu}$. Let $s \in W^P$ satisfy 3.3(1), (2). Then*

(1)
$$H^q(\mathfrak{g}, K; I \otimes F_\lambda) = 0 \quad \text{for } q \notin [q_0(^0M) + l(s), q_0(^0M) + l(s) + l_0(^0M) + \dim A].$$

By 3.3, there exists a finite dimensional representation L_s of M such that $H^*(\mathfrak{g}, K; I \otimes F_\lambda)$ is equal to the tensor product of $H^*(^0\mathfrak{m}, K \cap M; H \otimes L_s)$ by $\Lambda \mathfrak{a}_c^*$, up to a shift of degrees by $l(s)$. By 5.3, the first factor has cohomology concentrated in the interval $[q_0(^0M), q_0(^0M) + l_0(^0M)]$, whence our assertion.

7. Appendix: C^∞ vectors in certain induced representations

The purpose of this appendix is to make precise the relationship between the notion of induced representations in §2 of this chapter and the more common in-duction procedures (cf. [**113**], Chapter 5). The results of this appendix will also be useful in VII and VIII.

7.1. If M is a C^∞ manifold, then, in this appendix, a vector bundle over M will mean a continuous vector bundle in the usual sense, except that we allow the fibers to be infinite dimensional. A C^∞ *vector bundle* will mean a continuous vector bundle that is also a C^∞ manifold (possibly infinite dimensional) locally C^∞ isomorphic with a trivial vector bundle. An *Hermitian vector bundle* will mean a vector bundle with Hilbert spaces as fibers and a continuously varying inner product (giving the topology) on each fiber. If G is a Lie group acting on M, then a G-vector bundle (C^∞ G-vector bundle) will mean a (C^∞) vector bundle that is a (C^∞) G-space so that the projection is G-equivariant and the maps from fiber to fiber are given by linear maps.

7.2. Let G be a Lie group and let M be an orientable C^∞ manifold such that G acts on M. Let $E \xrightarrow{p} M$ be a G-vector bundle over M with an Hermitian structure $\langle\,,\,\rangle$. Fix a volume form ω on M. We say that $(E, \langle\,,\,\rangle)$ is *admissible* if for each compact subset Ω of G there is a constant $L_\Omega < \infty$ so that

$$(1) \qquad \langle g^{-1}v, g^{-1}v\rangle_x \le L_\Omega \langle v, v\rangle_{g \cdot x} \qquad (x \in M, g \in \Omega).$$

If M is compact, then every Hermitian G-vector bundle is admissible. We note that $(g^*\omega)_x = c(g, x)\omega_x$ with $c\colon G \times M \to \mathbf{R}$ of class C^∞. If η is another volume form on M, then $\eta = u\omega$, $u \in C^\infty(M)$, u nowhere 0. If $(g^*\eta)_x = d(g, x)\eta$, then

$$(2) \qquad d(g, x) = u(g \cdot x)c(g, x)u(x)^{-1}.$$

7.3. Let $(E, \langle\,,\,\rangle)$ be an admissible G-vector bundle. Let $\Gamma_c E$ denote the space of continuous cross-sections of E with compact support. If ω and $c(g, x)$ are as above, define for $f \in \Gamma_c E$

$$(1) \qquad (\pi(g)f)(x) = |c(g^{-1}, x)|^{1/2} g \cdot f(g^{-1} \cdot x).$$

If $f_1, f_2 \in \Gamma_c E$, set

$$(2) \qquad \langle f_1, f_2 \rangle = \int_M \langle f_1(x), f_2(x) \rangle_x \omega.$$

Set $H(E, \omega)$ equal to the Hilbert space completion of $\Gamma_c E$ relative to $\langle\,,\,\rangle$. Then, for $\Omega \subset G$ compact

$$(3) \qquad \|\pi(g)f\| \le L_\Omega^{1/2}\|f\| \quad \text{if } g \in \Omega,$$

with L_Ω as in 1.(1).

It is shown in [**107**], 2.4.6, that $(\pi, H(E, \omega))$ defines a continuous representation of G. If η is another volume form on M and $\eta = u\omega$ as in 7.2, let $\widetilde{\pi}$ be the corresponding action of G on $\Gamma_c E$. Define

$$(4) \qquad (Tf)(x) = |u(x)|^{1/2} f(x).$$

An obvious calculation shows that

$$(5) \qquad T \circ \widetilde{\pi}(g) = \pi(g) \circ T \quad \text{for } g \in G.$$

Furthermore, it is clear that T extends to a bijective unitary operator from $H(E, \eta)$ to $H(E, \omega)$.

We may thus assign to an admissible G-vector bundle $(E, \langle\,,\,\rangle)$ an equivalence class of representations π_E of G. We will abuse notation and let π_E denote $(\pi, H(E, \omega))$ when necessary.

We note that π_E is unitary if $(E, \langle\,,\,\rangle)$ is a unitary G-vector bundle.

7.4. We now specialize our considerations to the case where $M = G/H$ and H is a closed subgroup of G so that G/H is orientable. If $(E, \langle \, , \, \rangle)$ is an admissible G-vector bundle over M, then H acts on $E_{1 \cdot H}$ by a representation of H. Let (σ, E_1) denote this representation. Then $E = G \times_H E_1$ is a G-vector bundle.

Let E_1^∞ denote the space of C^∞ vectors of (σ, E_1). Then $E^\infty = G \times_H E_1^\infty$ is a C^∞ G-vector bundle over G/H with Fréchet spaces as fibers.

7.5. THEOREM. *Let $M = G/H$ be compact and let E be a G-vector bundle over M with Hermitian structure $\langle \, , \, \rangle$. The space of C^∞ vectors for $(\pi_E, H(E, \omega))$ is the space $\Gamma^\infty E^\infty$ of C^∞ cross-sections of E^∞ with the C^∞ topology.*

In this proof, we write π and V for π_E and $H(E, \omega)$.

$\mathfrak{g}$ acts on $\Gamma^\infty E^\infty$ (as usual) by

$$(X \cdot f)(y) = \frac{d}{dt}(e^{tX} \cdot |c(e^{-tX}, y)|^{1/2} f(e^{-tX}y))\big|_{t=0} \qquad (X \in \mathfrak{g}, \ f \in \Gamma^\infty E^\infty).$$

(1) $\qquad \Gamma^\infty E^\infty \subset V^\infty$ and $\pi(X)f = X \cdot f$ $\qquad (f \in \Gamma^\infty E^\infty, \ X \in \mathfrak{g})$.

It is enough to show that
(a) $g \mapsto \pi(g)f$ is of class C^1 as a map of G into V,
(b) $\frac{d}{dt}\pi(e^{tX})f\big|_{t=0} = X \cdot f$ $(X \in \mathfrak{g}, \ f \in \Gamma^\infty E^\infty)$.
Since G/H is compact, both (a) and (b) follow easily from Taylor's theorem. Moreover,

(2) $$V^\infty \subset \Gamma^\infty E^\infty.$$

This follows from Sobolev's lemma (cf. e.g. Yosida, *Functional Analysis*, p. 174) and from the fact that a weakly C^∞ function with values in E^∞ is C^∞.

We recall that the topology on $\Gamma^\infty E^\infty$ is defined by the semi-norms

(3) $\qquad q_u(f) = \sup\{\|u \cdot f(x)\| \mid x \in G/H\}$ $\qquad (u \in U(\mathfrak{g}), \ f \in \Gamma^\infty E^\infty)$,

while the topology on V^∞ is defined by the semi-norms

(4) $\qquad p_u(f) = \|\pi(u)f\|$ $\qquad (u \in U(\mathfrak{g}))$.

G/H is compact, hence there exists a constant $c > 0$ such that

(5) $\qquad p_u(f) \le c q_u(f)$ $\qquad (f \in \Gamma^\infty E^\infty, \ u \in U(\mathfrak{g}))$.

The definition of $H(E, \omega) = V$ implies that the natural map $i \colon \Gamma^\infty E^\infty \to V$ is injective. By (1) and (2) it induces a bijection of $\Gamma^\infty E^\infty$ onto V^∞. It is continuous by (5), hence an isomorphism by the open mapping theorem.

7.6. Let G be a reductive Lie group $(0, \S 3)$. Let $K \subset G$ be a maximal compact subgroup. Let $P \subset G$ be a parabolic subgroup with Levi decomposition MN. Then $G/P = K/K_P$, where as usual $K_P = K \cap P$. Let (σ, H_σ) be a continuous representation of M, with H_σ a Hilbert space so that $\sigma|_{K_P}$ is unitary. We extend σ to P by $\sigma(mn) = \sigma(m)$. Set $E = G \times H_\sigma = K \times_{K_P} H_\sigma$. Since $\sigma|_{K_P}$ is unitary, we give E the Hermitian structure coming from $\langle \, , \, \rangle$ on H_σ. We use for a volume element on G/P the normalized volume element on K/K_P. Then Theorem 7.5 applies.

We reformulate this situation. Let $f \in \Gamma E$. Put $f(g) = (\pi(g)f)(1 \cdot P)$. Let $I_{ct}(\sigma)$ be the space of all $f \in \Gamma E$.

Then $I_{\mathrm{ct}}(\sigma)$ is precisely the space of all $h\colon G \to H_\sigma$ such that

(1)
$$h \text{ is continuous and}$$
$$h(pg) = \delta(p)^{1/2}\sigma(p)f(g), \qquad p \in P,\ g \in G.$$

Also, if $f, g \in I_{\mathrm{ct}}(\sigma)$, then

(2)
$$\langle f, g \rangle = \int_K \langle f(k), g(k) \rangle\, dk.$$

We define $I_P^G(\sigma)$ to be the space of all measurable functions $f\colon G \to H_\sigma$ such that

(3)
$$f(pg) = \delta(p)^{1/2}\sigma(p)f(g), \qquad p \in P,\ g \in G$$

and

(4)
$$\int_K \|f(k)\|^2\, dk < \infty.$$

We set $(\pi_{P,\sigma}(g)f)(x) = f(xg)$ for $g, x \in G$.

Then $\pi_{P,\sigma}$ is equivalent with π_E by the above.

7.7. COROLLARY. *The space of C^∞ vectors for $I_P^G(\sigma)$ is precisely* $\mathrm{Ind}_P^G(\delta^{1/2}\sigma_\infty)$ *with the C^∞ topology (see 2.2), where $(\sigma_\infty, H_\sigma^\infty)$ is the smooth representation of M on the C^∞ vectors of (σ, H) with the C^∞ topology.*

7.8. We now assume that G is as in 7.6. Let $\Gamma \subset G$ be a cocompact discrete subgroup of G. Let $M = G/\Gamma$. Then M is a compact C^∞ manifold. We take ω to be the push-down of dg. Let (σ, H_σ) be a unitary representation of Γ, and $E = G \times_\Gamma H_\sigma$. We give E the Hermitian structure corresponding to $\langle\,,\,\rangle$ on H_σ. Arguing as in 7.6, we find that π_E is equivalent with the representation $I_\Gamma^G(\sigma)$ defined as follows:

(1) $I_\Gamma^G(\sigma)$ is the space of all $f\colon G \to H_\sigma$, f measurable and $f(\gamma g) = \sigma(\gamma)f(g)$, $\gamma \in \Gamma$, $g \in G$.

(2)
$$\int_{\Gamma\backslash G} \|f(g)\|^2\, dg < \infty$$

Here $(\pi_{\Gamma,\sigma}(x)f)(g) = f(gx)$.

7.9. COROLLARY. *The space of C^∞ vectors of $I_\Gamma^G(\sigma)$ is the space $\mathrm{Ind}_\Gamma^G(\sigma)$ of 2.1 with the C^∞ topology.*

7.10. In Chapter V we will need a version of 7.5 (and its corollaries) for continuously induced modules. We set up the relevant results in a more general context. Let G be a Lie group and let $H \subset G$ be a closed subgroup. Let (σ, W) be a continuous representation of H on a Fréchet space. Let

(1) $I_{\mathrm{ct}}(\sigma) = \{f\colon G \to W \mid f \text{ continuous},\ f(hg) = \sigma(h)f(g)\ (h \in H,\ g \in G)\}.$

Set $(\pi_\sigma(x)f)(g) = f(gx)$ as usual $(x, g \in G)$. We topologize $I_{\mathrm{ct}}(\sigma)$ using the topology of uniform convergence on compacta mod H. Then $I_{\mathrm{ct}}(\sigma)$ is a Fréchet G-module (cf. [**22**, X21, Cor. to Prop. 21]; recall that G is countable at infinity by our conventions).

7.11. PROPOSITION. $I_{\mathrm{ct}}(\sigma)^\infty$ *is topologically G-isomorphic with $\mathrm{Ind}_H^G(W^\infty)$ (as defined in 2.1).*

If $f \in I_{ct}(\sigma)$ has compact support mod H and if $\phi \in C_c^\infty(G)$, then it is easy to see that

$$\pi_\sigma(\phi)f \in \operatorname{Ind}_H^G(W^\infty)$$

and that the linear span $^\infty I_{ct}(G)$ of such functions is dense in $I_{ct}(\sigma)^\infty$ and in $\operatorname{Ind}_H^G(W^\infty)$.

The semi-norms on $\operatorname{Ind}_H^G(W^\infty)$ are defined by 7.5(4), but with $\| \ \|$ replaced by a semi-norm. Those of $I_{ct}(\sigma)^\infty$, by 7.5(3), but where the right-hand side is the sup on a compactum of a semi-norm. Therefore the topologies on $^\infty I_{ct}(\sigma)$ stemming from $I_{ct}(\sigma)^\infty$ and $\operatorname{Ind}_H^G(W^\infty)$ are the same. Hence the identity map of $^\infty I_{ct}(\sigma)$ extends to an isomorphism of $I_{ct}(\sigma)^\infty$ onto $\operatorname{Ind}_H^G(W^\infty)$.

CHAPTER IV

The Langlands Classification and Uniformly Bounded Representations

The main purpose of this chapter is to prove 4.13 and 5.2, which will play an essential role in the proof of the main vanishing theorem for relative Lie algebra cohomology in Chapter V. Their proof goes to the heart of the Langlands classification of irreducible admissible representations [76] in that it uses many of Langlands' preliminary results and some theorems of Harish-Chandra [50]. So we have also included a complete proof of the Langlands classification (in fact for reductive groups in the sense of 0, 3.1). Another reason to include it here is that it makes use of the modules $V/V(\mathfrak{n})$, which are the real analogues of the Jacquet modules in the p-adic case, so that it transcribes easily to the p-adic case by using the p-adic counterparts of the pertinent lemmas. This will also lead to a p-adic version of 4.13 and 5.2 (see XI).

Theorem 5.2 is a generalization of a result of Roger Howe (see Theorem 5.4). Howe's theorem says that the matrix entries of a non-trivial, irreducible, unitary representation of a real, simple, algebraic group vanish at infinity. Theorem 5.2 for unitary representations can be derived from Howe's theorem using Theorem 1.5. Similar results can be found in Trombi [103].

1. Some results of Harish-Chandra

1.1. Let G be a real, reductive Lie group as in (0, 3.1), and let $K \subset G$, θ, A_0, B be as in (0, §3). We fix a minimal parabolic pair (P_0, A_0). A p-pair (P, A) is then standard if $A \subset A_0$ and $P \supset P_0$. If (P, A) is a p-pair, then $P = MN$ is the standard Levi decomposition of P and $\Phi(P, A)$ the set of roots of P with respect to A. If (P, A) is standard, then we set

$$^*P = {}^0M \cap P_0, \qquad {}^*A = {}^0M \cap A_0$$

(see 0, 1.6 for ^0M). Then $(^*P, {}^*A)$ is a minimal p-pair in ^0M. Furthermore,

$$A_0 = {}^*A \times A, \qquad N_0 = {}^*N \ltimes N, \quad \text{where } {}^*N = R_u{}^*P.$$

If (P, A) is a p-pair, we set for $t > 0$, $\eta > 0$,

$$\mathfrak{a}^+_{P,t,\eta} = \{H \in \mathfrak{a} \mid \alpha(H) \geq \max(t, \eta\|H\|) \text{ for } \alpha \in \Phi(P, A)\}.$$

Here $\|H\| = B(H, H)^{1/2}$. If P is understood we will use the notation $\mathfrak{a}^+_{t,\eta}$. We also set

$$\mathfrak{a}^+_t = \{H \in \mathfrak{a} \mid \alpha(H) \geq t \ (\alpha \in \Phi(P, A))\}, \qquad \mathfrak{a}^+ = \bigcup_{t>0} \mathfrak{a}^+_t.$$

In this chapter, ρ_0 will stand for ρ_{P_0} (0, 3.0). Continuous representations of G will be on Hilbert spaces and unitary with respect to K.

1.2. THEOREM (Harish-Chandra; cf. [**114**], Chapter 9). *Let (π, H) be an admissible finitely generated representation of G, and H_0 the space of K-finite vectors in H. Then there exist a countable set $E(P_0, \pi)$ of elements of $(\mathfrak{a}_0)_c^*$ and a collection of non-zero functions $P_\Lambda \colon \mathfrak{a}_0 \times H_0 \times H_0 \to \mathbf{C}$, $\Lambda \in E(P_0, \pi)$, satisfying the following properties:*

1) If $h \in \mathfrak{a}_0$ and $\Lambda \in E(P_0, \pi)$, then $P_\Lambda(h; v_1, v_2)$ is linear in v_1 and conjugate linear in v_2.

2) If $W \subset H_0$ is a finite dimensional subspace, there is $d_W \in \mathbf{N}$ such that for any $v_1, v_2 \in W$ the function $h \mapsto P_\Lambda(h; v_1, v_2)$ is polynomial, of degree less than or equal to d_W.

3) If $h \in (\mathfrak{a}_0)^+$ and $v_1, v_2 \in H_0$, then

$$\langle \pi(\exp h) v_1, v_2 \rangle = \sum_{\Lambda \in E(P_0, \pi)} e^{\Lambda(h)} P_\Lambda(h; v_1, v_2),$$

with convergence uniform and absolute on the sets $\mathfrak{a}_{P_0, t, \eta}^+$ for $t > 0$, $\eta > 0$.

Actually 3) can be refined to uniform and absolute convergence in the sets $\mathfrak{a}_{P_0, t}^+$ for $t > 0$ (see [**106**]). However, we will not use this fact.

1.3. THEOREM (Harish-Chandra; cf. [**114**], Chapter 9). *There is a finite subset $E^0(P_0, \pi)$ of $E(P_0, \pi)$ satisfying:*

1) If $\Lambda \in E(P_0, \pi)$, there is $\mu \in E^0(P_0, \pi)$ so that $\mu - \Lambda$ is a sum of elements of $\Phi(P_0, A_0)$.

2) If $\Lambda_1, \Lambda_2 \in E^0(P_0, \pi)$, then $\Lambda_1 - \Lambda_2$ is not a sum of elements of $\Phi(P_0, A_0)$.

1.4. The set $E(P_0, \pi)$ will be called the *set of exponents* of π. If $\Lambda_1, \Lambda_2 \in (\mathfrak{a}_0)_c^*$, we say that $\Lambda_1 > \Lambda_2$ if $\Lambda_1 - \Lambda_2$ is a sum of (not necessarily distinct) elements of $\Phi(P_0, A_0)$.

$E^0(P_0, \pi)$ is called the *set of leading exponents* of π relative to P_0.

The next property of the asymptotic expansion of matrix entries of π which we will need is the following result of Harish-Chandra:

1.5. THEOREM (Harish-Chandra [**138**]). *Let (P, A) be a standard p-pair. Then there exist a countable subset $E(P, \pi) \subset \mathfrak{a}_c^*$ and a collection $\{q_{\mu, P}\}_{\mu \in E(P, \pi)}$ of non-zero functions $q_{\mu, P} \colon {}^*A \times \mathfrak{a} \times H_0 \times H_0 \to \mathbf{C}$ with the following properties:*

1) $q_{\mu, P}(a; h; v_1, v_2)$ is linear in v_1 and conjugate linear in v_2, and, for fixed v_1, v_2, it is analytic in $a \in {}^*A_1$ and a polynomial in h.

2) If $a \in {}^*A$ is fixed and $h \in \mathfrak{a}^+$, then

$$\langle \pi(a \exp h) v_1, v_2 \rangle = \sum_{\mu \in E(P, \pi)} e^{\mu(h)} q_{\mu, P}(a; h; v_1, v_2),$$

with convergence uniform and absolute on $\mathfrak{a}_{P, t, \eta}^+$ for $t > 0$, $\eta > 0$.

3) If $v_1, v_2 \in H_0$, ${}^*h \in {}^*\mathfrak{a}$, $h \in \mathfrak{a}$ and ${}^*h + h \in (\mathfrak{a}_0)^+$, then

$$e^{\mu(h)} q_{\mu, P}(\exp {}^*h; h; v_1, v_2) = \sum_{\substack{\Lambda \in E(P_0, \pi) \\ \Lambda\big|_{\mathfrak{a}} = \mu}} e^{\Lambda(h)} P_\Lambda(h; v_1, v_2).$$

In particular, $E(P, \pi) = \{\Lambda\big|_{\mathfrak{a}} \mid \Lambda \in E(P_0, \pi)\}$.

1.6.   Let Z be the split component of G. As is well known, G/ZK has the structure of a Riemannian symmetric space. For $x \in G$, define $\sigma(x)$ to be the distance in G/ZK from $1.ZK$ to xZK. Then $\sigma(xy) \leq \sigma(x) + \sigma(y)$. We can fix the Riemannian structure on G/ZK so that if $H \in \mathfrak{a}_0$ and $B(H, \mathfrak{z}) = 0$ ($\mathfrak{z}$ the Lie algebra of Z, as usual), then $\sigma(\exp H)^2 = B(H, H)$. We note that $\sigma(k_1 g k_2) = \sigma(g)$, $k_1, k_2 \in K$, and $\sigma(zg) = \sigma(g)$, $z \in Z$, $g \in G$.

The modular function δ_0 of P_0 is extended to G as usual by the rule

$$\delta_0(pk) = \delta_0(p) \qquad (k \in K;\ p \in P_0).$$

Harish-Chandra's function Ξ is defined by

$$\Xi(g) = \int_K \delta_0(kg)^{1/2}\, dk, \quad \text{where} \quad \int_K dk = 1.$$

It satisfies the rule $\Xi(zg) = \Xi(g)$ ($z \in Z$; $g \in G$). It is well known (see Harish-Chandra [49]) that there is d so that if $H \in \mathfrak{a}_0^+$ and $\alpha(H) \geq 0$ for $\alpha \in \Phi(P_0, A_0)$, then:

1) $e^{-\rho_0(H)} \leq \Xi(\exp H) \leq (1 + \sigma(\exp H))^d\, e^{-\rho_0(H)}$.

2) There is an e so that $(1 + \sigma)^{-e}\Xi \in L^2(Z\backslash G)$.

Let $\alpha_1, \ldots, \alpha_n$ be the simple roots in $\Phi(P_0, A_0)$. Let $\beta_1, \ldots, \beta_n \in \mathfrak{a}_0^*$ be defined by

a) $\beta_i(\mathfrak{z}) = 0$, $i = 1, \ldots, n$.

b) $\langle \alpha_i, \beta_j \rangle = \delta_{ij}$,

$\langle\ ,\ \rangle$ the dual form to $B\big|_{\mathfrak{a}_0 \times \mathfrak{a}_0}$.

1.7. THEOREM (Harish-Chandra [50]). *Let (π, H) be an irreducible admissible representation of G. Assume that $\mathrm{Re}\langle \Lambda + \rho_0, \beta_i \rangle < 0$, $i = 1, \ldots, n$, for each $\Lambda \in E^0(P_0, \pi)$. Then, if $v, w \in H_0$ and $d > 0$, there is a constant $C_{d,v,w}$ depending on d, v, w so that*

$$|\langle \pi(g)v, w \rangle| \leq C_{d,v,w}(1 + \sigma(g))^{-d}\Xi(g) \quad \text{for all } g \in {}^0G.$$

1.8.   If (σ, H) is a representation of ${}^0G$ and if $\nu \in \mathfrak{z}_c^*$, we denote by σ_ν the representation of G given by $\sigma_\nu(zg) = z^\nu \sigma(g)$ ($z \in \mathcal{Z}(G)$, $g \in G$).

The following lemma is well known. We include a proof since it is usually stated in the literature slightly differently.

1.9. LEMMA. *Suppose that (π, H) is an irreducible admissible representation of G such that for each $v, w \in H_0$, $g \mapsto \langle \pi(g)v, w \rangle$ is in $L^2({}^0G)$. Then there is an irreducible unitary representation (σ, W) of ${}^0G$ and a $\nu \in \mathfrak{z}_c^*$ so that π is infinitesimally equivalent with σ_ν. Furthermore, (σ, W) can be chosen to be an irreducible subrepresentation of the left regular representation of ${}^0G$ on $L^2({}^0G)$.*

Fix $v \in H_0$, $v \neq 0$. Define $A(w)(g) = \langle \pi(g^{-1})w, v \rangle$. Then $A(w) \in L^2({}^0G)$, for all $w \in H_0$, by hypothesis. If $f \in C_c^\infty({}^0G)$, define $\pi(f)w = \int_{{}^0G} f(g)\pi(g)w\, dg$. If $X \in {}^0\mathfrak{g}$, define $(l_X f)(g) = \frac{d}{dt}f(\exp(-tX)g)\big|_{t=0}$. If $f \in C_c^\infty({}^0G)$, set $\tilde{f}(g) = f(g^{-1})$. Then it is easy to see that if $f \in C_c^\infty({}^0G)$ and $w \in H_0$, then

$$\pi((l_X f)\tilde{\ })w = \pi(\tilde{f})\pi(X)w \quad \text{for } x \in {}^0\mathfrak{g}.$$

This implies that if $w \in H_0$, then

$$\text{(1)} \qquad \int_{{}^0G} l_X f(g)A(w)(g)\, dg = \int_{{}^0G} f(g)A(\pi(X)w)(g)\, dg$$

for $X \in {}^0\mathfrak{g}$ and $f \in C_c^\infty({}^0G)$.

Iterating (1), we see that $A(w)$ has weak derivatives of all orders in $L^2({}^0G)$. Hence A is a $(\mathfrak{g}, K)$-morphism of H_0 into the space $L^2({}^0G)_\infty$ of C^∞ vectors of the left regular representation.

Using K-finiteness and the Casimir operator of ${}^0G$, we see that each $A(w)$ satisfies an analytic elliptic differential equation (cf. [10]). But then $A(H_0)$ consists of weakly analytic vectors for the left regular representation of ${}^0G$. This implies that the L^2-closure W of $A(H_0)$ is stable under the left regular representation of ${}^0G$. Take σ to be the restriction of the left regular representation of ${}^0G$ to W. Since (π, H) is admissible, it is an easy matter to see that $A(H_0)$ is precisely the space of K-finite vectors of W. The result now follows, since the irreducibility and admissibility of (π, H) imply that there is $\nu \in \mathfrak{z}_c^*$ so that if $z \in \mathcal{Z}(G)$, then $\pi(z) = z^\nu I$.

2. Some ideas of Casselman

2.1. We retain the notation of section 1. We fix (π, H) to be an admissible, finitely generated representation of G.

For (P, A) a standard p-pair, let $(\overline{P}, A)$ denote the *opposite* p-pair $(\overline{P} = \theta(P))$. Then if $P = MN$, $\overline{P} = M\overline{N}$ with $\overline{N} = \theta(N)$.

We denote by (π^*, H) the *conjugate dual* representation of G. That is, $\pi^*(g)$ is defined by $\langle \pi(g)v, \pi^*(g)w \rangle = \langle v, w \rangle$ for $g \in G$. Then (π^*, H) is an admissible representation of G.

We will use the notation $\pi(X)v$ (for $X \in \mathfrak{g}$, $v \in H_0$) for the action of $\mathfrak{g}$ on H_0. We note that H_0 is also the space of K-finite vectors for π^* (indeed, $\pi^*(k) = \pi(k)$ for $k \in K$). We have

$$\langle \pi(X)v, w \rangle = -\langle v, \pi^*(X)w \rangle$$

for $X \in \mathfrak{g}$, $v, w \in H_0$.

2.2. LEMMA (Casselman; cf. Miličić [85]). *If $Y \in \overline{\mathfrak{n}}$, $X \in \mathfrak{n}$ and $\Lambda \in E^0(P, \pi)$, then $P_\Lambda(h; \pi(Y)v_1, v_2) = P_\Lambda(h; v_1, \pi^*(X)v_2) = 0$ for $v_1, v_2 \in H_0$, $h \in \mathfrak{a}_0$.*

If $Y \in \overline{\mathfrak{n}}$, then $Y = \sum Y_{-\alpha}$ (the sum over $\Phi(P_0, A_0)$) and $\mathrm{Ad}(a)Y_{-\alpha} = a^{-\alpha}Y_{-\alpha}$ for $a \in A_0$. If $h \in \mathfrak{a}_0^+$, $Y \in \overline{\mathfrak{n}}$, and $Y = Y_{-\alpha}$ for some $\alpha \in \Phi(P_0, A_0)$, then

$$\langle \pi(\exp h)\pi(Y)v_1, v_2 \rangle = e^{-\alpha(h)} \langle \pi(Y)\pi(\exp h)v_1, v_2 \rangle$$
$$= -e^{-\alpha(h)} \langle \pi(\exp h)v_1, \pi^*(Y)v_2 \rangle$$
$$= -e^{-\alpha(h)} \sum_{\mu \in E(P_0, \pi)} e^{\mu(h)} P_\mu(h; v_1, \pi^*(Y)v_2).$$

This implies that the only exponentials that occur in the expansion of $\langle \pi(\exp h)\pi(Y)v_1, v_2 \rangle$ are of the form $\mu - \alpha$, $\mu \in E(P_0, \pi)$. The definition of $E^0(P_0, \pi)$ implies that $P_\Lambda(h; \pi(Y)v_1, v_2) = 0$, $h \in \mathfrak{a}_0$, $\Lambda \in E^0(P_0, \pi)$. Since $Y \in \overline{\mathfrak{n}}$ is of the form $\sum Y_{-\alpha}$, we have shown that

$$P_\Lambda(h; \pi(Y)v_1, v_2) = 0 \quad \text{for } h \in \mathfrak{a}_0, \ Y \in \overline{\mathfrak{n}}, \ v_1, v_2 \in H_0.$$

If $X \in \mathfrak{n}$ and $a \in A_0$, then

$$\langle \pi(a)v_1, \pi^*(X)v_2 \rangle = -\langle \pi(X)\pi(a)v_1, v_2 \rangle = -\langle \pi(a)\pi(\mathrm{Ad}(a)^{-1}X)v_1, v_2 \rangle.$$

Now argue as above to complete the proof of the lemma.

2.3.   Let V be a finitely generated $(\mathfrak{g}, K)$-module. If P is a parabolic subgroup and $\mathfrak{n}$ the Lie algebra of $R_u P$, then we let $\mathfrak{n} \cdot V = V(\mathfrak{n})$ be the subspace spanned by the vectors $n \cdot v$ $(n \in \mathfrak{n}, v \in V)$, and $V_{\mathfrak{n}} = V/\mathfrak{n} \cdot V$. If (P, A) is standard, it follows directly from this definition that we have

$$(1) \qquad\qquad V_{\mathfrak{n}_0} = V_{\mathfrak{n}}/((V_{\mathfrak{n}})(^*\mathfrak{n})),$$

where $\mathfrak{n}_0$ (resp. $^*\mathfrak{n}$) is the Lie algebra of $N_0 = R_u P_0$ (resp. $^*N = R_u{}^*P$) (cf. 1.1). If $V \to W$ is a surjective morphism of $(\mathfrak{g}, K)$-modules, then $V(\mathfrak{n}) \to W(\mathfrak{n})$ is surjective.

2.4. THEOREM. *Let V be a finitely generated admissible $(\mathfrak{g}, K)$-module, and (P, A) a standard p-pair. Then $V_{\mathfrak{n}}$ is a non-zero finitely generated admissible $(\mathfrak{m}, K_M)$-module. In particular, $V_{\mathfrak{n}_0}$ is finite dimensional.*

We note that $\mathfrak{g} = \mathfrak{n} + \mathfrak{m} + \mathfrak{k}$. Let $\sigma_1, \ldots, \sigma_r \in \mathcal{E}(K)$ be such that $\pi(U(\mathfrak{g}_c))\sum_{i=1}^r V_{\sigma_1} = V$. Then

$$\pi(U(\mathfrak{m}_c)) \left(\sum_1^r V_{\sigma_i} \right) + V(\mathfrak{n}) = V.$$

It follows that $V_{\mathfrak{n}}$ is a finitely generated $(\mathfrak{m}, K \cap M)$-module.

Let $\mathfrak{q}_P \colon Z(\mathfrak{g}) \to Z(\mathfrak{m})$ be defined by $\mathfrak{q}_P(z) \equiv z \bmod \mathfrak{n} \cdot U(\mathfrak{g})$. Then the Harish-Chandra isomorphism of $Z(\mathfrak{g})$ with the Weyl group invariants in the enveloping algebra of a Cartan subalgebra implies that

$$Z(\mathfrak{m}) = \sum_{i=1}^s u_i \cdot \mathfrak{q}_P(Z(\mathfrak{g})), \quad \text{for suitable } u_1, \ldots, u_s \in Z(\mathfrak{m}).$$

This implies that if $v \in V_{\mathfrak{n}}$, then $\dim Z(\mathfrak{m}) \cdot v \leq s \cdot r_\pi$, where $r_\pi = \sum_i \dim V_{\sigma_i}$. It follows (cf. [**110**], 5.3) that $V_{\mathfrak{n}}$ is admissible as an $(\mathfrak{m}, K_M)$-module; hence it is finite dimensional if P is minimal.

It remains to show that $V \neq V(\mathfrak{n})$. Assume first that V is the space of K-finite vectors in a finitely generated admissible G-module (π, H). Let $\Lambda \in E^0(\overline{P}_0; \pi)$. Since $P_\Lambda(h; v_1, v_2)$ is not identically zero, there exists $v_1 \in V$, not in $V(\mathfrak{n})$, by 2.2.

In the general case, V has an irreducible quotient W. The latter is the space of K-finite vectors in an irreducible admissible G-module [**77**]; hence $W \neq W(\mathfrak{n})$. But then, $V \neq V(\mathfrak{n})$.

REMARK. It is known that V itself is the space of K-finite vectors in an admissible finitely generated G-module. However, the previous theorem is used to prove this result; therefore we have preferred not to invoke it.

2.5. LEMMA (Casselman, cf. Miličić [**85**]). *Let $\Lambda \in E^0(\overline{P}_0, \pi)$. Then there exists $v \neq 0$ in $H_0/H_0(\mathfrak{n}_0)$ such that*

$$(1) \qquad\qquad h \cdot v = \Lambda(h) \cdot v \quad \text{for all } h \in \mathfrak{a}_0.$$

Let $t > 0$, $\eta > 0$. If $h \in \mathfrak{a}^+_{P_0, t, \eta}$ and $v, w \in H_0$, then

$$\langle \pi(\exp h)v, w \rangle = \sum_{\mu \in E(\overline{P}_0, \pi)} e^{\mu(h)} P_\mu(h; v, w).$$

The absolute and uniform convergence allows us to differentiate term by term, and we find that if $h_1 \in \mathfrak{a}_0$, then

$$\langle \pi(\exp h)\pi(h_1)v, w \rangle$$

$$= \sum_{\mu \in E(\overline{P}_0, \pi)} e^{\mu(h)} \left\{ \mu(h_1) P_\mu(h; v, w) + \frac{d}{dt} P_\mu(h + th_1; v, w)\big|_{t=0} \right\}.$$

We set

$$\partial_X Q(h) = \frac{d}{dt} Q(h + tX)\big|_{t=0} \qquad (Q \in C^\infty(\mathfrak{a}_0), \ X \in \mathfrak{a}_0).$$

Then

(2)
$$P_\Lambda(h; \pi(x)v, w) = \Lambda(x) P_\Lambda(h; v, w) + \partial_x P_\Lambda(h; v, w)$$

$$(\Lambda \in E^0(\overline{P}_0, \pi), \ x \in \mathfrak{a}_0).$$

We also note that if V and W are finite dimensional subspaces of H_0 such that $W + H_0(\overline{\mathfrak{n}}_0) = H_0$ and $V + \pi^*(\mathfrak{n}_0)H_0 = H_0$, then $\deg_h P_\Lambda(h; v, w) \le d_{V+W}$ by 1.2(2).

Let $w_0 \in H_0$ be fixed so that $\mathfrak{q}(h; v) = P_\Lambda(h; v, w_0) \not\equiv 0$. Let $\mathfrak{q}(h; v) = \sum_{j=0}^d \mathfrak{q}_j(h; v)$, with $\mathfrak{q}_j(h; v)$ homogeneous in h of degree j and $\mathfrak{q}_d(h; v) \not\equiv 0$. Then, comparing terms of degree d in (2), we see that

(3)
$$\mathfrak{q}_d(h; \pi(x)v) = \Lambda(x)\mathfrak{q}_d(h; v) \qquad (h \in \mathfrak{a}_0, \ x \in \mathfrak{a}_0, \ v \in H_0).$$

Fix $h \in \mathfrak{a}_0$ so that $\mu(v) = \mathfrak{q}_d(h; v) \not\equiv 0$. Then $\mu(\pi(\mathfrak{n}_0)H_0) = 0$ and $\mu(\pi(x)v) = \Lambda(x)\mu(v)$ for $x \in \mathfrak{a}_0$. This proves the lemma.

2.6. Let (P, A) be a p-pair, $P = MN$. If (σ, H_σ) is a representation of ${}^0M$ and if $\nu \in \mathfrak{a}_c^*$, set σ_ν equal to the representation of M given by $\sigma_\nu(ma) = a^\nu \sigma(m)$, $m \in {}^0M$, $a \in A$.

2.7. LEMMA. *Let V be an admissible finitely generated $(\mathfrak{g}, K)$-module. Suppose that σ is an irreducible finite dimensional representation of ${}^0M_0$ and $\nu \in (\mathfrak{a}_0)_c^*$ is such that σ_ν occurs as an $(\mathfrak{m}_0, K \cap M_0)$-module subquotient of $V_{\mathfrak{n}_0}$. Then $\operatorname{Hom}_{\mathfrak{g}, K}(V, I_{P_0, \sigma, \nu - \rho_0}) \ne 0$ (see III, 3.2 for the definition of $I_{P_0, \sigma, \nu}$).*

Let W denote the $\sigma|_{K \cap M_0}$ isotypic component of $V_{\mathfrak{n}_0}$. Then $\mathfrak{a}_0 \cdot W \subset W$; hence W is an $(\mathfrak{m}_0, K \cap M_0)$-module direct summand. The hypothesis of the lemma implies that

$$W_\nu = \{w \in W \mid (h - \nu(k))^k w = 0 \text{ for some } k, \text{ all } h \in \mathfrak{a}_0\} \ne (0).$$

Thus W_ν has σ_ν as a quotient. Since W_ν is a direct summand of $V_{\mathfrak{n}_0}$ we see that there is an $(\mathfrak{m}_0, K \cap M_0)$-module homomorphism $\widetilde{\mathfrak{q}} \colon V_{\mathfrak{n}_0} \to (\sigma_\nu, H_\sigma)$. For $v \in V$, let $\mathfrak{q}(v) = \widetilde{\mathfrak{q}}(v + V(\mathfrak{n}_0))$. Then $\mathfrak{q} \colon V \to (\sigma_\nu, H_\sigma)$ is a $(\mathfrak{p}_0, K \cap P_0)$-module homomorphism (here σ_ν is extended to P_0 by setting $\sigma_\nu(n) = I$, $n \in \mathbf{N}_0$).

Define $A(v)(k) = \mathfrak{q}(kv)$ for $v \in V$. Then

$$A(k_1 \cdot v)(k_2) = A(v)(k_2 k_1) \ (k_1, k_2 \in K)$$

and

$$A(v)(mk) = \sigma(m) A(v)(k) \qquad (k \in K, \ m \in M_0 \cap K).$$

Define

$$A(v)(p_0 k) = \sigma_\nu(p_0) A(v)(k) \qquad (p_0 \in P_0; \ k \in K).$$

Then we have:

1) $A\colon V \to I_{P_0,\sigma,\nu-\rho_0}$ is linear and non-zero.

2) $A(k \cdot v) = \pi(k)A(v)$ $(k \in K,\ v \in V,\ \pi = \pi_{P_0,\sigma,\nu-\rho_0})$.

We must show that $A(X \cdot v) = \pi(X)A(v)$ for all $X \in \mathfrak{g}$. We have

$$A((k \cdot X) \cdot (k \cdot v))(1) = A(k \cdot (X \cdot v))(1) = A(X \cdot v)(k).$$

Thus, it suffices to show that

$$A(X \cdot v)(1) = (\pi(X)A(v))(1) \qquad (x \in \mathfrak{g}).$$

Now $\mathfrak{g} = \mathfrak{k} + \mathfrak{a} + \mathfrak{n}_0$. If $X \in \mathfrak{k}$, then $A(X \cdot v)(1) = \pi(X)A(v)(1)$ by 2). If $X \in \mathfrak{a}_0 + \mathfrak{n}_0$, then $A(X \cdot v)(1) = \mathfrak{q}(X \cdot v) = \sigma_\nu(X)\mathfrak{q}(v)$. Also, $(X \cdot A(v))(1) = \sigma_\nu(X)A(v)(1)$ by the definition of $I_{P_0,\sigma,\nu-\rho_0}$. The lemma now follows.

3. The Langlands classification (first step)

3.1. Let V be an admissible $(\mathfrak{g}, K)$-module. Let (P, A) be a standard p-pair, $P = MN$. Then we have seen that $V_\mathfrak{n}$ is an admissible $(\mathfrak{m}, K \cap M)$-module. Thus $\dim U(\mathfrak{a}_c) \cdot v < \infty$ for $v \in V_\mathfrak{n}$. This implies that $V_\mathfrak{n}$ is the direct sum of the subspaces

$$V_{\mathfrak{n},\nu} = \{v \in V_\mathfrak{n} \mid (H - \nu(H))^k v = 0$$
$$\text{for all } H \in \mathfrak{a} \text{ and some } k\} \qquad (\nu \in \mathfrak{a}_c^*).$$

Set $e(P, V) = \{\nu \in \mathfrak{a}_c^* \mid V_{\mathfrak{n},\nu} \neq 0\}$. If (π, H) is an admissible representation of G, then set $e(P, \pi) = e(P, H_0)$.

3.2. LEMMA. *Let (π, H) be an admissible finitely generated representation of G. Then $E^0(\overline{P}_0, \pi) \subset e(P_0, V)$.*

This is just a restatement of Lemma 2.5.

3.3. Let $\{\alpha_i\}$ and $\{\beta_j\}$ $(1 \leq i, j \leq l)$ be as in 1.6. Let $\mathcal{F}_c = \sum \mathbf{C}\alpha_i = \sum \mathbf{C}\beta_i$. If $\lambda \in \mathfrak{z}_c^*$, extend λ to $(\mathfrak{a}_0)_c^*$ by $\lambda|_{\mathfrak{a}_0 \cap [\mathfrak{g},\mathfrak{g}]} = 0$. Then $(\mathfrak{a}_0)_c^* = \mathfrak{z}_c^* \oplus \mathcal{F}_c$. Set $\mathcal{F} = \sum \mathbf{R}\alpha_i = \sum \mathbf{R}\beta_i$. Then $\mathfrak{a}_0^* = \mathfrak{z}^* \oplus \mathcal{F}$. Let $^0\lambda$ denote the projection of $\lambda \in (\mathfrak{a}_0)_c^*$ onto $\mathcal{F}_c$. If $\lambda \in \mathfrak{a}_0^*$, then $^0\lambda \in \mathcal{F}$.

If $\nu \in (\mathfrak{a}_0)_c^*$, let $\mathrm{Re}\,\nu \in \mathfrak{a}_0^*$ be given by $\mathrm{Re}\,\nu(h) = \mathrm{Re}(\nu(h))$, $h \in \mathfrak{a}_0$. Clearly, $\mathrm{Re}\,^0\nu = {}^0\mathrm{Re}\,\nu$.

If (P, A) is a standard p-pair, then we have $\mathfrak{a}^* = \mathfrak{z}^* \oplus \mathcal{F}_\mathfrak{p}$, where $\mathcal{F}_\mathfrak{p} = \{\nu \in \mathfrak{a}^* \mid \nu(\mathfrak{z}) = 0\}$ and $\mathfrak{a}_c^* = \mathfrak{z}_c^* \oplus \mathcal{F}_{\mathfrak{p},c}$, as above. If $\nu \in \mathfrak{a}_c^*$, we denote by $^0\nu$ the projection of ν onto $\mathcal{F}_{\mathfrak{p},c}$. If $\nu \in \mathfrak{a}_c^*$, we extend ν to $\mathfrak{a}_0$ by $\nu|_{*\mathfrak{a}} = 0$. We remark that if $\nu \in \mathfrak{a}_c^*$, then $^0\nu$ extended to $\mathfrak{a}_0$ is the same as ^{0}(extension of ν to $\mathfrak{a}_0$).

If $\nu, \mu \in \mathcal{F}$, we say $\nu \geq \mu$ if $\langle \nu - \mu, \beta_i \rangle \geq 0$ for all i. If $\nu \in \mathcal{F}$, we set $\nu_0 = \sum_{i \notin F} x_i \beta_i$ if

$$\nu \in S_F = \left\{ \lambda \in \mathcal{F} \mid \lambda = \sum_{i \notin F} x_i \beta_i - \sum_{i \in F} y_i \alpha_i \ (x_i > 0,\ y_i \geq 0) \right\}.$$

(See 6.6, 6.11, 6.12.)

If $F \subset \{1, \ldots, n\}$, set $\mathfrak{a}_F = \sum_{i \notin F} \mathbf{R}\beta_i + \mathfrak{z}^*$. Let M_F be the centralizer in G of $A_F = \exp \mathfrak{a}_F$. Let (P_F, A_F) the corresponding standard p-pair, and let $\mathfrak{n}_F = \bigoplus \mathfrak{n}_\alpha$ (the sum over those $\alpha \in \Phi(P_0, A_0)$ with $\alpha|_{\mathfrak{a}_F} \neq 0$).

REMARK. It was observed by J. Carmona that ν_0 is the projection of ν onto the cone $(\mathfrak{a}^*)^+$ [**131**].

3.4. LEMMA. *Let (π, H) be an irreducible admissible representation of G. Let $\nu \in e(P_0, \pi)$ be such that ${}^0\mathrm{Re}\,\nu$ is minimal relative to $>$. Let $F = F(-{}^0\mathrm{Re}\,\nu + \rho_0)$ (see 6.11). Let $(P, A) = (P_F, A_F)$. Then there exists an irreducible, admissible representation (σ, H_σ) of ${}^0M$ such that*

1) (π, H_0) is equivalent with a subrepresentation of $I_{P,\sigma,\nu}\big|_{\mathfrak{a} - \rho_P}$,

2) if $\mu \in e({}^*P, \sigma)$ is extended to $\mathfrak{a}_0$ by $\mu(\mathfrak{a}) = 0$, then $\mathrm{Re}^0\,\mu - \rho_{*P} \geq 0$.

Let V denote (π, H_0) as a $(\mathfrak{g}, K)$-module. Then by 2.3(1), there exists $\xi \in e(P, \pi)$ so that $\xi = \nu|_{\mathfrak{a}}$. Let W be an irreducible quotient of $V_\mathfrak{n}$. Lemma 2.7 implies the existence of an irreducible representation (σ, H_σ) of ${}^0M$ such that W is equivalent with $(\sigma_\xi, (H_\sigma)_0)$ as an $(\mathfrak{m}, K \cap {}^0M)$-module. Let $j: V/V(\mathfrak{n}) \to (H_\sigma)_0$ be the corresponding $(\mathfrak{m}, K \cap {}^0M)$-module homomorphism. Let $q: V \to (H_\sigma)_0$ be given by $q(V) = j(v + V(\mathfrak{n}))$.

For $v \in V$, define $A(v)(pk) = \sigma_\xi(p)q(k \cdot v)$, $p \in P$, $k \in K$. The argument of the proof of Lemma 2.7 implies that $A: V \to I_{P,\sigma,\xi - \rho_P}$ is a $(\mathfrak{g}, K)$-module homomorphism. Since $A \neq (0)$ by construction and V is irreducible, A is injective.

To complete the proof we must show that σ satisfies 2). Let $\mu \in e({}^*P, \sigma)$. Then $\mathrm{Re}\,\mu - \rho_{*P} = \sum_{i \in F} x_i \alpha_i$, $x_i \in \mathbf{R}$. We must show $x_i \geq 0$. We note that

$$\mu + \xi \in e(P_0, \pi).$$

This is clear from the definitions.
Moreover,

$$ {}^0\mathrm{Re}(\mu + \xi - \rho_0) = \sum_{i \in F} x_i \alpha_i + {}^0\mathrm{Re}\,\xi - \rho_P $$

by the definitions, and

$$ {}^0\mathrm{Re}\,\nu - \rho_0 = \sum_{i \in F} z_i \alpha_i - \sum_{i \notin F} y_i \beta_i $$

with $z_i \geq 0$ and $y_i > 0$, by the definition of F. Also

$$ {}^0\mathrm{Re}(\xi - \rho_P) = -\sum_{i \notin F} y_i \beta_i. $$

Hence

$$ {}^0\mathrm{Re}(\mu + \xi - \rho_0) = \sum_{i \in F} x_i \alpha_i - \sum_{i \notin F} y_i \beta_i. $$

Let $F = F_1 \cup F_2$ with $x_i \geq 0$, $i \in F_1$, $x_i < 0$, $i \in F_2$. Then

$$ -{}^0\mathrm{Re}(\mu + \xi - \rho_0) \geq -\sum_{i \in F_1} x_i \alpha_i + \sum_{i \notin F} y_i \beta_i. $$

Hence (see 6.12)

$$ (-{}^0\mathrm{Re}(\mu + \xi - \rho_0))_0 \geq \sum_{i \notin F} y_i \beta_i = (-{}^0\mathrm{Re}(\nu - \rho_0))_0. $$

But 6.13 implies that $(-{}^0\mathrm{Re}(\nu - \rho_0))_0 \geq (-{}^0\mathrm{Re}(\mu + \xi - \rho_0))_0$, since ν was chosen so that ${}^0\mathrm{Re}\,\nu$ is minimal. Hence we see that

$$(-{}^0\mathrm{Re}(\mu + \xi - \rho_0)) = (-{}^0\mathrm{Re}\,\nu - \rho_0)).$$

But then $F(-{}^0\mathrm{Re}(\mu + \xi - \rho_0)) = F$. Hence $F_2 = \varnothing$.   Q.E.D.

3.5. LEMMA. *Let (π, H) be an irreducible admissible representation of G such that if $\nu \in e(P_0, \pi)$, then ${}^0\mathrm{Re}(\nu - \rho_0) \geq 0$. Then there exists a standard p-pair (P, A), and also $\sigma \in \mathcal{E}_d({}^0M)$ and $\mu \in \mathfrak{a}_c^*$, so that ${}^0\mu \in i\mathfrak{a}^*$ and (π, H_0) is equivalent with a $(\mathfrak{g}, K)$-module direct summand of $I_{P,\sigma,\mu}$ (note that $I_{P,\sigma,\mu}$ is a unitary representation of ${}^0G$). Moreover, σ can be chosen so that if $\mu \in e(^*P, \sigma)$, then ${}^0(\mathrm{Re}\,\mu - \rho_{*P}) = \sum x_\alpha \alpha$ with $x_\alpha > 0$ (the sum over $\alpha \in \Delta(^*P, ^*A)$).*

If for each $\nu \in e(P_0, \pi)$ we have $\langle {}^0\mathrm{Re}\,\nu - \rho_0, \beta_i \rangle > 0$ for $i = 1, \ldots, n$, then for each $i = 1, \ldots, n$ and $\nu \in E^0(\overline{P}_0, \pi)$ we see that $\langle {}^0\mathrm{Re}\,\nu - \rho_0, \beta_i \rangle > 0$ by 3.2. Hence the result follows from Theorem 1.7, Lemma 1.9 and the definition of $\mathcal{E}_d({}^0G)$.

Let $\nu \in e(P_0, \pi)$ be such that

$$F_\nu = \{ i \mid \langle {}^0\mathrm{Re}\,\nu - \rho_0, \beta_i \rangle > 0 \}$$

has minimal order. Let $(P, A) = (P_F, A_F)$, $F = F_\nu$. The argument in the proof of 3.8 1) shows that there is an irreducible representation (σ, H_σ) of ${}^0M$ so that (π, H_0) is equivalent with a subrepresentation of $I_{P,\sigma,\xi-\rho_P}$, where $\xi = \nu|_\mathfrak{a}$. We note that $\mathrm{Re}\langle \xi - \rho_P, \beta_i \rangle = 0$ if $i \notin F$ by the definition of F.

To complete the proof we must show that if $\mu \in e(^*P, \sigma)$, then $\langle \mathrm{Re}\,\mu - \rho_{*P}, \beta_i \rangle > 0$ for $i \in F$.

If $\mu \in e(^*P, \sigma)$, then $\mu + \xi \in e(P_0, \pi)$ (as in the proof of 3.3), and $\mu + \xi - \rho_0 = \mu - \rho_{*P} + \xi - \rho_P$. Hence ${}^0\mathrm{Re}(\mu + \xi - \rho_0) = \mathrm{Re}(\mu - \rho_{*P}) + {}^0\mathrm{Re}(\xi - \rho_P)$. But ${}^0\mathrm{Re}(\xi - \rho_P) = 0$ by definition of P. Hence ${}^0\mathrm{Re}(\mu + \xi - \rho_0) = \mathrm{Re}(\mu - \rho_{*P})$. If $\langle \mathrm{Re}(\mu - \rho_{*P}), \beta_i \rangle = 0$ for some $i \in F$ and if $\delta = \mu + \xi$, then $F_\delta \subsetneq F_\nu$, which contradicts the definition of ν. This completes the proof.

3.6.   Let (π, H) be an irreducible admissible representation of G. We say that π is *tempered* if for each $v, w \in H_0$ there is a constant C such that

$$|\langle \pi(g)v, w \rangle| \leq C\Xi(g),$$

for $g \in {}^0G$.

3.7. PROPOSITION. *Let (π, H) be an irreducible admissible representation of G. The following conditions are equivalent:*

(1) *(π, H) is tempered.*

(2) *If $\nu \in e(P_0, \pi)$, then ${}^0\mathrm{Re}\,\nu \geq \rho_0$.*

(3) *There exist a standard p-pair (P, A), $\sigma \in \mathcal{E}_d({}^0M)$ and $\nu \in i\mathfrak{a}^*$ such that (π, H_0) is equivalent with a $(\mathfrak{g}, K)$-module summand of $I_{P,\sigma,\nu}$.*

That (2) implies (3) is 3.5.

We now show that (3) implies (1). Since $\sigma \in \mathcal{E}_d({}^0M)$, if $x, y \in (H_\sigma)_0$, then

$$|\langle \sigma(m)x, y \rangle| \leq C\Xi_0(m)   \text{ for } m \in {}^0M;$$

here $\Xi_{{}^0M}$ is defined for ${}^0M$ in the same way as Ξ is defined for G. Extend $\Xi_{{}^0M}$ to G by the rule $\Xi_{{}^0M}(mank) = a^{\rho_P}\Xi_{{}^0M}(m)$ $(m \in {}^0M, k \in K, a \in A, n \in N)$. Then

$$\int_K \Xi_{{}^0M}(kg)\,dk = \Xi(g)   (g \in G).$$

This, combined with an obvious computation, shows that (3) implies (1).

To complete the proof we show that not (2) implies not (1). Suppose that (π, H_0) does not satisfy (2). Let P, σ, ν be as in 3.4 for (π, H_0). Then $P \neq G$ by hypothesis. Since σ satisfies (2) and (2) implies (1) (since (2) implies (3) implies (1) has already been proven), σ is tempered. 3.4 and 1.6(1) now imply that (π, H_0) does not satisfy (1).

4. The Langlands classification (second step)

4.1. If (P, A) is a p-pair for G, we normalize the Haar measure $d\overline{n}$ on $\overline{N}$ $(\overline{P} = M\overline{N})$ by $\int_{\overline{N}} \delta_P(\overline{n}) \, d\overline{n} = 1$. This can be done, since $P \cap K \backslash K = P \backslash G$ and if $d\overline{n}$ is a Haar measure on $\overline{N}$, then $\int_K \phi(k) \, dk = \int_{\overline{N}} \phi(\overline{n}) \delta_P(\overline{n}) \, d\overline{n}$ for ϕ integrable on $P \cap K \backslash K = P \backslash G$.

4.2. LEMMA (Harish-Chandra [**54**, Lemma 10.2]). *Let (P, A) be a p-pair. Extend Ξ_{0M} to G by $\Xi_{0M}(mank) = \delta_P(a)^{1/2}\Xi_{0M}(m)$, where $m \in {}^0M$, $a \in A$, $n \in N$, $k \in K$. If $\nu \in \mathfrak{a}_c^*$, define*

$$\Xi_{0M,\nu}(mank) = \Xi_{0M}(mank)a^\nu \qquad (k \in K, \ m \in M, \ a \in A, \ n \in N).$$

If $\nu \in \mathfrak{a}_c^*$ and $\operatorname{Re}\langle \nu, a \rangle > 0$ for $\alpha \in \Phi(P, A)$, then the integral

$$\int_{\overline{N}} \Xi_{0M,\nu}(\overline{n}g) \, d\overline{n}$$

converges absolutely and uniformly on any compact subset of G.

4.3. Let (P, A) be a standard p-pair. Let (σ, H_σ) be an irreducible, tempered representation of ${}^0M$. Let $\nu \in \mathfrak{a}_c^*$ be such that $\operatorname{Re}\langle \nu, \alpha \rangle > 0$ for $\alpha \in \Phi(P, A)$.

We define for $f \in I_{P,\sigma,\nu}$

$$(1) \qquad\qquad (j(\nu)f)(g) = \int_{\overline{N}} f(\overline{n}g) \, d\overline{n}.$$

Lemma 4.2, combined with 3.6, implies that the integral defining $j(\nu)$ converges absolutely and uniformly for g in a compact set. It is easy to see that $j(\nu)f \in I_{\overline{P},\sigma,\nu}$ and, more precisely,

$$(2) \qquad j(\nu) \colon I_{P,\sigma,\nu} \to I_{\overline{P},\sigma,\nu} \text{ is a homomorphism of } (\mathfrak{g}, K) - modules.$$

If $f \in C^\infty(A)$ we define $\lim_{a \to \infty P} f(a) = c$ to mean that $\lim f(\exp H)$ $(H \to \infty$, $H \in \mathfrak{a}_{P,t,\eta}^+ \cap \mathfrak{z}^\perp)$ exists and equals c for each $\eta > 0$ and $t > 0$ (see 1.1 for $\mathfrak{a}_{P,t,\eta}^+$).

4.4. LEMMA (Langlands [**76**]; cf. [**151**, 5.3.4]). *Let P, σ, ν be as in 4.3. Then*
1) $\lim_{a \to \infty P} a^{\rho_P - \nu} \langle \pi(am)f, g \rangle = \langle \sigma_\nu(m)(j(\nu)f)(1), g(1) \rangle$ for $f, g \in I_{P,\sigma,\nu}$.
2) $j(\nu) \colon I_{P,\sigma,\nu} \to I_{\overline{P},\sigma,\nu}$ is not identically zero.

PROOF (SKETCH). 1) Use the integration formulae on p. 46 of Harish-Chandra [**55**] to compute $\langle \pi(am)f, g \rangle$ as an integral over $\overline{N}$. Now use Lemma 20.1 on p. 49 of [**55**] to interchange integration and limits.

2) If $j(\nu) = 0$ on $I_{P,\sigma,\nu}$, then $\int_{\overline{N}} f(\overline{n}) \, d\overline{n} = 0$ for all $f \in C^\infty(G; H_\sigma)$ such that $f(pg) = \sigma_\nu(p)\delta_P(p)^{1/2}f(g)$ for $g \in G$, $p \in P$. Let $\phi \in C_c^\infty(\overline{N})$ be such that $\int_{\overline{N}} \phi(\overline{n}) \, d\overline{n} \neq 0$, and let $\nu \in H_\sigma$, $v \neq 0$. Define $f(p\overline{n}) = \sigma_\nu(p)\delta_P(p)^{1/2}\phi(\overline{n})v$, $f(g) = 0$ if $g \notin P\overline{N}$. Then $f \in C^\infty(G; H_\sigma)$ and satisfies the above properties. Moreover,

$$(j(\nu)f)(1) = \left(\int_{\overline{N}} \phi(n) \, dn \right) v \neq 0.$$

4.5. LEMMA (Miličić [**85**], Langlands). *Let P, σ, ν be as in 4.3.*

(1) *$j(\nu)I_{P,\sigma,\nu}$ is irreducible.*

(2) *If $f \notin \operatorname{Ker} j(\nu)$, then f is cyclic for $I_{P,\sigma,\nu}$.*

PROOF. Clearly, (1) follows from (2). To prove (2) it suffices to show that if $g \in I_{P,\sigma,\nu}$ is such that $\langle \pi(U(\mathfrak{g}))\pi(K)f, g \rangle = 0$, then $g = 0$. By real analyticity of K-finite vectors we see that

(3) $\langle \pi(k_1 \times k_2)f, g \rangle = 0$ for $k_1, k_2 \in K$, $x \in G$.

(3) combined with Lemma 4.4 1) implies

(4) $\langle \sigma_\nu(m)(j(\nu)\pi(k_1)f)(1), (\pi(k_2)g)(1) \rangle = 0$ for $m \in M$, $k_1, k_2 \in K$.

Since $(j(\nu)\pi(k_1)f)(1) = (j(\nu)f)(k_1^{-1})$, we see that there is $k_1 \in K$ so that $(j(\nu)\pi(k_1)f)(1) \neq 0$. Thus there is $w \in (H_\sigma)_0$, $w \neq 0$, so that

(5) $\langle \sigma_\nu(m)w, (\pi(k)g)(1) \rangle = 0$ for $k \in K$, $m \in M$.

Since σ_ν is irreducible, this implies that $(\pi(k)g)(1) = 0$ for all $k \in K$. Hence $g = 0$. This concludes the proof of 2), hence of the lemma.

4.6. COROLLARY (Miličić [**85**]). *Let P, σ, ν be as in 4.3. Then $I_{P,\sigma,\nu}$ has a unique non-zero irreducible quotient, $J_{P,\sigma,\nu}$. Furthermore, $J_{P,\sigma,\nu}$ is equivalent with $j(\nu)I_{P,\sigma,\nu}$.*

Suppose $W \subset I_{P,\sigma,\nu}$ is an invariant subspace. If $j(\nu)W \neq 0$, then 4.5(2) implies that $W = I_{P,\sigma,\nu}$. Thus if $W \neq I_{P,\sigma,\nu}$, then $W \subset \operatorname{Ker} j(\nu)$. This proves the corollary.

4.7. COROLLARY. *Let P, σ, ν be as in 4.3. If $W \subset I_{\overline{P},\sigma,\nu}$ is an irreducible non-zero $(\mathfrak{g}, K)$-submodule, then $W = j(\nu)I_{P,\sigma,\nu} \equiv J_{P,\sigma,\nu}$.*

3.4 and 3.5 show that we may assume (σ, H_σ) to be a unitary representation of ^0M. We first note

1) If $\pi_\xi = \pi_{P,\sigma,\xi}(\xi \in \mathfrak{a}_c^*)$, then $\langle f_1, f_2 \rangle = \langle \pi_\xi(g)f_1, \pi_{-\overline{\xi}}(g)f_2 \rangle$ for $g \in G$, where $\overline{\xi}$ is defined by $\overline{\xi}(H) = \overline{\xi(H)}$ $(H \in \mathfrak{a})$.

Since $\operatorname{Re}\langle -\nu, \alpha \rangle < 0$ for $\alpha \in \Phi(P, A)$, 4.6 applies to $I_{\overline{P},\sigma,-\overline{\nu}}$. That is, $I_{\overline{P},\sigma,-\overline{\nu}}$ has a unique non-zero irreducible quotient. But then $I_{\overline{P},\sigma,\nu}$ has a unique non-zero irreducible subrepresentation by (1). Since $(0) \neq j(\nu)I_{P,\sigma,\nu} \subset I_{\overline{P},\sigma,\nu}$, the corollary follows.

REMARK. Implicit in 1) above is the fact that the conjugate dual representation to $I_{P,\sigma,\nu}$ is $I_{P,\sigma,-\overline{\nu}}$ for σ unitary. Similarly, if σ is admissible and $\widetilde{\sigma}$ is the admissible dual of σ, then the admissible dual of $I_{P,\sigma,\nu}$ is $I_{P,\widetilde{\sigma},-\nu}$. Both assertions follow from the following integration formula (cf. [**107**], 7.6.6):

$$\int_{K_p \backslash K} f(kg)\delta_P(kg)\, dk = \int_{K_p \backslash K} f(k)\, dk.$$

Here $K_p \backslash K = P \backslash G$ and $\delta_P(pk) = \delta_P(p)$ for $p \in P$, $k \in K$.

4.8. LEMMA. *Let P, σ, ν be as in 4.3 and $\lambda \in E(P_0, I_{P,\sigma,\nu})$. Then $^0\operatorname{Re}\lambda + \rho_0 \leq {}^0\operatorname{Re}\nu$.*

Let $f_1, f_2 \in I_{P,\sigma,\nu}$. Set $\pi = \pi_{P,\sigma,\nu}$. If $a \in A_0$, then

$$\langle \pi(a)f_1, f_2 \rangle = \int_K \langle f_1(ka), f_2(k) \rangle\, dk.$$

Now $|\langle f_1(ka), f_2(k) \rangle| \leq C \Xi_{^0M,\operatorname{Re}\nu}(ka)$ (for notation see 4.2 and 3.6). Thus

$$|\langle \pi(a)f_1, f_2 \rangle| \leq C' \int_K \Xi_{^0M,\operatorname{Re}\nu}(ka)\, dk.$$

But

$$\int_K \Xi_{0M,\mathrm{Re}\,\nu}(ka)\,dk = \int_K e^{(\rho_0+\mathrm{Re}\,\nu)(H(ka))}\,dk = \phi_{\mathrm{Re}\,\nu}(a)$$

(here $g = n\exp H(g)k(g)$, $k(g) \in K$, $H(g) \in \mathfrak{a}_0$, $n \in N_0$). This can be seen, for instance, by using induction in stages. We now note

(i) If $a = \exp H$, $H \in \mathfrak{a}_t^+$, $t > 0$ and $\langle H, \mathfrak{z}\rangle = 0$, then $\phi_{\mathrm{Re}\,\nu}(a) \leq a^{\mathrm{Re}\,\nu}\Xi(a)$.

Indeed, let $\mu \in \mathfrak{a}^*$ be such that $\langle \mu, \alpha\rangle \geq 0$, $\alpha \in \Phi(P_0, A_0)$. Then, with $\rho = \rho_0$,

$$\phi_\mu(a) = \int_K e^{(\rho+\mu)(H(ka))}\,dk = \int_{\overline{N}_0} e^{(\rho-\mu)(H(\overline{n}))}e^{(\rho+\mu)(H(\overline{n}a))}\,d\overline{n};$$

here we use the facts that $k(\overline{n}) \in N\exp(-H(\overline{n}))\overline{n}$ and

$$\int_{\overline{N}} e^{2\rho(H(\overline{n}))}\phi(k(\overline{n}))\,d\overline{n} = \int_{M\backslash K} \phi(Mk)\,d(Mk).$$

After a change of variables we find that

$$\phi_\mu(a) = a^{\mu-\rho}\int_{\overline{N}_0} e^{(\rho+\mu)(H(\overline{n}))}e^{(\rho-\mu)(H(a\overline{n}a^{-1}))}\,d\overline{n}.$$

But $a = \exp H$, $H \in \mathfrak{a}_t^+$, $t > 0$. Hence $\mu(H(\overline{n}) - H(a\overline{n}a^{-1})) \leq 0$ (cf. [**107**], 8.13.7). Thus

$$\phi_\mu(a) \leq a^{\mu-\rho}\int_{\overline{N}} e^{\rho(H(\overline{n}))}e^{\rho(H(a\overline{n}a^{-1}))}\,d\overline{n} = a^\mu\phi_0(a) = a^\mu\Xi(a).$$

This proves (i). Combined with 1.6(1), it implies

ii) If $v, w \in I_{P,\sigma,\nu}$, then

$$\lim_{a\to\infty P} a^{\rho-\mathrm{Re}\,\nu-\varepsilon\rho}|\langle\pi(a)v, w\rangle| = 0,$$

for each $\varepsilon > 0$.

Now let $H \in \mathfrak{a}_t^+$ for some $t > 0$. Set $\phi(t) = \langle\pi(\exp tH)v, w\rangle$. Then ϕ has an expansion as in Lemma 7.2 with the $\lambda_i = \Lambda(H)$, $\Lambda \in {}^0E(P_0, \pi)$. ii) combined with Lemma 7.2 implies that if $\Lambda \in E(P_0, \pi)$ and $H \in \mathfrak{a}^+$, then $({}^0\mathrm{Re}\,\Lambda + \rho)(H) < {}^0\mathrm{Re}\,\nu(H) + \varepsilon\rho(H)$ for $\varepsilon > 0$. The result follows by taking the limit as $\varepsilon \to 0$.

4.9. LEMMA (Langlands [**76**]). *Let* (P, A) *and* (P', A') *be standard p-pairs. Let* σ *(resp.* σ'*) be an irreducible tempered representation of* ${}^0M$ *(resp.* ${}^0M'$*). Let* $\nu \in \mathfrak{a}_c^*$ *(resp.* $\nu' \in (\mathfrak{a}')_c^*$*) be such that* $\mathrm{Re}\langle\nu, \alpha\rangle > 0$ *for* $\alpha \in \Phi(P, A)$ *(resp.* $\mathrm{Re}\langle\nu', \alpha\rangle > 0$ *for* $\alpha \in \Phi(P', A')$*). If* $J_{P,\sigma,\nu}$ *is equivalent with* $J_{P',\sigma',\nu'}$*, then* $P = P'$*,* $\nu = \nu'$*, and* σ *is infinitesimally equivalent with* σ'*.*

Let π denote $J_{P,\sigma,\nu}$.

(1) There exists $\lambda \in E(P_0, \pi)$ such that $\mathrm{Re}^0\lambda\big|_{\mathfrak{a}} = \mathrm{Re}^0\nu - \rho_P$.

Indeed, if $\lambda \in E(P, \pi)$ and $t > 0$, then we have seen that $({}^0\mathrm{Re}\,\lambda + \rho_P)(H) \leq {}^0\mathrm{Re}\,\nu(H)$, $H \in \mathfrak{a}_{P,t}^+$. Suppose $\mathrm{Re}^0\Lambda \neq {}^0\nu - \rho_P$ for any $\Lambda \in E(P, \pi)$. Set $S_\Lambda = \{H \in \mathfrak{a}^+ \mid ({}^0\mathrm{Re}\,\Lambda - \rho_P)(H) = {}^0\mathrm{Re}\,\nu(H)\}$. S_Λ has measure zero in $\mathfrak{a}^+$ if $\Lambda \in E(P, \pi)$. Since $E(P, \pi)$ is countable, $\bigcup S_\Lambda$ has measure zero. Hence there is $H \in \mathfrak{a}^+$ so that $({}^0\mathrm{Re}\,\Lambda - {}^0\mathrm{Re}\,\nu + \rho_P)(H) < 0$. Applying Lemma 7.2, we get a contradiction to Lemma 4.4.

We assert that $({}^0\mathrm{Re}\,\lambda + \rho_0)_0 = {}^0\mathrm{Re}\,\nu$. Indeed, ${}^0\mathrm{Re}\,\lambda + \rho_0 - {}^0\mathrm{Re}\,\nu\big|_{\mathfrak{a}} = 0$ and ${}^0\mathrm{Re}\,\lambda + \rho_0 - {}^0\mathrm{Re}\,\nu \leq 0$ (4.8). Hence ${}^0\mathrm{Re}\,\lambda + \rho_0 - {}^0\mathrm{Re}\,\nu = -\sum_{i\in F} y_i\alpha_i$, $y_i \geq 0$ (here $(P, A) = (P_F, A_F)$). We therefore see that $({}^0\mathrm{Re}\,\lambda + \rho_0)_0 = {}^0\mathrm{Re}\,\nu$ and $({}^0\mathrm{Re}\,\lambda' + \rho_0) =$

$^0\mathrm{Re}\,\nu'$. Now $^0\mathrm{Re}\,\lambda + \rho_0 \leq {}^0\mathrm{Re}\,\nu' = ({}^0\mathrm{Re}\,\nu')_0$. Hence $({}^0\mathrm{Re}\,\nu)_0 = ({}^0\mathrm{Re}\,\lambda + \rho_0)_0 \leq ({}^0\mathrm{Re}\,\nu')_0$ (6.13). Similarly, $({}^0\mathrm{Re}\,\nu') \leq ({}^0\mathrm{Re}\,\nu)_0$. Hence $^0\mathrm{Re}\,\nu = {}^0\mathrm{Re}\,\nu'$. But then $P = P'$. Furthermore

$$\lim_{a \to \infty P} a^{\rho_P - \nu}\langle \pi(a)v, w \rangle = L(v, w),$$

$$\lim_{a \to \infty P} a^{\rho_P - \nu'}\langle \pi(a)v, w \rangle = L'(v, w).$$

Since L and L' are not identically 0, we see that $\lim_{a \to \infty P} a^{\nu - \nu'}$ exists. Since $\mathrm{Re}\,\nu = \mathrm{Re}\,\nu'$, this can occur only if $\nu = \nu'$.

Finally we see that σ is infinitesimally equivalent with σ', since

$$\lim_{a \to \infty P} a^{\rho_0 - \nu}\langle \pi(ma)v, w \rangle$$

is a matrix entry of both σ and σ'.

4.10. If (P, A) is a standard p-pair and $P = MN$ the standard Levi decomposition of P, σ an irreducible tempered representation of ^0M and $\nu \in \mathfrak{a}^*$ such that $\mathrm{Re}\langle \nu, \alpha \rangle > 0$ for $\alpha \in \Delta(P, A)$, then we refer to P, σ, ν as *Langlands data*. If P, σ, ν are Langlands data, then $J_{P,\sigma,\nu}$ will be called the corresponding *Langlands quotient or representation*. With these definitions in mind we can state the *Langlands classification*.

4.11. THEOREM (Langlands [**76**]). *Let (π, H) be an irreducible admissible representation of G. Then there exist a unique set of Langlands data P, σ, ν such that (π, H_0) is equivalent with $J_{P,\sigma,\nu}$.*

The existence follows from 3.4 (with P_0 replaced by $\overline{P}_0$) and 4.7, the uniqueness from 4.8.

4.12. Let (π, H) be an irreducible admissible representation of G. Let P, σ, ν be as in 4.10. Let $\lambda_\pi = {}^0\mathrm{Re}\,\nu$. Then λ_π is called the *Langlands parameter* associated with π and P_0.

4.13. PROPOSITION. *Let P, σ, ν be Langlands data. If (π, H_0) is isomorphic to a constituent of $I_{P,\sigma,\nu}$, then $\lambda_\pi \leq {}^0\mathrm{Re}\,\nu$, and equality holds if and only if (π, H_0) is isomorphic to $J_{P,\sigma,\nu}$.*

By 4.4(1) there exists $\mu \in E(P_0, I_{P,\sigma,\nu})$ with $\lambda_\pi = ({}^0\mathrm{Re}\,\mu + \rho_0)_0$. Now use 4.8, 6.13 and 4.4(1).

5. A necessary condition for uniform boundedness

In this section we assume that G is a connected, simple Lie group with finite center.

5.1. A representation (π, H) of G is said to be *uniformly bounded* if there is a constant C so that $\|\pi(g)v\| \leq C\|v\|$ for $g \in G$, $v \in H$. It is clear that a unitary representation is uniformly bounded.

We denote by $\Pi_\infty(G)$ the set of all equivalence classes of irreducible admissible representations that contain either a tempered representation or a Langlands quotient $J_{P,\sigma,\nu}$ with

$$(1) \qquad (\mathrm{Re}\,\nu - \rho_P)(h) < 0 \quad \text{for } h \in \mathrm{Cl}(\mathfrak{a}^+) - \{0\}, \ P \neq G.$$

5.2. THEOREM. *If (π, H) is an irreducible non-trivial uniformly bounded representation of G, then (π, H_0) is in $\Pi_\infty(G)$.*

Suppose (π, H_0) is not tempered. Then there exist Langlands data P, σ, ν, $P \neq G$, such that (π, H_0) is equivalent with $J_{P,\sigma,\nu}$.

Let $h \in \mathfrak{a}^+$. Set $a_t = \exp th$ $(t \in \mathbf{R})$ and $\rho = \rho_P$. Then Lemma 4.4 1) and the definition of $J_{P,\sigma,\nu}$ imply

$$(1) \qquad \lim_{t \to +\infty} e^{t(\rho-\nu)(h)} \langle \pi(a_t)v_1, v_2 \rangle = L(v_1, v_2) \quad (v_1, v_2 \in J_{P,\sigma,\nu}),$$

where L is linear in v_1, anti-linear in v_2 and not identically zero.

Since π is uniformly bounded, $|\langle \pi(a_t)v_1, v_2 \rangle| \leq C\|v_1\|\,\|v_2\|$. Combined with (1), this implies

$$(2) \qquad \text{If } h \in \mathfrak{a}^+, \text{ then } \mathrm{Re}(\rho - \nu)(h) \geq 0.$$

Suppose $h \in \mathfrak{a}^+$ and $\mathrm{Re}(\rho - \nu)(h) = 0$. Then (1) and the uniform boundedness imply

$$(3) \qquad |L(v_1, v_2)| \leq C\|v_1\|\,\|v_2\|, \qquad v_1, v_2 \in H_0.$$

As a consequence, L extends to a bounded sesquilinear form on H. Thus $L(v_1, v_2) = \langle Bv_1, v_2 \rangle$, with B bounded.

Set $ic = (\rho - \nu)(h)$, $c \in \mathbf{R}$. Then (1) can be written

$$(4) \qquad \lim_{t \to +\infty} e^{itc} \langle \pi(a_t)v_1, v_2 \rangle = L(v_1, v_2).$$

Using (4), we see easily that

$$(5) \qquad B \circ \pi(a_t) = \pi(a_t) \circ B = e^{-ict}B \qquad (t \in \mathbf{R}).$$

We now want to prove that

$$(6) \qquad \pi(n) \cdot B = B \qquad (n \in N).$$

Using (5), we get

$$\pi(n)B = e^{ict} \cdot \pi(n)\pi(a_t)B = e^{ict} \cdot \pi(a_t) \cdot \pi(a_t^{-1} \cdot n \cdot a_t)B \qquad (t \in \mathbf{R}; \ n \in N).$$

Let $v \in H$. If $\varepsilon > 0$ is given, there exists $T > 0$ such that

$$\|\pi(a_t^{-1} \cdot n \cdot a_t)Bv - Bv\| < \varepsilon \quad \text{for } t \geq T.$$

Hence,

$$\|\pi(n)Bv - Bv\| = \|e^{ict} \cdot \pi(a_t) \cdot \pi(a_t^{-1}na_t)Bv - Bv\| \leq C \cdot \varepsilon \qquad (t \geq T).$$

Since ε and v are arbitrary, this implies (6). Moreover, we also have

$$e^{-ict} \cdot \pi(\overline{n}) \cdot \pi(a_t^{-1}) \cdot B = \pi(\overline{n}) \cdot B \qquad (\overline{n} \in N; \ t \in \mathbf{R}).$$

Therefore the same argument yields

$$(7) \qquad \pi(\overline{n}) \cdot B = B \qquad (\overline{n} \in \overline{N}).$$

By assumption, $P \neq G$; therefore N and $\overline{N}$ generate G. But then (6) and (7) yield

$$(8) \qquad \pi(g)B = B, \qquad g \in G.$$

Since $B \neq 0$ and (π, H) is non-trivial, (8) is a contradiction. We have proven

$$(9) \qquad \begin{array}{l} \text{If } h \in \mathrm{Cl}(\mathfrak{a}^+), \text{ then } \mathrm{Re}(\rho - \nu)(h) \geq 0. \text{ If } \mathrm{Re}(\rho - \nu)(h) = 0, \\ \text{then there is } \alpha \in \Phi(P, A) \text{ so that } \alpha(h) = 0. \end{array}$$

If $h \in \mathrm{Cl}(\mathfrak{a}^+)$ and $h \neq 0$, then there is a proper standard p-pair (P_1, A_1) so that $P_1 \supset P$, $A_1 \subset A$ and $h \in (\mathfrak{a}_1)^+$.

We apply Theorem 1.5 to both (P, A) and (P_1, A_1). Let (Q, B) be a standard p-pair. Let $^*Q = {}^0M_Q \cap P_0$, as usual. $(^*Q, {}^*B)$ is a minimal p-pair for ^0M_Q. By Theorem 1.5

$$\langle \pi(a \exp h)v_1, v_2 \rangle = \sum_{\mu \in E(Q,\pi)} e^{\mu(h)} q_{\mu, Q}(a; h; v_1, v_2)$$

(10)
$$(a \in {}^*B, \ h \in \mathfrak{b}_{Q,t,\eta}^+ \ (t > 0, \ \eta > 0)),$$

with convergence and $q_{\mu, Q}$ as in Theorem 1.5.

(1) and $\mu \neq \nu - \rho$ combined with Lemma 4.8 imply that if $\mu \in E(P, \pi)$, then $\mu = \nu - \rho - d_\mu$ with $\mathrm{Re}\, d_\mu(h) > 0$ for $h \in \mathfrak{a}^+$. This implies

$$q_{\nu - \rho}(1; h; v_1, v_2) = L(v_1, v_2) \qquad (h \in \mathfrak{a}^+). \tag{11}$$

If $\mu \in E(P_1, \pi)$, then $\mu = \xi \mid \mathfrak{a}_1$ for some $\xi \in E(P, \pi)$ (see 1.9). If $\xi \neq \nu - \rho$, then $\xi = \nu - \rho - d_\xi$ as above, $d_\xi(h) > 0$ for $h \in \mathfrak{a}^+$. If $d_\xi|_{(\mathfrak{a}_1)^+} \neq 0$, then $d_\xi(h) > 0$ for $h \in (\mathfrak{a}_1)^+$. Hence if $\mu \in E(P_1, \pi)$ and $\mu \neq (\nu - \rho)|_{\mathfrak{a}_1}$, and if $\mu_0 = (\nu - \rho)|_{\mathfrak{a}_1}$, then $\mu = \mu_0 - e_\mu$ with $e_\mu(h) > 0$ for $h \in \mathfrak{a}_1^+$. Applying Lemma 7.2, we see that if $h \in \mathfrak{a}_1^+$ and $a \in {}^*A_1$, then

$$\lim_{t \to +\infty} (e^{-t\mu_0(h)} \langle \pi(a \exp th)v_1, v_2 \rangle - q_{\mu_0, P_1}(a; th; v_1, v_2)) = 0. \tag{12}$$

Suppose $h \in (\mathfrak{a}_1)^+$ and $\mathrm{Re}(\nu - \rho_P)(h) = \mathrm{Re}\, \mu_0(h) = 0$. Then (12) and the relation

$$|\langle \pi(g)v_1, v_2 \rangle| \leq C\|v_1\|\, \|v_2\|$$

imply that

$$q_{\mu_0, P_1}(a; th; v_1, v_2) \qquad (a \in {}^*A_1, \ v_1, v_2 \in H_0)$$

is independent of t. Set $q(a; v_1, v_2) = q_{\mu_0; P_1}(a; th; v_1, v_2)$. If $q \equiv 0$, then Theorem 1.5 shows that $L \equiv 0$. Hence there are $a \in {}^*A_1$ and $v_1, v_2 \in H_0$ so that $q(a; v_1, v_2) \neq 0$. We also have

$$\lim_{t \to +\infty} e^{-t\mu_0(h)} < \pi(\exp th)\pi(a)v_1, v_2 \rangle = q(a; v_1, v_2) \tag{13}$$

for $v_1, v_2 \in H_0$, h, $a \in {}^*A_1$ as above.

Arguing as above, we find that $|q(a; v_1, v_2)| \leq C\|v_1\|\, \|v_2\|$ for $v_1, v_2 \in H_0$. Thus $q(a; \cdot, \cdot)$ extends to a continuous sesquilinear form on $H \times H$. Using an "$\varepsilon/3$" argument, it is easy to see that (13) is now true for all $v_1, v_2 \in H$. We therefore have

$$q(a; v_1, v_2) = M(\pi(a)v_1, v_2) \quad \text{with } M \text{ a continuous}$$

(14)
$$\text{sesquilinear form on } H \times H.$$

We can now apply to $M(\ ,\ )$ the same arguments as to $L(\ ,\ )$ above. There exists then a bounded operator T on H such that

$$M(v_1, v_2) = \langle Tv_1, v_2 \rangle \qquad (v_1, v_2 \in H),$$
$$T\pi(e^{th}) = \pi(e^{th})T = e^{\mu_0(h)t} \cdot T \qquad (t \in \mathbf{R}),$$

and then (see the proof of (8)) $\pi(g)T = T$ $(g \in G)$. Since $T \neq 0$, this is a contradiction. The proof of the theorem is now complete.

5.3. LEMMA. *Let G be connected and simple, and (π, H) an element of $\Pi_\infty(G)$. There is $0 < t < \infty$ so that if $v, w \in H_0$, then*

$$|\langle \pi(g)v, w\rangle| \le C\Xi(g)^t \qquad (g \in G)$$

for some constant C.

If (π, H) is tempered, this follows from 1.6, 3.6. Otherwise, $H_0 = J_{P,\sigma,\nu}$ with Langlands data P, σ, ν, and $\mathrm{Re}(\rho_P - \nu)(h) > 0$ for $h \in \mathrm{Cl}(\mathfrak{a}^+)$, $h \ne 0$. It is shown in the first part of the proof of 4.8 that if $v, w \in H_0$ and $h \in \mathrm{Cl}((\mathfrak{a}_0)^+)$, then

(1) $$|\langle \pi(\exp h)v, w\rangle| \le Ce^{\mathrm{Re}\,\nu(h)}\Xi(\exp h).$$

On the other hand, there is $0 < \eta < 1$ so that

$$\mathrm{Re}\,\nu(h) \le \eta\rho_0(h) \quad \text{for } h \in \mathrm{Cl}((\mathfrak{a}_0)^+).$$

Applying 1.6(1), we therefore find that if $0 < t < \eta$, then

(2) $$|\langle \pi(\exp h)v, w\rangle| \le C'\Xi(\exp h)^t$$

for $v, w \in H_0$.

The result now follows from the fact that $G = K \exp \mathrm{Cl}((\mathfrak{a}_0)^+)K$, and from (2).

5.4. THEOREM. *If (π, H_0) is in $\Pi_\infty(G)$ (in particular, if (π, H) is uniformly bounded and non-trivial), then*

(1) The matrix entries of (π, H) vanish at infinity.

(2) There is $p \in (0, \infty)$ such that every K-finite matrix entry of π is of class L^p on G.

(2) follows from 5.2, 5.3 and 1.6(2).

We divide the proof of (1) into three steps.

(a) Let $v, w \in H_0$ and $\varepsilon > 0$. There is N so that if $h \in \mathrm{Cl}((\mathfrak{a}_0)^+)$ and $\|h\| \ge N$, then $|\langle \pi(\exp h)v, w\rangle| < \varepsilon$.

This follows from 5.3 and 1.6(1).

(b) Let $v, w \in H$ and $\varepsilon > 0$. There exists N so that if $h \in \mathrm{Cl}(\mathfrak{a}_0^+)$ and $\|h\| \ge N$, then $|\langle \pi(\exp h)v, w\rangle| < \varepsilon$.

We may assume $\|v\| = \|w\| = 1$. There exist $v_0, w_0 \in H_0$ so that $\|v - v_0\| < \varepsilon/3C$ $\|w - w_0\| < \varepsilon/3C$ (C as in 5.1). Let N be chosen so that if $h \in \mathrm{Cl}(\mathfrak{a}_0^+)$ and $\|h\| \ge N$, then

$$|\langle \pi(\exp h)v_0, w_0\rangle| < \varepsilon/3.$$

We have

$$|\langle \pi(\exp h)v, w\rangle| \le \|\pi(\exp h)(v - v_0)\|\,\|w\|$$
$$+ \|\pi(\exp h)^*(w - w_0)\|\,\|v\| + |\langle \pi(\exp h)v_0, w_0\rangle|;$$

hence

$$|\langle \pi(\exp h)v, w\rangle| < \varepsilon \quad \text{if } \|h\| \ge N, \text{ by (a)}.$$

(c) We can now prove (1). Let $B_N = \{g \in G \mid \sigma(g) \le N\}$. Then B_N is compact. Also

$$B_N = \{k_1(\exp h)k_2 \mid k_1, k_2 \in K \text{ and } h \in \mathrm{Cl}(\mathfrak{a}_0^+), \|h\| \le N\}.$$

Clearly, (b) implies that if $v, w \in H$ and $\varepsilon > 0$, then there is N so that if $x \notin B_N$, then $|\langle \pi(x)v, w\rangle| < \varepsilon$. This implies that $c_{v,w}$ vanishes at infinity, as asserted.

Theorem 5.4 in the case when (π, H) is unitary is precisely Howe's theorem ([**63**]).

5.5. PROPOSITION. *Let (π, H) be an irreducible admissible representation of G on a Hilbert space. Suppose that for each $v, w \in H_0$ and $h \in \mathrm{Cl}(\mathfrak{a}_0^+)$, $h \neq 0$,*

$$\lim_{t \to +\infty} \langle \pi(\exp th)v, w \rangle = 0.$$

Then (π, H) is in $\Pi_\infty(G)$.

We first note that
(1) If $v, w \in H$, then

$$\lim_{t \to +\infty} \langle \pi(\exp th)v, w \rangle = 0, \qquad h \in \mathrm{Cl}((\mathfrak{a}_0)^+), \ h \neq 0.$$

This is proved in the same way as (b) in the proof of 5.4(1).

Suppose that $h \in \mathrm{Cl}((\mathfrak{a}_0)^+)$, $h \neq 0$. Let (P_1, A_1) be a standard p-pair such that $h \in \mathfrak{a}_1^+$. As usual, $A_0 = {}^*A_1 A_1$. Let $v, w \in H_0$. Then 1.5 implies
(2) If $a \in {}^*A_1$ is fixed, then

$$\phi(t) = \langle \pi(\exp th)\pi(a)v, w \rangle = \sum_{\mu \in E(P_1, \pi)} e^{t\mu(h)} q_{\mu, P_1}(a; th; v, w),$$

with convergence as in 1.5.

Let

$$\{\mu_1, \ldots, \mu_r\} = \{\Lambda\big|_{\mathfrak{a}_1} \mid \Lambda \in E^0(P_0, \pi)\}.$$

(1) implies that $\lim_{t \to +\infty} \phi(t) = 0$. Since $E(P_1, \pi) = \{\Lambda\big|_{\mathfrak{a}_1}, \Lambda \in E(P_0, \pi)\}$ and $\mu \in E(P_1, \pi)$ is of the form $\mu = \mu_i - \xi$ $(1 \leq i \leq r)$, where ξ is a positive integral linear combination of elements of $\Delta(P_1, A_1)$, Lemma 7.2 implies
(3) If $\Lambda \in E^0(P_0, \pi)$ and $h \in \mathrm{Cl}((\mathfrak{a}_0)^+)$, $h \neq 0$, then $\mathrm{Re}\,\Lambda(h) < 0$.

We now prove the proposition. Suppose that (π, H_0) is not tempered. Then there exist Langlands data (P, σ, ν) so that (π, H_0) is equivalent with $J_{P,\sigma,\nu}$.

(1) in the proof of 4.9 says that there is $\Lambda \in E(P_0, \pi)$ such that $\mathrm{Re}\,\Lambda\big|_{\mathfrak{a}} = \mathrm{Re}\,\nu - \rho_P$. (3) now implies
(4) If $h \in \mathrm{Cl}(\mathfrak{a}^+)$ and $h \neq 0$, then $(\mathrm{Re}\,\nu - \rho_P)(h) < 0$.

This proves the result.

6. Appendix: Langlands' geometric lemmas

6.1. Let $(V, \langle\ ,\ \rangle)$ be an n-dimensional inner product space over $\mathbf{R}$. We fix a basis $\{\alpha_1, \ldots, \alpha_n\}$ of V so that $\langle \alpha_i, \alpha_j \rangle \leq 0$ for $i \neq j$. Let $\beta_1, \ldots, \beta_n \in V$ be defined by $\langle \beta_i, \alpha_j \rangle = \delta_{ij}$.

6.2. LEMMA. $\langle \beta_i, \beta_j \rangle \geq 0$ *for $1 \leq i, j \leq n$, and $\beta_i = \sum e_{ji} \alpha_j$ with $e_{ji} \geq 0$ for all $1 \leq i, j \leq n$.*

This lemma is an easy exercise, and is left to the reader.

6.3. If $F \subset \{1, \ldots, n\}$, we set $V_F = \sum_{i \notin F} \mathbf{R}\beta_i$. If $U \subset V$ is a subspace, we denote by $U^\perp$ the orthogonal complement of U in V. Then $V_F^\perp = \sum_{i \in F} \mathbf{R}\alpha_i$.

Let $\beta_i^F = \beta_i$ if $i \notin F$, and let β_i^F be the projection of β_i on $V_F^\perp$ if $i \in F$. Define α_i^F, $i = 1, \ldots, n$, by $\langle \alpha_i^F, \beta_j^F \rangle = \delta_{ij}$.

6.4. LEMMA. 1) $\alpha_i^F = \alpha_i$ if $i \in F$.

2) If $i \notin F$, then $\alpha_i^F = \alpha_i + \sum_{j \in F} c_{ji}\alpha_j$ with $c_{ji} \geq 0$ for $j \in F$, $i \notin F$.

3) $\langle \beta_i^F, \beta_j^F \rangle \geq 0$ for $1 \leq i, j \leq n$.

4) $\langle \alpha_i^F, \alpha_j^F \rangle \leq 0$ for $i \neq j$.

PROOF. If $i \in F$ and $j \notin F$, then $\langle \alpha_i, \beta_j^F \rangle = \langle \alpha_i, \beta_j \rangle = 0$. If $j \in F$, then $\langle \alpha_i, \beta_j^F \rangle = \langle \alpha_i, \beta_j \rangle = \delta_{ij}$. Hence $\alpha_i^F = \alpha_i$. This proves 1).

If $i, j \in F$ we note that $\langle \beta_i^F, \alpha_j \rangle = \delta_{ij}$, and hence 6.2 implies that $\langle \beta_i^F, \beta_j^F \rangle \geq 0$ for $i, j \in F$. If $i \notin F$, then $\langle \alpha_i^F, V_F^\perp \rangle = 0$. Hence $\alpha_i^F = \alpha_i + \sum_{j \in F} c_{ji}\alpha_j$. If $j \in F$, then $0 = \langle \alpha_i^F, \beta_j^F \rangle = c_{ji} + \langle \alpha_i, \beta_j^F \rangle$. Using 6.2, we see that if $j \in F$ and $i \notin F$, then $\langle \alpha_i, \beta_j^F \rangle \leq 0$. Hence $c_{ji} \geq 0$. This proves 2).

We observe that we have already shown that $\langle \beta_i^F, \beta_j^F \rangle \geq 0$ for $i, j \in F$. If $i \notin F$ and $j \in F$, then $\langle \beta_i^F, \beta_j^F \rangle = 0$. If $i, j \notin F$, then $\langle \beta_i, \beta_j \rangle = \langle \beta_i^F, \beta_j^F \rangle$. This proves 3).

If $i, j \in F$, then $\langle \alpha_i, \alpha_j \rangle = \langle \alpha_i^F, \alpha_j^F \rangle$ by 1). Hence if $i, j \in F$, $i \neq j$, then $\langle \alpha_i^F, \alpha_j^F \rangle \leq 0$. If $i \notin F$ and $j \in F$, then $\langle \alpha_i^F, \alpha_j^F \rangle = \langle \alpha_i^F, \alpha_j \rangle = 0$. If $i, j \notin F$, then $\langle \alpha_i^F, \alpha_j^F \rangle = \langle \alpha_i, \alpha_j^F \rangle = \langle \alpha_i, \alpha_j + \sum_{k \in F} c_{kj}\alpha_k \rangle \leq 0$ if $i \neq j$, by 2).

6.5. LEMMA. 1) $\langle \beta_i, \beta_j^F \rangle \geq 0$ for all i, j, F.

2) $\langle \beta_j, \alpha_i^F \rangle \geq 0$ for all i, j, F.

3) $\langle \alpha_i^F, \alpha_j \rangle \leq 0$ for all $i \neq j$, F.

PROOF. If $i \notin F$, 1) is clear. If $i \in F$, then $\beta_i^F = \sum_{k \in F} b_{k,i}\alpha_k$ and $b_{k,i} \geq 0$ by 6.2. This implies 1).

If $i \in F$, then 2) follows from 6.4(1). If $i \notin F$, then by 6.4(2) $\alpha_i^F = \alpha_i + \sum_{k \in F} c_{ki}\alpha_k$ with $c_{ki} \geq 0$. This implies 2).

If $i \in F$, then 3) follows from 6.4(1). If $i, j \notin F$, $i \neq j$, then $\langle \alpha_i^F, \alpha_j \rangle \leq 0$ by 6.4(2). If $i \notin F$ and $j \in F$, then $\langle \alpha_i^F, \alpha_j \rangle = \langle \alpha_i^F, \alpha_j^F \rangle \leq 0$ by 6.4(1) and (4).

6.6. If $F \subset \{1, \ldots, n\}$, then $\{\beta_i\}_{i \notin F} \cup \{\alpha_i\}_{i \in F}$ is a basis of V. Let

$$S_F = \left\{ \lambda \in V \mid \lambda = \sum_{i \notin F} x_i\beta_i - \sum_{j \in F} y_j\alpha_j \mid x_i > 0,\ y_j \geq 0 \right\}.$$

6.7. LEMMA. If $F \subset \{1, \ldots, n\}$, then $S_F = \{\lambda \in V \mid \langle \lambda, \alpha_i^F \rangle > 0$ for $i \notin F$, $\langle \lambda, \beta_i^F \rangle \leq 0$ for $i \in F\}$.

PROOF. Denote by S_F' the right-hand side of the assertion of the lemma. If $\lambda \in S_F$, then $\lambda = \sum_{i \notin F} x_i\beta_i - \sum_{i \in F} y_j\alpha_j$, $x_i > 0$, $y_j \geq 0$. $\langle \lambda, \alpha_i^F \rangle = x_i$ if $i \notin F$ and $\langle \lambda, \beta_i^F \rangle = y_i$ if $i \in F$ by the definition of β_i^F, α_i^F and 6.4(1). Hence $S_F \subset S_F'$.

If $\lambda \in S_F'$, then $\lambda = \sum_{i \notin F} x_i\beta_i - \sum_{i \in F} y_i\alpha_i$ (see 6.6). Now reverse the reasoning of the above argument to see that $x_i > 0$, $y_i \geq 0$. Hence $S_F' \subset S_F$.

6.8. LEMMA. Let $F, F' \subset \{1, \ldots, n\}$. Then $S_F \cap S_{F'} \subset S_{F \cap F'}$.

PROOF. Set $G = F \cap F'$. Suppose $\lambda \in S_F \cap S_{F'}$. Then

$$\lambda = \sum_{i \notin G} x_i\beta_i - \sum_{i \in G} y_i\alpha_i, \qquad x_i, y_i \in \mathbf{R},$$

$$\lambda = \sum_{i \notin F} a_i\beta_i - \sum_{i \in F} b_i\alpha_i, \qquad a_i > 0,\ b_i \geq 0.$$

Thus, if $i \notin F$,

$$\langle \lambda, \alpha_i^G \rangle = \sum_{j \notin F} a_j \langle \beta_j, \alpha_i^G \rangle - \sum_{j \in F} b_j \langle \alpha_j, \alpha_i^G \rangle \geq \sum_{j \notin F} a_j \langle \beta_j, \alpha_i^G \rangle = a_i > 0$$

(here we use 6.5(3)). Similarly, if $i \notin F'$, then $\langle \lambda, \alpha_i^G \rangle > 0$. Hence if $i \notin F'$ or $i \notin F$, then $\langle \lambda, \alpha_i^G \rangle > 0$. This implies that if $i \notin G$, then $\langle \lambda, \alpha_i^G \rangle > 0$. Thus $x_i > 0$, $i \notin G$.

If $i \in G$, then

$$\langle \lambda, \beta_i^F \rangle = \sum_{j \notin G} x_j \langle \beta_j, \beta_i^F \rangle - \sum_{j \in G} y_j \langle \alpha_j, \beta_i^F \rangle \geq -y_i$$

(here we use 6.5(1)). But $\langle \lambda, \beta_i^F \rangle \leq 0$, $i \in F \supset G$. Thus $-y_i \leq 0$, $i \in G$. This proves the lemma.

6.9. LEMMA. *If $F \neq F'$, then $S_F \cap S_{F'} = \varnothing$.*

PROOF. Let $\lambda \in S_F \cap S_{F'}$.

1) $F = \varnothing$. Then $\langle \lambda, \alpha_i \rangle > 0$, $i = 1, \dots, n$. But $F' \neq F$ implies there is $j \in F'$ with $\langle \lambda, \beta_j^{F'} \rangle \leq 0$. But $\beta_j^{F'} = \sum_{i \in F'} d_{ij} \alpha_i$ with $d_{ij} \geq 0$ and $\sum_i d_{ij} > 0$. Thus $\langle \lambda, \beta_j^{F'} \rangle > 0$. This contradiction proves the lemma in this case.

2) $F' \supset F$. For $\mu \in V$ let μ_F be the orthogonal projection of μ on V_F. Since $\lambda \in S_F$,

$$\lambda_F = \sum_{i \notin F} x_i (\beta_i)_F = \sum_{i \notin F} x_i \beta_i, \qquad x_i > 0.$$

But $\lambda \in S_{F'}$; hence

$$\lambda = \sum_{i \notin F'} a_i \beta_i - \sum_{j \in F'} b_j \alpha_j, \qquad a_i > 0, \ b_j \geq 0.$$

Thus

$$\lambda_F = \sum_{i \notin F'} a_i (\beta_i)_F - \sum_{i \in F' i \notin F} b_i (\alpha_i)_F.$$

If $i \notin F'$, then $i \notin F$ and hence $(\beta_i)_F = \beta_i$. If $i \notin F$, then $(\alpha_i)_F = \alpha_i^F$. Thus

$$\lambda_F = \sum_{i \notin F'} a_i \beta_i - \sum_{i \in F' - F} b_i \alpha_i^F, \qquad a_i > 0, \ b_i \geq 0.$$

We are now in situation 1) using V_F, α_i^F, β_i, $i \notin F$. Thus this case follows from 1).

3) If $F' \neq F$, then $S_{F'} \cap S_F \subset S_{F' \cap F}$. Thus we are reduced to case 2).

6.10. LEMMA. *If $\lambda \in V$, then $\lambda \in S_F$ for some $F \subset \{1, \dots, n\}$.*

PROOF. By induction on n. If $\dim V = 1$, the result is clear. Assuming it true for $n-1 \geq 1$, we now prove it for n. If $\lambda \notin S_\varnothing$, then we may, by relabeling $\alpha_1, \dots, \alpha_n$, assume that $\langle \lambda, \alpha_n \rangle \leq 0$. Set $E = \{n\}$. Let λ_E be the orthogonal projection of λ on V_E. Then $\alpha_1^E, \dots, \alpha_{n-1}^E$, $\beta_1^E, \dots, \beta_{n-1}^E$ satisfy all of the conditions assumed for $V, \alpha_1, \dots, \alpha_n$. Hence there is $F' \subset \{1, \dots, n-1\}$ so that

$$\lambda_E = \sum_{i \notin F' \cup E} x_i' \beta_i - \sum_{i \in F'} y_i' \alpha_i^E, \qquad x_i' > 0, \ y_i' \geq 0.$$

We have $\alpha_i^E = \alpha_i + c_i \alpha_n$, $c_i \geq 0$ for $i < n$, by 6.4(2). Also $\lambda_E = \lambda - (\langle \lambda, \alpha_n \rangle / \langle \alpha_n, \alpha_n \rangle) \alpha_n$. Hence

$$\lambda = \lambda_E + (\langle \lambda, \alpha_n \rangle / \langle \alpha_n, \alpha_n \rangle) \alpha_n$$
$$= \sum_{i \notin F' \cup E} x_i' \beta_i - \sum_{i \in F'} y_i' \alpha_i - \sum_{i \in F'} y_i' c_i \alpha_n + (\langle \lambda, \alpha_n \rangle / \langle \alpha_n, \alpha_n \rangle) \alpha_n.$$

But $\langle \lambda, \alpha_n \rangle \leq 0$. Thus $\lambda \in S_{F' \cup \{n\}}$.   Q.E.D.

Lemmas 6.9 and 6.10 imply the following lemma of Langlands.

6.11. LEMMA (Langlands [**76**]). *If $\lambda \in V$, then there exists a unique $F \subset \{1, \ldots, n\}$, to be denoted $F(\lambda)$, such that $\lambda \in S_F$.*

6.12. If $\lambda \in V$ and $\lambda \in S_F$, set $\lambda_0 = \sum_{i \notin F} x_i \beta_i$ if $\lambda = \sum_{i \notin F} x_i \beta_i - \sum_{i \in F} y_i \alpha_i$.

6.13. LEMMA (Langlands [**76**]). *If $\lambda, \mu \in V$ and $\langle \lambda, \beta_i \rangle \geq \langle \mu, \beta_i \rangle$ for $i = 1, \ldots, n$, then $\langle \lambda_0, \beta_i \rangle \geq \langle \mu_0, \beta_i \rangle$ for $i = 1, \ldots, n$.*

If $i \notin F(\mu)$, then $\langle \lambda_0, \beta_i \rangle \geq \langle \lambda, \beta_i \rangle \geq \langle \mu, \beta_i \rangle = \langle \mu_0, \beta_i \rangle$. If $i \in F(\mu)$, then $\langle \alpha_i, \lambda_0 - \mu_0 \rangle = \langle \alpha_i, \lambda_0 \rangle \geq 0$. Hence $\langle \beta_i^{F(\mu)}, \lambda_0 - \mu_0 \rangle \geq 0$. Also $\beta_i^{F(\mu)} = \beta_i - \sum_{j \notin F(\mu)} a_j \beta_j$. We assert that $a_j \geq 0$. Indeed, if $j \notin F(\mu)$, then (1.2) implies that

$$-a_j = \langle \beta_i^{F(\mu)}, \alpha_j \rangle \leq 0.$$

Hence $\beta_i = \beta_i^{F(\mu)} + \sum_{j \notin F(\mu)} a_j \beta_j$ with $a_j \geq 0$. Thus

$$\langle \beta_i, \lambda_0 - \mu_0 \rangle \geq \langle \beta_i^{F(\mu)}, \lambda_0 - \mu_0 \rangle \geq 0.$$

7. Appendix: A lemma on exponential polynomial series

7.1. As usual on $\mathbf{R}^n$, set $\langle x, y \rangle = \sum x_i y_i$. If $m = (m_1, \ldots, m_n) \in \mathbf{N}^n$, set $|m| = m_1 + \cdots + m_n$.

7.2. LEMMA. *Let $\lambda_1, \ldots, \lambda_k \in \mathbf{C}$ be distinct. Let $\mu = (\mu_1, \ldots, \mu_n) \in \mathbf{R}^n$, $\mu_j > 0$, $j = 1, \ldots, n$. Let $p_{i,m}(t) \in \mathbf{C}[t]$ for $m \in \mathbf{N}$, $i = 1, \ldots, k$. Suppose that*

$$\phi(t) = \sum_{i=1}^{k} e^{\lambda_i t} \sum_{m \in \mathbf{N}^n} e^{-\langle m, \mu \rangle t} p_{i,m}(t),$$

with convergence absolute and uniform for $t \geq 1$. Suppose also that $p_{i,0} \not\equiv 0$ for $i = 1, \ldots, k$. Then $\lim_{t \to +\infty} \phi(t) = 0$ if and only if $\operatorname{Re} \lambda_i < 0$, $i = 1, \ldots, k$.

Set

$$\psi_i(t) = \sum_{m \in \mathbf{N}^n, |m| > 0} e^{-\langle m, \mu \rangle t} p_{i,m}(t) \qquad (i = 1, \ldots, k).$$

1) $\lim_{t \to +\infty} \psi_i(t) = 0$, $i = 1, \ldots, k$.

To prove 1) we note that if $\varepsilon > 0$ is given, then there is M so that

(a) $$\sum_{|m| \geq M} e^{-\langle m, \mu \rangle t} |p_{i,m}(t)| < \varepsilon \quad \text{for } t \geq 1.$$

Also, since $p_{i,m}$ is a polynomial, there is a constant C so that

(b) $$\sum_{0 < |m| \leq M} e^{-\langle m, \mu \rangle t} |p_{i,m}(t)| \leq C \sum_{0 < |m| \leq M} e^{-1/2 \langle m, \mu \rangle t}.$$

(b) implies that there is $T > 1$ so that if $t \geq T$, then

$$\sum_{0 < |m| \leq M} e^{-\langle m, \mu \rangle t} |p_{i,m}(t)| \leq \varepsilon.$$

Hence if $t \geq T$, then $|\psi_i(t)| \leq 2\varepsilon$. This proves 1).
Now

$$\phi(t) = \sum_{i=1}^{k} e^{\lambda_i t}(p_{i,0}(t) + \psi_i(t)).$$

If $\operatorname{Re} \lambda_i < 0$ for $i = 1, \ldots, k$, then

$$\lim_{t \to +\infty} e^{\lambda_i t} = 0, \qquad \lim_{t \to +\infty} e^{\lambda_i t} p_{i,0}(t) = 0.$$

Thus, by 1), $\lim_{t \to +\infty} \phi(t) = 0$.

Thus to complete the proof we need only show that if $\lim_{t \to +\infty} \phi(t) = 0$, then $\operatorname{Re} \lambda_i < 0$, $i = 1, \ldots, k$. If not, then after renumbering we may assume that $\operatorname{Re} \lambda_1 \geq \operatorname{Re} \lambda_2 \geq \cdots \geq \operatorname{Re} \lambda_k$ and $\operatorname{Re} \lambda_1 \geq 0$. We have

$$\left| e^{-\lambda_1 t} \phi(t) - \sum_{i=1}^{k} e^{(\lambda_i - \lambda_1)t} p_{i,0}(t) \right| = \left| \sum_{i=1}^{k} e^{(\lambda_i - \lambda_1)t} \psi_i(t) \right|.$$

Since $\operatorname{Re} \lambda_1 \geq 0$, we see that

$$\lim_{t \to +\infty} e^{-\lambda_1 t} \phi(t) = 0.$$

Hence by 1) we have

$$\lim_{t \to +\infty} \left| \sum_{i=1}^{k} e^{(\lambda_i - \lambda_1)t} p_{i,0}(t) \right| = 0.$$

Let $\operatorname{Re} \lambda_1 = \operatorname{Re} \lambda_2 = \cdots = \operatorname{Re} \lambda_{k_0} > \operatorname{Re} \lambda_{k_0+1}$. Then

$$\lim_{t \to +\infty} \left| \sum_{i > k_0} e^{(\lambda_i - \lambda_1)t} p_{i,0}(t) \right| = 0.$$

We therefore have

i) $\lim_{t \to +\infty} |\sum_{i=1}^{k_0} e^{(\lambda_i - \lambda_1)t} p_{i,0}(t)| = 0$.

Now $p_{i,0}(t) = \sum_{j=0}^{q} a_{i,j} t^j$, $i = 1, \ldots, k_0$, with $a_{i,q} \neq 0$ for at least one i. Multiplying through in i) by t^{-q}, we see that

ii) $\lim_{t \to +\infty} |\sum_{i=1}^{k_0} e^{(\lambda_i - \lambda_1)t} a_{i,q}| = 0$.

Now Lemma A.3.2.1 on p. 428 of [**114**] implies $a_{i,q} = 0$ for $i \leq k_0$, whence a contradiction.

CHAPTER V

Cohomology with Coefficients in $\Pi_\infty(G)$

In this chapter we prove some results on the cohomology with coefficients in certain admissible $(\mathfrak{g}, K)$-modules with $\mathfrak{g}$ semisimple. We shall proceed by induction, starting from the results of III on induced representations and using Langlands' classification (Chapter IV). Although we are mainly interested in unitary $(\mathfrak{g}, K)$-modules, we consider more generally those $(\mathfrak{g}, K)$-modules whose coefficients satisfy the necessary conditions for unitarizability from IV, 5.2, and denote by $\Pi_\infty(G)$ the set of infinitesimal equivalence classes of such representations (see §2). In §3 it is shown that if $H \in \Pi_\infty(G)$ and if F is a finite dimensional $(\mathfrak{g}, K)$-module, then

$$H^q(\mathfrak{g}, K; H \otimes F) = 0 \quad \text{for } q < \mathrm{rk}_{\mathbf{R}}\, G, \ q > \dim(G/K) - \mathrm{rk}_{\mathbf{R}}\, G.$$

The vanishing of $H^q(\mathfrak{g}, K; H)$ below the $\mathbf{R}$-rank has also been proved by G. Zuckerman (see [119]).

In §4 we study the cohomology of a particular $(\mathfrak{g}, K)$-module that is a real analogue of the Steinberg module for p-adic reductive groups or finite groups. We use a partial determination of its cohomology to show that the vanishing theorem for $\Pi_\infty(G)$ is best possible (4.6).

In §5 we show how the results of this chapter and of II can be used to derive some results of Delorme on the relationship between H^1 and the topology of the unitary dual of G.

§6 gives a vanishing theorem for $H^1(\mathfrak{g}, K; H \otimes F)$ when $\mathfrak{g}$ is simple of real rank one (6.1), which is a representation-theoretic analogue of a result of Raghunathan (see 6.9 and VII).

1. Preliminaries

The notation of Chapter III *is freely used.*

1.1. Let r_0 be the restriction mapping from $\mathfrak{h}_c^*$ to $(\mathfrak{a}_0)_c^*$ and from $X(H)$ to $X(A_0)$. We fix compatible orderings on $\mathfrak{h}^*$ and $\mathfrak{a}_0^*$. Let Δ (resp. $_{\mathbf{R}}\Delta$) be the corresponding set of simple roots in $\Phi(\mathfrak{g}_c, \mathfrak{h}_c)$ (resp. $_{\mathbf{R}}\Phi = \Phi(\mathfrak{g}_c, \mathfrak{a}_{0c})$). We have then

(1) $$_{\mathbf{R}}\Delta \subset r_0(\Delta) \subset {}_{\mathbf{R}}\Delta \cup \{0\}.$$

Let

(2) $$\Delta = \Delta_0 \cup \bigcup_{\beta \in {}_{\mathbf{R}}\Delta} \Delta_\beta,$$

where

(3) $$\Delta_0 = \{\alpha \in \Delta \mid r_0\alpha = 0\}, \qquad \Delta_\beta = \{\alpha \in \Delta \mid r_0\alpha = \beta\} \quad (\beta \in {}_{\mathbf{R}}\Delta).$$

In particular,

$$(4) \qquad\qquad \Delta_0 = \Delta_M$$

is the set of simple roots of $\Phi(\mathfrak{m}_c, \mathfrak{h}_c) = \Phi({}^0\mathfrak{m}_c, \mathfrak{b}_c)$.

1.2. For the standard parabolic subgroups of G (resp. G_c) we use the usual indexing by subsets of ${}_{\mathbf{R}}\Delta$ (resp. Δ) (see [113, 1.2]). If P is a standard parabolic subgroup of G, there is a unique subset $J = J(P)$ of ${}_{\mathbf{R}}\Delta$ such that $P = P_J$. Then A_P is the intersection of the kernels of the $\alpha \in J$. The complexification P_c of P, viewed as a standard parabolic subgroup of G_c, is then $P_{\tilde{J}}$, where

$$(1) \qquad\qquad \tilde{J} = r_0^{-1}(J) \cap \Delta = \Delta_0 \cap \bigcup_{\beta \in J(P)} \Delta_\beta = \Delta_{M_P}.$$

Let (P, A) be a standard p-pair, and $r_P \colon X(A_0) \to X(A)$ the restriction mapping. Then

$$(2) \qquad\qquad \Delta(P, A) \subset r_P({}_{\mathbf{R}}\Delta) \subset \Delta(P, A) \cup \{0\}.$$

More precisely,

$$(3) \qquad\qquad r_P(J) = 0; \qquad r_P \colon {}^c J \simeq \Delta(P, A) \text{ is a bijection.}$$

In particular,

$$(4) \qquad\qquad \operatorname{prk}(P) = \dim A_P = \operatorname{Card} {}^c J.$$

1.3. Weyl chambers. On $\mathfrak{a}$ and $\mathfrak{a}^*$ we use the scalar product induced by the Killing form. We put

$$(1) \qquad \mathfrak{a}^+ = \{a \in \mathfrak{a} \mid \beta(a) > 0 \ (\beta \in \Delta(P, A))\}, \qquad A^+ = \exp \mathfrak{a}^+,$$

$$(2) \qquad\qquad \mathfrak{a}^{*+} = \{\lambda \in \mathfrak{a}^* \mid (\lambda, \beta) > 0 \ (\beta \in \Delta(P, A))\},$$

$$(3) \qquad\qquad {}^+\mathfrak{a}^* = \left\{ \lambda \in \mathfrak{a}^* \mid \lambda = \sum_{\beta \in \Delta(P,A)} x_\beta \cdot \beta \ (x_\beta > 0 \text{ for all } \beta) \right\}.$$

If

$$(4) \qquad\qquad \operatorname{Cl}(\mathfrak{a}^+) = \{a \in \mathfrak{a} \mid \beta(a) \geq 0 \ (\beta \in \Delta(P, A))\},$$

then

$$(5) \qquad\qquad {}^+\mathfrak{a}^* = \{\lambda \in \mathfrak{a}^* \mid \lambda(a) > 0 \text{ for all } a \in \operatorname{Cl}(\mathfrak{a}^+) - \{0\}\}.$$

As is well known,

$$(6) \qquad \mathfrak{a}^{*+} \subset {}^+\mathfrak{a}^*, \qquad {}^+\mathfrak{a}^* = \{\lambda \in \mathfrak{a}^* \mid (\lambda, \mu) > 0 \text{ for all } \mu \in \mathfrak{a}^{*+}\}.$$

1.4. Let (P, A) be a standard p-pair, $P = M \cdot N$ the standard Levi decomposition of P. Let w_G (resp. w_M) be the longest element in W (resp. W_M). Then $s \mapsto s' = w_M \cdot s \cdot w_G$ is an involution of W^P, and we have

$$(1) \qquad l(s) + l(s') = \dim N \qquad (s \in W^P).$$

The proof is elementary, and is left to the reader. As is well known, the lengths of w_G and w_M are respectively equal to the number of positive roots in Φ and Φ_M. As a consequence, $l(s)$ takes all values between 0 and $\dim N$ when s ranges through W^P. The longest element is $w_M \cdot w_G$.

We have $w_G \rho = -\rho$; therefore

$$(2) \qquad s\rho\big|_A + s'\rho\big|_A = 0 \qquad (s \in W^P).$$

Let $\mathfrak{b} = \mathfrak{h} \cap {}^0\mathfrak{m}$ (cf. III, 1.2). Let $s \in W^P$. We have

$$(3) \qquad s'\rho = -w_M s\rho, \qquad (s'\rho - \rho)\big|_{\mathfrak{b}} = -w_M(s\rho - \rho)_{\mathfrak{b}}.$$

The automorphism $-w_M$ of $\mathfrak{b}_c^*$ transforms the highest weight of an irreducible representation of ${}^0\mathfrak{m}_c$ into that of the contragredient one [**28**, VIII, §7, n° 5]. In particular, *the irreducible representations of ${}^0\mathfrak{m}_c$ with highest weights $(s\rho - \rho)\big|_{\mathfrak{b}}$ and $(s'\rho - \rho)\big|_{\mathfrak{b}}$ are contragredient to one another.*

We note that if we replace the given ordering on Φ by the opposite one, then W^P and the length function on W^P are unchanged.

The group W_M acts trivially on A; therefore, if $\lambda \in \mathfrak{h}_c^*$, then $s\lambda\big|_A = t\lambda\big|_A$ whenever $t \in W_M s$ $(s, t \in W)$. Hence

$$(4) \qquad \{s\lambda\big|_A\}_{s \in W} = \{s\lambda\big|_A\}_{s \in W^P} \qquad (\lambda \in \mathfrak{h}_c^*).$$

1.5. PROPOSITION. *Let (P, A) be a standard p-pair, $P = M \cdot N$ the standard Levi decomposition of P, and $(\overline{P}, A)$ the p-pair opposite to (P, A). Let (σ, H_σ) be a unitary representation of ${}^0M$, and let $\nu \in \mathfrak{a}_c^*$. Let $I = I_{P,\sigma,\nu}$, $I' = I_{\overline{P},\sigma,\nu}$ (III, 2.2). Then $H^q(\mathfrak{g}, K; I) = H^{n-q}(\mathfrak{g}, K; I')$ for all q's, where $n = 2q(G)$ (III, 4.3).*

If ν is not real, then both cohomology groups are zero (III, 3.3), so assume $\nu \in \mathfrak{a}^*$. Let $\widetilde{\rho}$ be the sum of the positive roots for the order opposite to the given one on Φ. Then $\widetilde{\rho} = -\rho$, and $(\overline{P}, A)$ is standard for that new ordering. Let $s \in W^P$. Then the conditions

$$(1) \qquad s\rho\big|_A + \nu = 0, \qquad \chi_\sigma = \chi_{-s(\rho)|\mathfrak{b}},$$

are equivalent to

$$(2) \qquad s'\widetilde{\rho}\big|_A + \nu = 0, \qquad \chi_\sigma = \chi_{-s'(\widetilde{\rho})|\mathfrak{b}},$$

as follows immediately from 1.4. Also, the representations L_s, $L_{s'}$ of M with highest weights $s\rho - \rho$ (in the given ordering) and $s'\widetilde{\rho} - \widetilde{\rho}$ (in the opposite ordering) have equivalent restrictions to ${}^0M$. We have then, by III, 3.3,

$$(3) \qquad \begin{aligned} H^{q+l(s)}(\mathfrak{g}, K; I) &= H^{q+l(s')}(\mathfrak{g}, K; I') \\ &= (H^*({}^0\mathfrak{m}; K_p; H \otimes L_s) \otimes \Lambda \mathfrak{a}_c)^q \qquad (q \in \mathbf{N}). \end{aligned}$$

But it follows from II, 3.4 that the first factor on the right-hand side satisfies Poincaré duality, the top dimension being $2q({}^0M) + \dim A$. Since $l(s) + l(s') = \dim N$ and $2q(G) = 2q({}^0M) + \dim A + \dim N$, our assertion follows.

2. The class $\Pi_\infty(G)$

2.1.   We let $\Pi_\infty(G)$ denote the infinitesimal equivalence classes of irreducible admissible smooth representations (π, V) of G which are either tempered or represented by a Langlands quotient $J_{p,\omega,\nu}$ (see IV, 4.6), where (P, A) is a standard p-pair and

(1) $$\operatorname{Re}\nu \in \mathfrak{a}^{*^+}, \qquad \rho_P - \operatorname{Re}\nu \in {}^+\mathfrak{a}^*.$$

Often we shall say that (π, V) belongs to $\Pi_\infty(G)$ if its infinitesimal equivalence class does. Let G be simple and non-compact. As is shown in IV, 5.2, all non-trivial unitarizable (in fact uniformly bounded) Langlands quotients belong to $\Pi_\infty(G)$; therefore $\Pi_\infty(G)$ contains all non-trivial simple unitarizable representations of G. If $G = G' \times G''$, then $\Pi_\infty(G) = \Pi_\infty(G') \times \Pi_\infty(G'')$, via tensor product. It follows that, in general, a simple unitarizable representation of G belongs to $\Pi_\infty(G)$ if and only if it has a compact kernel.

2.2. PROPOSITION. *Let (P, A) be a p-pair, ω an irreducible tempered representation of ${}^0M_P$, and $\nu \in \mathfrak{a}_c^*$. Assume that 2.1(1) is satisfied. Then all constituents of the induced representation $I_{P,\omega,\nu}$ (III, 3.2) belong to $\Pi_\infty(G)$.*

After conjugation, we may assume (P, A) to be semi-standard. Let (P', A) be the standard p-pair associated to (P, A). Then $I_{P,\omega,\nu}$ and $I_{P',\omega,\nu}$ have the same character [**56**, §21, Lemma 3], hence the same constituents. We may therefore assume (P, A) to be standard. Moreover, it suffices to prove 2.2 for G simple. But this is just IV, 4.13.

2.3.   Let (P, A) be a standard p-pair, $(\overline{P}, A)$ the opposite p-pair. Let $\nu \in \mathfrak{a}_c^*$ be such that $\operatorname{Re}\nu \in \mathfrak{a}^{*^+}$, and let ω be as in 2.2. Then there is an intertwining operator

(1) $$A \colon I_{P,\omega,\nu} \to I_{\overline{P},\omega,\nu},$$

whose image is the Langlands quotient $J_{P,\omega,\nu}$. We have therefore two exact sequences of admissible finitely generated G-modules

(2) $$0 \to U \to I_{P,\omega,\nu} \to J_{P,\omega,\nu} \to 0,$$

(3) $$0 \to J_{P,\omega,\nu} \to I_{\overline{P},\omega,\nu} \to U' \to 0.$$

3. A vanishing theorem for the class $\Pi_\infty(G)$

3.1. LEMMA. *Let (P, A) be a standard p-pair, $J = J(P)$, ${}^cJ = {}_{\mathbf{R}}\Delta - J$, and $\lambda \in \mathfrak{h}^*$ a dominant weight of $\mathfrak{g}_c$. Let $\nu \in \mathfrak{a}^*$ be such that $\rho_P + \nu \in {}^+\mathfrak{a}^*$. Let $s \in W^P$ be such that $s(\rho + \lambda)\big|_A + \nu = 0$. Then $l(s) \geq \dim A$. More precisely, if $s = s_{\alpha_1} \cdots s_{\alpha_m}$ ($m = l(s)$) is a reduced decomposition of s, then $\{\alpha_i\}_{1 \leq i \leq m}$ contains at least one element of each set Δ_β ($\beta \in {}^cJ$) (cf. 1.1(2)).*

We may write

(1) $$\lambda = \sum c_\alpha \varpi_\alpha \qquad (c_\alpha \in \mathbf{N}).$$

The ϖ_α are positive linear combinations of the simple roots; therefore (1.1, 1.2)

(2) $$\lambda\big|_A \in \operatorname{Cl}({}^+\mathfrak{a}^*) = \left\{ \mu \in \mathfrak{a}^* \mid \mu = \sum_{\beta \in \Delta(P,A)} y_\beta \cdot \beta \ (y_\beta \geq 0) \right\}.$$

Since $\rho\big|_A = \rho_P$, our assumption on ν then implies

$$(3) \qquad (\rho + \lambda)\big|_A + \nu \in {}^+\mathfrak{a}^*.$$

On the other hand, $s(\rho + \lambda)$ is a weight of the finite dimensional irreducible representation of G_c with highest weight $\rho + \lambda$. Therefore

$$(4) \qquad s(\rho + \lambda) = \rho + \lambda - \sum_{\alpha \in \Delta} m_\alpha(s)\alpha, \quad \text{with } m_\alpha(s) \in \mathbf{N},$$

and hence

$$(5) \qquad s(\rho + \lambda)\big|_A + \nu = (\rho + \lambda)\big|_A + \nu - \sum_{\beta \in {}^cJ} r_P(\beta)\left(\sum_{\alpha \in \Delta_\beta} m_\alpha(s)\right).$$

The left-hand side of (5) is zero by assumption; therefore (3) implies

$$(6) \qquad \sum_{\alpha \in \Delta_\beta} m_\alpha(s) > 0 \quad \text{for every } \beta \in {}^cJ.$$

Now, if $\gamma \in \Delta$ and $\mu \in \mathfrak{h}_c^*$, then $s_\gamma(\mu) - \mu$ is an integral multiple of γ. Therefore, since the $\gamma \in \Delta$ are linearly independent, we see that if a reduced decomposition of s does not contain s_γ, then $m_\gamma(s) = 0$. The lemma then follows from (6).

3.2. LEMMA. *Let (P, A) and λ be as in 3.1, and let $\nu \in \mathfrak{a}^*$ be such that $\rho_P - \nu \in {}^+\mathfrak{a}^*$. Let $s \in W^P$ be such that $s(\rho + \lambda)\big|_A + \nu = 0$. Then $l(s) \leq \dim N - \dim A$.*

We shall reduce this to 3.1 by using the involution $t \mapsto t'$ of W^P introduced in 1.4.

Let $\lambda' = -w_G(\lambda)$. It is also a dominant weight. We have

$$s'(\rho + \lambda') = w_M s w_G(\rho + \lambda') = w_M s(-\rho - \lambda) = -w_M s(\rho + \lambda).$$

Therefore, since W_M acts trivially on A,

$$s'(\rho + \lambda')\big|_A = -s(\rho + \lambda)\big|_A = \nu.$$

Thus, s', λ' and $\nu' = -\nu$ satisfy the conditions of 3.1. Hence $l(s') \geq \dim A$. But then $l(s) = \dim N - l(s') \leq \dim N - \dim A$.

3.3. THEOREM. *Let (σ, F) be a finite dimensional representation of G. Let (π, V) be an irreducible admissible representation whose class belongs to $\Pi_\infty(G)$. Then*

$$(1) \qquad H^q(\mathfrak{g}, K; V \otimes F) = 0 \quad \text{for } q < \mathrm{rk}_{\mathbf{R}}\, G \text{ and } q > 2q(G) - \mathrm{rk}_{\mathbf{R}}\, G.$$

In this proof, we shall write $H^q(U)$ instead of $H^q(\mathfrak{g}, K; U)$, if U is a $(\mathfrak{g}, K)$-module.

(a) Let $j \in \mathbf{N}$. Assume that $H^j(U \otimes F) = 0$ for all $U \in \Pi_\infty(G)$. Then we have $H^j(U \otimes F) = 0$ for every admissible $(\mathfrak{g}, K)$-module of finite length whose constituents belong to $\Pi_\infty(G)$.

In fact, if

$$(2) \qquad 0 \to U' \to U \to U'' \to 0$$

is an exact sequence of G-modules, then the long exact sequence associated to the exact sequence

$$(3) \qquad 0 \to U' \otimes F \to U \otimes F \to U'' \otimes F \to 0$$

yields the exact sequence

(4) $$H^j(U' \otimes F) \to H^j(U \otimes F) \to H^j(U'' \otimes F).$$

Therefore, if the two extreme terms are zero, so is the middle one. Our assertion then follows by induction on the length of U.

(b) If V is tempered, then our theorem follows from III, 5.1. It therefore remains to consider the case where $V = J_{P,\omega,\nu}$ is a Langlands quotient with ν satisfying 2.1(1).

(c) We now prove the vanishing below the split rank by induction on q. It is obvious for $q < 0$, so let $q < \mathrm{rk}_{\mathbf{R}}\, G$, $q \geq 0$, and assume our assertion proved for $q - 1$. The exact sequence 2.3(3) gives rise to the exact sequence

(5) $$0 \to V \otimes F \to I_{\overline{P},\omega,\nu} \otimes F \to U' \otimes F \to 0,$$

whence an exact sequence

(6) $$H^{q-1}(U' \otimes F) \to H^q(V \otimes F) \to H^q(I_{\overline{P},\omega,\nu} \otimes F).$$

The constituents of U' all belong to $\Pi_\infty(G)$ by 2.2; hence the first term of (6) is zero by (a). In view of III, 6.1,

(7) $$H^j(I_{\overline{P},\omega,\nu} \otimes F) = 0 \quad \text{for } j < q_0(^0M) + l(s),$$

where $s \in W^P$ is such that

(8) $$s(\rho + \lambda)\big|_{A_{\overline{P}}} + \nu = 0,$$

and the ordering on $\mathfrak{h}_c^*$ is such that $(\overline{P}, A_{\overline{P}})$ is standard. We have

(9) $$A_P = A_{\overline{P}}, \qquad {}^+\mathfrak{a}_P^* = -{}^+\mathfrak{a}_{\overline{P}}^*, \qquad \rho_P = -\rho_{\overline{P}};$$

therefore the condition $\rho_P - \mathrm{Re}\,\nu \in {}^+\mathfrak{a}_P^*$ of 2.1(1) can be written

(10) $$\rho_{\overline{P}} + \mathrm{Re}\,\nu \in {}^+\mathfrak{a}_{\overline{P}}^*.$$

But then 3.1 holds for $\overline{P}$ and shows that $l(s) \geq \dim A_P$. Since $q_0(^0M) \geq \mathrm{rk}_{\mathbf{R}}\, {}^0M$ (III, 4.4) and $\mathrm{rk}_{\mathbf{R}}\, G = \mathrm{rk}_{\mathbf{R}}\, {}^0M + \dim A_P$, it follows from (7) that the last term of (6) is also zero. But then so is the second one.

(d) The second part of (1) will be proved similarly by descending induction on q. It is trivial for $q > 2q(G)$, so we let $q > 2q(G) - \mathrm{rk}_{\mathbf{R}}\, G$ and assume our assertion is true for $q + 1$. We now consider the exact sequence

(11) $$0 \to U \otimes F \to I_{P,\omega,\nu} \otimes F \to V \otimes F \to 0$$

associated to 2.3(2) and the exact sequence

(12) $$H^q(I_{P,\omega,\nu} \otimes F) \to H^q(V \otimes F) \to H^{q+1}(U \otimes F).$$

By (a), 2.2 and the induction assumption, the last term is zero. By III, 4.4 and 6.1, we have

(13) $$H^j(I_{P,\omega,\nu} \otimes F) = 0 \quad \text{for } j > 2q(^0M) - \mathrm{rk}_{\mathbf{R}}\, {}^0M + \dim A_P + l(s),$$

where $s \in W^P$ satisfies the condition

(14) $$s(\rho + \lambda)\big|_{A_P} + \nu = 0.$$

In view of 2.1(1), we can apply 3.2; hence $l(s) \leq \dim N_P - \dim A_P$. We have then

(15) $$2q(^0M) - \mathrm{rk}_{\mathbf{R}}\, {}^0M + \dim A_P + l(s) \leq 2q(^0M) + \dim N_P - \mathrm{rk}_{\mathbf{R}}\, {}^0M.$$

But

$$2q(^0M) + \dim N_P + \dim A_P = 2q(G),$$

and $\mathrm{rk}_{\mathbf{R}}\, G = \mathrm{rk}_{\mathbf{R}}\, {}^0M + \dim A_P$. Therefore the right-hand side of (15) is equal to $2q(G) - \mathrm{rk}_{\mathbf{R}}\, G$, and so, by (13), the first term of (12) is also zero, and our assertion follows.

REMARK. The second part also follows from the first one and (I, 7.6).

3.4. COROLLARY. *Let* (π, V) *be an irreducible unitary representation of* G *with compact kernel. Then* $H^q(\mathfrak{g}, K; V \otimes F) = 0$ *for* $q < \mathrm{rk}_{\mathbf{R}}\, G$ *and* $q > 2q(G) - \mathrm{rk}_{\mathbf{R}}\, G$.

In fact, the equivalence classes of such representations all belong to $\Pi_\infty(G)$, as remarked in 2.1.

4. Cohomology with coefficients in the Steinberg representation

In this section, G *is connected, linear and semi-simple. 1 also denotes the class of the trivial representation of a group.*

4.1. Let P be a parabolic subgroup of G. The representation space for $I^\infty_{P,1,-\rho_P}$ (resp. $I_{P,1,-\rho_P}$) is the space $C^\infty(P\backslash G)$ (resp. the space of K-finite vectors in $C^\infty(P\backslash G)$), with G (resp. $\mathfrak{g}$) acting by right translations (resp. differentiations). Similarly, let $I^c_{P,1,-\rho_P} = C(P\backslash G)$ be the space of continuous complex valued functions on $P\backslash G$, acted upon by right translations. We note that the space of K-finite vectors in $I^c_{P,1,-\rho_P}$ consists of smooth functions, hence is equal to $I_{P,1,-\rho_P}$. In fact, since $G = K \cdot P$, the space $I^c_{P,1,-\rho_P}$ may be identified, as a K-module, with $C((K \cap P)\backslash K)$, for which this is clear.

If $Q \supset P$, then the projection $\pi_{P,Q} \colon P\backslash G \to Q\backslash G$ induces injections

$$I_{Q,1,-\rho_Q} \to I_{P,1,-\rho_P}, \quad I^\infty_{Q,1,-\rho_Q} \to I^\infty_{P,1,-\rho_P}, \quad I^c_{Q,1,-\rho_Q} \to I^c_{P,1,-\rho_P},$$

all to be denoted $i_{P,Q}$.

We now consider standard parabolic subgroups P_J $(J \subset \Delta = \Delta(P_0, A_0))$, put

$$\rho_J = \rho_{P_J}, \quad I_J = I_{P_J,1,-\rho_J}, \quad I^\infty_J = I_{P_J,1,-\rho_P}, \quad I^c_J = I^c_{P_j,1,-\rho_J},$$

and write π_{IJ}, i_{IJ} for $\pi_{P,Q}$, $i_{P,Q}$ if $P = P_I$, $Q = P_J$ $(I \subset J)$. Let $l = |\Delta|$, and

$$D_j = \bigoplus_{|J|=l-j} I_J \quad (0 \le j \le l).$$

Define D^∞_j and D^c_j similarly, using I^∞_J and I^c_J. Then the direct sum D (resp. D^∞, resp. D^c) of the D_j (resp. D^∞_j, resp. D^c_j) is made into a complex with a differential of degree 1, given in degree q by

$$d_q f = \sum (-1)^j i_{I-i_j,I}(f)$$

$$(I = \{i_1, \ldots, i_q\}, \quad f \in I_I \text{ (resp. } I^\infty_I, \text{ resp. } I^c_I))$$

[**17**, 3.1].

4.2. PROPOSITION. (a) *The sequence of* $(\mathfrak{g}, K)$-*modules*

(1) $$0 \to D_0 \to D_1 \to \cdots \to D_{l-1} \to D_l$$

is exact.

(b) *The two-sequences of G-modules*

(2) $$0 \to D_0^c \to D_1^c \to \cdots \to D_{l-1}^c \to D_l^c,$$

(3) $$0 \to D_0^\infty \to D_1^\infty \to \cdots \to D_{l-1}^\infty \to D_l^\infty$$

are exact.

The exactness of (2) is stated in [**17**, 3.10] with some indications on the proof. Since we need this result, we shall give some more details, using freely the notation of [**17**]. Let w_m, $C(w_m)$ and E_m be as in [**17**, 2.3], and let $E_{m,I} = \pi_I(E_m)$ $(1 \le m \le N)$. It follows from [**17**, 2.4] that

(4)
$$E_{m,I} = E_{m-1,I} \quad \text{if } I \not\subset I_m,$$
$$E_{m,I} - E_{m-1,I} = C(w_m) \quad \text{if } I \subset I_m \qquad (1 < m \le N).$$

Let

$$F_{m,I} = \{f \in C(P_I \backslash G), \ f \text{ is zero on } E_{m-1,I}\} \qquad (I \subset \Delta; \ 1 \le m \le N),$$

with the understanding that $E_{-1,I}$ is the empty set. The direct sum F_m of the $F_{m,I}$, $I \subsetneq \Delta$, is a subcomplex of $D_{(1)}^c = \bigoplus_{i \ge 1} D_i^c$. The F_m's form a decreasing filtration of D^c, and $F_1 = D_{(1)}^c$. We set $F_{N+1} = \{0\}$. Lemma 2.4 in [**17**] implies that

$$H^*(F_m/F_{m+1}) = H_c^*(L_m) \otimes A_m,$$

where A_m is the space of continuous functions on $C(w_m)$ which tend to zero at infinity $(1 < m \le N)$. But [**17**, p. 216]

$$H_c^*(L_m) = 0 \quad (1 < m < N),$$
$$H_c^q(L_N) = 0 \quad (q \ne l), \qquad H_c^l(L_N) = \mathbf{Z}$$

(taking into account that our grading in (2) differs by one from the one in [**17**]). Therefore

$$H^*(F_m/F_{m+1}) = 0 \qquad (1 < m \le N),$$
$$H^*(F_2) = H^*(F_N), \qquad H^q(F_2) = 0 \quad (q \ne l).$$

Moreover, it is clear that F_1/F_2 has the cohomology of the simplex spanned by Δ, with coefficients in $\mathbf{C}$. Since $D_0 = \mathbf{C}$, our assertion follows.

If $\delta \in \widehat{K}$, then the functor which assigns to a topological G-module V the isotypic space V_δ of type δ is obviously an exact functor. Hence the exactness of (2) implies that of (1).

The spaces I_J^c are Fréchet spaces. Moreover, I_J^∞ may be viewed as the space of C^∞-vectors of I_J^c (III, 7.11). The exactness of (3) then follows from that of (2) and from the fact that $V \mapsto V^\infty$ is an exact functor on Fréchet G-modules (see IX, 6.5(iii)).

REMARK. The exactness of (3) answers a question raised in [**17**, footnote, p. 218] for the C^∞-case. The analogous sequence with analytic functions is also exact [**146**].

4.3.   We set

$$\widetilde{S} = D_l/d(D_{l-1}) = I_\varnothing / \sum_{I, I \neq \varnothing} i_{\varnothing, I}(I_I).$$

Let

$$B_j = \bigoplus_{|I|=l-j} I_{P_1, 1, \rho_I}.$$

In particular,

$$B_0 = \mathbf{C}, \qquad B_l = I_{P_0, 1, \rho_{P_0}}.$$

Let δ_j be the dual map to d_{j-1}, and $i : S \to I_{P_0, 1, \rho_{P_0}}$ the inclusion map. Then 4.2 implies the exactness of the (g, K)-module sequence

$$(1) \qquad 0 \longrightarrow S \xrightarrow{\;i\;} B_l \xrightarrow{\;\delta_l\;} B_{l-1} \longrightarrow \cdots \xrightarrow{\;\delta_1\;} B_0 \longrightarrow 0.$$

4.4. PROPOSITION. *We have*

$$H^i(\mathfrak{g}, K; S) = (0) \quad (i < l), \qquad H^l(\mathfrak{g}, K; S) \neq (0).$$

We use the notation $H^i(U)$ for $H^i(\mathfrak{g}, K; U)$. We compute $H^i(I_{P, 1, \rho_P})$ using Theorem III, 3.3. The discussion in 1.4 implies that the $s \in W^P$ of Theorem III, 3.3(i) is $w_M w_G$, and $l(s) = \dim N_P$. Let $r = r(\mathfrak{g}_c)$ be the rank of $\mathfrak{g}_c$. Since any Cartan subalgebra of $\mathfrak{p}$ acts faithfully on $\mathfrak{n}_P$, if $P \neq G$, we have $\dim N_P \geq r$. III, 3.3 implies

$$(1) \qquad\qquad\qquad H^i(I_{P, 1, \rho_P}) = 0, \qquad i < r.$$

Let $S_j = \delta_j(B_j)$ $(1 \leq j \leq l)$ and $S_{l+1} = i(S) \simeq S$. The short exact sequences

$$(2) \qquad\qquad 0 \to S_j \to B_{j-1} \to S_{j-1} \to 0 \qquad (1 < j \leq l+1),$$

combined with the long exact sequence of cohomology, give rise to the exact sequences

$$(3)$$
$$H^i(B_{j-1}) \to H^i(S_{j-1}) \to H^{i+1}(S_j) \to H^{i+1}(B_{j-1}) \qquad (1 < j \leq l+1;\ i \in \mathbf{N}).$$

Using (1) we see that

$$(4) \qquad\qquad H^i(S_{j-1}) = H^{i+1}(S_j) \qquad (i \leq r-2;\ 2 \leq j \leq l+1),$$

$$(5) \qquad\qquad \dim H^{i+1}(S_j) \geq \dim H^i(S_{j-1}) \qquad (i \leq r-1;\ 2 \leq j \leq l+1).$$

Since $S_1 = B_0 = \mathbf{C}$ and $S_{l+1} = S$, this gives

$$H^i(S) = H^{i-l}(\mathbf{C}) \quad (i < r), \qquad \dim H^r(S) \geq \dim H^{r-l}(\mathbf{C}),$$

and 4.4 follows, since $r \geq l$.

4.5. LEMMA. *Every constituent of S is in $\Pi_\infty(G)$.*

Let $\rho_{P_0} = \rho_0$ and $\pi_0 = \pi_{P_0, 1, \rho_0}$. It suffices (see IV, 5.5) to show that

$$\lim_{t \to \infty} \langle \pi_0(\exp tH) f, g \rangle = 0,$$

$$(1)$$
$$\text{for all } f \in S,\ g \in I_{P_0, 1, \rho_0} \text{ and } H \in \mathrm{Cl}(\mathfrak{a}_0^+) - \{0\}.$$

Let $^*\Phi = \{\alpha \in \Phi(P_0, A_0) \mid \alpha(H) = 0\}$ and $\mathfrak{n} = \bigoplus \mathfrak{n}_\lambda$, where the sum is over the $\lambda \in \Phi(P_0, A_0)$ for which $\lambda(H) > 0$. Let $M = \{g \in G \mid \mathrm{Ad}(g) \cdot H = H\}$, $N = \exp \mathfrak{n}$. Then $P = MN$ is a standard parabolic subgroup of G. Set $^*P = M \cap P_0$. Then

$^*P = {}^0M_0A_0{}^*N$ and $\overline{N}_0 = {}^*\overline{N} \cdot \overline{N}$ ($\overline{P} = M\overline{N}$ the parabolic subgroup opposite to P containing M).

Set $a_t = \exp tH$. If $g \in G$, write

$$g = n(g)a(g)k(g) \qquad (n(g) \in N_0,\ a(g) \in A_0,\ k(g) \in K).$$

By definition

$$\langle \pi_0(a_t)f, g \rangle = \int_K f(ka_t)\overline{g(k)}\, dk = \int_K a(ka_t)^{2\rho_0} f(k(ka_t))\overline{g(k)}\, dk$$

$$= \int_K f(k)\overline{g(k(ka_t^{-1}))}\, dk = \int_{\overline{N}_0} f(k(\overline{n}_0))\overline{g(k(\overline{n}_0a_t^{-1}))}a(\overline{n}_0)^{2\rho_0}\, d\overline{n}_0.$$

Here we have used standard integration formulae (cf. [**107**], 7.6.6 and 7.6.8). We therefore have

$$(1) \qquad \langle \pi_0(a_t)f, g \rangle = \int_{\overline{N}_0} f(k(\overline{n}_0))\overline{g(k(a_t\overline{n}_0a_t^{-1}))}a(\overline{n}_0)^{2\rho_0}\, d\overline{n}_0.$$

Since $\int_{\overline{N}_0} a(n_0)^{2\rho_0}\, dn_0$ exists, we can use the Lebesgue dominated convergence theorem to see that

$$(2) \qquad \lim_{t\to+\infty} \langle \pi_0(a_t)f, g \rangle = \int_{{}^*\overline{N}\times N} f(k({}^*\overline{n}\,\overline{n}))\overline{g(k({}^*\overline{n}))}a({}^*\overline{n}\,\overline{n})^{2\rho_0}\, d{}^*\overline{n}\, d\overline{n}.$$

We now compute the right-hand side of (2). First note that ${}^*\overline{n} \in [M, M]$ and ${}^*\overline{n} = n({}^*\overline{n})a({}^*\overline{n})k({}^*\overline{n})$ with $n({}^*\overline{n}) \in {}^*N$, $a({}^*\overline{n}) \in [M, M] \cap A_0$, $k({}^*\overline{n}) \in K_P$. Set I equal to the right-hand side of (2). Then

$$I = \int_{{}^*\overline{N}\times N} f(k(k({}^*\overline{n})\overline{n}))\overline{g(k({}^*\overline{n}))}a({}^*\overline{n})^{2\rho_0}a(k({}^*\overline{n})\overline{n})^{2\rho_0}\, d{}^*\overline{n}\, d\overline{n}$$

$$= \int_{{}^*\overline{N}\times N} f(k(k({}^*\overline{n})\overline{n}k({}^*\overline{n})^{-1})k({}^*\overline{n}))\overline{g(k({}^*\overline{n}))} \cdot a({}^*\overline{n})^{2\rho_0}a(k({}^*\overline{n})\overline{n}k({}^*\overline{n})^{-1})^{2\rho_0}\, d{}^*\overline{n}\, d\overline{n}.$$

Since the action of ${}^0M$ on $\overline{N}$ under conjugation preserves the measure $d\overline{n}$, we have

$$(3) \qquad \lim_{t\to+\infty} \langle \pi_0(a_t)f, g \rangle = \int_{{}^*\overline{N}\times N} f(k(\overline{n})k({}^*\overline{n}))\overline{g(k({}^*\overline{n}))}a({}^*\overline{n})^{2\rho_0}a(\overline{n})^{2\rho_0}\, d{}^*\overline{n}\, d\overline{n}.$$

(3) implies that to prove the lemma we need only show that if $f \in S$, then

$$(4) \qquad I_P(f) = \int_{\overline{N}} f(k(\overline{n}))a(\overline{n})^{2\rho_0}\, d\overline{n} = 0.$$

To prove (4) we use the following easy observation:

$$(5) \qquad f \in S \quad \text{if and only if} \quad \int_{K_Q} f(kx)\, dk = 0 \text{ for all } x \in K,\ Q \supset P_0.$$

Set $\widetilde{N}_0 = {}^*N\overline{N}$; then $M_0\widetilde{N} = \widetilde{P}_0$ is a minimal parabolic subgroup of G with split component A_0. Hence $\widetilde{N}_0 = sN_0s^{-1}$ with $s \in W(A_0)$.

The integral in (4) can be reinterpreted as

$$(6) \qquad \int_{sN_0s^{-1}\cap N_0\backslash sN_0s^{-1}} f(\dot{n})\, d\dot{n} = I_P(f).$$

Set

$$A_s(f)(x) = \int_{sN_0s^{-1}\cap N_0\backslash sN_0s^{-1}} f(\dot{n}x)\, d\dot{n}.$$

The *Gindikin-Karpelevič formula* (cf. [**107**], 8.10.11) implies that there is $\alpha \in \Delta(P_0, A_0)$ so that $A_s(f) = A_{s'}(A_{s_\alpha}(f))$ with $s' \in W(A_0)$, $s's_\alpha = s$. Now

$$A_{s_\alpha}(f)(x) = \int_{s_\alpha N_0 s_\alpha^{-1} \cap N_0 \backslash s_\alpha N_0 s_\alpha^{-1}} f(\dot{n}x) \, d\dot{n}.$$

Setting

$$\overline{\mathfrak{n}}^\alpha = \overline{\mathfrak{n}}_{-\alpha} + \overline{\mathfrak{n}}_{-2\alpha} \quad \text{and} \quad \overline{N}^\alpha = \exp(\overline{\mathfrak{n}}^\alpha),$$

we have

$$(N_0 \cap s_\alpha N_0 s_\alpha^{-1}) \cdot \overline{N}^\alpha = s_\alpha N_0 s_\alpha^{-1}.$$

Hence

$$A_{s_\alpha}(f)(x) = \int_{\overline{N}^\alpha} f(\overline{n}^\alpha x) \, d\overline{n}^\alpha = \int_{\overline{N}^\alpha} f(k(\overline{n}^\alpha)x) a(\overline{n}^\alpha)^{2\rho_0} \, d\overline{n}^\alpha.$$

If $P' = (P_0)_\alpha$, then

$$\int_{{}^0 M_0 \backslash K_{P'}} f(\dot{k}x) \, d\dot{k} = \int_{\overline{N}^\alpha} f(k(\overline{n}^\alpha)x) a(\overline{n}^\alpha)^{2\rho_0} \, d\overline{n}^\alpha$$

(cf. [**107**], 7.6.8). Hence (5) implies that $A_{s_\alpha}(f) = 0$. But then $A_s(f) = 0$. (6) now implies $I_P(f) = 0$. We have therefore proved (4).

4.6. COROLLARY. *Let $l = \mathrm{rk}_{\mathbf{R}}(G)$. Then there exists $H \in \Pi_\infty(G)$ such that $H^l(\mathfrak{g}, K; H) \neq (0)$.*

This follows from 4.4, 4.5 and the cohomology sequence. 4.6 implies that Theorem 3.3 is a best possible vanishing theorem in $\Pi_\infty(G)$.

5. H^1 and the topology of $\mathcal{E}(G)$

5.1.   We denote by $\mathcal{E}(G)$ the set of all equivalence classes of irreducible unitary representations of G. If $\omega \in \mathcal{E}(G)$, let $\mathcal{P}(\omega)$ be the set of matrix coefficients of ω of the form $c_{v,v}$, with $(\pi, H) \in \omega$ and $v \in H - \{0\}$. If $S \subset \mathcal{E}(G)$, set

$$\mathcal{P}(S) = \bigcup_{\omega \in S} \mathcal{P}(\omega).$$

Let $C(G)$ denote the space of all complex valued continuous functions on G with the topology of uniform convergence on compacta.

If $S \subset \mathcal{E}(G)$ and $\omega \in \mathcal{E}(G)$, we say that $\omega \in \overline{S}$ (the closure of S) if

$$\mathcal{P}(\omega) \cap \overline{\mathcal{P}(S)} \neq 0,$$

where $\overline{\mathcal{P}(S)}$ is the closure of $\mathcal{P}(S)$ in $C(G)$. It can be shown that if $\omega \in \overline{S}$, then $\mathcal{P}(\omega) \subset \overline{\mathcal{P}(S)}$ (see Dixmier [**37**], 18.1.4, 18.1.5). This notion of closure defines a topology on $\mathcal{E}(G)$. We will denote by $\mathcal{E}(G)$ this topological space. Let $1 \in \mathcal{E}(G)$ denote the class of the trivial representation.

The following theorem is due to Delorme [**36**]. The proof below is new.

5.2. THEOREM. *Suppose that G is a connected semi-simple Lie group with finite center. If 1 is not isolated in $\mathcal{E}(G)$, then there exists $\omega \in \mathcal{E}(G)$ such that if H is the corresponding $(\mathfrak{g}, K)$-module, then $H^1(\mathfrak{g}, K; H) \neq (0)$.*

Suppose that 1 is not isolated in $\mathcal{E}(G)$. Then 1 is in the closure of $\mathcal{E}(G) - \{1\}$. This implies that there is a sequence $\omega_j \in \mathcal{E}(G) - \{1\}$ and a $\phi_j \in \mathcal{P}(\omega_j)$ such that $\lim_{j \to \infty} \phi_j = 1$ uniformly on compacta. If $f \in C(G)$, define

$$^0 f(g) = \int_{K \times K} f(k_1 g k_2) \, dk_1 \, dk_2.$$

Let $(\pi_j, H_j) \in \omega_j$. Let $E_j \colon H_j \to H_j$ be defined by

$$E_j = \int_K \pi_j(k) \, dk,$$

and let $v_j \in H_j$ be such that

$$\phi_j(g) = \langle \pi(g) v_j, v_j \rangle, \qquad j = 1, 2, \cdots, \quad g \in G.$$

Hence $^0\phi_j(g) = \langle \pi(g) E_j v_j, E_j v_j \rangle$, $j = 1, 2, \cdots$, $g \in G$. This implies that $^0\phi_j \in \mathcal{P}(\omega_j)$. It is also clear that

$$\lim_{j \to \infty} {}^0\phi_j = 1 \quad \text{uniformly on compacta.}$$

This implies that $^0\phi_j \not\equiv 0$ for j large. We may therefore assume that $\phi_j \not\equiv 0$ for all j. Hence $E_j H_j \neq (0)$, $j = 1, 2, \cdots$. Harish-Chandra's parametrization of zonal spherical functions (cf. Helgason [58], Chapter 10) implies

$$^0\phi_j(g) = c_j \langle \pi_{P_0, 1, \nu_j}(g) 1, 1 \rangle \qquad (g \in G),$$

with $\nu_j \in (\mathfrak{a}_0)^*_c$, $c_j \in \mathbf{R}$, $c_j > 0$ and $\operatorname{Re}\langle \nu_j, \alpha \rangle \geq 0$, $\alpha \in \Phi(P_0, A_0)$. Since $\lim_{j \to \infty} {}^0\phi_j = 1$, we may assume $c_j = 1$ for all j.

Since $\omega_j \in \mathcal{E}(G)$, IV, 5.2 implies that $\langle \operatorname{Re} \nu_j, \operatorname{Re} \nu_j \rangle \leq \langle \rho_0, \operatorname{Re} \nu_j \rangle$. Hence $\| \operatorname{Re} \nu_j \| \leq \| \rho_0 \|$. We may therefore assume that $\lim_{j \to \infty} \operatorname{Re} \nu_j = \mu_0$ exists.

If $f \in C_c^\infty(\mathfrak{a}_0)$, define

$$\widehat{f}(\nu) = \int_{\mathfrak{a}_0} f(h) e^{-i\nu(h)} \, dh$$

(dh is Euclidean measure on $\mathfrak{a}_0$).

If $f \in C_c^\infty(G)$, define

$$F_f(h) = e^{\rho_0(h)} \int_{N_0} f(n \exp h) \, dn, \quad h \in \mathfrak{a}_0.$$

(Here dn is normalized so that

$$\int_{N_0} a(\theta(n_0))^{2\rho_0} \, dn_0 = 1$$

where for $g \in G$, we have $g = n(g) a(g) k(g)$, $n(g) \in N_0$, $a(g) \in A_0$, $k(g) \in K_0$.)

It is well known (cf. [114, 9.2.2.3]) that if $f \in C_c^\infty(G)$, then

$$\int_G {}^0\phi_j(g) f(g) \, dg = (F_{0_f})\widehat{}(-i\nu_j).$$

Suppose that $f \in C_c^\infty(G)$ and $\int_G f(g) \, dg = 1$. Then

$$\lim_{j \to \infty} \int_G {}^0\phi_j(g) f(g) \, dg = \int_G f(g) \, dg = 1.$$

Hence

$$\lim_{j \to \infty} (F_{0_f})\widehat{}(-i\nu_j) = 1.$$

If $\|\operatorname{Im} \nu_j\|$ were not bounded, then the Paley-Wiener theorem (for the Euclidean Fourier transform) would imply that there is a subsequence of the ν_j so that $(F_{0_f})^\wedge(-i\nu_j) \to 0$ (since $\operatorname{Re} \nu_j \to \mu_0$). We may thus replace ν_j by a subsequence and assume that $\lim_{j\to\infty} \nu_j = \nu_0 \in (\mathfrak{a}_0)_c^*$ exists.

The equality $\lim_{j\to\infty} {}^0\phi_j(g) = 1$ implies that $\langle \pi_{P_0,1,\nu_0}(g)1, 1\rangle = 1$. Since $\langle \operatorname{Re} \nu_0, \alpha\rangle \geq 0$ for $\alpha \in \Phi(P_0, A_0)$, $\nu_0 = \rho$, we conclude that

$$\lim_{j\to\infty} \nu_j = \rho. \tag{1}$$

(1) implies that if j is large, then $\operatorname{Re}\langle \nu_j, \alpha\rangle > 0$ for $\alpha \in \Phi(P_0, A_0)$. We may take a subsequence and assume that this is true for all j. If $\nu \in (\mathfrak{a}_0)_c^*$ and $\operatorname{Re}\langle \nu, \alpha\rangle > 0$ for $\alpha \in \Phi(P_0, A_0)$, then Harish-Chandra's c-function is non-zero at ν (cf. Wallach [107], 8.10.16). This implies that $(H_j)_0$ is isomorphic with $J_{P_0,1,\nu_j}$.

Let $Z_j \subset I_{P_0,1,\nu_j}$ be such that $J_{P_0,1,\nu_j} = I_{P_0,1,\nu_j}/Z_j$. Then, as a K-module, $J_{P_0,1,\nu_j}$ is isomorphic with $Z_j^\perp$. Let

$$\alpha_j \colon \operatorname{Hom}_K(\mathfrak{p}, Z_j^\perp) \otimes \mathfrak{p} \to Z_j^\perp$$

be defined by $\alpha_j(A \otimes X) = A(X)$. Since $\omega_j \neq 1$ for all j, $U_j = \operatorname{Im} \alpha_j \neq (0)$. This last inequality follows from the easy observation that $\alpha_j = 0$ implies that $\pi_{P_0,1,\nu_j}(\mathfrak{g}) \cdot 1 \subset Z_j$. Let $\langle\ ,\ \rangle$ be the inner product on $I_{P_0,1,\nu_j}$ (i.e. the L^2-inner product on $L^2(K_{P_0}\backslash K)$). Then the inner product on U_j corresponding to ω_j is given by

$$(v, w) = \langle B_j v, w\rangle$$

with $B_j \colon U_j \to U_j$ self-adjoint and positive non-degenerate.

Hence if $v_j \in U_j$, then

$$h_j(g) = \langle \pi_{P_0,1,\nu_j}(g)v_j, B_j v_j\rangle$$

is in $\mathcal{P}(\omega_j)$. Since B_j is positive non-degenerate, there is $v_j \in U_j$ with $\langle v_j, v_j\rangle = 1$ and $\lambda_j > 0$ so that $B_j v_j = \lambda_j v_j$. Hence

$$\psi_j(g) = \langle \pi_{P_0,1,\nu_j}(g)v_j, v_j\rangle$$

is in $\mathcal{P}(\omega_j)$.

Observing that as a K-module (and a Hilbert space) $I_{P_0,1,\nu_j} = I_{P_0,1,\rho_0}$, we see that the v_j are contained in the unit sphere of a finite dimensional subspace of $I_{P_0,1,\rho_0}$. Hence we may assume that $\lim_{j\to\infty} v_j = v_0$ exists.

If $\psi_0(g) = \langle \pi_{P_0,1,\rho_0}(g)v_0, v_0\rangle$. Then

$$\lim_{j\to\infty} \psi_j = \psi_0 \quad \text{uniformly on compacta.}$$

ψ_0 is thus a positive definite function on G, transforming under K by a sum of K-representations contained in $(\operatorname{Ad}\big|_K, \mathfrak{p}_c)$.

We therefore see that $I_{P_0,1,\rho_0}$ contains a unitarizable subquotient $H_0 \in \omega_0$ such that $\operatorname{Hom}_K(\mathfrak{p}, H_0) \neq (0)$. Since $\pi_{P_0,1,\rho}(C) = 0$ (C the Casimir operator of $\mathfrak{g}$) we see that $H^1(\mathfrak{g}, K; H_0) \neq (0)$.

5.3. COROLLARY. *If G is simple and* $\operatorname{rk}_{\mathbf{R}} G > 1$ *or if G is a $\mathbf{R}$-rank one real form of $\mathbf{F}_4$ or $\mathbf{C}_n$ ($n \geq 3$), then 1 is isolated in $\mathcal{E}(G)$.*

This follows from 3.4 and the results in II, 7.8.

5.4. The above result is due to Kazhdan [**69**] and Wang [**111**] for $\mathrm{rk}_{\mathbf{R}}(G) > 1$. The result for $\mathrm{rk}_{\mathbf{R}} G = 1$ and G a real form of $\mathbf{C}_n$, $n \geq 3$, or $\mathbf{F}_4$ is essentially due to Kostant [**73**].

5.5. The above result can be rephrased to say that if G is simple and 1 is not isolated in $\mathcal{E}(G)$, then G is locally isomorphic with $\mathbf{SO}(n, 1)$ or $\mathbf{SU}(n, 1)$.

6. A more detailed examination of first cohomology

6.1. THEOREM. *Let G be connected and simple. Let V be an irreducible, unitary $(\mathfrak{g}, K)$-module. Let F be an irreducible, finite dimensional $(\mathfrak{g}, K)$-module.*

1) If $\mathrm{rk}_{\mathbf{R}}(G) > 1$ or if $\mathfrak{g}_c$ is of type $\mathbf{C}_n$ or $\mathbf{F}_4$, then $H^1(\mathfrak{g}, K; V \otimes F^*) = (0)$.

2) Let G be locally isomorphic with $\mathbf{SO}(n, 1)$, $n \geq 3$. Let Λ_0 be the highest weight of the standard representation of $\mathbf{SO}(n, 1)$ on $\mathbf{C}^{n+1}$. If the highest weight of F is not of the form $k\Lambda_0$, $k = 0, 1, 2, \cdots$, then $H^1(\mathfrak{g}, K; V \otimes F^*) = (0)$.

3) Let G be locally isomorphic with $\mathbf{SU}(n, 1)$, $n \geq 1$. Let Λ_0 (resp. Λ_0^*) denote the highest weight of the standard representation of $\mathbf{SU}(n, 1)$ on $\mathbf{C}^{n+1}$ (resp. $(\mathbf{C}^{n+1})^*$). If the highest weight of F is not of the form $k\Lambda_0$ or $k\Lambda_0^*$ for $k = 0, 1, \cdots$, then $H^1(\mathfrak{g}, K; V \otimes F^*) = (0)$.

If $\mathrm{rk}_{\mathbf{R}}(G) > 1$, then (1) follows from Corollary 3.4. We may thus assume $\mathrm{rk}_{\mathbf{R}}(G) = 1$. If $\mathfrak{g}_c$ is of type $\mathbf{F}_4$ or $\mathbf{C}_n$ and F is the trivial representation, then the vanishing is implied by II, 7.8. To complete the proof we must examine the groups of $\mathbf{R}$-rank 1 in more detail. We need the following lemma.

6.2. LEMMA. *Assume that $\mathrm{rk}_{\mathbf{R}}(G) = 1$. Let $J = J_{P_0, \sigma, \nu}$, $J \in \Pi_\infty(G)$. Let F be the irreducible, finite dimensional $(\mathfrak{g}, K)$-module with highest weight λ. If*

$$H^1(\mathfrak{g}, K; J \otimes F^*) \neq (0),$$

then there is $\alpha \in \Delta$ so that $\alpha\big|_{\mathfrak{a}_0} \neq 0$ and $\nu = s_\alpha(\rho + \lambda)\big|_{\mathfrak{a}_0}$.

We consider the exact sequence 2.3(3)

$$0 \to J \to I' \to U' \to 0,$$

$I' = I_{\overline{P}_0, \sigma, \nu}$.

This gives rise to the exact sequence

(1) $H^0(U' \otimes F^*) \to H^1(J \otimes F^*) \to H^1(I' \otimes F^*) \to H^1(U' \otimes F^*)$.

If V is a constituent of U', then V is either tempered or of the form $J' = J_{P_0, \sigma', \nu'}$ with $\nu' < \nu$ relative to the partial order of IV. If V is tempered, then the results of III, §5 imply that $H^0(V \otimes F^*) = 0$. Suppose $H^0(J' \otimes F^*) \neq (0)$. Then we must have $H^0(I_{\overline{P}, \sigma', \nu'} \otimes F^*) \neq 0$. Theorem III, 3.3 implies that $\nu' = (\rho + \lambda)\big|_{\mathfrak{a}_0}$. Since $\nu > \nu'$, this contradicts the assumption $J \in \Pi_\infty(G)$. Hence $H^0(U' \otimes F^*) = 0$. The relation 1) now implies that $H^1(I' \otimes F^*) \neq (0)$.

We have $\rho_{\overline{P}} = -\rho_P = -\rho\big|_P$. Moreover, $-\lambda$ is the highest weight of F_λ^* with respect to the opposite ordering to the given one. In view of this, the lemma follows from III, 3.3.

6.3.   We now continue the proof of 6.1. We first look at the case where $\mathrm{rk}_{\mathbf{R}}\, G = 1$ and $\mathfrak{g}_c$ is of type $\mathbf{C}_{n+1}$, $n \geq 2$. Then G is locally isomorphic with $\mathbf{Sp}(n,1)$. We order $\Delta = \{\alpha_1, \ldots, \alpha_{n+1}\}$ relative to the Dynkin diagram

There is a unique $\alpha \in \Phi^+$ so that $\theta\alpha = -\alpha$, given by

$$\alpha = \alpha_1 + 2\sum_{i=2}^{n} \alpha_i + \alpha_{n+1}.$$

If $\alpha_j\big|_{\mathfrak{a}_0} \neq 0$, then $j = 2$.

We suppose that $H^1(J \otimes F^*) \neq (0)$, $J = H_{P_0,\sigma,\nu}$, $J \in \Pi_\infty(G)$ and F has highest weight λ. Then $\nu = s_{\alpha_2}(\lambda + \rho)\big|_{\mathfrak{a}_0}$, and we must have $(\nu - \rho_0, \alpha) < 0$. Thus we must have

$$(\lambda + \rho, s_{\alpha_2}\alpha) < (\rho, \alpha).$$

Since α is short, $2(\rho, \alpha)/(\alpha, \alpha) = 2n + 1$ and $2(\rho, s_{\alpha_2}\alpha)/(s_{\alpha_2}\alpha, s_{\alpha_2}\alpha) = 2n$. Thus we must have

$$2(\lambda, s_{\alpha_2}\alpha)/(s_{\alpha_2}\alpha, s_{\alpha_2}\alpha) < 1.$$

But $s_{\alpha_2}\alpha = \alpha_1 + \alpha_2 + 2\sum_{i=3}^n \alpha_i + \alpha_{n+1}$ if $n \geq 3$, and $S_{\alpha_2}\alpha = \alpha_1 + \alpha_2 + \alpha_3$ if $n = 2$. Hence if $J \in \Pi_\infty(G)$ and $H^1(J \otimes F^*) \neq (0)$, then $\lambda = 0$. Applying Theorem II, 7.8, we see that there is no unitary, irreducible $(\mathfrak{g}, K)$-module V so that $H^1(V) \neq (0)$.

If V is tempered and $H^1(V \otimes F^*) \neq 0$, then, since the infinitesimal character of F is regular and G has discrete series, we see that V must be in the discrete series. But then $\dim G/K = 2$ by II, 5.4, which is absurd. Since we have exhausted all possibilities, 6.1 1) is proved for $\mathfrak{g}_c$ of type $\mathbf{C}_n$.

6.4.   We now look at the case where $\mathrm{rk}_{\mathbf{R}}\, G = 1$ and $\mathfrak{g}_c$ is of type $\mathbf{F}_4$. We take $\Delta = \{\alpha_1, \alpha_2, \alpha_3, \alpha_4\}$ according to

If $\alpha_j\big|_{\mathfrak{a}_0} \neq 0$, then $j = 4$. The unique $\alpha \in \Phi^+$ such that $\theta\alpha = -\alpha$ is $\alpha = \alpha_1 + 2\alpha_2 + 3\alpha_3 + 2\alpha_4$. It is short, and $s_{\alpha_4}\alpha = \alpha_1 + 2\alpha_2 + 3\alpha_3 + \alpha_4$. We have $2(\rho, \alpha)/(\alpha, \alpha) = 11$ and $2(\rho, s_{\alpha_4}\alpha)/(s_{\alpha_4}\alpha, s_{\alpha_4}\alpha) = 10$. Thus the above proof of 6.1(1) in the case of $\mathbf{C}_{n+1}$ $(n \geq 2)$ goes over to the present case.

6.5.   We now begin the proof of 6.1(2). It is reasonable to look at the cases $n = 3$, $n = 2k + 1$, $k > 1$, and $n = 2k$, $k \geq 2$, separately. In the cases n odd we will be using results of II, §6. These results are stated in terms of the highest weights of finite dimensional representations of $\mathfrak{g}$ relative to an order such that the set of positive roots is θ-stable. Thus to apply the results to the case at hand the conditions on the highest weights must be properly interpreted for the system of simple roots Δ. The interpretations are routine, and are left to the reader. Let $n = 3$; then $\Delta = \{\alpha_1, \alpha_2\}$ and $-\theta\alpha_1 = \alpha_2$. If λ is the highest weight of F and there is a unitary irreducible $(\mathfrak{g}, K)$-module V such that $H^1(V \otimes F^*) \neq (0)$, then $-\theta\lambda = \lambda$ (see II, 6.12). Hence if λ_1, λ_2 are the basic highest weights, $2(\lambda_i, \alpha_j)/(\alpha_j, \alpha_j) = \delta_{ij}$, then $\lambda = k(\lambda_1 + \lambda_2)$. Since $\lambda_0 = \lambda_1 + \lambda_2$, 6.1(2) follows in this case.

6.6. We now assume G to be locally isomorphic to $\mathbf{SO}(n,1)$, $n = 2k+1$, $k \geq 2$. We take $\Delta = \{\alpha_1, \ldots, \alpha_{k+1}\}$ according to

If $\alpha_j\big|_{\mathfrak{a}_0} \neq 0$, then $j = 1$. If V is an irreducible unitary $(\mathfrak{g}, K)$-module such that $H^1(V \otimes F^*) \neq (0)$, then V cannot be tempered by III, §5; hence $V = J_{P_0, \sigma, \nu}$, with $\nu = s_{\alpha_1}(\rho + \lambda)\big|_{\mathfrak{a}_0}$.

Let $\lambda_1, \ldots, \lambda_{k+1}$ be the basic highest weights. If $\lambda = \sum n_i \lambda_i$ and there exists an irreducible unitary $(\mathfrak{g}, K)$-module W with $H^*(W \otimes F^*) \neq (0)$, then (II, 6.12) easily implies that $n_k = n_{k+1}$.

Also, $\alpha_1 - \theta\alpha_1 = 2\alpha$, $\Phi(P_0, A_0) = \{\alpha\}$, $\alpha_1 - \theta\alpha_1 = 2\alpha_1 + \cdots + 2\alpha_{k-1} + \alpha_k + \alpha_{k+1}$ and $(\alpha_1, \theta\alpha_1) = 0$. We now compute

$$2(\nu - \rho_0, \alpha)/(\alpha, \alpha) = 2(s_{\alpha_1}(\lambda + \rho), \alpha)/(\alpha, \alpha) - 2(\rho, \alpha)/(\alpha, \alpha)$$

$$= (\lambda + \rho, -\alpha_1 - \theta\alpha_1)/(\alpha, \alpha) - k$$

$$= \sum_{i=2}^{k-1} n_i + (n_k + n_{k+1})/2 + 1.$$

Hence if $V \in \Pi_\infty(G)$, then $\sum_{i=2}^{k-1} n_i + (n_k + n_{k+1})/2 < 1$. But $n_k = n_{k+1}$; hence $n_i = 0$, $i = 2, \ldots, k+1$. This implies that $\lambda = n\lambda_1$. Since $\lambda_0 = \lambda_1$, this proves 6.1(2) for $n = 2k+1$.

6.7. We now consider the case $n = 2k$, $k \geq 2$. We label $\Delta = \{\alpha_1, \ldots, \alpha_k\}$ as follows:

If $\alpha_i\big|_{\mathfrak{a}_0} \neq 0$, then $j = 1$. The unique $\alpha \in \Phi^+$ such that $-\theta\alpha = \alpha$ is $\alpha = \alpha_1 + \cdots + \alpha_k$.

As in the previous cases, we need only consider $J = J_{P_0, \sigma, \nu}$, $\nu = s_{\alpha_1}(\lambda + \rho)\big|_{\mathfrak{a}_0}$. We write $\lambda = \sum n_i \lambda_i$. Noting that α is short, we see that

$$(\nu - \rho, \alpha)/(\alpha, \alpha) = \sum_{i=2}^{k-1} n_i + n_k/2 - 1.$$

Thus $J \in \Pi_\infty(G)$ if and only if $n_i = 0$, $i = 2, \ldots, k-1$, and $n_k = 0$ or 1. We now use III, 3.3 i 2) to see that if $n_k = 1$, σ cannot be trivial on the center of $^0M_0$. Hence Proposition 55 on p. 561 in [**71**] implies that if $n_k = 1$, J cannot be unitarizable. But then we have $\lambda = n\lambda_1$ and $\lambda_1 = \lambda_0$, and the proof of 6.1(2) is complete.

6.8. Assume now that G is locally isomorphic with $\mathbf{SU}(n,1)$, $n \geq 2$. We label $\Delta = \{\alpha_1, \ldots, \alpha_n\}$ in accordance with the diagram

If $\alpha_j\big|_{\mathfrak{a}_0} \neq 0$, then $j = 1$ or n. The root $\alpha = \alpha_1 + \cdots + \alpha_n$ is the unique element of Φ^+ such that $-\theta\alpha = \alpha$. If F has highest weight λ and V is unitary and such that $H^1(V \otimes F^*) \neq (0)$, then $V = J_{P_0,\sigma,\nu}$, $\nu = s_{\alpha_i}(\lambda + \rho)\big|_{\mathfrak{a}_0}$, $i = 1$ or n. Also $2(\nu - \rho, \alpha)/(\alpha, \alpha) = \sum_{j \neq i} n_i - 1$. Thus if $V \in \Pi_\infty(G)$ and $\nu = s_{\alpha_i}(\lambda + \rho)\big|_{\mathfrak{a}_0}$, $i = 1$ or n, then $\lambda = k\lambda_1$ or $k\lambda_n$. This proves 6.1 3) in this case.

If $n = 1$, then *every* irreducible finite dimensional representation of G has highest weight of the form $k\lambda_0$. The proof of the theorem is now complete.

6.9.   Theorem 6.1 for $F \neq \mathbf{C}$ is a representation theoretic analogue of a result of Raghunathan [**92**] (see Chapter VII for the relationship with $H^1(\Gamma; F)$).

6.10.   The results in 6.1(2),(3) are best possible. For $\mathbf{SO}(n, 1)$ this follows from [**139**], and for $\mathbf{SU}(n, 1)$ it follows from VIII, 2.12.

CHAPTER VI

The Computation of Certain Cohomology Groups

0. Translation functors

0.1. In sections 1–4 of this chapter we will be studying $\text{Ext}^*_{\mathfrak{g},K}(\mathbf{C}, V) = H^*(\mathfrak{g}, K; V)$ for V admissible and in $\mathcal{C}(\mathfrak{g}, K)$. In this section we will show how the translation functors can be used to translate these results to computations of the spaces $\text{Ext}^*_{\mathfrak{g},K}(F, V) = H^*(\mathfrak{g}, K; V \otimes P^*)$ with F an irreducible $(\mathfrak{g}, K)$-module. We begin with some generalities.

0.2. Let $\mathfrak{g}$ be a reductive Lie algebra over $\mathbf{C}$ and $\mathfrak{h}$ a Cartan subalgebra of $\mathfrak{g}$. Let W be the Weyl group of $\mathfrak{g}$ with respect to $\mathfrak{h}$. Let $\mathfrak{b} \supset \mathfrak{h}$ denote a Borel subalgebra of $\mathfrak{g}$ and let Φ^+ be the corresponding system of positive roots in $\Phi(\mathfrak{g}, \mathfrak{h})$. Let $\mathfrak{n}^{\pm} = \bigoplus_{\alpha \in \Phi^+} \mathfrak{g}_{\pm \alpha}$ ($\mathfrak{g}_\alpha$ the α root space). Then $\mathfrak{g} = \mathfrak{n}^- \oplus \mathfrak{h} \oplus \mathfrak{n}^+$. We therefore have a decomposition

$$U(\mathfrak{g}) = U(\mathfrak{h}) \oplus (\mathfrak{n}^- U(\mathfrak{g}) + U(\mathfrak{g}) \mathfrak{n}^+).$$

Let $p \colon U(\mathfrak{g}) \to U(\mathfrak{h})$ be the corresponding projection. Let $\rho \in \mathfrak{h}^*$ be, as usual, $\frac{1}{2} \sum_{\alpha \in \Phi^+} \alpha$. We set $\mu_\rho(H) = H - \rho(H)$. We will use the same notation for its extension to $S(\mathfrak{h}) = U(\mathfrak{h})$. We write $\delta(z) = \mu_\rho(p(z))$ for $z \in Z(\mathfrak{g})$. Harish-Chandra has shown that δ defines an algebra isomorphism of $Z(\mathfrak{g})$ onto $U(\mathfrak{h})^W$ (cf. [**151**], 3.2). This implies that if $\chi \colon Z(\mathfrak{g}) \to \mathbf{C}$ is an algebra homomorphism, then there exists $\lambda \in \mathfrak{h}^*$ such that $\chi(z) = \lambda(\delta(z))$ ($z \in Z(\mathfrak{g})$). We will thus write $\chi = \chi_\lambda$. We note that $\chi_\lambda = \chi_\mu$ if and only if there exists $s \in W$ such that $s\mu = \lambda$.

0.3. For the remainder of this section we take G to be a connected semisimple Lie group with finite center, and let K be a maximal compact subgroup of G. The complexification of the Lie algebra of G will be denoted $\mathfrak{g}$ (rather than $\mathfrak{g}_c$) only in this section. We maintain the notation of subsection 0.2. If $\chi \colon Z(\mathfrak{g}) \to \mathbf{C}$ is an algebra homomorphism and if $V \in \mathcal{C}(\mathfrak{g}, K)$, then V is said to have *generalized infinitesimal character* χ if for each $v \in V$ there exists r such that $(z - \chi(z))^r v = 0$ for all $z \in Z(\mathfrak{g})$. We denote by $\mathcal{C}_\chi(\mathfrak{g}, K)$ the full subcategory of $\mathcal{C}(\mathfrak{g}, K)$ consisting of those V with generalized infinitesimal character χ.

Harish-Chandra's finiteness theorem implies that if $V \in \mathcal{C}_\chi(\mathfrak{g}, K)$ and if $S \subset V$ is finite dimensional, then the span of $U(\mathfrak{g})S$ is admissible (cf. [**151**], 3.4.7).

If $V \in \mathcal{C}(\mathfrak{g}, K)$ is admissible and if we do the obvious analysis on each K-isotypic component, then we see that V is a direct sum of submodules $V_\chi \in \mathcal{C}_\chi(\mathfrak{g}, K)$ for appropriate χ. If $V \in \mathcal{C}_\chi(\mathfrak{g}, K)$ and F is a finite dimensional $(\mathfrak{g}, K)$-module and $v \in V$, then $U(\mathfrak{g})(v \otimes F)$ is admissible. Thus if μ is another homomorphism of $Z(\mathfrak{g})$ to $\mathbf{C}$, then we have a natural $(\mathfrak{g}, K)$-module projection $P_{\mu,\chi,F} \colon V \otimes F \mapsto (V \otimes F)_\mu$. If $T \colon V \to V_1$ is a morphism of $\mathcal{C}_\chi(\mathfrak{g}, K)$, then

$$P_{\mu,\chi,F} \circ (T \otimes I) \colon P_{\mu,\chi,F}(V \otimes F) \to P_{\mu,\chi,F}(V_1 \otimes F)$$

is a morphism in $\mathcal{C}_\mu(\mathfrak{g}, K)$. We have therefore defined a functor

$$\Psi_{\mu,\chi,F} : \mathcal{C}_\chi(\mathfrak{g}, K) \to \mathcal{C}_\mu(\mathfrak{g}, K)$$

by

$$\Psi_{\mu,\chi,F}(V) = P_{\mu,\chi,F}(V \otimes F) \qquad (V \in \mathcal{C}(\mathfrak{g}, K))$$

and

$$\Psi_{\mu,\chi,F}(T) = P_{\mu,\chi,F} \circ (T \otimes I)$$

for T a morphism in $\mathcal{C}_\chi(\mathfrak{g}, K)$.

The following is an easy exercise.

LEMMA. $\Psi_{\mu,\chi,F}$ *defines an exact functor from* $\mathcal{C}_\chi(\mathfrak{g}, K)$ *to* $\mathcal{C}_\mu(\mathfrak{g}, K)$.

0.4. We will only need the simplest form of the results of Zuckerman [**155**]. Let B denote the Killing form of $\mathfrak{g}$. Let $(\ ,\ )$ denote the dual form on $\mathfrak{h}^*$. We say that λ is *dominant regular* if $\mathrm{Re}(\lambda, \alpha) > 0$ for all $\alpha \in \Phi^+$. If F is an irreducible, finite dimensional $(\mathfrak{g}, K)$-module, then λ_F will denote its highest weight with respect to Φ^+. We set $\Phi_\lambda^{\lambda+\lambda_F} = \Psi_{x_{\lambda_F+\lambda}, \chi_\lambda, F}$.

PROPOSITION. *Assume that* λ *is dominant regular and that* F *is an irreducible, finite dimensional* $(\mathfrak{g}, K)$-module. *Then*

$$\Phi_\lambda^{\lambda+\lambda_F} : \mathcal{C}_{\chi_\lambda}(\mathfrak{g}, K) \to \mathcal{C}_{\chi_{\lambda+\lambda_F}}(\mathfrak{g}, K)$$

is an equivalence of categories.

For a proof the reader may consult [**155**] (cf. [**151**], 6.A.3.9).

This proposition combined with I, 5.5 implies

0.5. PROPOSITION. *If* $V \in \mathcal{C}_{\chi_\rho}(\mathfrak{g}, K)$ *and if* F *is an irreducible finite dimensional* $(\mathfrak{g}, K)$-module, *then there is a natural isomorphism between* $\mathrm{Ext}_{\mathfrak{g},K}^*(\mathbf{C}, V)$ *and* $\mathrm{Ext}_{\mathfrak{g},K}^*(F, \Phi_\rho^{\rho+\lambda_F}(V))$.

We note that I, 5.3 implies that if $V \in \mathcal{C}_{\chi_\rho}(\mathfrak{g}, K)$, then

$$\mathrm{Ext}_{\mathfrak{g},K}^*(F, V \otimes F) = \mathrm{Ext}_{\mathfrak{g},K}^*(F, \Phi_\rho^{\rho+\lambda_F}(V)).$$

0.6. The above proposition has the following immediate consequence

COROLLARY. *Let* $V \in \mathcal{C}_{\chi_{\lambda_F}}(\mathfrak{g}, K)$ *and assume that* $\mathrm{Ext}_{\mathfrak{g},K}^*(F, V) \neq 0$. *Then there exists* $V_1 \in \mathcal{C}_{\chi_\rho}(\mathfrak{g}, K)$ *such that* $V = \Phi_\rho^{\rho+\lambda_F}(V_1)$. *Furthermore,* $\mathrm{Ext}_{\mathfrak{g},K}^*(\mathbf{C}, V_1) = \mathrm{Ext}_{\mathfrak{g},K}^*(F, V)$.

0.7. Further properties of these functors are established in [**151**], 6.A.3—in particular, the fact that they preserve square integrability and the class of induced representations appearing in Chapter IV. A comprehensive account can be found in [**140**], Chapter VII.

1. Cohomology with respect to minimal non-tempered representations. I

1.1.   In this section and the next one, G is a connected linear semisimple Lie group. If V is a $(\mathfrak{g}, K)$-module, we write $H^*(V)$ for $H^*(\mathfrak{g}, K; V)$.

Let $\Pi^\rho(G)$ denote the set of equivalence classes of irreducible admissible $(\mathfrak{g}, K)$-modules with trivial central character and the same infinitesimal character χ_ρ as the trivial representation. For $V \in \Pi^\rho(G)$, let λ_V be the Langlands parameter of V (see IV, 4.12). We fix $J \in \Pi^\rho(G)$ such that λ_J is $\neq 0$ and minimal with those properties: if $W \in \Pi^\rho(G)$, then either $\lambda_W = 0$, or λ_W is not comparable with λ_J with respect to the partial ordering of $\mathfrak{a}_0^*$ (see IV, 3.3), or $\lambda_W \geq \lambda_J$.

There are Langlands data (P, A), δ, ν, where (P, A) is a standard p-pair, such that $J = J_{P,\delta,\nu}$. Set $I = I_{P,\delta,\nu}$ and $\overline{I} = I_{\overline{P},\delta,\nu}$, where $(\overline{P}, A)$ is the p-pair opposite to (P, A). Let $U \subset I$ be such that $I/U = J$, and let $\overline{U} = \overline{I}/J$. Then U and $\overline{U}$ have the same constituents, and we have the exact sequences

(1)
$$0 \to U \to I \to J \to 0,$$

(2)
$$0 \to J \to \overline{I} \to \overline{U} \to 0.$$

If W is a constituent of U (hence of $\overline{U}$), then $\lambda_W < \lambda_J$ (IV, 4.13), and hence $\lambda_W = 0$ implies that every constituent of U or $\overline{U}$ is tempered.

1.2.   We now consider the $(\mathfrak{g}, K)$-cohomology with respect to I and $\overline{I}$. We note first that there is a unique $s \in W^P$ such that

(1)
$$s\rho\big|_{\mathfrak{a}} = \nu,$$
(2)
$$\chi_\delta = \chi_{s\rho}\big|_{\mathfrak{b}_c},$$

where $\mathfrak{b}$ is as in III, 3.3.

Let $s' = w_M \cdot s \cdot w_G$ (see V, 1.4). Then

(3)
$$l(s) + l(s') = \dim N.$$

If $t \in W^P$, let L_t be the irreducible finite dimensional representation of M_c with highest weight $(t\rho - \rho)\big|_{\mathfrak{b}}$. Then III, 3.3 implies

(4)
$$H^{q+l(s')}(I) = (H^*(^0m, K_P; H_\delta \otimes L_{s'}) \otimes \Lambda^* a)^q \qquad (q \in \mathbf{Z}),$$

(5)
$$H^{q+l(s)}(\overline{I}) = (H^*(^0\mathfrak{m}, K_P; H_\delta \otimes L_{s'}) \otimes \Lambda^* \mathfrak{a})^q \qquad (q \in \mathbf{Z}).$$

1.3.   We now assume that $H^*(I) \neq 0$; hence, by 1.2(4),

(1)
$$H^*(^0\mathfrak{m}, K_P; H_\delta \otimes L_{s'}) \neq 0.$$

This and III, 5.1 imply that δ is a direct summand of an induced representation $I_{*Q,\omega,0}$, where (Q, A_Q) is a p-pair dominated by (P, A) and $^*Q = {}^0M \cap Q$, and $\omega \in \mathcal{E}_d(^0(^*M_Q))$, where $^*M_Q = {}^0M \cap M_Q$ is the standard Levi subgroup of $^*Q$. It follows also from III, 5.1 that $(^*Q, A_Q \cap {}^0M)$ is a fundamental p-pair for ^0M; hence $\dim R_u {}^*Q = {}^*N_Q$ is even (III, 4.2), and therefore

(2)
$$\dim N \equiv \dim N_Q \mod 2.$$

In [56] it is shown that, under those circumstances, $I_{*Q,\omega,0}$ is irreducible. We extend ν to $\mathfrak{a}_Q$ by setting $\nu(\mathfrak{a}_Q \cap {}^0\mathfrak{m}) = 0$. Then, induction in stages implies that

$I = I_{Q,\omega,\nu}$, $\overline{I} = I_{\overline{Q},\omega,\nu}$. We now apply III, 5.1 to this description of I and $\overline{I}$ rather than to the initial one. We get

(3) $$\dim H^q(I) = \binom{\dim \mathfrak{a}_Q}{q - q(^0M_Q) - l(t')} \qquad (q \in \mathbf{Z}),$$

(4) $$\dim H^q(\overline{I}) = \binom{\dim \mathfrak{a}_Q}{q - q(^0M_Q) - l(t)} \qquad (q \in \mathbf{Z}),$$

where t and t' are the analogues of s and s'.

1.4. Until the end of §1, we assume that G has a compact Cartan subgroup. Then $q(G)$ is an integer. A tempered representation of G with regular infinitesimal character is square integrable (see [**52**]); hence (II, 5.3) implies that if $W = U, \overline{U}$, then

(1) $$\dim H^q(W) = \begin{cases} 0, & \text{if } q \neq q(G), \\ r, & \text{if } q = q(G), \end{cases}$$

where r is the number of constituents of W. From this and the cohomology sequences associated to 1.1(1), (2), we get

(2) $$H^q(J) = H^q(I) \quad \text{for } q \neq q(G) - 1, \ q(G),$$

(3) $$H^q(J) = H^q(\overline{I}) \quad \text{for } q \neq q(G), \ q(G) + 1.$$

1.5. LEMMA. *The conditions $H^*(I) = 0$ and $H^*(J) = 0$ are equivalent and imply that $I = J$.*

Assume $H^*(I) = 0$. Then $H^*(\overline{I}) = 0$ by 1.2(4), (5); the exact cohomology sequences associated to 1.1(1), (2) yield

(1) $$H^q(J) = H^{q+1}(U) = H^{q-1}(\overline{U}), \quad \text{for all } q \in \mathbf{Z}.$$

But this contradicts 1.4(1) unless $r = 0$, i.e., $U = (0)$, $I = J$, and then $H^*(J) = 0$. Assume now that $H^*(J) = 0$. Then $H^*(I) = H^*(U)$ in view of 1.1(1). It follows then from 1.2(4) and 1.4(1) that $U = 0$ and $I = J$; hence $H^*(I) = 0$.

1.6. We now derive some results about $l(t)$ and $H^*(J)$. First,

(1) $$l(t) + q(^0M_Q) < q(G) \quad \text{and} \quad l(t) + q(^0M_Q) + \dim \mathfrak{a}_Q \geq q(G).$$

If $l(t) + q(^0M) \geq q(G)$, then 1.3(4) and 1.4(3) imply that $H^q(J) = 0$ for $q < q(G)$. But J is admissible and irreducible; hence

(2) $$H^{2q(G)-q}(J) = H^q(J) \qquad (q \in \mathbf{Z})$$

(I, 7.6). We would then have $H^q(J) = 0$ for $q \neq q(G)$, but this is not compatible with 1.3(4) and 1.4(3). The proof of the second inequality is similar.

(3) $$\text{If } H^{q(G)+2}(J) \neq 0, \text{ then } l(t) = (\dim N_Q)/2 \text{ and } \dim \mathfrak{a}_Q \geq 4.$$

By 1.3(3) and 1.6(2) we have

(4) $$H^{q(G)-m}(\overline{I}) = H^{q(G)-m}(J) = H^{q(G)+m} = H^{q(G)+m}(\overline{I}) \qquad (m = 2, 3, \cdots).$$

If these groups are non-zero for some $m \geq 2$, then, by 1.3(4), $\dim \mathfrak{a}_Q \geq 4$ and

(5) $$2 \cdot q(^0M_Q) + 2 \cdot l(t) + \dim \mathfrak{a}_Q = 2 \cdot q(G).$$

But (III, 4.4(4))

(6) $$2q(G) = 2q(^0M_Q) + \dim \mathfrak{a}_Q + \dim \mathfrak{n}_Q,$$

and the first assertion of (3) follows. We now want to prove

(7) $\quad$ *If $H^{q(G)+2}(J) = 0$, then either $\dim \mathfrak{a}_Q = 1$ and $2l(t) = \dim N_Q - 1$*
$$\text{or } \dim \mathfrak{a}_Q = 2 \text{ and } 2l(t) = \dim N_Q.$$

If $H^{q(G)+2}(J) = 0$, then (1), (4) and 1.3(4) show that

(8) $\quad q(G) - 1 \leq l(t) + q(^0M_Q) \quad$ and $\quad q(G) + 1 \geq l(t) + q(^0M_Q) + \dim \mathfrak{a}_Q,$

hence $\dim \mathfrak{a}_Q = 1, 2$. If $\dim \mathfrak{a}_Q = 2$, then we have equalities in (8) and our assertion follows from (6). If $\dim \mathfrak{a}_Q = 1$, then (8) yields

$$2q(G) = 2l(t) + 2q(^0M_Q) + 1$$

and (6) gives the first part of (7).

1.7. THEOREM. *Let J, P, A, δ, ν be as in 1.1, and $P = M \cdot N$ the standard Levi decomposition of P.*

(i) *If $H^*(J) = 0$, then I is irreducible and $I = J$.*

(ii) *Let $\dim N$ be odd. Then $\dim A = 1$; if $H^*(J) \neq 0$, then $\delta \in \mathcal{E}_d(^0M)$, and*

(1) $$H^q(J) = \begin{cases} 0, & \text{if } q \neq q(G) - 1, \ q(G) + 1, \\ \mathbf{C}, & \text{if } q = q(G) - 1, \ q(G) + 1. \end{cases}$$

If $s \in W^P$ is as in 1.2(5), then $2l(s) = \dim N - 1$.

(iii) *Assume that $\dim N$ is even. If $H^*(J) \neq 0$, then $2l(s) = \dim N$ and*

(2) $$H^q(J) = H^q(I) = H^q(\overline{I}), \quad \text{for } q \neq q(G),$$

(3) $$\dim H^{q(G)}(J) + r = \binom{\dim \mathfrak{a}}{(\dim \mathfrak{a})/2},$$

where r is the number of constituents of U, (1.4).

We recall that G is assumed to have a compact Cartan subgroup, cf. 1.3. (i) follows from 1.5. We prove (iii). The space N_Q (cf. 1.3) is even dimensional since N is (1.3(2)). By 1.2(3) (for Q) and 1.6(3), (7), we have

(4) $$2l(t) = 2l(t') = \dim N_Q.$$

Therefore, 1.3(3), (4) imply

(5) $$H^q(I) = H^q(\overline{I}) \qquad (q \in \mathbf{Z}),$$

and (2) follows from 1.4(2), (3). In view of this and 1.4(1), the cohomology sequence associated to 1.1(2) yields the exact sequence

(6) $$0 \to H^q(U) \to H^q(I) \to H^q(J) \to 0, \quad \text{for } q = q(G).$$

By (2) and 1.6(2), $H^*(I)$ satisfies Poincaré duality. Now the first factor on the right-hand side of 1.2(4) or 1.2(5) also satisfies Poincaré duality. In view of (5), and 1.6(5) for P, we get

$$l(s) = l(s') = (\dim N)/2.$$

The remaining part of (iii) then follows from (6) and 1.3(3), (4).

(ii) By 1.3(2), N_Q is odd dimensional. Then 1.6(3), (7) show that $\dim A_Q = 1$. Therefore $Q = P$. Then 1.3(3),(4) and the results of 1.6 imply

$$
\begin{aligned}
H^q(J) &= 0 \qquad (q \neq q(G) \pm 1, q(G)), \\
H^{q(G)\pm 1}(J) &= \mathbf{C}, \qquad 2l(s) = \dim N - 1.
\end{aligned}
$$
(7)

Since $H^q(U) = 0$ for $q \neq q(G)$, 1.1(1) yields the exact sequence

(8) $\qquad 0 \to H^{q-1}(I) \to H^{q-1}(J) \to H^q(U) \to H^q(I) \to H^q(J) \to 0$

for $q = q(G)$. Since $H^{q(G)-1}(I) = 0$, this and (7) imply the exactness of

(9) $\qquad\qquad 0 \to \mathbf{C} \to H^{q(G)}(U) \to \mathbf{C} \to H^q(J) \to 0.$

This shows that the number of constituents of U is 1 or 2 (cf. 1.4(1)). But the character identity in Hecht-Schmid [57] corresponding to a non-compact simple root shows that $r \geq 2$. Indeed, their results imply that under our circumstances, I has three constituents. But then (9) shows that $H^q(J) = 0$ for $q = q(G)$, which completes the proof of 1.7.

2. Cohomology with respect to minimal non-tempered representations. II

In this section, we keep the assumptions and notation of 1.1, 1.2, 1.3 and assume moreover that G has no compact Cartan subgroup. We let $q_0 = q_0(G)$, $l_0 = l_0(G)$ (cf. III, 4.3). Then $2q(G) = 2q_0 + l_0$.

2.1. Let V be a constituent of U (or $\overline{U}$). It is tempered (1.1) and has infinitesimal character χ_ρ. There is a p-pair (P_1, A_1) such that V is an irreducible summand of $I_{P_1,\omega,i\mu}$ with $\omega \in \mathcal{E}_d(^0M_1)$, $\mu \in \mathfrak{a}_1^*$. Since $\chi_V = \chi_\rho$, we see that $\mu = 0$, and P_1 is fundamental if $H^*(V) \neq 0$ (III, 5.1). It is well known that if P_1 is fundamental and $\omega \in \mathcal{E}_d(^0M_1)$, then $I_{P_1,\omega,0}$ is irreducible [56]. Moreover, it can be shown that there is $\omega_0 \in \mathcal{E}_d((^0M_1)^0)$ such that $I_{P_1,\omega,0} = I_{P_1^0,\omega_0,0}$. By III, 5.1 we have then

(1) $\qquad\qquad \dim H^q(V) = \begin{pmatrix} l_0 \\ q - q_0 \end{pmatrix} \qquad (q \in \mathbf{Z}).$

This being true for every constituent of U such that $H^*(V) \neq 0$, we get

(2) $\qquad H^q(U) = H^q(\overline{U}) = (0), \quad \text{if } q \notin [q_0, q_0 + l_0],$

(3) $\qquad H^q(U) \neq 0, \ H^q(\overline{U}) \neq 0, \quad \text{if } U \neq (0) \text{ and } q = q_0, \ q_0 + l_0.$

By the cohomology sequences associated to 1.1(1), (2), this implies

(4) $\qquad\qquad H^q(J) = H^q(\overline{I}) \quad \text{if } q \notin [q_0, q_0 + l_0 + 1],$

(5) $\qquad\qquad H^q(J) = H^q(I) \quad \text{if } q \notin [q_0 - 1, q_0 + l_0].$

2.2. LEMMA. *If $H^*(I) = (0)$, then $H^*(J) = (0)$ and $I = J$ is irreducible.*

If $H^*(I) = 0$, then $H^*(\overline{I}) = 0$ by 1.2(4), (5), and 1.5(1) again holds. But this contradicts 2.1(2), (3) if $U \neq (0)$.

2.3.   We now assume $H^*(I) \neq 0$ and derive some results on $l(t)$ and $H^*(J)$.

(1) $$If \ U = (0), \ then \ l(t) = l(t') = (\dim N_Q)/2.$$

Indeed, then $H^*(I) = H^*(\overline{I}) = H^*(J)$; hence $H^*(I)$ satisfies Poincaré duality, and (1) follows from 1.3(3) and III, 4.4(4).

(2) $$l(t) + q(^0M) \leq q_0(G), \ with \ equality \ if \ and \ only \ if \ P \ is$$
$$fundamental, \ and \ then \ 2l(t) = \dim N_Q.$$

Assume $l(t) + q(^0M) \geq q_0$. Then

(3) $$H^q(\overline{I}) = H^q(I) = 0 \quad for \ q < q_0,$$

by 1.3(3),(4). It follows then from 2.1(4) and Poincaré duality that

(4) $$H^q(J) = 0 \quad for \ q \notin [q_0, q_0 + l_0].$$

Therefore, by (3) and 2.1(5),

(5) $$H^q(I) = 0 \quad for \ q \notin [q_0, q_0 + l_0].$$

If $l(t) + q(^0M) > q_0$, then by 1.3(3),

(6) $$H^q(I) = 0 \quad for \ q = q_0.$$

by 2.1(3), the relation (5) (resp. (6)) implies that

(7) $$\dim \mathfrak{a}_Q = l_0 \quad (\text{resp. } \dim \mathfrak{a}_Q < l_0).$$

Since ^0M has a compact Cartan subgroup, the strict inequality is impossible, and the equality implies that Q is fundamental. In the latter case, the equality $l(t) = (\dim N_Q)/2$ follows from

(8) $$2q_0 + l_0 = 2q(G) = 2q(^0M_Q) + \dim \mathfrak{a}_Q + \dim \mathfrak{n}_Q$$

(cf. III, §4). This proves (2).

(9) $$Assume \ H^{q_0-2}(J) \neq 0. \ Then \ 2l(t) = \dim N_Q$$
$$and \ \dim \mathfrak{a}_Q \geq l_0 + 4.$$

We have

(10) $$H^{q_0-m}(I) = H^{q_0-m}(J) = H^{q_0+l_0+m}(J) = H^{q_0+l_0+m}(I) \qquad (m = 2, 3, \cdots)$$

by 2.1(5) and Poincaré duality for $H^*(J)$. In view of (2) and 1.3(4) we have

$$\dim \mathfrak{a}_Q - (q_0 - 2 - q(^0M) - l(t)) = q_0 + l_0 + 2 - q(^0M) - l(t).$$

Together with (8), this implies that $2 \cdot l(t) = \dim N_Q$. Also, since $H^q(I) \neq 0$ for $q = q_0 + l_0 + 2$, 1.3(4) yields the second assertion of (9).

(11) $$If \ H^{q_0-2}(J) = 0, \ then \ either \ Q \ is \ fundamental \ and \ 2l(t) = \dim N_Q,$$
$$or \ \dim \mathfrak{a}_Q \leq l_0 + 2 \ and \ 2l(t) = \dim N_Q + \dim \mathfrak{a}_Q - l_0 - 2.$$

Under our assumption, we have, by (10) and 2.1(4),

(12) $$q(^0M) + l(t) \geq q_0 - 1.$$

Then (2) shows that either $q(^0M) + l(t) = q_0$, Q is fundamental and $2l(t) = \dim N_Q$, or

(13) $$q(^0M) + l(t) = q_0 - 1.$$

Since $H^{q_0+l_0+2}(J) = 0$ by Poincaré duality, we see from (4) and 1.3(3) that the cohomology with respect to I is concentrated in the interval $[q_0 - 1, q_0 + l_0 + 1]$; hence $\dim \mathfrak{a}_Q \leq l_0 + 2$. The last equality of (11) follows from (13) and (8).

2.4. THEOREM. *Let (P, A), J, δ, ν be as in 1.1 and Q as in 1.3.*
(i) *If $H^*(I) = 0$, then I is irreducible, $I = J$, and $H^*(J) = 0$.*
Assume $H^*(I) \neq 0$.
(ii) *If $\dim N$ is odd or if $\dim N$ is even and $2 \cdot l(t) \neq \dim N_Q$, then*

$$H^q(J) = \begin{cases} 0, & \text{for } q \notin [q_0 - 1, q_0 + l_0 + 1], \\ \mathbf{C}, & \text{for } q = q_0 - 1, \ q_0 + l_0 + 1. \end{cases}$$

(iii) *If $\dim N$ is even and $2 \cdot l(t) = \dim N_Q$, then $H^q(J) = H^q(I)$ for $q \notin [q_0, q_0 + l_0]$.*

(i) is just 2.2.
(ii) We recall that $\dim N$ and $\dim N_Q$ have the same parity (1.3(2)). Then, by 2.3(9), (11), either assumption of (ii) implies that we have 2.3(13) and that $H^q(J) = 0$ for $q \notin [q_0 - 1, q_0 + l_0 + 1]$. Furthermore, by 1.3(3), (4) and 2.1(4), (5), we have

$$H^q(J) = H^q(\bar{I}) = \mathbf{C}, \quad \text{for } q = q_0 - 1,$$
$$H^q(J) = H^q(I) = \mathbf{C}, \quad \text{for } q = q_0 + l_0 + 1.$$

Hence (ii) is proved.
(iii) If $2l(t) = \dim N_Q$, then $l(t) = l(t')$; hence $H^*(I) = H^*(\bar{I})$ by 1.3(3), (4). Then (iii) follows from 2.1(4), (5).

3. Semi-simple Lie groups with R-rank 1

3.1. In this section we assume that G is connected, simple, linear, and that $\mathrm{rk}_{\mathbf{R}} G = 1$. We also assume that if (P, A) is a fixed minimal (hence unique proper standard) p-pair then ^0M is connected. This is really no assumption, since we will be studying only elements of $\Pi^p(G)$ (see 1.1, 2.1).

To describe $\Pi^p(G)$ it is convenient to consider two cases.
I) (P, A) is fundamental.
II) (P, A) is not fundamental (i.e. G has a compact Cartan subgroup).
We first look at case I). In this case we may assume that $G = \mathbf{SO}(2k + 1, 1)^0$ (the identity component of the group of all linear transformations of $\mathbf{R}^{2k+2}$ leaving invariant the quadratic form $\sum_{i=1}^{2k+1} x_i^2 - x_{2k+2}^2$).
We take $K = G \cap \mathbf{O}(2k + 2)$. Then K is isomorphic with $\mathbf{SO}(2k + 1)$. We let A be the subgroup of G leaving the hyperplane spanned by $e_1, \ldots, e_{2k}$ pointwise fixed (here $\{e_i\}$ is the standard basis of $\mathbf{R}^{2k+2}$).
^0M is isomorphic with $\mathbf{SO}(2k)$. We fix a Cartan subalgebra $\mathfrak{h} \supset \mathfrak{a}$ of $\mathfrak{g}$ and an order on $\Phi = \Phi(\mathfrak{g}_c, \mathfrak{h}_c)$ compatible with $\Phi(P, A) = \{\beta\}$. Then, using the labeling of simple roots in [**27**, p. 256], we have

(1) $$\Delta = \{\alpha_1, \ldots, \alpha_{k+1}\},$$
(2) $$\Delta_M = \{\alpha_2, \ldots, \alpha_{k+1}\}.$$

Set $s_0 = 1$ and $s_j = s_{\alpha_1} \cdots s_{\alpha_j}$, $1 \leq j \leq k$. Let $t_k = s_{k-1} s_{\alpha_{k+1}}$. For $s \in W^P$, let $s' = w_M s w_G$. Then $W^P = \{1, s_1, \ldots, s_k, t_k, s'_0, \ldots, s'_{k-1}\}$. Since $\mathrm{Card}\, W^P =$

$2(k+1)$, s_i and t_k are easily seen to be in W^P and

$$(3) \qquad\qquad l(s_i) = i, \qquad l(t_k) = k, \qquad l(s_i') = 2k - i.$$

It is easily checked that

$$(4) \qquad\qquad s_j\rho = \rho - \alpha_j - 2\alpha_{j-1} - \cdots - j\alpha_1 \qquad (1 \le j \le k),$$

$$(5) \qquad\qquad t_k\rho = \rho - \alpha_{k+1} - 2\alpha_{k-1} - \cdots - k\alpha_1.$$

$$(6) \qquad\qquad \text{Set } \nu_j = s_j\rho\big|_{\mathfrak{a}} = (m-j)\beta, \qquad 0 \le j \le k.$$

For $s \in W^P$, let $\nu_s = s\rho\big|_{\mathfrak{a}}$, and let δ_s be equal to the irreducible representation of ${}^0M$ with highest weight $(s\rho - \rho)\big|_{\mathfrak{b}_c}$. Clearly $\nu_j = \nu_{s_j}$. Set $I_s = I_{P,\delta_s,\nu_s}$.

We note that $\nu_k = 0$ and $\nu_{t_k} = 0$. We also note that if $m \in N_K(A)$, $m \notin {}^0M$, then $\delta_{t_k}^m(x) = \delta_{t_k}(m^{-1}xm)$ defines a representation equivalent with δ_{s_k}. Thus we see that

$$(7) \qquad\qquad I_{t_k} \text{ is equivalent with } I_{s_k}.$$

We set $I_j = I_{s_j}$ and $\delta_j = \delta_{s_j}$, $0 \le j \le k$.

The Langlands classification now implies the following result.

3.2. THEOREM. (i) *If (P, A) is fundamental and if $\dim N = 2k$, then*

$$(1) \qquad\qquad \Pi^\rho(G) = \{J_0, J_1, \ldots, J_{k-1}\} \cup \{I_{P,\delta_k,0}\}$$

with J_0 the trivial representation, $I_{P,\delta_k,0}$ the unique tempered representation in $\Pi^\rho(G)$,

$$(2) \qquad\qquad J_i = J_{P,\delta_i,\nu_i} \qquad (0 \le i \le k-1),$$

where δ_i has highest weight $(s_i\rho - \rho)\big|_{\mathfrak{b}}$, $\nu_i = s_i\rho\big|_{\mathfrak{a}} = (k-i)\beta$ and $l(s_i) = i$.

(ii) *If V is an irreducible $(\mathfrak{g}, K)$-module such that $H^*(V) \ne (0)$, then V is equivalent with an element of $\Pi^\rho(G)$.*

3.3. We now look at case II), that is, G has a compact Cartan subgroup. Let $\mathfrak{h} \subset \mathfrak{g}$ be a Cartan subalgebra containing $\mathfrak{a}$. Let Φ^+ be a system of positive roots for $\Phi = \Phi(\mathfrak{g}_c, \mathfrak{h}_c)$ compatible with $\Delta(P, A) = \{\beta\}$. Then there exists a unique element $\alpha_0 \in \Phi^+$ such that $\alpha_0(\mathfrak{h}_c \cap \mathfrak{k}) = (0)$. Clearly, $\alpha_0(\mathfrak{h}) = \mathbf{R}$.

Let ${}^PW = \{s \in W^P \mid s^{-1}\alpha_0 \in \Phi^+\}$. Then

$$(1) \qquad\qquad {}^PW = \{s \in W^P \mid (s\rho\big|_{\mathfrak{a}}, \beta) > 0\}.$$

If $s \in W^P$, set $\nu_s = s\rho\big|_{\mathfrak{a}}$ and let δ_s be the irreducible representation of ${}^0M$ with highest weight $(s\rho - \rho)\big|_{\mathfrak{b}}$. Set $I_s = I_{P,\delta_s,\nu_s}$, and if $s \in {}^PW$ set $J_s = J_{P,\delta_s,\nu_s}$.

We note that

$$(2) \qquad\qquad W^P = {}^PW \cup w_M \cdot {}^PW \cdot w_G, \qquad \text{a disjoint union.}$$

The Langlands classification now implies

3.4. THEOREM. *Let G satisfy 3.1 II). Then*

$$\Pi^\rho(G) = \{J_s \mid s \in {}^PW\} \cup (\mathcal{E}_d(G) \cap \Pi^\rho(G)).$$

If V is an irreducible $(\mathfrak{g}, K)$-module such that $H^*(V) \ne (0)$, then $V \in \Pi^\rho(G)$.

3.5.   We partially order ^PW as follows: if $s, t \in {}^PW$, we say $s \to t$ if there is $\alpha \in \Phi^+ - \Phi_M^+$, simple in $s\Phi^+$ ($\Phi_M^+ = \Phi^+ \cap \Phi(^0\mathfrak{m}_c, \mathfrak{b}_c)$), so that $t = s_\alpha s$ and $l(t) = l(s) + 1$. If $s, t \in {}^PW$, then $s < t$ if there exist $u_1, \ldots, u_k \in {}^PW$ such that $s \to u_1 \to \cdots \to u_k \to t$.

3.6. LEMMA. *Let $s \in {}^PW$.*
   1) *If $t \in {}^PW$ and $s < t$, then $2(s\rho, \alpha_0)/(\alpha_0, \alpha_0) > 2(t\rho, \alpha_0)/(\alpha_0, \alpha_0)$.*
   2) *We have $s > 1$.*
   3) *Each of the two conditions:*
       a) $2l(s) = (\dim N) - 1$,
       b) $2(s\rho, \alpha_0)/(\alpha_0, \alpha_0) = 1$,
is equivalent to s being maximal.

To prove 1) it is enough to look at the case $s \to t$. Then $t = s_\alpha s$, $\alpha \in \Phi^+ - \Phi_M^+$, $l(s_\alpha s) = l(s) + 1$, and $2(s\rho, \alpha)/(\alpha, \alpha) = 1$. But then $t\rho = s\rho - \alpha$. Hence

$$2(t\rho, \alpha_0)/(\alpha_0, \alpha_0) = 2(s\rho, \alpha_0)/(\alpha_0, \alpha_0) - 2(\alpha, \alpha_0)/(\alpha_0, \alpha_0).$$

Since $\alpha \in \Phi^+ - \Phi_M^+$, we have $\alpha|_\mathfrak{a} = r\alpha_0|_\mathfrak{a}$ with $r > 0$. (In fact, $r = 1/2$, 1 or 2.) Hence 1).

We now prove 2). Let $s \in {}^PW$. Then if $s \neq 1$ there is $\beta \in s\Phi^+$ with β simple in $s\Phi^+$ and $\beta \notin \Phi^+$. We assert that $s_\beta s \in {}^PW$. We note that $\beta \notin \Phi_M$, since $\Phi_M^+ \subset s\Phi^+$. Hence $(\beta, \alpha_0) < 0$. Now

$$(s_\beta s\rho, \alpha_0) = (s\rho, s_\beta \alpha_0) = (s\rho, \alpha_0) - (2(\beta, \alpha_0)/(\beta, \beta))(s\rho, \beta) > 0.$$

Hence $s_\beta s \in {}^PW$ as asserted, and by (1) $s_\beta s \to s$. Set $t_1 = s_\beta s$. If $t_1 \neq 1$, then we can argue as above to find $t_2 \to t_1$, $t_2 \in {}^PW$. Arguing recursively we find t_j so that if $t_j \neq 1$ there is $t_{j+1} \to t_j$. If $t_k \neq 1$ the process continues. Since ^PW is finite, there exists k such that

$$t_{k+1} = 1, \qquad t_{k+1} \to t_k \to \cdots \to t_1 \to s;$$

hence $1 < s$.

We now prove 3). Suppose that $s \in {}^PW$ is maximal and α_0 is not simple in $s\Phi^+$. If $s\Delta = \{\gamma_1, \ldots, \gamma_n\}$, then $\alpha_0 = \sum n_i \gamma_i$ and $n_i \geq 0$. After relabeling we may assume that $n_1 > 0$ and $(\gamma_1, \alpha_0) > 0$. Also, since α_0 is not simple in $s\Phi^+$, $s_{\gamma_1} \alpha_0 \in s\Phi^+$. This implies that $(s_{\gamma_1} s\rho, \alpha_0) = (s\rho, s_{\gamma_1}\alpha_0) > 0$. Hence $s_{\gamma_1} s \in {}^PW$. Clearly $s \to s_{\gamma_1} s$. This contradicts the maximality of s, and proves (a).

It follows that if s is maximal, then J_s satisfies the hypotheses of Theorem 1.7. Since P is not fundamental and $\dim A = 1$, $\dim N$ is odd. Hence 1.7(ii) shows that (b) holds.

If s satisfies (b), it is maximal by (1). If (a) holds, then $l(s) \geq l(t)$ for every $t \in {}^PW$. Hence s is maximal.

3.7. LEMMA. *If $s, t \in {}^PW$ and $l(s) < l(t)$, then $(s\rho, \alpha_0) \geq (t\rho, \alpha_0)$.*

Let p (resp. q) be the dimension of the β (resp. 2β) root space relative to $\mathfrak{a}$. Then $\rho = \frac{1}{2}(p + 2q)\beta$ and $\dim N = p + q$.

Let δ be a maximal element of ^PW. Then 3.6, 2) implies that we have $s_0 = 1 \to s_1 \to s_2 \to \cdots \to s_r = \delta$, with $s_i = s_{\gamma_i} s_{i-1}$, $i = 1, \ldots, r$. Furthermore, $2r = p + q - 1$ by 3.6, 3). Since $s_{\gamma_i} s_{i-1}\rho = s_{i-1}\rho - \gamma_i$, we see that

(1) $1 = 2(\delta\rho, \alpha_0)/(\alpha_0, \alpha_0) = 2(\rho, \alpha_0)/(\alpha_0, \alpha_0) - \displaystyle\sum_{i=1}^{r} 2(\gamma_i, \alpha_0)/(\alpha_0, \alpha_0).$

Using (1), we see that

(2)   *If $\alpha_0 = \beta$, then $q = 0$.*

Indeed $\gamma_i\big|_{\mathfrak{a}} = m_i\beta$, $m_i = 1$ or 2. Thus $1 = (p + 2q) - \sum_{i=1}^{r} 2m_i$. If $q \neq 0$, then p is even. This contradiction proves (2).

We first prove the lemma under the assumption that $q = 0$. Then $\alpha_0 = \beta$. Let $s \in {}^P W$. Let $1 \to s_1 \to \cdots \to s_u = s$, with $s_i = s_{\beta_i} s_{i-1}$ as above. Then $\beta_i\big|_{\mathfrak{a}} = \beta$. Hence

$$2(s\rho, \alpha_0)/(\alpha_0, \alpha_0) = 2(\rho, \alpha_0)/(\alpha_0, \alpha_0) - \sum_{i=1}^{u} 2(\beta_i, \alpha_0)/(\alpha_0, \alpha_0)$$
$$= p - 2u = p - 2l(s).$$

This clearly implies the lemma in this case.

We now assume that $q \neq 0$; then $\alpha_0 = 2\beta$. This implies that if $\gamma \in \Phi^+ - \Phi_M^+$, then

(3) $2(\gamma, \alpha_0)/(\alpha_0, \alpha_0) = \begin{cases} 1, & \gamma\big|_{\mathfrak{a}} = \beta, \\ 2, & \gamma\big|_{\mathfrak{a}} = 2\beta. \end{cases}$

Let δ be a maximal element of ${}^P W$. Let $s_0 = 1 \to s_1 \to \cdots \to s_r = \delta$ be as above. Let r_1 be the number of i such that $\gamma_i\big|_{\mathfrak{a}} = \beta$, and r_2 the number of i such that $\gamma_i\big|_{\mathfrak{a}} = 2\beta$. Then

(4) $2 = (p + 2q) - 2(r_1 + 2r_2), \qquad 2(r_1 + r_2) = p + q - 1.$

The two equations in (4) imply

(5) $r_2 = (q - 1)/2, \qquad r_1 = p/2.$

There are three possibilities for $q \neq 0$,

                    I) $q = 1$,   II) $q = 3$,   III) $q = 7$.

If $q = 1$, then $r_2 = 0$. This implies that if $s \to t$, $t = s_\gamma s$, then $\gamma\big|_{\mathfrak{a}} = \beta$. Hence we have

        In case I), if $s \in {}^P W$ then $2(s\rho, \alpha_0)/(\alpha_0, \alpha_0) = (p + 2)/l(s) - 2$.

This implies the lemma when I) is satisfied.

If II) is satisfied and if $s \in {}^P W$, then (5) implies that

$$2(s\rho, \alpha_0)/(\alpha_0, \alpha_0) = (p/2) + 3 - l(s),$$

or

$$2(s\rho, \alpha_0)/(\alpha_0, \alpha_0) = (p/2) + 3 - (l(s) + 1),$$

whence the lemma in this case.

If III) is satisfied, then $\mathfrak{g}$ is the **R**-rank 1 form of F_4. We leave it to the reader to check that in this case we have

(6) $\operatorname{Card} {}^P W = 12, \qquad \dim N = 15,$

(7)   *For each $l(s)$, the possible values of $2(s\rho, \alpha_0)/(\alpha_0, \alpha_0)$*
                          are given by the following table:

$l(s)$	$2(s\rho, \alpha_0)/(\alpha_0, \alpha_0)$
0	11
1	10
2	9
3	7
4	5, 6
5	4, 5
6	2, 3
7	1, 1

(8)

This completes the proof of the lemma.

3.8. THEOREM. *Assume that G is as in 3.1. Let $^PW = \{s \in W^P \mid (s\rho\big|_{\mathfrak{a}}, \beta) > 0\}$. Let δ_s and ν_s be as in 3.1 or 3.3. If J is non-tempered and $H^*(J) \neq (0)$, then J must be one of the $J_s = J_{P,\delta_s,\nu_s}$.*
  1) $H^q(J_s) = 0$ *for $q < l(s)$ and $q > 2q(G) - l(s)$.*
  2) *If $s \in {}^PW$, $l(s) = k$ and $(s\rho\big|_{\mathfrak{a}}, \beta) \leq (t\rho\big|_{\mathfrak{a}}, \beta)$ for all $t \in {}^PW$ with $l(t) = k$, then $H^{l(s)}(J_s) = \mathbf{C}$.*
  3) *If s is a maximal element of ^PW, then $l(s) = q_0(G) - 1$ and*

$$H^q(J_s) = \begin{cases} 0, & q \neq l(s), \quad 2q(G) - l(s), \\ \mathbf{C}, & q = l(s), \quad 2q(G) - l(s). \end{cases}$$

We first note that 3) follows from 1.7(ii) if G satisfies 3.1 II).

If G satisfies 3.1 I), then 2.4(ii) will imply 3) if we prove that $H^{q_0(G)}(J_s) = 0$. Set $J = J_s$. Then taking $G = \mathbf{SO}(2k+1, 1)^0$ (which we can without loss of generality), we have $J = J_{k-1}$. The s in 1.7 is s_{k-1}. The results of 2.4(ii) imply that $\overline{I}/J \neq (0)$ $(\overline{I} = I_{\overline{P},\delta_{k-1},\nu_{k-1}})$. Since s is maximal, $\overline{I}/J$ has all of its subquotients isomorphic with $I_{P,\delta_k,0}$. Frobenius reciprocity implies that I is multiplicity free as a representation of K. Hence $\overline{I}/J = I_{P,\delta_k,0}$.

It is classical that $\Lambda^k(\mathfrak{g}_c/\mathfrak{k}_c)$ is an irreducible representation of K. Since $\overline{I}$ is multiplicity free, we see that

$$\dim \mathrm{Hom}_K(\Lambda^k(\mathfrak{g}/\mathfrak{k}), \overline{I}) \leq 1.$$

But

$$\dim \mathrm{Hom}_K(\Lambda^k(\mathfrak{g}/\mathfrak{k}), I_{\overline{P},\delta_k,0}) \geq 1$$

since $H^k(\overline{I}_{\overline{P},\delta_k,0}) \neq 0$. Hence

$$\mathrm{Hom}_K(\Lambda^k(\mathfrak{g}/\mathfrak{k}), J) = 0.$$

This clearly implies $H^k(J) = 0$. This completes the proof of (3).

We prove (1) and (2) by downward induction on $l(s)$. If $l(s)$ is maximal for $s \in {}^PW$, the result has been proved. Suppose that the result has been proved for $l(t) > l(s)$, $t \in {}^PW$. Let $l(s) = q$, and let $s_1, \ldots, s_p$ be the elements of ^PW such that $l(s_i) = q$. We also assume that

$$2(s_1\rho, \beta)/(\beta, \beta) \leq 2(s_2\rho; \beta)/(\beta, \beta) \leq \cdots \leq 2(s_p\rho, \beta)/(\beta, \beta).$$

Now $l(s_i') > q$ by our assumptions. Hence $H^u(I_{s_i}) = 0$ for $u \leq q$. Let $U_{s_i} \subset I_{s_i}$ be such that the sequence

(4) $$0 \to U_{s_i} \to I_{s_i} \to J_{s_i} \to 0$$

is exact.

We first look at s_1. Then the constituents of U_{s_1} must be either tempered or of the form J_t with $l(t) > q$. Thus $H^r(U_{s_1}) = 0$ for $r \leq q$ by the induction hypothesis. The long exact sequence of cohomology now implies that

$$H^r(I_{s_1}) = H^r(J_{s_1}) \quad \text{for } r < q.$$

This implies (1) in this case.

We also have the exact sequence

(5) $$0 \to J_{s_i} \to \overline{I}_{s_i} \to \overline{U}_{s_i} \to 0.$$

Since $H^r(\overline{U}_{s_1}) = 0$ for $r \leq q$, we find that

$$H^r(J_{s_1}) = H^r(\overline{I}_{s_1})$$

for $r \leq q$. Since $H^q(I_{s_1}) = \mathbf{C}$, this proves (2).

Suppose that we have shown that $H^r(J_{s_i}) = 0$ for $1 \leq i \leq r-1$ and $r < q$. The constituents of $\overline{U}_{s_r}$ must be either tempered, or of the form J_t with $l(t) > q$, or of the form J_{s_i} with $1 \leq i \leq r-1$. Then

$$H^r(\overline{U}_{s_r}) = 0 \quad \text{for } r < q.$$

The sequence (5) now implies that if $r < q$, then

$$H^r(J_{s_r}) = H^r(\overline{I}_{s_r}) = 0,$$

since $l(s_r) = q$. This completes the proof of (1).

3.9. In the special case that the 2β root space has dimension at most 1, a great deal more can be said. We study this case in more detail in the next section.

4. The groups $\mathbf{SO}(n,1)$ and $\mathbf{SU}(n,1)$

In this section we show how the results of §3 can be made very precise for $\mathbf{SO}(n,1)$ and $\mathbf{SU}(n,1)$.

4.1. We first look at the case $G = \mathbf{O}(n,1)^0$ (here $\mathbf{O}(n,1)$ is the group of all endomorphisms of $\mathbf{R}^{n+1}$ preserving $\sum_{i=1}^{n} x_i^2 - x_{n+1}^2$). We let $e_1, \ldots, e_{n+1}$ be the standard basis of $\mathbf{R}^{n+1}$, and $K = \mathbf{O}(n+1) \cap G$. Then

$$K = \{g \in G \mid g \cdot e_{n+1} = e_{n+1}\} = \mathbf{SO}(n).$$

We let A be the group of $g \in G$ such that $g \cdot e_i = e_i$, $i = 1, \ldots, n-1$, and $P = {}^0MAN$ the stabilizer of the hyperplane $\bigoplus_{i=1}^{n-1} \mathbf{R}e_i$.

We have

(1) $$2q(G) = \dim G - \dim K = n,$$
$$\dim N = n - 1, \qquad {}^0M = \mathbf{SO}(n-1).$$

(2) $$\Phi(P, A) = \Delta(P, A) = \{\beta\}, \qquad \rho_P = ((n-1)/2)\beta.$$

We fix a Cartan subalgebra $\mathfrak{h}$ of $\mathfrak{g}$ so that $\mathfrak{h} \supset \mathfrak{a}$, and an ordering on $\Phi = \Phi(\mathfrak{g}_c, \mathfrak{h}_c)$ compatible with the one on ${}_{\mathbf{R}}\Phi = \{\pm\beta\}$. We use the labeling of the simple roots of [**27**, p. 252, 256]:

(3) $$\Delta = \{\alpha_1, \ldots, \alpha_r\}, \qquad r = [(n+1)/2].$$

(4) $$\Delta_0 = \Delta_M = \{\alpha_2, \ldots, \alpha_r\}, \qquad \beta = \alpha_1\big|_A.$$

We first assume that n is odd. We are then in the situation of 3.2. We use the notation of Theorem 3.2.

4.2. THEOREM. *Suppose* $G = \mathbf{O}(n,1)^0$, $n = 2k+1$. *Let* J_i, $i = 1, \ldots, k-1$, *and* $I_{P,\delta_k,0} = I_k$ *be as in 3.2. If* V *is an irreducible* $(\mathfrak{g}, K)$-*module such that* $H^*(V) \neq (0)$, *then* V *is one of the* J_i *or* $V = I_k$. *Furthermore,*

$$(1) \qquad\qquad H^q(I_k) = \begin{cases} 0 & \text{if } q \neq k \text{ or } k+1, \\ \mathbf{C} & \text{if } q = k, \ k+1. \end{cases}$$

$$(2) \qquad\qquad H^q(J_i) = \begin{cases} 0 & \text{if } q \neq i, \ 2k+1-i, \\ \mathbf{C} & \text{if } q = i, \ 2k+1-i. \end{cases}$$

The first assertion and (1) have already been proved.

As a representation of K, $\Lambda^q(\mathfrak{g}/\mathfrak{k})$ is just $\Lambda^q \mathbf{R}^n$ with $K = \mathbf{SO}(n)$ acting as usual. Let us denote this representation by τ_q.

We compute δ_i as a representation of $^0M = \mathbf{SO}(n-1)$. Using the notation of [**27**, p. 256, 257], it is an exercise to show that the highest weight λ_i of δ_i on $\mathfrak{b}$ is given as follows:

$$(3) \qquad\qquad \lambda_i = \begin{cases} 0, & i = 0, \\ \varepsilon_1 + \cdots + \varepsilon_i, & i = 1, \ldots, k-1. \end{cases}$$

Thus, as a representation of ^0M, δ_i is just the complexification of $\Lambda^i \mathbf{R}^{n-1}$, with $\mathbf{R}^{n-1}$ the standard representation of $^0M = \mathbf{SO}(n-1)$. Clearly

$$(4) \qquad\qquad \tau_q = \delta_q \oplus \delta_{q-1}, \quad \text{as a representation of } ^0M.$$

We now prove (2) by downward induction on i. If $i = k-1$, (2) is contained in 3.8(3). Suppose it to be true for $i < j \leq k-1$. Then we have a $(\mathfrak{g}, K)$-module exact sequence

$$(5) \qquad\qquad 0 \to U_i \to I_i \to J_i \to 0.$$

U_i can have as subquotients only $J_{i+1}, \ldots, J_{k-1}$ or I_k. But $H^j(J_j) = \mathbf{C}$ for all j by 3.8(2). Hence $\text{Hom}_K(\Lambda^j(\mathfrak{g}/\mathfrak{k}), J_j) \neq 0$.

Since it is well known that the K-isotypic components of $I_{P,\delta,\nu}$ are all of multiplicity 1, (4) and Frobenius reciprocity imply

$$(6) \qquad\qquad U_i = J_{i+1} \quad \text{or} \quad U_i = (0).$$

However, (5) and the fact that $l(s') = 2k+1-i$ show that

$$(7) \qquad\qquad H^q(J_i) = H^{q+1}(U_i), \qquad q < 2k+1-i.$$

Hence $H^{i+1}(U_i) \neq 0$. It follows that $U_i = J_{i+1}$. Using (7), we find that

$$(8) \qquad\qquad H^q(J_i) = H^{q+1}(J_{i+1}).$$

Combined with the induction hypothesis, this yields

$$(9) \qquad\qquad H^q(J_i) = 0 \ \text{ if } q < i, \ i < q \leq k, \qquad H^i(J_i) = \mathbf{C}.$$

This, combined with the fact that $H^{2k+1-q}(J_i) = H^q(J_i)$, completes the induction.

4.3. We note that in the course of the proof of 4.2 we have shown that there is a $(\mathfrak{g}, K)$-module exact sequence

$$(1) \qquad\qquad 0 \to J_{i+1} \to I_i \to J_i \to 0 \qquad (i \leq k-2).$$

Using Zuckerman's translation principle [**118**], it is not too hard to derive the composition series of the full analytic continuation of the principal series from (1).

4.4. We now look at the case when $n = 2k$, $k > 0$, $k \in \mathbf{Z}$. First,

$$(1) \qquad\qquad s_i = s_{\alpha_1} \cdots s_{\alpha_i} \quad (1 \leq i \leq k-1), \qquad s_0 = 1.$$

Then (see 3.3)

$$(2) \qquad\qquad {}^P W = \{s_0, s_1, \ldots, s_{k-1}\}.$$

Set $I_i = I_{s_i}$, $J_i = J_{s_i}$ $(i = 1, \ldots, k-1)$. As is well known, there are two elements D_1, D_2 in $\Pi^p(G) \cap \mathcal{E}_d(G)$.

Let $\delta_i = \delta_{s_i}$, $i = 1, \ldots, k-1$, and let τ_j be the complexification of the representation of K on $\Lambda^j(\mathfrak{g}/\mathfrak{k})$ $(j = 1, \ldots, k-1)$. Just as in the proof of 4.2, we find that

$$(3) \qquad \begin{aligned} \tau_j\big|_{{}_0 M} &= \delta_j \oplus \delta_{j-1} \quad \text{for } j = 1, \ldots, k-1, \\ &\text{and } \tau_j \text{ is irreducible in this range.} \end{aligned}$$

In addition,

$$(4) \qquad \tau_k = \tau_k^+ \oplus \tau_k^-, \quad \text{with } \tau_k^\pm \text{ irreducible and } \tau_k^\pm\big|_{{}_0 M} = \delta_{k-1}.$$

4.5. THEOREM. *Suppose* $G = \mathbf{O}(n,1)^0$, *with* $n = 2k$, $k > 0$, $k \in \mathbf{Z}$.

(1) *If* V *is an irreducible* $(\mathfrak{g}, K)$-*module such that* $H^*(V) \neq 0$, *then* $V \in \{J_0, J_1, \ldots, J_{k-1}, D_1, D_2\}$.

(2) *We have*

$$(a) \quad H^q(D_i) = \begin{cases} 0 & \text{if } q \neq k, \\ \mathbf{C} & \text{if } q = k, \end{cases} \qquad (i = 1, 2, \ q \in \mathbf{N});$$

$$(b) \quad H^q(J_i) = \begin{cases} 0 & \text{if } q \neq i, \ k+i, \\ \mathbf{C} & \text{if } q = i, \ k+i, \end{cases} \qquad (1 \leq i < k; \ q \in \mathbf{N}).$$

(1) is just 3.8 in this case. (2)(a) has been proved in Chapter II, §5, (2)(b) for $i = k-1$ is 3.8(3), and for $i \leq k-1$ the proof of (2)(b) by downward induction is identical with the proof of 4.2(2).

4.6. It should be noted that the arguments in 4.2(2) imply that there are exact sequences

$$0 \to D_1 \oplus D_2 \to I_{k-1} \to J_{k-1} \to 0,$$
$$0 \to J_{i+1} \to I_i \to J_i \to 0 \qquad (0 \leq i \leq k-2).$$

The implications are the same as in 4.3.

4.7. For the remainder of this section we assume that $G = \mathbf{SU}(n,1)$, $n \geq 2$. Recall that $\mathbf{SU}(n,1)$ is the group of elements in $\mathbf{SL}(n+1,\mathbf{C})$ leaving invariant the Hermitian form

$$\sum_{i=1}^{n} |z_i|^2 - |z_{n+1}|^2.$$

We take $K = \mathbf{U}(n+1) \cap G$, and A to be the subgroup of G consisting of matrices of the form

$$\begin{bmatrix} a & 0 & b \\ 0 & I_{n-1} & 0 \\ b & 0 & a \end{bmatrix}, \quad \text{where } a^2 - b^2 = 1, \ a,b \in \mathbf{R}.$$

We fix a parabolic subgroup $P = {}^0MAN$ with $C_G(A) = {}^0MA$. Then we leave it to the reader to check that

(1) $$2q(G) = \dim G - \dim K = 2n,$$

(2) $$\dim N = 2n - 1,$$

(3) $${}^0M \text{ is the subgroup of } K \text{ consisting of the matrices}$$

$$\begin{bmatrix} e^{i\theta} & 0 & 0 \\ 0 & u & 0 \\ 0 & 0 & e^{i\theta} \end{bmatrix}, \quad \text{with } \det u = e^{-2i\theta} \ (\theta \in \mathbf{R}).$$

(4) $$\Phi(P,A) = \{\beta, 2\beta\}, \qquad \Delta(P,A) = \{\beta\}, \qquad \rho_P = n\beta.$$

Fix a Cartan subalgebra $\mathfrak{h}$ of $\mathfrak{g}$ containing $\mathfrak{a}$. Order $\Phi(\mathfrak{g}_c, \mathfrak{h}_c) = \Phi$ compatibly with the order of ${}_{\mathbf{R}}\Phi = \{\pm\beta, \pm 2\beta\}$. Using the numbering of the simple roots $\Delta = \Delta(\mathfrak{g}_c, \mathfrak{h}_c)$ as in [**27**, p. 250], we have

(5) $$\Delta = \{\alpha_1, \ldots, \alpha_n\}.$$

(6) $$\Delta_0 = \Delta_M = \varnothing \quad \text{if } n = 2,$$
$$\Delta_0 = \{\alpha_2, \ldots, \alpha_{n-1}\} \quad \text{if } n \geq 3.$$

(7) $$\beta = \alpha_1\big|_A = \alpha_n\big|_A.$$

(8) $$\alpha_0 = \alpha_1 + \cdots + \alpha_n, \quad \alpha_0\big|_{\mathfrak{a}} = 2\beta \quad (\text{see 3.3 for } \alpha_0).$$

4.8. As in [**27**, p. 250], we write $\alpha_j = \varepsilon_j - \varepsilon_{j+1}$ for $1 \leq j \leq n$. Then Φ^+ corresponds to the Weyl chamber $\varepsilon_1 > \cdots > \varepsilon_{n+1}$.

For $i+j \leq n-1$, $i,j \geq 0$, let Φ_{ij}^+ be the system of positive roots corresponding to the Weyl chamber $\varepsilon_{r_1} > \cdots > \varepsilon_{r_{n+1}}$ with $r_{i+1} = 1$, $r_{n+1-j} = n+1$ and $\varepsilon_2 > \cdots > \varepsilon_n$.

Let $s_{ij} \in W$ be defined by $s_{ij}\Phi^+ = \Phi_{ij}^+$. Then

(1) $${}^PW = \{s_{ij} \mid i+j \leq n-1, (i,j \geq 0)\}. \quad \text{Furthermore, } l(s_{ij}) = i+j.$$

In our notation we have

(2) $$\alpha_0 = \varepsilon_1 - \varepsilon_{n+1}. \quad \text{Hence } \mathfrak{b}_c = \{H \in \mathfrak{h} \mid \varepsilon_1(H) = \varepsilon_{n+1}(H)\}.$$

Set $J_{ij} = J_{s_{ij}}$ and $I_{ij} = I_{s_{ij}}$, $i, j \geq 0$, $i + j \leq n - 1$ (see 3.3). Then

$$\Pi^p(G) = \{J_{ij} \mid i, j \geq 0, \ i + j \leq n - 1\} \cup \{D_0, \dots, D_n\}.$$

Here, as is well known, $\Pi^p(G) \cap \mathcal{E}_d(G)$ consists of $n + 1$ elements $D_0, \dots, D_n$.

We will need a particular labeling of the D_i. For this we must analyze the decomposition of $\Lambda^q(\mathfrak{g}_c/\mathfrak{k}_c)$ as a representation of $K = \mathbf{U}(n)$.

Let τ be the standard representation of $\mathbf{U}(n)$ on $\mathbf{C}^n$. Set $\tau_1 = (\det) \otimes \tau$. Then

(3)   *As a representation of K, $\mathfrak{g}_c/\mathfrak{k}_c$ is $\tau_1 \oplus \tau_1^*$.*

This is an easy computation, which is left to the reader.

Let $\{e_1, \dots, e_n\}$ be the standard basis of $\mathbf{C}^n$ and let $\{e_1^*, \dots, e_n^*\}$ be the dual basis of $(\mathbf{C}^n)^*$. Set

$$\omega = \sum e_i \wedge e_i^* \in \Lambda^2(\mathfrak{g}_c/\mathfrak{k}_c).$$

Define, as in II, 4.6,

(4) $L \colon \Lambda^q(\mathfrak{g}_c/\mathfrak{k}_c) \to \Lambda^{q+2}(\mathfrak{g}_c/\mathfrak{k}_c)$ by $L\eta = \omega \wedge \eta.$

As is well known,

(5) $\Lambda^q \tau_1$ *and* $\Lambda^q \tau_1^*$ *are irreducible representations of K.*

As a representation of K,

$$\Lambda^q(\mathfrak{g}_c/\mathfrak{k}_c) = \bigoplus_{r+s=q} (\Lambda^r \mathbf{C}^n) \otimes \Lambda^s(\mathbf{C}^n)^* = \bigoplus_{r+s=q} \Lambda^{r,s},$$

with K acting on $\Lambda^{r,s}$ by $\Lambda^r \tau_1 \otimes \Lambda^s \tau_1^*$.

Clearly,

$$L \colon \Lambda^{r,s} \to \Lambda^{r+1,s+1}.$$

Fix a maximal torus $T \subset K$ so that $\mathfrak{t} \supset \mathfrak{b}$. We fix an order on the roots of $(\mathfrak{k}_c, \mathfrak{t}_c)$ compatible with the order on $\Phi(^0\mathfrak{m}_c, \mathfrak{b}_c)$ giving Φ_M^+.

4.9. LEMMA. *Let λ_p be the highest weight of $\Lambda^p \tau_1$ and let λ_q^* be the highest weight of $\Lambda^q \tau_1^*$. Let $F_{p,q}$ be the irreducible component of $\Lambda^{p,q}$ with highest weight $\lambda_p + \lambda_q^*$. Suppose that $p + q \leq n$, $p, q \geq 0$. Then*

(1) $\Lambda^{p,q} = L\Lambda^{p-1,q-1} \oplus F_{p,q}$ *(here $\Lambda^{-1,p} = \Lambda^{p,-1} = 0$ if $p \in \mathbf{Z}$).*

(2) $L \colon \Lambda^{p,q} \to \Lambda^{p+1,q+1}$ *is injective if $p + q \leq n - 1$.*

The assertion (2) is contained in II, 4.6.

For $\lambda \in (i\mathfrak{t})^*$, let F_λ be zero if λ is not $\Phi^+(\mathfrak{k}_c, \mathfrak{t}_c) = \Phi_K^+$-dominant integral, and let F_λ be the irreducible finite dimensional representation of K with highest weight λ if λ is Φ_K^+-dominant integral.

An easy computation using the Weyl character formula implies that

(3) $$F_\lambda \otimes F_\mu = \sum m_{\lambda,\mu}(\xi) F_{\lambda+\xi},$$

where the sum is over the weights ξ of F_μ relative to $\mathfrak{t}_c$ and $m_{\lambda,\mu}(\xi)$ is less than or equal to the dimension of the ξ weight space in F_μ.

Let $p + q \leq n$, $p \geq 0$, $q \geq 0$. Let $\eta_1 > \eta_2 > \cdots > \eta_n$ be the weights of τ. Then

(4) $$\lambda_p = p(\eta_1 + \cdots + \eta_n) + \sum_{i \leq p} \eta_i,$$

(5)
$$\lambda_q^* = -q(\eta_1 + \cdots + \eta_n) - \sum_{i \geq n+1-q} \eta_i.$$

The weights of $F_{\lambda_q^*}$ are of the form

(6) $\xi = -q(\eta_1 + \cdots + \eta_n) - \eta_{j_1} - \cdots - \eta_{j_q}, \qquad 1 \leq j_1 < \cdots < j_q \leq n,$

and each weight has multiplicity 1.

Using (4), (5), (6) and (3), we find that if $p + q \leq n$, then

(7) $$\Lambda^{p,q} = F_{\lambda_p} \otimes F_{\lambda_q^*} = \sum_{j=0}^{\min(p,q)} m_{p,q;j} F_{p-j,q-j}, \quad \text{and} \quad m_{p,q;j} \leq 1.$$

We assume that (1) is true for $0 \leq p + q \leq j - 1$, $p \geq 0$, $q \geq 0$, $p + q \leq n - 1$. If $j \leq n - 1$, then (7) combined with the inductive hypothesis implies (1) for $p + q = j$.

Since we have already observed that (1) is true for $\Lambda^{p,0}$ and $\Lambda^{0,q}$, the proof is complete.

4.10.   If $D \in \Pi^p(G) \cap \mathcal{E}_d(G)$, then

(1) $$H^q(D) = \operatorname{Hom}_K(\Lambda^q(\mathfrak{g}_c/\mathfrak{k}_c), D).$$

Since

(2) $$H^q(D) = \begin{cases} 0, & q \neq n, \\ \mathbf{C}, & q = n, \end{cases}$$

we see that the only $F_{p,q}$ such that $\operatorname{Hom}_K(F_{p,q}, D) \neq (0)$ are of the form $F_{i,n-i}$.

From the results used in II, 5.3, we see that if $\operatorname{Hom}_K(F_{i,n-i}, D) \neq (0)$, then $F_{i,n-i}$ is the lowest K-type of D. We may thus label D_i by the unique $F_{i,n-i}$ it contains. That is, D_i, $0 \leq i \leq n$, is determined by

(3) $$\operatorname{Hom}_K(F_{i,n-i}, D_i) \neq (0).$$

Let $\delta_{ij} = \delta_{s_{ij}}$ for $i, j \geq 0$, $i + j \leq n - 1$ (see 3.3). Using the formula for $2s_{ij}\rho$ [**27**, p. 251] and the classical branching rules for $\mathbf{U}(n)$ to $\mathbf{U}(n-1)$ (cf. [**4**]), it is an exercise to prove that if $\tau_{p,q}$ is the representation of K on $F_{p,q}$, then

(4) $$\tau_{i,n-i}\big|_{0M} = \begin{cases} \delta_{0,n-1}, & i = 0, \\ \delta_{i-1,n-i-1} \oplus \delta_{i-1,n-i} \oplus \delta_{i,n-i-1}, & 0 < i < n, \\ \delta_{n-1,0}, & i = n. \end{cases}$$

If $p + q \leq n - 1$, $p \geq 0$, $q \geq 0$, then

(5) $$\tau_{p,q}\big|_{0M} = \delta_{p,q} \oplus \delta_{p-1,q} \oplus \delta_{p,q-1} \oplus \delta_{p-1,q-1}.$$

Here $\delta_{-1,p} = \delta_{p,-1} = 0$ for $p \in \mathbf{Z}$.

We are now ready to prove the analogue of Theorem 4.5 for $\mathbf{SU}(n,1)$.

4.11.   THEOREM.   *Let $G = \mathbf{SU}(n,1)$.*

(1) *If V is an irreducible $(\mathfrak{g}, K)$-module such that $H^*(V) \neq (0)$, then V is one of the J_{ij} or the D_i.*

(2) $$H^q(D_i) = \begin{cases} 0 & \text{if } q \neq n, \\ \mathbf{C} & \text{if } q = n; \end{cases}$$

(3)
$$H^q(J_{ij}) = \begin{cases} \mathbf{C}, & \text{if } q = i + j + 2l \ (0 \le l \le n - i - j), \\ 0, & \text{otherwise.} \end{cases}$$

As above, (1), (2), and (3) for $i + j = n - 1$ have already been proven. Frobenius reciprocity implies

(4)
$$\begin{aligned} I_{p,q} \text{ contains } F_{p,q} \oplus F_{p+1,q} \oplus F_{p,q+1} \\ \text{and no other } F_{ij} \text{ if } p + q = n - 1, \end{aligned}$$

(5)
$$\begin{aligned} \text{If } p + q < n - 1, \text{ then } I_{p,q} \text{ contains} \\ F_{p,q} \oplus F_{p+1,q} \oplus F_{p,q+1} \oplus F_{p+1,q+1} \quad \text{and no other } F_{ij}. \end{aligned}$$

Let $U_{ij} \subset I_{ij}$ be the $(\mathfrak{g}, K)$-module such that the following sequence is exact:

(6)
$$0 \to U_{ij} \to I_{ij} \to J_{ij} \to 0.$$

As usual, if $i + j \le n - 1$

(7)
$$H^q(I_{ij}) = \begin{cases} \mathbf{C}, & \text{if } q = 2n - i - j, 2n - i - j + 1, \\ 0, & \text{otherwise.} \end{cases}$$

(7), combined with (6), implies

(8)
$$\begin{aligned} \text{If } q \le n - 1 \text{ and } i + j \le n - 2, \text{ then} \\ H^q(J_{ij}) = H^{q+1}(U_{ij}). \end{aligned}$$

We prove by downward induction on $i + j$ the following assertion:

(9)
$$J_{i,j} \text{ contains } F_{i,j} \text{ and no other } F_{p,q}.$$

To start the induction we must prove (9) for $J_{i,n-1-i}$, $0 \le i \le n - 1$.

(4) combined with the proof of 3.8(3) and 4.10(3) implies that $I_{i,n-1-i}$ contains D_i and D_{i+1} for $i = 0, \ldots, n-1$. Thus $U_{i,n-1-i}$ has a composition series consisting of D_i and D_{i+1}. (4) now implies that $J_{i,n-i-1}$ contains $F_{i,n-i-1}$ and no other $F_{p,q}$. Thus the first step in the induction has been proven.

To continue the induction we need a weak form of a result of Kraljević [75]:

(10)　　　　$$\text{If } i + j \le n - 2, \text{ then } I_{ij} \text{ has four non-zero subquotients.}$$

We now look at the case $i + j = n - 2$. That is, $i, n - 2 - i$, $i = 0, \ldots, n - 2$. The only possible subquotients of $I_{i,n-2-i}$ are $J_{i,n-2-i}$, $J_{i,n-1-i}$, $J_{i+1,n-2-i}$ and D_{i+1}, by (4), (5), and (9) for $i + j = n - 1$. Hence (10) implies that they all occur. Since $H^{n-2}(J_{i,n-2-i}) = \mathbf{C}$ by 3.8(2), we see that

$$\operatorname{Hom}_K(\Lambda^{n-2}(\mathfrak{g}/\mathfrak{k}), J_{i,n-2-i}) \ne 0.$$

The only constituent of $\Lambda^{n-2}(\mathfrak{g}/\mathfrak{k})$ contained in $I_{i,n-2-i}$ is $F_{i,n-2-i}$. Thus $F_{i,n-2-i}$ is contained in $J_{i,n-2-i}$. Since $F_{i,j}$ is contained in $J_{i,j}$ for $i + j = n - 1$ and $F_{i,n-i} \in D_i$ for $i = 0, \ldots, n$, we see that (9) is true for $i + j = n - 2$.

Assume (9) for $0 \le p \le i + j \le n - 2$. Then if $0 \le i \le p$, (4), (5) and (9) for $i + j > p$ imply that the only possible constituents of $I_{i,p-i}$ are $J_{i,p+1-i}$, $J_{i+1,p-i}$ and $J_{i+1,p+1-i}$. Since $H^p(J_{i,p-i}) = \mathbf{C}$, the argument above proves (9) for p. Thus (9) is true.

Now (9) combined with 4.9 implies that if $i + j \le n - 2$, then

(11)　　　$$\dim \operatorname{Hom}_K(\Lambda^q(\mathfrak{g}/\mathfrak{k}), J_{i,j}) = \begin{cases} 1, & \text{if } q = i + j + 2l, \ 0 \le l \le n - i - j, \\ 0, & \text{otherwise.} \end{cases}$$

Since $H^q(J_{ij})$ is the cohomology of the complex $\text{Hom}_K(\Lambda^q(\mathfrak{g}/\mathfrak{k}), J_{ij})$, the theorem now follows.

4.12. THEOREM. (1) *Let* $G = \mathbf{SO}(n,1)$. *Then the representations* J_i, $i \le [n/2] - 1$, *are unitary.*

(2) (Kraljević [**75**]). *The representations* J_{ij}, $i + j \le n - 1$, *are unitary.*

This theorem can be derived from results in Knapp-Stein [**71**]. We note that in case (1), [**86**] shows the existence of cocompact discrete subgroups $\Gamma \subset \mathbf{SO}(n,1)^0$ such that $H^q(\Gamma; \mathbf{C}) \ne 0$ for $q = 0, 1, \ldots, n$. In view of 4.2, 4.5 and VII, 3.2, this also proves (1).

5. The Vogan-Zuckerman theorem

The purpose of this section is to describe the Vogan-Zuckerman theorem that gives a complete classification of irreducible unitary $(\mathfrak{g}, K)$ modules V such that there exists a finite dimensional $(\mathfrak{g}, K)$-module with $\text{Ext}^*_{\mathfrak{g}, K}(F, V) \ne 0$. Since they also calculate the corresponding Ext groups, their theorem thereby calculates $\text{Ext}^*_{\mathfrak{g}, K}(F, V)$ for all finite dimensional F and all irreducible unitary V. In particular, one sees that the vanishing theorem in II.10 is best possible.

5.1. We will use the notation of II.10. In addition we assume that $\mathfrak{g}$ is simple over $\mathbf{R}$. We now introduce the (substantial) additional notation that is necessary to state the main theorems. Let $\mathfrak{q}$ be a θ-stable parabolic subalgebra of $\mathfrak{g}_c$. Fix $\mathfrak{b}_k$, a Borel subalgebra of $\mathfrak{k}_c$. Up to the action of $\text{Ad}(K)$ we may assume that $\mathfrak{q} \cap \mathfrak{k}_c \supset \mathfrak{b}_k$. Fix, $\mathfrak{t} \subset \mathfrak{b}_k$, a Cartan subalgebra of $\mathfrak{k}_c$. Let $\mathfrak{h}$ be the centralizer of $\mathfrak{t}$ in $\mathfrak{g}_c$. Then $\mathfrak{h}$ is a Cartan subalgebra of $\mathfrak{g}_c$. Let $\mathfrak{m}(\mathfrak{q}) = \mathfrak{q}/\mathfrak{u}(\mathfrak{q})$, and let $p: \mathfrak{q} \to \mathfrak{m}(\mathfrak{q})$ be the canonical projection. Since $\theta\mathfrak{u}(\mathfrak{q}) = \mathfrak{u}(\mathfrak{q})$, the map θ induces an involutive automorphism of $\mathfrak{m}(\mathfrak{q})$. The projection p restricted to $\mathfrak{h}$ defines an isomorphism onto its image. We will identify $\mathfrak{h}$ with $p(\mathfrak{h})$. We set $K_M = \{k \in K \mid \text{Ad}(k)\mathfrak{q} \subset \mathfrak{q}\}$. Then p is injective on $\text{Lie}(K_M)$. We also identify $p(\text{Lie}(K_M))$ with $\text{Lie}(K_M)$. If $k \in K_M$, then $\text{Ad}(k)\mathfrak{u}(\mathfrak{q}) = \mathfrak{u}(\mathfrak{q})$, and so K_M acts as a group of automorphisms on $\mathfrak{m}(\mathfrak{q})$. We may thus speak of $(\mathfrak{m}(\mathfrak{q}), K_M)$-modules.

Let F be a finite dimensional irreducible $(\mathfrak{g}, K)$-module. We say that F is θ-*compatible* if the highest weight of F with respect to a θ-stable Borel subalgebra is fixed by θ. Let $\mathcal{P}_0(F) = \{\mathfrak{q} \in \mathcal{P}(F)|_{\mathfrak{q}} \supset \mathfrak{b}_k\}$. Let $\mathfrak{q} \in \mathcal{P}_0(F)$. Let s_K denote the longest element of the Weyl group of K with respect to $\mathfrak{t}$ corresponding to the choice of $\mathfrak{b}_k$. Let s_M be the longest element of the Weyl group of K_M with respect to $\mathfrak{t}$ corresponding to $\mathfrak{b}_k \cap \text{Lie}(K_M)_c$. Let $s_0 = s_M s_K$, and fix $k \in K$ such that $\text{Ad}(k)|_{\mathfrak{t}} = s_0$. Set $\mathfrak{q}' = \text{Ad}(k)^{-1}\mathfrak{q}$. Let $q = \dim \mathfrak{u}(\mathfrak{q})$. Put K'_M equal to $k^{-1}K_M k$, and let Z denote the 1-dimensional $(\mathfrak{q}', K'_M)$-module that is given by $F^{\mathfrak{u}(\mathfrak{q})} \otimes \Lambda^q \mathfrak{u}(\mathfrak{q})$ twisted by k^{-1}. That is, if σ is the action of $\mathfrak{q}$, then the twisted action by k^{-1} is $\sigma \circ k$. Set $N_{\mathfrak{q}}(F) = U(\mathfrak{g}_c) \otimes_{U(\mathfrak{q}')} Z$, which we look upon as an element of $\mathcal{C}(\mathfrak{g}, K'_M)$ in the usual way. Let $n = \frac{1}{2} \dim(K/K_M)$. We will also use the notation in 1.8.

5.2. THEOREM ([**148**]; cf. [**151**], 6.10.3, 9.5.9). (1) $R^i\Gamma^{\mathfrak{g}, K}_{\mathfrak{g}, K'_M}(N_{\mathfrak{q}}(F)) = 0$ for $i \ne n$.

(2) *If* F *is* θ-*compatible, then* $R^n\Gamma^{\mathfrak{g}, K}_{\mathfrak{g}, K'_M}(N_{\mathfrak{q}}(F))$ *is an irreducible* $(\mathfrak{g}, K)$-*module, to be denoted* $A_{\mathfrak{q}}(F)$, *that admits a positive non-degenerate inner product with respect to which it is unitary.*

We will only give a brief description of a proof of this theorem. The first step in the proof of (1) is to show that $N_{\mathfrak{q}}(F)$ has a $(\mathfrak{k}, K_M)$-module filtration

$$0 = N^0 \subset N^1 \subset N^2 \subset \cdots$$

with $\bigcup_i N^i = N_{\mathfrak{q}}(F)$ and $N^i/N^{i-1} \cong U(\mathfrak{k}) \otimes_{U(\mathfrak{q}_k)} W_i$ $(i \geq 0)$, where W_i is an irreducible finite dimensional $(\mathfrak{q}_k, K_M)$-module (cf. [**151**], 6.4.4). Thus $N_{\mathfrak{q}}(F) \otimes \mathcal{H}(K)$ has the $(\mathfrak{k}, K_M)$-module filtration

$$0 = N^0 \otimes \mathcal{H}(K) \subset N^1 \otimes \mathcal{H}(K) \subset N^2 \otimes \mathcal{H}(K) \subset \cdots$$

with

$$(N^i \otimes \mathcal{H}(K))/(N^{i-1} \otimes \mathcal{H}(K)) \cong (U(\mathfrak{k}) \otimes_{U(\mathfrak{q}_k)} W_i) \otimes \mathcal{H}(K)$$
$$\cong U(\mathfrak{k}) \otimes_{U(\mathfrak{q}_k)} (W_i \otimes H(K)).$$

The last equation follows from I, 8.5 (ii). One then observes the general fact that if $\mathfrak{k} \supset \mathfrak{l} \supset \mathfrak{k}_M$ and if T is an $(\mathfrak{l}, K_M)$-module, then

$$H^i(\mathfrak{k}, K_M; U(\mathfrak{k}) \otimes_{U(\mathfrak{l})} T) = 0 \qquad (i \leq \dim \mathfrak{k}/\mathfrak{l})$$

(cf. [**151**], 6.A.1.5). Using our definition of the Zuckerman functors, this vanishing assertion, and the long exact sequence of cohomology applied to this filtration, we have

$$R^i \Gamma_{\mathfrak{g}, K'_M}^{\mathfrak{g}, K}(N_{\mathfrak{q}}(F)) = 0, \qquad i < n.$$

The reverse inequality is less formal and involves concepts to be used in the proof of 2). Let $\check{Z}$ denote the conjugate dual of Z in $\mathcal{C}(\mathfrak{m}(\mathfrak{q}'), K'_M)$. We extend this module to a $\mathfrak{q}'$-module by letting $\mathfrak{u}(\mathfrak{q}')$ act by 0. Set $V = U(\mathfrak{g}_c) \otimes_{U(\mathfrak{q}')} \check{Z}$. We now describe a general method of defining a sesquilinear pairing between modules of the form of $N_{\mathfrak{q}}(F)$ and V. Applying the Poincaré-Birkhoff-Witt theorem, we see that

$$U(\mathfrak{g}_c) = U(\mathfrak{m}(\mathfrak{q}')) \oplus (\overline{\mathfrak{u}(\mathfrak{q}')}U(\mathfrak{g}) + U(\mathfrak{g})\mathfrak{u}(\mathfrak{q}')),$$

with the bar indicating complex conjugation in $\mathfrak{g}_c$ with respect to $\mathfrak{g}$. Let p denote the corresponding linear projection of $U(\mathfrak{g}_c)$ onto $U(\mathfrak{m}(\mathfrak{q}'))$. Set

(1) $\langle g \otimes z, h \otimes \check{z} \rangle = \langle p(h^*g)z, \check{z} \rangle$ $(g, h \in U(\mathfrak{g}_c), z \in Z, \check{z} \in \check{Z}).$

Here $g \mapsto g^*$ is the anti-involution of $U(\mathfrak{g}_c)$ defined by

$$1^* = 1,$$

$$(uv)^* = v^* u^* \qquad (u, v \in U(\mathfrak{g}_c)),$$

$$X^* = -\overline{X} \qquad (X \in \mathfrak{g}_c).$$

It is easily seen that the sesquilinear pairing defined in (1) pushes down to $N_{\mathfrak{q}}(F) \times V$. The only non-formal part of the proof is the assertion that it is non-degenerate (cf. [**151**], 6.4.5, 6.4.6). Now Theorem I, 8.11, combined with the argument above applied to V, implies the vanishing assertion in 1) for $i > n$.

If F is θ-compatible, then Z is equivalent with $\check{Z}$. Thus $N_{\mathfrak{q}}(F)$ is endowed with a non-degenerate sesquilinear form. Applying I, 8.11, we therefore have a non-degenerate sesquilinear form on $R^n \Gamma_{\mathfrak{g}, K'_M}^{\mathfrak{g}, K}(N_{\mathfrak{q}}(F))$. If we multiply this form by an appropriate power of $\sqrt{-1}$, we may assume that it is Hermitian. To prove 2) one must show that the product is definite. The argument in [**151**], 6.7 proving this relies heavily on Vogan's idea of *signature character*. We refer the reader to that reference and to the treatment in [**140**].

5.3. THEOREM. 1) *If F is not θ-compatible, then $H^*(\mathfrak{g}, K; V \otimes F^*) = 0$.*

2) ([**149**]; cf. [**151**], 9.6.6). *If F is θ-compatible and V is an irreducible unitary $(\mathfrak{g}, K)$-module such that $H^*(\mathfrak{g}, K; V \otimes F^*) \neq 0$, then there exists $\mathfrak{q} \in \mathcal{P}_0(F)$ such that V is $(\mathfrak{g}, K)$-equivalent with $A_\mathfrak{q}(F)$. Furthermore,*

$$H^*(\mathfrak{g}, K; A_\mathfrak{q}(F) \otimes F^*) = H^*(m(\mathfrak{q}), K_M; C)[-r]$$

with $r = \dim \mathfrak{u}(\mathfrak{q}) \cap \mathfrak{p}$.

The first assertion is a restatement of II, 6.12 1). The second is the theorem of Vogan and Zuckerman. The proof is quite complicated (cf. [**151**], 9.7). However, to prove that II, 10.1 is best possible it is enough to observe the following special case of the formula in 2).

5.4. If $h \in \mathfrak{t}$, then set $\rho_n(\mathfrak{q})(h) = \frac{1}{2} \text{tr}(\text{ad}(h)|_{\mathfrak{u}(\mathfrak{q}) \cap \mathfrak{p}})$. Let V_λ denote the $(\mathfrak{k}, K)$-module with highest weight λ with respect to the choice of $\mathfrak{b}_k$. Using II, 8.8, it is not hard to show that

$$\dim \text{Hom}_K(V_{\lambda + 2\rho_n(\mathfrak{q})}, A_\mathfrak{q}(F)) = 1.$$

This, combined with II, 3.1 (b) and the argument in II, 7.2, implies that

$$\dim H^r(\mathfrak{g}, K; A_\mathfrak{q}(F) \otimes F^*) \geq 1.$$

5.5. REMARK. If F is a finite dimensional $(\mathfrak{g}, K)$-module and if $\mathfrak{q} \in \mathcal{P}(F)$, then, in the notation of §0,

$$\Phi_\rho^{\rho + \lambda \, F}(A_\mathfrak{q}(\mathbf{C})) = A_\mathfrak{q}(F)$$

(cf. [**151**], 6.6.3; [**140**], VIII, 5).

CHAPTER VII

Cohomology of Discrete Subgroups and Lie Algebra Cohomology

In this chapter, we consider the cohomology spaces $H^*(\Gamma; E)$ of a discrete subgroup Γ of a Lie group G with finitely many connected components, with coefficients in a finite dimensional complex Γ-module (ρ, E), and we express them in terms of relative Lie algebra cohomology. This is first done in general in §2 and yields an isomorphism

$$(1) \qquad\qquad H^*(\Gamma; E) = H^*(\mathfrak{g}, K; I^\infty(E)),$$

where K is a maximal compact subgroup of G and

$$(2) \quad I^\infty(E) = I_\Gamma^G(E) = \{f \in C^\infty(G, E) \mid f(\gamma \cdot g) = \rho(\gamma) \cdot f(g) \ (\gamma \in \Gamma; g \in G)\}$$

(see 2.5). In the most important case for us, where (ρ, E) is in fact a G-module, this takes the form

$$(3) \qquad\qquad H^*(\Gamma; E) = H^*(\mathfrak{g}, K; C^\infty(\Gamma \backslash G) \otimes E)$$

(see 2.7). From §3 on, we assume Γ to be cocompact, and E to be either a unitary Γ-module (§§3, 4) or a G-module (§§5, 6). The right-hand side of (1) or (3) then decomposes into a finite direct sum of cohomology algebras of the type considered in the earlier chapters (see 3.2, 3.4, 5.2, 6.1). When G^0 is semi-simple with finite center, the results of II, V, VI translate into properties of $H^*(\Gamma; E)$ which are discussed in §§4, 6.

1. Manifolds

In this section we review some familiar material on manifolds, mainly to fix our notation. For more details, see for instance [**112**].

1.1. Unless otherwise stated, manifolds are C^∞. Smooth is used synonymously with C^∞. Let M be a manifold, $L = \mathbf{R}$ or $\mathbf{C}$, and E a finite dimensional vector space over L. Then $T(M)_m$ is the tangent space at $m \in M$, $C^\infty(M; E)$ the space of smooth functions with values in E, $A^q(M; E)$ the space of smooth E-valued differential q-forms on M ($q = 0, 1, \cdots$), $A_1(M)$ the space of smooth vector fields on M, and $A_1(M; L) = A_1(M) \otimes_{\mathbf{R}} L$. If $E = L$ and L is clear from the context, it will often be omitted from the notation. We have

$$(1) \qquad C^\infty(M; E) = A^0(M; E), \qquad A^q(M; E) = A^q(M; L) \otimes_L E.$$

Let $\omega \in A^q(M; E)$. It associates to each $m \in M$ an element of $\mathrm{Hom}(\Lambda^q T(M)_m, E)$. The value of ω on a q-vector y at m will sometimes be denoted $\omega(m; y)$. Often,

ω will be viewed as a $C^\infty(M; L)$-multilinear alternating map on $A_1(M; L)$, with values in $C^\infty(M; E)$. If $x \in A_1(M, L)$, the interior product $i_x \omega$ is defined by

$$(2) \qquad i_x\omega(x_1, \ldots, x_{q-1}) = \omega(x, x_1, \ldots, x_{q-1}) \qquad (x_1, \ldots, x_{q-1} \in A_1(M; L)).$$

The exterior differential $d \colon A^q(M; E) \to A^{q+1}(M; E)$ is given by

$$(3) \qquad \begin{aligned} d\omega(x_0, \ldots, x_q) &= \sum (-1)^i y_i \cdot \omega(x_0, \ldots, \widehat{x}_i, \ldots, x_q) \\ &\quad + \sum_{i<j} (-1)^{i+j} \omega([x_i, x_j], x_0, \ldots, \widehat{x}_i, \ldots, \widehat{x}_j, \ldots, x_q), \end{aligned}$$

where $[\ ,\ ]$ refers to the bracket of vector fields, and $\widehat{}$ means omission of the corresponding argument.

1.2. If N is a manifold and $\pi \colon M \to N$ a smooth map, then $d\pi_m \colon T(M)_m \to T(N)_{\pi(m)}$ is the differential of π at m. The map π induces a homomorphism ${}^t\pi \colon \omega \mapsto \omega \circ \pi$ of $A^p(N; E)$ into $A^p(M; E)$, given by

$$(1) \qquad (\omega \circ \pi)(m, y) = \omega(\pi(m), d\pi_m(y)) \qquad (m \in M;\ y \in \Lambda^q T(M)_m).$$

1.3. Let $\widetilde{E}$ be the local system of coefficients on M associated to a representation on E of the fundamental group of M. Then, similarly, $C^\infty(M; \widetilde{E})$ denotes the space of $\widetilde{E}$-valued C^∞-functions on M, i.e. of smooth cross-sections of $\widetilde{E}$, and $A^q(M; \widetilde{E})$ the space of smooth $\widetilde{E}$-valued q-forms on M. Since the transition functions of $\widetilde{E}$ are locally constant, the exterior differentiation still makes sense on $A^q(M; \widetilde{E})$ and 1.1(3) remains valid.

1.4. Lie derivative. Let $x \in A_1(M; L)$. Then θ_x denotes the Lie derivative in the direction x [**112**, 2.24]. In particular,

$$(1) \qquad \theta_x f = x \cdot f \qquad (f \in C^\infty(M; E)),$$

$$(2) \qquad \theta_x y = [x, y] \qquad (y \in A_1(M)),$$

$$(3) \qquad \begin{aligned} (\theta_x \omega)(x_1, \ldots, x_q) &= \theta_x(\omega(x_1, \ldots, x_q)) - \sum_i \omega(x_1, \ldots, [x, x_i], \ldots, x_q) \\ &\qquad (x, x_1, \ldots, x_q \in A_1(M);\ \omega \in A^q(M; E)). \end{aligned}$$

The vector field x defines (locally) a one-parameter group of transformations $\{\phi_t\}$ (t in a neighborhood of the origin in $\mathbf{R}$), and we have

$$(4) \qquad \theta_x f(m) = \frac{d}{dt} f(\phi_t(m))\big|_{t=0},$$

$$(5) \qquad \theta_x(y)(m) = \frac{d}{dt} d\phi_{-t}(y_{\phi_t(m)})\big|_{t=0},$$

$$(6) \qquad (\theta_x \omega)(m, y) = \frac{d}{dt} \omega(\phi_t(m), d\phi_t(y(m)))\big|_{t=0}.$$

The operators d, i_x, θ_x are related by

$$(7) \qquad d \cdot i_x + i_x \cdot d = \theta_x.$$

1.5. Let G be a group. Assume that it operates by diffeomorphisms on M and via a linear representation ρ on E. Then we let G operate on $A^q(M; E)$ by

$$(g \circ \omega)(m, x_1, \ldots, x_p) = \rho(g)(\omega(g^{-1}m, g^{-1}x_1, \ldots, g^{-1}x_q))$$

$$(m \in M; \ x_1, \ldots, x_q \in T(M)_m; \ g \in G).$$

The space of invariant q-forms is denoted $A^q(M; E)^G$. Thus

$$\omega \in A^q(M; E)^G \Leftrightarrow \rho(g) \circ \omega = \omega(g \cdot x_1, \ldots, g \cdot x_q)$$

$$(g \in G; \ x_1, \ldots, x_q \in A_1(M)).$$

1.6. We now assume $M = G$ to be a Lie group. We let l_g (resp. r_g) denote left (resp. right) translation by g. In particular,

(1)
$$l_g f(x) = f(g^{-1} \cdot x), \qquad r_g f(x) = f(x \cdot g)$$

$$(f \in C^\infty(M; E); \ g, x \in G).$$

If H is a closed subgroup, then G/H (resp. $H\backslash G$) is the space of left (resp. right) cosets $x \cdot H$ (resp. $H \cdot x$) ($x \in G$). By definition, the Lie algebra $\mathfrak{g}$ of G is the Lie algebra of left-invariant vector fields. As usual, the tangent space $T(G)_1$ is identified to $\mathfrak{g}$ by assigning to $x \in T(G)_1$ the unique left-invariant vector field which is equal to x at 1. The one-parameter group $\{\phi_t\}$ associated to $x \in \mathfrak{g}$ is the group of right translations by the elements e^{tx} ($t \in \mathbf{R}$). In particular,

$$xf(g) = \frac{d}{dt} f(g \cdot e^{tx})\big|_{t=0}.$$

2. Discrete subgroups

From now on, G is a Lie group with finitely many connected components, G^0 its identity component, K a maximal compact subgroup of G, $X = G/K$, Γ a discrete subgroup of G, and (ρ, E) a finite dimensional real or complex linear representation of Γ.

We recall that the maximal compact subgroups of G are conjugate and that X is homeomorphic to Euclidean space. If G is connected, this is the well-known Cartan-Iwawasa-Malcev theorem. The extension to groups with finite component group is due to G. D. Mostow [**87**].

2.1. Let M be any compact subgroup of G. Then Γ acts properly on G/M by left translations (i.e., for every compact set C, $\{\gamma \in \Gamma \mid \gamma C \cap C \neq \varnothing\}$ is finite). If Γ has no torsion, then it acts freely (no $\gamma \neq 1$ has a fixed point). Conversely, if Γ acts freely, then its elements of finite order act trivially, hence are contained in the intersection of all the conjugates of K. If G is connected, these elements belong to the center of G.

2.2. THEOREM. *The space $H^*(\Gamma; E)$ is canonically isomorphic to*

$$H^*(A(X; E)^\Gamma).$$

This is well known. However, since it is basic for us, we recall the proof. Assume first that Γ acts freely. Then $\Gamma \backslash X$ is a smooth manifold and, since X is contractible, it is also an Eilenberg-MacLane space $K(\Gamma, 1)$. Then we have

$$(1) \qquad H^*(\Gamma; E) = H^*(\Gamma \backslash X; \widetilde{E}),$$

where $\widetilde{E}$ is the local system on $\Gamma \backslash X$ defined by (ρ, E). Let $\pi: X \to \Gamma \backslash X$ be the canonical projection. Then it is immediate that $\omega \mapsto \omega \circ \pi$ defines an isomorphism of $A(\Gamma \backslash X; \widetilde{E})$ with $A(X; E)^\Gamma$. Our assertion in this case follows then from de Rham's theorem (with a locally constant sheaf of coefficients).

Assume now that Γ has a torsion-free normal subgroup Γ' of finite index. Then Γ / Γ' acts on $H^*(\Gamma'; E)$, and we have

$$(2) \qquad H^*(\Gamma; E) = (H^*(\Gamma'; E))^{\Gamma/\Gamma'},$$

as follows e.g. from the Hochschild-Serre spectral sequence. On the other hand,

$$(3) \qquad A(X; E)^\Gamma = (A(X; E)^{\Gamma'})^{\Gamma/\Gamma'}.$$

Since taking invariants under a finite group is an exact functor in characteristic zero, this gives

$$(4) \qquad H^*(A(X; E)^\Gamma) = H^*(A(X; E)^{\Gamma'})^{\Gamma/\Gamma'},$$

and (2), (4) provide a reduction to the first case considered.

This suffices for our needs. To be complete, we treat the general case too. For $q \in \mathbf{N}$, let $\mathfrak{F}^q$ be the sheaf on $\Gamma \backslash X$ associated to the presheaf $U \mapsto A^q(\pi^{-1}(U); E)^\Gamma$ (U open in $\Gamma \backslash X$). Since the isotropy groups of Γ on X are finite, it follows by a simple averaging process from the Poincaré lemma on X that $\{\mathfrak{F}^q\}$ is a resolution of the constant sheaf $(\Gamma \backslash X) \times E$ on $\Gamma \backslash X$. Using a partition of unity, one sees moreover that $\mathfrak{F}^q$ is a fine sheaf. Since $A^q(X; E)^\Gamma$ is just the space of global cross-sections of $\mathfrak{F}^q$, this gives

$$(5) \qquad H^*(\Gamma \backslash X; \widetilde{E}) = H^*(A(X; E)^\Gamma).$$

On the other hand, since the isotropy groups Γ_x ($x \in X$) of Γ on X are finite, the groups $H^i(\Gamma_x; E) = 0$ are all zero for $i > 0$. By general principles, [**47**, p. 204], (1) is still valid, and our assertion follows from (1) and (5).

2.3. The quotient $G \times_\Gamma E$ of $G \times E$ by the equivalence relation $(g, e) \sim (\gamma \cdot g, \rho(\gamma) \cdot e)$ ($g \in G$; $e \in E$; $\gamma \in \Gamma$) is the total space of a vector bundle $\widetilde{E}$ over $\Gamma \backslash G$ with typical fiber E and structure group Γ. We let $I^\infty(E) = C^\infty(G, E)^\Gamma$ be the space of its smooth cross-sections, i.e.

$$(1) \qquad I^\infty(E) = \{f \in C^\infty(G; E) \mid f(\gamma \cdot g) = \rho(\gamma) \cdot f(g) \ (\gamma \in \Gamma; \ g \in G)\}.$$

Otherwise said, $I^\infty(E)$ is the space of the representation $I_\Gamma^G(E)$ induced from (ρ, E) to G, in the C^∞ sense.

Assume now that (ρ, E) is the restriction to Γ of a representation of G, still denoted in the same way. Then the map $f \mapsto F$ of $C^\infty(G; E)$ into itself, given by

$$(2) \qquad F(g) = \rho(g)^{-1} \cdot f(g) \qquad (g \in G; \ f \in C^\infty(G; E)),$$

is immediately seen to yield an isomorphism of G-modules

$$(3) \qquad I^\infty(E) \overset{\sim}{\to} C^\infty(\Gamma \backslash G; L) \otimes_L E,$$

where the G-module structure on the right-hand side is the tensor product of the right regular representation on $C^\infty(\Gamma\backslash G; L)$ and of ρ.

2.4. For $g \in G$, the left translation by g^{-1} provides a canonical isomorphism of $T(G)_g$ with $\mathfrak{g} = T(G)_1$, whence an identification

$$(1) \qquad \iota\colon A^q(G; E) = \operatorname{Hom}(\Lambda^q\mathfrak{g}, C^\infty(G; E)) = C^q(\mathfrak{g}; C^\infty(G; E)) \qquad (q \in \mathbf{N}).$$

Let $\omega \in A^q(G; E)^\Gamma$. Then, for $y \in \Lambda^q\mathfrak{g}$, we have

$$(2) \qquad\qquad\qquad \omega(\gamma \cdot g, y) = \rho(\gamma) \cdot \omega(g, y);$$

hence ω is identified to an element of $\operatorname{Hom}(\Lambda^q\mathfrak{g}, I^\infty(E))$. The converse is clear, so that we get an isomorphism, also to be denoted ι,

$$(3) \qquad\qquad\qquad \iota\colon A^q(G; E)^\Gamma \xrightarrow{\sim} C^q(\mathfrak{g}; I^\infty(E)).$$

It follows from 1.1(3) and I, §1 that the isomorphisms ι commute with the differentials, hence give rise to an isomorphism

$$(4) \qquad\qquad\qquad \iota^*\colon H^*(A(G; E)^\Gamma) \xrightarrow{\sim} H^*(\mathfrak{g}; I^\infty(E)).$$

Let $\widetilde{E}$ be the local system on $\Gamma\backslash G$ defined by (ρ, E). Then

$$(5) \qquad\qquad\qquad A(G; E)^\Gamma \cong A(\Gamma\backslash G; \widetilde{E}),$$

so that the left-hand side of (4) can be viewed as the cohomology of $\Gamma\backslash G$ with coefficients in the locally constant sheaf defined by $\widetilde{E}$.

If now (ρ, E) comes from a representation of G, then, by 2.3, $\omega \mapsto \omega^0$, where $\omega^0(g) = \rho(g)^{-1}\omega(g)$ $(g \in G)$, yields an isomorphism

$$(6) \qquad\qquad\qquad A(G; E)^\Gamma \xrightarrow{\sim} C^*(\mathfrak{g}; C^\infty(\Gamma\backslash G; L) \otimes E),$$

whence also

$$(7) \qquad\qquad\qquad H^*(\Gamma\backslash G; \widetilde{E}) \xrightarrow{\sim} H^*(\mathfrak{g}; C^\infty(\Gamma\backslash G; L) \otimes E).$$

We now want to divide by K on the right and relate similarly the cohomology of Γ and relative Lie algebra cohomology.

2.5. PROPOSITION. *Let $\pi\colon G \to X = G/K$ be the canonical projection. Then $^t\pi\colon \omega \mapsto \omega \circ \pi$ induces an isomorphism of graded complexes of $A(X; E)^\Gamma$ onto $C^*(\mathfrak{g}, K; I^\infty(E))$. In particular, $H^*(\Gamma; E)$ is canonically isomorphic to $H^*(\mathfrak{g}, K; I^\infty(E))$.*

The map $^t\pi$ clearly commutes with left translations, hence sends $A(X; E)^\Gamma$ into $A(G; E)^\Gamma$. Let A_0 be its image. Since π is constant along the left K-cosets, A_0 consists of the elements of $A(G; E)^\Gamma$ which are right invariant under K and annihilated by the interior products i_x $(x \in \mathfrak{k})$. It then follows from 2.4 and the definitions that $\iota \circ {}^t\pi$ induces an isomorphism

$$(1) \qquad\qquad\qquad A(X; E)^\Gamma = C^*(\mathfrak{g}, K; I^\infty(E)).$$

Our assertion now follows from 2.2.

2.6. REMARK. If we associate to $e \in E^\Gamma$ the constant function on G equal to e, then we get a map $E^\Gamma \to I^\infty(E)^G$, which is readily seen to be bijective. The inclusion $I^\infty(E)^G \subset I^\infty(E)$ then yields a canonical homomorphism

$$(1) \qquad\qquad j^*: H^*(\mathfrak{g}, K; E^\Gamma) \to H^*(\mathfrak{g}, K; I^\infty(E)).$$

2.7. COROLLARY. *Assume that (ρ, E) extends to a representation of G. Then the map which associates to $\omega \in A(X; E)^\Gamma$ the form $\omega^0 \colon g \mapsto \rho(g^{-1}) \cdot (\omega \circ \pi)(g)$ induces an isomorphism of $A(X; E)^\Gamma$ onto $C^*(\mathfrak{g}, K; C^\infty(\Gamma \backslash G; L) \otimes E)$ and an isomorphism of $H^*(\Gamma; E)$ onto $H^*(\mathfrak{g}, K; C^\infty(\Gamma \backslash G; L) \otimes E)$.*

By 2.3, the map $f \mapsto F$ given by $F(g) = \rho(g^{-1}) \cdot f(g)$ induces a G-equivariant isomorphism of $I^\infty(E)$ onto $C^\infty(\Gamma \backslash G; L) \otimes E$. The corollary then follows from the proposition.

We note further that if we go back to the definitions, we see that the image of $A(X; E)^\Gamma$ in $A(G; E)$ under the map $\omega \mapsto \omega^0$ consists of all the $\eta \in A(G; E)$ which satisfy the three following conditions

$$(1) \qquad \begin{aligned} l_\gamma \circ \eta &= \eta & (\gamma \in \Gamma), \\ r_k \circ \eta &= \rho(k)^{-1} \cdot \eta & (k \in K), \\ i_x \eta &= 0 & (x \in \mathfrak{k}). \end{aligned}$$

2.8. REMARK. We have now an identification $e \mapsto 1 \otimes e$ of E onto the space of constant E-valued functions on $\Gamma \backslash G$, whence a natural homomorphism

$$(1) \qquad\qquad j^*: H^*(\mathfrak{g}, K; E) \to H^*(\mathfrak{g}, K; C^\infty(\Gamma \backslash G; L) \otimes E).$$

2.9. As remarked in [**82**, §3], the case of 2.5, when E is a unitary Γ-module, could be subsumed to that of 2.7 by adding a compact factor to G.

2.7 is in substance proved in [**82**, **83**], although stated there under narrower assumptions.

2.10. Assume now that G is semi-simple and G/K carries an invariant complex structure. We take the notation of II, §4, and let $A^{p,q}$ denote the space of forms of type p, q. Then the isomorphism of 2.5 induces isomorphisms

$$A^{p,q}(X; E)^\Gamma \xrightarrow{\sim} C^{p,q}(\mathfrak{g}, K; E) \qquad (p, q \in \mathbf{N}),$$

and the cohomology of $A^*(X; E)^\Gamma$ is naturally bigraded. We let $H^{p,q}(\Gamma; E)$ be the space of classes represented by cocycles of type (p, q). We have then

$$H^{p,q}(\Gamma; E) = H^{p,q}(\mathfrak{g}, K; I^\infty(E)),$$

$$H^*(\Gamma; E) = \bigoplus_{p,q} H^{p,q}(\Gamma; E).$$

If Γ is torsion free, $\Gamma \backslash X$ is a Kaehlerian manifold and $H^{p,q}(\Gamma; E)$ is the (p, q)-part of the cohomology of X with coefficients in the local system $\widetilde{E}$.

3. Γ cocompact, E a unitary Γ-module

3.1. We keep the notation and assumptions of §2, and moreover assume Γ to be cocompact, and E to be a unitary Γ-module, $L = \mathbf{C}$. The group G is then necessarily unimodular. Let dx denote a Haar measure on G, and the associated measure on $\Gamma\backslash G$, and let $(\ ,\ )_E$ denote the scalar product on E. If $u, v \in I^\infty(E)$, then

$$(1) \qquad (u(\gamma \cdot x), v(\gamma \cdot x))_E = (u(x), v(x))_E \qquad (x \in G;\ \gamma \in \Gamma).$$

Hence this scalar product $f_{u,v}$ is a function on $\Gamma\backslash G$, and we can define a global scalar product (u, v) by integrating it over $\Gamma\backslash G$. The completion of $I^\infty(E)$ under this scalar product is the space $I_2(E)$ of square integrable cross-sections of the bundle $G \times_\Gamma E \to \Gamma\backslash G$ (see 2.3). The space $I_2(E)$ is a unitary G-module with respect to right translations. It follows from (III, 7.9) that $(I_2(E))^\infty = I^\infty(E)$, topologically.

By a theorem of Gelfand and Piatetski-Shapiro [**42**, 1, §2], $I_2(E)$ decomposes into a discrete Hilbert direct sum with finite multiplicities of irreducible G-modules. We can write

$$(2) \qquad I_2(E) = \widetilde{\bigoplus_{\pi \in \widehat{G}}} m(\pi, \Gamma, E) H_\pi,$$

where the $m(\pi, \Gamma, E)$ are natural numbers. (2) and the above imply

$$(3) \qquad I^\infty(E) = \left(\widetilde{\bigoplus_{\pi \in \widehat{G}}} m(\pi, \Gamma, E) H_\pi \right)^\infty.$$

3.2. THEOREM. *We have*

$$(1) \qquad H^*(\Gamma, E) = \bigoplus_{\pi \in \widehat{G}} m(\pi, \Gamma, E) H^*(\mathfrak{g}, K; H_{\pi,0}),$$

where the sum is finite and may be restricted to the $\pi \in \widehat{G}$ which have trivial infinitesimal and central characters. The natural homomorphism $j^* \colon H^*(\mathfrak{g}, K; E^\Gamma) \to H^*(\Gamma, E)$ of 2.6 is injective. Its image is the contribution of the trivial representation π_0 of G to (1), and we have $m(\pi_0, \Gamma, E) = \dim E^\Gamma$.

By 2.5 and 3.1(3), we have

$$(2) \qquad H^*(\Gamma; E) = H^* \left(\mathfrak{g}, K; \left(\widehat{\bigoplus} m(\pi, \Gamma, E) H_\pi \right)^\infty \right).$$

The main point of the proof is to show that the coefficients on the right-hand side can be replaced by the algebraic direct sum of the $m(\pi, \Gamma, E) H_\pi^\infty$. For $\pi \in \widehat{G}$ and $q \in \mathbf{N}$, let

$$(3) \qquad C_\pi^q = C^q(\mathfrak{g}, K; m(\pi, \Gamma, E) H_\pi^\infty), \qquad C_\pi^* = \bigoplus_q C_\pi^q.$$

Let $S \subset \widehat{G}$ be finite; put

$$(4) \qquad C_S^* = \bigoplus_{\pi \in S} C_\pi^*, \qquad C_{S'}^* = C^* \left(\mathfrak{g}, K; \left(\widetilde{\bigoplus_{\pi \in \widehat{G}-S}} m(\pi, \Gamma, E) H_\pi \right)^\infty \right).$$

Then $C^*(\mathfrak{g}, K; I^\infty(E)) = C_S^* \oplus C_{S'}^*$, and hence

$$(5) \qquad H^*(\Gamma, E) = \bigoplus_{\pi \in S} m(\pi, \Gamma, E) H^*(\mathfrak{g}, K; H_\pi^\infty) \oplus H^*(C_{S'}^*).$$

The space $H^*(\Gamma; E)$ is the cohomology of $\Gamma \backslash X$ with coefficients in a local system. The space $\Gamma \backslash X$ is compact and locally contractible (in fact, it may be triangulated); hence

$$(6) \qquad\qquad \dim H^*(\Gamma; E) < \infty.$$

In view of (5), there exists then a finite set $S \subset \widehat{G}$ such that $H^*(C_\pi^*) = 0$ for $\pi \notin S$ with $m(\pi, \Gamma, E) \neq 0$. Assuming S to be so chosen, we want to show that $H^*(C_{S'}^*) = 0$. This will prove (1). The second assertion then follows from 2.6 and I, 5.3, and, in view of 2.6; the third one is clear. Note that $H^*(C_{S'}^*)$ is finite dimensional by (5) and (6). Therefore the vanishing of $H^*(C_{S'}^*)$ follows from the following lemma.

3.3. LEMMA. *Let T be a countable set of irreducible unitary representations (π, H_π) of G, and V the Hilbert direct sum of the H_π's. Assume that $H^*(\mathfrak{g}, K; H_\pi^\infty) = 0$ for all $\pi \in T$ and that $H^*(\mathfrak{g}, K; V)$ is finite dimensional. Then $H^*(\mathfrak{g}, K; V^\infty) = 0$.*

Let $C^*(V) = C^*(\mathfrak{g}, K; V^\infty)$. We view it as a topological direct sum of finitely many copies of V^∞. The map $d \colon C^{q-1}(V^\infty) \to C^q(V^\infty)$ is continuous. This follows directly from its definition (I, §1) and the definition of the topology on V^∞. Therefore

$$(7) \qquad\qquad Z^q = C^q(V^\infty) \cap \ker d$$

is closed. We have an exact sequence

$$(8) \qquad\qquad 0 \to dC^{q-1}(V^\infty) \to Z^q \to H^q(C^*(V^\infty)) \to 0.$$

Since $H^q(C^*(V^\infty))$ is finite dimensional, $dC^{q-1}(V^\infty)$ has finite codimension in Z^q; hence it has a closed complement. Since these spaces are Fréchet spaces, it follows that $dC^{q-1}(V^\infty)$ is closed (see e.g. Cor. 1 on p. 25 in [**23**]).

For $S \subset T$ finite, let pr_S be the projection of V^∞ onto the sum of the H_π^∞ ($\pi \in S$), with kernel $(\widehat{\bigoplus_{\pi \in T-S} H_\pi})^\infty$. It defines a projection, denoted in the same way:

$$(9) \qquad \mathrm{pr}_S \colon C^*(V^\infty) \to \bigoplus_{\pi \in S} C_\pi^*, \quad \text{with kernel } C^*\left(\left(\widehat{\bigoplus_{\pi \in T-S} H_\pi}\right)^\infty\right).$$

It follows from the definition of the topology of $C^*(V^\infty)$ that an element $x \in C^*(V^\infty)$ is the limit of the $\mathrm{pr}_S x$, as S tends to T. Now let $z \in Z^q$. By assumption $\mathrm{pr}_S z$ is a coboundary for every finite S. Since z is the limit of the $\mathrm{pr}_S z$, it is then in the closure of $dC^{q-1}(V^\infty)$. But we have seen that this space is closed. Hence $z \in dC^{q-1}(V^\infty)$, and the lemma is proved.

3.4. COROLLARY. *Assume G to be reductive $(0, 3)$. Then*

$$H^q(\Gamma; E) = \bigoplus_{\pi \in \widehat{G}, \chi_\pi = \chi_0, \omega_\pi = \omega_0} m(\pi, \Gamma, E) \operatorname{Hom}(\Lambda^q(\mathfrak{g}/\mathfrak{k}), H_{\pi,0}) \qquad (q \in \mathbf{N}).$$

Indeed, we have $H_{\pi,K} = H_{\pi,0}$. The corollary then follows from 3.2 and (II, 3.1).

3.5. REMARKS. (1) If G is connected, and $E = \mathbf{C}$, this relation is due to Y. Matsushima [**80**], except for the fact that the sum in [**80**] is over the π which map the Casimir element into zero.

(2) The proof in 3.2 shows that d has closed image in $C^*_{S'}$, hence also in C^*, since each C^*_π is finite dimensional (II, 3.4); it also applies to the dual operator ∂ (II, 2.3). We have therefore a Hodge decomposition

$$C^q = C^q(\mathfrak{g}, K; I^\infty(E)) = \mathcal{H}^q \oplus dC^{q-1} \oplus \partial C^{q+1}, \quad \text{where } \mathcal{H}^q = \ker \partial \cap \ker d \cap C^q,$$

as in the case of an admissible module (II, 3.4), so that $H^q(\Gamma, E)$ may be identified to the space of harmonic q-forms in C^q. In this case, the isomorphisms of 2.4 identify harmonic forms in $C^*(g, K; I^\infty(E))$, in the sense of (II, §2), with $\widetilde{E}$-valued harmonic forms in $\Gamma \backslash X$ (say if Γ acts freely; otherwise one has to invoke the theory of harmonic forms on V-manifolds). Thus the above yields a proof of the Hodge theorem in this case.

(3) Assume (ρ, E) is irreducible. Then the center $C(\Gamma)$ of Γ acts by scalars. If it does not act trivially, then $H^*(\Gamma; E) = 0$. This follows by the argument used in §4 of I, in the category of modules over the group algebra of Γ. If N is a finite central subgroup of Γ which acts trivially on E, then the Hochschild-Serre spectral sequence of $\Gamma \bmod N$ shows that $H^*(\Gamma; E) = H^*(\Gamma/N; E)$.

Now assume G to be connected. Then the formula of 3.4 effectively involves only representations of G which are trivial on the center $C(G)$ of G, i.e., only representations of the adjoint group $\mathrm{Ad}\,\mathfrak{g}$ of G. If $C(G)$ is finite, the computation of $H^*(\Gamma; E)$ may therefore be reduced to the case where G is its own adjoint group.

3.6. The complex case. Assume G/K to be Hermitian symmetric. Then it follows from 2.10 that 3.2, 3.4 and their proofs remain valid if the degree is replaced by the bidegree and $\Lambda^q(\mathfrak{g}/\mathfrak{k})$ by $\Lambda^p\mathfrak{p}_c^+ \otimes \Lambda^q\mathfrak{p}_c^-$.

4. G semi-simple, Γ cocompact, E a unitary Γ-module

We assume now that G^0 is semi-simple with finite center. Γ and (ρ, E) are as in §3.

4.1. We say that G^0 *has no compact factor* if it has no infinite normal compact subgroup. A discrete subgroup L of G is said to be *irreducible* if the image of $L \cap G^0$ under any surjective morphism $f \colon G^0 \to G'$ with non-trivial image and non-compact kernel is non-discrete. If G/L has finite invariant volume, and G^0 has no compact factor, then this condition implies in fact that $f(L)$ is dense in G' [**5**].

4.2. LEMMA. *Assume that G is connected with no compact factor, and has a direct product decomposition $G = G_1 \times \cdots \times G_t$, and that Γ is irreducible in G. Let (π, H) be a unitary irreducible representation of G which occurs in $I_2(E)$, and $\pi = \pi_1 \widehat{\otimes} \cdots \widehat{\otimes} \pi_t$ its canonical decomposition. If π is not trivial, then no π_i is.*

If E is a direct sum of unitary Γ-modules, then $I_2(E)$ decomposes accordingly, so we may assume E to be simple. Assume that one of the π_i's, say π_1, is trivial. We have to show that π is trivial, too. Since π_1 is trivial, H_π^∞ consists of functions which are right-invariant under G_1. Since G_1 is normal in G, they are also left-invariant under G_1. Let $G' = G_2 \times \cdots \times G_t$, $\sigma \colon G \to G'$ the natural projection and $\Gamma' = \sigma(\Gamma)$. Then Γ' is dense in G' (see above). By definition, H_π^∞ consists of smooth functions $f \colon G \to E$ such that

$$(1) \qquad\qquad f(\gamma \cdot g) = \rho(\gamma) \cdot f(g) \qquad (\gamma \in \Gamma;\ g \in G).$$

If $\gamma \in \Gamma \cap G_1$, then $f(\gamma \cdot g) = f(g)$; hence $\rho(\gamma)$ is the identity on any element $e \in E$ of the form $f(g)$ for some $g \in G$ and $f \in H_\pi^\infty$. Since E is assumed to be irreducible, (1) implies that the set of such e's spans E; hence $\rho(\gamma) = \mathrm{Id}$, and ρ may be viewed as a representation of Γ'. Assume that $\gamma_n \in \Gamma$ is a sequence such that $\sigma(\gamma_n) \to 1$. Then for $g \in G'$ and $f \in H_\pi^\infty$ we have $f(\gamma_n \cdot g) \to f(g)$ by continuity and left G_1-invariance. By (1), this shows that $\rho(\gamma_n) \cdot f(g) \to f(g)$. Since the $f(g)$'s span E, we see that $\rho(\gamma_n) \to 1$, i.e., the representation ρ of Γ' is continuous for the topology induced by that of G'. But then ρ extends to a finite dimensional unitary representation of G', hence is trivial. By (1), the elements of H_π^∞ are then left-invariant under Γ, hence under $G_1 \cdot \Gamma = G_1 \times \Gamma'$, which is dense in G. Thus H_π^∞ is the space of constant functions.

4.3. PROPOSITION. *Assume that G is connected and has no compact factor. Let $\mathfrak{g} = \mathfrak{g}_1 \times \cdots \times \mathfrak{g}_t$ be the decomposition of $\mathfrak{g}$ into simple ideals. Assume Γ to be irreducible in G. Then the natural homomorphism $j^* \colon H^q(\mathfrak{g}, K; E^\Gamma) \to H^q(\Gamma; E)$ (see 2.6) is an isomorphism for $q < \sum_i (M(\mathfrak{g}_i) + 1)$ (where $M(\mathfrak{g}_i)$ is as in II, 9.1), in particular, for $q < \mathrm{rk}_{\mathbf{R}}\, \mathfrak{g}$.*

(We recall that $M(\mathfrak{g}_i)$ is the greatest integer such that

$$H^q(\mathfrak{g}_i, \mathfrak{k}_i; V) = 0 \quad \text{for} \quad q \leq M(\mathfrak{g}_i)$$

and any non-trivial irreducible admissible unitary $(\mathfrak{g}_i, \mathfrak{k}_i)$-module V. In particular, $M(\mathfrak{g}_i) \geq \mathrm{rk}_{\mathbf{R}}\, \mathfrak{g}_i - 1$ (V, 3.4) and $M(\mathfrak{g}_i) \geq m(\mathfrak{g}_i)$, where $m(\mathfrak{g}_i)$ is Matsushima's constant (II, 8.2).)

Using 3.5(3), we see that it suffices to prove 4.3 when $G = \mathrm{Ad}\, \mathfrak{g}$. By Theorem 3.4, $H^q(\Gamma; E)$ is the sum of $H^q(\mathfrak{g}, \mathfrak{k}; E^\Gamma)$ and of the spaces $H^q(\mathfrak{g}, \mathfrak{k}; H_{\pi, K})$, with $\pi \in \widehat{G}$, π non-trivial. Since $G = \mathrm{Ad}\, \mathfrak{g}$, the decomposition of $\mathfrak{g}$ into simple ideals corresponds to one of G as a product of simple groups and 4.2 obtains; we can therefore apply II, 9.4, which shows that those groups vanish in the range indicated.

4.4. COROLLARY. a) *If (ρ, E) is irreducible and non-trivial, then $H^q(\Gamma; E) = 0$ for $q < \sum_i (M(\mathfrak{g}_i) + 1)$, in particular for $q < \mathrm{rk}_{\mathbf{R}}\, \mathfrak{g}$.*

b) *The homomorphism j^* is an isomorphism of $H^q(\mathfrak{g}, \mathfrak{k}; \mathbf{C})$ onto $H^q(\Gamma; \mathbf{C})$ for $q < \sum_i (M(\mathfrak{g}_i) + 1)$, in particular for $q < \mathrm{rk}_{\mathbf{R}}\, \mathfrak{g}$.*

These are in fact special cases of 4.3. Since $M(\mathfrak{g}_i) \geq m(\mathfrak{g}_i)$, we see in particular that if $E = \mathbf{C}$ is the trivial Γ-module and $\mathfrak{g}$ is simple, then j^* is an isomorphism at least up to $m(\mathfrak{g})$, a result due to Y. Matsushima [**80**, Thm. 1].

4.5. The space $H^1(\mathfrak{g}, K; L)$ is zero for any finite dimensional $(\mathfrak{g}, K)$-module L. Thus, in particular, $H^1(\Gamma; E) = 0$ for any E if $t \geq 2$. Assume now that $t = 1$, i.e. $\mathfrak{g}$ is simple non-compact. Then 3.2, (II, §7) and (V, §3) imply that $H^1(\Gamma; E) = 0$ if $\mathfrak{g}$ is not of type $\mathfrak{so}(n, 1)$ or $\mathfrak{su}(n, 1)$, in particular if the split rank $\mathrm{rk}_{\mathbf{R}}\, \mathfrak{g}$ of $\mathfrak{g}$ is > 1. This proves the first assertion of the following corollary:

4.6. COROLLARY. *Let G and Γ be as in 4.3. Assume that $\mathrm{rk}_{\mathbf{R}}\, \mathfrak{g} \geq 2$ or that $\mathfrak{g}$ is not isomorphic to $\mathfrak{su}(n, 1)$ or $\mathfrak{so}(n, 1)$ for any $n \geq 1$. Then $H^1(\Gamma; E) = 0$. Let Q be a compact connected Lie group. Then, up to inner automorphisms of Q, there are only finitely many homomorphisms of Γ into Q.*

Since Γ is finitely generated, the second assertion is a consequence of the first and of the following lemma. (See [**8**, 1.1] for a similar proof.)

4.7. LEMMA. *Let L be a finitely generated group and Q a compact connected Lie group. Assume that for every finite dimensional unitary representation (ρ, E)*

of L, the group $H^1(L; E)$ is zero. Then, up to inner automorphisms of Q, there are only finitely many homomorphisms of L into Q.

The space $R(L, Q)$ of homomorphisms of L into Q may be viewed as the set of real points of an affine algebraic variety defined over $\mathbf{R}$, namely the space $R(L, Q_c)$ of homomorphisms of L into the complexification of Q_c of Q (see [**117**]). Let $f \in \text{Hom}(L, Q)$, and let $\rho = \text{Ad} \circ f$ be the representation of L into the Lie algebra $\mathfrak{q}$ of Q defined by composing f with the adjoint representation of Q. Our assumption insures that $H^1(L; \mathfrak{q}) = 0$. Then we also have $H^1(L; \mathfrak{q}_c) = 0$. By [**117**], the irreducible component of $R(L, Q_c)$ passing through f is the orbit of Q_c, acting by inner automorphisms. Thus $R(L, Q)$ is contained in finitely many orbits of Q_c. But then it is also the union of finitely many orbits of Q [**15**, 6.4].

4.8. In particular, we see that, up to equivalence, Γ has only finitely many unitary representations of a given degree m. As is known, this is false if $G = \mathbf{SL}_2(\mathbf{R})$. In fact, if S is a compact Riemann surface of genus ≥ 2, then equivalence classes of certain holomorphic bundles on S correspond canonically to equivalence classes of unitary representations of a suitable Fuchsian group (see [**88**]). The results recalled above show that the only possible exceptions to 4.6 would occur when $\mathfrak{g} = \mathfrak{so}(n + 1, 1)$ or $\mathfrak{su}(n, 1)$. We do not know whether they do for $n \geq 2$.

4.9. PROPOSITION. *Let $\mathfrak{g} = \mathfrak{so}(n, 1)$ $(n \geq 2)$. Let D_m^+, D_m^-, J_q be as in* VI, *§4. Then*

(1) $\dim H^q(\Gamma; E) = \dim \text{Hom}_G(J_q; I_2(E)),$ *if $q < n/2$;*

(2) $\dim H^q(\Gamma; E) = \dim \text{Hom}_G(D_q^+ \oplus D_q^-, I_2(E)),$ *if $q = n/2$.*

This follows from 3.2 and VI, §4.

REMARK. [**86**] gives examples of arithmetic subgroups Γ for which $H^q(\Gamma; \mathbf{C}) \neq 0$ for all q between 0 and n. For $n \geq 4$ and $q \neq n/2$, this provides examples of non-tempered representations occurring in $L^2(\Gamma \backslash G)$.

4.10. Let $G = \mathbf{SU}(n, 1)$. Then $X = G/K$ is isomorphic to the open unit ball in $\mathbf{C}^n$. Assume Γ to have no non-central element of finite order. Then $Y = \Gamma \backslash X$ is a compact Kaehler manifold. Since $H^*(\Gamma; \mathbf{C})$ is canonically isomorphic with $H^*(Y; \mathbf{C})$, 3.4 and VI yield

4.11. PROPOSITION. *Let J_{ij}, D_i be as in* VI, *§4. Then*

(1) $\dim H^{p,q}(\Gamma \backslash X; \mathbf{C})_{\text{pr}} = \dim \text{Hom}_G(J_{p,q}, L^2(\Gamma \backslash G))$
                        $(0 \leq p, q \leq n, p + q \neq n),$

(2) $\dim H^{n-i,n+i}(\Gamma \backslash X; \mathbf{C})_{\text{pr}} = \dim \text{Hom}_G(D_i, L^2(\Gamma \backslash G))$ $(0 \leq i \leq n).$

5. Γ cocompact, E a G-module

5.1. In this section, Γ is a cocompact subgroup, and E a finite dimensional G-module. As a special case of 3.1(2)(3), we have discrete sum decompositions with finite multiplicities

$$(1) \qquad\qquad L^2(\Gamma\backslash G) = \widetilde{\bigoplus_{\pi\in\widehat{G}}} m(\pi,\Gamma)H_\pi,$$

$$(2) \qquad C^\infty(\Gamma\backslash G) = (L^2(\Gamma\backslash G))^\infty = \left(\widetilde{\bigoplus_{\pi\in\widehat{G}}} m(\pi,\Gamma)H_\pi\right)^\infty.$$

Moreover, the canonical isomorphism 2.3(3) yields

$$(3) \qquad\qquad I^\infty(E) \cong \left(\widetilde{\bigoplus_{\pi\in\widehat{G}}} m(\pi,\Gamma)H_\pi\right)^\infty \otimes E.$$

The summand corresponding to the trivial representation π_0 represents the constant E-valued functions on G. Obviously

$$(4) \qquad\qquad\qquad m(\pi_0,\Gamma) = 1.$$

5.2. THEOREM. *We have*

$$(1) \qquad\qquad H^*(\Gamma;E) = \bigoplus_{\pi\in\widehat{G}} m(\pi,\Gamma)H^*(\mathfrak{g},K;H_{\pi,0}\otimes E).$$

The natural homomorphism $j^*\colon H^*(\mathfrak{g},K;E) \to H^*(\Gamma;E)$ (see 2.8) is injective. Its image is the contribution of the trivial representation of G to (1).

By 2.6,

$$(2) \qquad\qquad H^*(\Gamma;E) = H^*\left(\mathfrak{g},K;\left(\widetilde{\bigoplus_{\pi}} m(\pi,\Gamma)H_\pi\right)^\infty \otimes E\right).$$

The proof that we can replace the topological sum on the right-hand side by an algebraic direct sum is then the same as in the case of 3.2, and will not be repeated. By (I, 2.2), we can substitute $H_{\pi,0}$ for H_π^∞. The last assertion is then obvious.

5.3. Assume G to have no compact factor and E to be a simple G-module. Then E is also a simple Γ-module [**5**], and the center $C(\Gamma)$ of Γ acts by scalars. If it acts non-trivially, then $H^*(\Gamma;E) = 0$ (3.4). So assume it acts trivially.

Let $C_\rho(G) = C(G)\cap\ker\rho$. Then 3.4 also shows that $H^*(\Gamma;E) = H^*(\Gamma';E)$, where $\Gamma' = \Gamma/(\Gamma\cap C_\rho(G))$. Thus we may replace G by G' ($G' = G/C_\rho(G)$) and Γ by Γ'. Now G' admits a faithful linear representation, namely the sum of its adjoint representation and of ρ. We may therefore assume G to be linear. Let G_0 be the analytic group generated by $\mathfrak{g}$ in the simply connected complex Lie group with Lie algebra the complexification $\mathfrak{g}_c$ of $\mathfrak{g}$. Then G is a quotient of G_0. Let $\alpha\colon G_0\to G'$ be the natural projection, and $\Gamma' = \rho' = \alpha^{-1}(\Gamma)$. We may view E as a G_0-module on which $\ker\alpha$ acts trivially. Therefore, $H^*(\Gamma;E) = H^*(\Gamma';E)$.

In conclusion, the computation of $H^*(\Gamma;E)$ may be reduced to the case where G is a real form of a simply connected complex semi-simple Lie group G_c. In particular G may be assumed to be linear, and to have a global direct product decomposition $G = G_1\times\cdots\times G_t$ corresponding to the decomposition of $\mathfrak{g}$ into simple ideals $\mathfrak{g}_i$ ($1\le i\le t$).

6. G semi-simple, Γ cocompact, E a G-module

In this section, G is connected and semi-simple with finite center and no compact factor. (ρ, E) and Γ are as in §5.

6.1. THEOREM. *Assume $\rho(E)$ to be irreducible. Let (ρ^*, E^*) be the contragredient representation to (ρ, E). Then, in the notation of 5.1, we have*

$$(1) \qquad H^q(\Gamma; E) = \bigoplus_{\pi \in \widehat{G}} m(\pi, \Gamma) H^q(\mathfrak{g}, K; H_{\pi, K} \otimes E) \qquad (q \in \mathbf{N}),$$

where the sum is finite and restricted to those π such that $\chi_\pi = \chi_{\rho^*}$ and $\omega_\pi = \omega_{\rho^*}$. For those π's, we have $H^q(\mathfrak{g}, K; H_{\pi, K} \otimes E) = \mathrm{Hom}_K(\Lambda^q(\mathfrak{g}/\mathfrak{k}), H_{\pi, K} \otimes E)$ $(q \in \mathbf{N})$.

Since π is admissible, $H_{\pi, K} = H_{\pi, K}^\infty$. The first assertion then follows from 5.2 and (I, 5.3), the second from the first and (II, 3.1). Note that since G is assumed to be connected, we could replace K by $\mathfrak{k}$.

6.2. The complex case. Assume that G/K is Hermitian symmetric. Then 6.1 remains valid for (p, q)-type, i.e., it holds true if q is replaced by (p, q) and $\Lambda^q(\mathfrak{g}/\mathfrak{k})$ by $\Lambda^p \mathfrak{p}_c^+ \otimes \Lambda^q \mathfrak{p}_c^-$ (in the notation of II, §4). This follows from 2.10.

6.3. Let $\mathfrak{g} = \mathfrak{g}_1 \oplus \cdots \oplus \mathfrak{g}_t$ be the decomposition of $\mathfrak{g}$ into simple ideals. Assume E to be a simple G-module. Write accordingly

$$(1) \qquad E = E_1 \otimes \cdots \otimes E_t, \qquad \rho = \rho_1 \otimes \cdots \otimes \rho_t,$$

where (ρ_i, E_i) is a simple $\mathfrak{g}_i$-module $(1 \leq i \leq t)$. Let $M(\mathfrak{g}_i, \rho_i)$ be as in II, 9.1 and 4.4(b).

6.4. PROPOSITION. *We keep the previous assumptions and notation, and moreover assume E to be non-trivial and Γ to be irreducible. Then*

$$(1) \qquad H^q(\Gamma; E) = 0 \quad \text{for } q < \sum_{1 \leq i \leq t} (M(\mathfrak{g}_i, \rho_i) + 1),$$

in particular for $q < \mathrm{rk}_\mathbf{R} G$.

By the reductions described in 5.3, we may assume that $G = G_1 \times \cdots \times G_t$, where G_i has Lie algebra $\mathfrak{g}_i$ $(1 \leq i \leq t)$. By 6.1,

$$(2) \qquad H^*(\Gamma; E) = H^*(\mathfrak{g}, \mathfrak{k}; E) \bigoplus {\bigoplus}' m(\pi, \Gamma) H^*(\mathfrak{g}, \mathfrak{k}; H_{\pi, K} \otimes E),$$

where ${\bigoplus}'$ extends over those π which have the same infinitesimal and central characters as ρ^*. In particular, π is not trivial. By II, 3.2,

$$(3) \qquad H^q(\mathfrak{g}, \mathfrak{k}; E) = 0.$$

Any $\pi \in \widehat{G}$ decomposes as $\pi = \pi_1 \widetilde{\otimes} \cdots \widetilde{\otimes} \pi_t$ $(\pi_i \in \widehat{G}_i, i = 1, \dots, t)$. By the Künneth rule (I, 1.3), we have, taking 6.3 into account,

$$(4) \qquad H^q(\mathfrak{g}, \mathfrak{k}; H_{\pi, K} \otimes E) = \bigoplus_{q_1 + \cdots + q_t = q} \left(\bigotimes_i H^{q_i}(\mathfrak{g}_i, \mathfrak{k}_i; H_{\pi_i, K_i} \otimes E_i) \right),$$

where $K_i = K \cap G_i$, $\mathfrak{k}_i = \mathfrak{k} \cap \mathfrak{g}_i$ $(1 \leq i \leq t)$. By 4.2, if π is non-trivial, then no π_i is trivial; therefore, for the left-hand side of (4) to be non-zero, it is necessary that $q_i > M(\mathfrak{g}_i, \rho_i)$ for all i. Since $M(\mathfrak{g}_i, \rho_i) \geq \mathrm{rk}_\mathbf{R} \mathfrak{g}_i$ by V, 3.2, the proposition is proved.

6.5. PROPOSITION (Raghunathan [**92**]). *Assume* Γ *to be irreducible and* (ρ, E) *to be simple non-trivial. Then* $H^1(\Gamma; E) = 0$ *except possibly when* $\mathfrak{g} = \mathfrak{so}(n+1, 1)$ *(resp.* $\mathfrak{g} = \mathfrak{su}(n, 1)$*) and the highest weight of* ρ *is a multiple of the highest weight of the standard representation of* $\mathfrak{so}(n+1, 1)$ *(resp. of the standard representation of* $\mathfrak{su}(n, 1)$ *or of its contragredient representation)* $(n \geq 1)$.

If $\mathfrak{g}$ is not simple, this is a consequence of 6.4. If $\mathfrak{g}$ is simple, it follows from (V, §6) and 6.1.

6.6. We now translate the results of II, §§6,7 into properties of the spaces $H^*(\Gamma; *)$. Let $\mathfrak{h}$ be a θ-stable fundamental Cartan subalgebra of $\mathfrak{g}$. Put

$$\mathfrak{h} = \mathfrak{h}_c \cap \mathfrak{k}_c, \qquad \Phi_k = \Phi(\mathfrak{k}_c, \mathfrak{h}_c \cap \mathfrak{k}_c), \qquad \Phi = \Phi(\mathfrak{g}_c, \mathfrak{h}_c).$$

Fix $\Phi_k^+ \subset \Phi_k$. Let Φ^+ be a system of positive roots for Φ compatible with Φ_k^+ (see II, 6.6).

6.7. THEOREM. *Let* F *be an irreducible finite dimensional* G-*module with highest weight* $\Lambda - \rho$. *If* $\theta\Lambda \neq \Lambda$, *then* $H^*(\Gamma; F) = (0)$.

This follows from 6.4(2),(3) and II, 6.12(1).

6.8. Let Λ be Φ^+-dominant integral and regular. Let W^1 be as in II, 6.9, and $\mathfrak{p}^+(t\Phi^+)$ as in II, 7.1 ($t \in W^1$). We say that Λ is *strongly* Φ_k^+-*dominant integral* if
   (1) $t\Lambda + t\rho - 2\rho_k - \xi$ and $t\Lambda - \rho_k - \xi$ are Φ_k^+-dominant integral for all weights ξ of $\Lambda\mathfrak{p}^+(t\Phi^+)$ and all $t \in W^1$.

6.9. THEOREM. *Let* l_0 *be as in* II, 6.9 *and* $q = \dim \mathfrak{p}_c^+$. *Let* F *be an irreducible finite dimensional* G-*module with highest weight* $\Lambda - \rho$. *If* Λ *is strongly* Φ_k^+-*dominant integral, then*

$$H^j(\Gamma; F) = \sum_{t \in W^1} \binom{l_0}{j - q} m(\pi_t, \Gamma) \qquad (j \in \mathbf{N}),$$

where π_t *is the element of the fundamental series (*[**38**]*) for* G *relative to* $t\Phi^+$ *with lowest* K-*type* $\tau_{t\Lambda + t\rho - 2\rho_k}$. *In particular,*

$$H^j(\Gamma; F) = (0) \quad \text{for } j < \dim \mathfrak{p}_c^+.$$

This is a consequence of 6.4(2),(3) and II, 7.3(1),(2).

6.10. REMARK. Assume that $G = {}^0G$ and $\mathfrak{h} \subset \mathfrak{k}$. Then the fundamental series is the discrete series. 6.9 in this case sharpens an unpublished result of Langlands.

Assume further that Γ is torsion free. Let Λ, F, π_t be as in 6.9. If dg is a fixed Haar measure on G, let d be the formal degree of π_t for $t \in W^1$. (It is independent of $t \in W^1$.) Then it is shown in [**61**] that

$$m(\pi_t, \Gamma) = d.\text{vol}(\Gamma\backslash G).$$

Thus 6.10 becomes

$$\dim H^j(\Gamma; F) = \begin{cases} |W^1|\, d.\text{vol}(\Gamma\backslash G), & \text{if } 2j = \dim G/K, \\ 0, & \text{otherwise.} \end{cases}$$

The Construction of Certain Unitary Representations and the Computation of the Corresponding Cohomology Groups

In this chapter the oscillator representation is used to construct non-trivial unitary representations V of $\mathbf{SU}(p,q)$ $(p \geq q > 0)$ such that $H^q(\mathfrak{g}, \mathfrak{k}; V) \neq (0)$. This is of interest since $q = \mathrm{rk}_{\mathbf{R}} \, \mathbf{SU}(p,q)$ (see V, 3.4). Using Weil's ideas on the relationship between theta functions and automorphic forms, we give in §5 a generalization of Kazhdan's theorem on the first Betti number of certain discrete cocompact subgroups of $\mathbf{SU}(n,1)$.

The material of this chapter is independent of most of the results in the preceding ones. It could be read after Chapter II.

1. The oscillator representation

1.1. We look upon $\mathbf{R}^{2n}$ as the space of all columns

(1)
$$\begin{bmatrix} x \\ y \end{bmatrix}, \qquad x, y \in \mathbf{R}^n.$$

We define

$$\beta \left(\begin{bmatrix} x \\ y \end{bmatrix}, \begin{bmatrix} x' \\ y' \end{bmatrix} \right) = \langle x, y' \rangle - \langle y, x' \rangle,$$

where $\langle x, y \rangle = \sum x_i y_i$ for $x = (x_1, \ldots, x_n)$, $y = (y_1, \ldots, y_n)$.

The *Heisenberg group* of dimension $2n + 1$ is the group with underlying space $\mathbf{R}^{2n} \times \mathbf{R}$ and multiplication given by

$$(z, t) \cdot (w, s) = \left(z + w, t + s + \tfrac{1}{2} \beta(z, w) \right).$$

We denote this Lie group by H_n.

It is easy to see that the 1-parameter subgroups of H_n are of the form $s \mapsto (sw, st)$. Thus the Lie algebra $\mathfrak{h}_n$ of H_n is $\mathbf{R}^{2n} \times \mathbf{R}$ with the bracket

$$[(x, t), (y, s)] = (0, \beta(x, y)).$$

1.2. The Stone-von Neumann theorem says that H_n has (up to dilation and duality) one infinite dimensional, irreducible, unitary representation $(\pi, L^2(\mathbf{R}^n))$ with

$$\left(\pi \left(\begin{bmatrix} x \\ y \end{bmatrix}, t \right) f \right)(z) = \exp \left(i \left(t + \langle x, z - \tfrac{1}{2} y \rangle \right) \right) f(z - y)$$

for $f \in L^2(\mathbf{R}^n)$, $x, y \in \mathbf{R}^n$, $t \in \mathbf{R}$.

Let the unitary group $U(L^2(\mathbf{R}^n)) = U$ of $L^2(\mathbf{R}^n)$ be given the strong operator topology (i.e., the weakest topology such that $U \times L^2(\mathbf{R}^n) \to L^2(\mathbf{R}^n)$, $u, \phi \mapsto u \cdot \phi$ is continuous.)

1.3. LEMMA. *The map*

$$\mathbf{R}^{2n} \to U(L^2(\mathbf{R}^n)) \quad \text{given by} \quad x \mapsto \pi(x, 0)$$

is a homeomorphism of $\mathbf{R}^{2n}$ *onto its image.*

A proof of this lemma can be found in [**65**] and in [**108**].

1.4. Let $\mathcal{S}(\mathbf{R}^n)$ denote the Schwartz space of $\mathbf{R}^n$ with the Schwartz topology. If $\phi \in \mathcal{S}(\mathbf{R}^n)$, define

$$(1) \qquad \mathcal{F}(\phi)(z) = (2\pi)^{-n/2} \int_{\mathbf{R}^n} \phi(x) e^{-i\langle x, z \rangle} \, dx.$$

Then $\mathcal{F}$ extends to a unitary operator on $L^2(\mathbf{R}^n)$, and an easy computation shows that

$$(2) \qquad \mathcal{F}\pi\left(\begin{bmatrix} x \\ y \end{bmatrix}, t\right) \mathcal{F}^{-1} = \pi\left(\begin{bmatrix} -y \\ x \end{bmatrix}, t\right),$$

for $x, y \in \mathbf{R}^n$, $t \in \mathbf{R}$. Let

$$G = \{g \in U(L^2(\mathbf{R}^n)) \mid g\pi(z, t)g^{-1} = \pi(z', t) \ (z \in \mathbf{R}^{2n}, \ t \in \mathbf{R})\}.$$

In this definition z' clearly depends on z and g. 1.3 implies that $z' = \nu(g)z$, with $\nu(g) \colon \mathbf{R}^{2n} \to \mathbf{R}^{2n}$ a homeomorphism.

Let $\mathbf{Sp}(n, \mathbf{R})$ denote the symplectic group. That is,

$$\mathbf{Sp}(n, \mathbf{R}) = \{g \in \mathbf{GL}(2n, \mathbf{R}) \mid \beta(g \cdot v, g \cdot w) = \beta(v, w) \text{ for } v, w \in \mathbf{R}^{2n}\}.$$

1.5. LEMMA. $\nu(g) \in \mathbf{Sp}(n, \mathbf{R})$ *for* $g \in G$.

This follows from the definition of G and the relation

$$\pi(z, t)\pi(w, s) = \pi\left(z + w, t + s + \tfrac{1}{2}\beta(z, w)\right).$$

1.6. PROPOSITION. *Let* $T^1 = \{e^{i\theta} \mid \theta \in \mathbf{R}\}$. *Then the following sequence is exact:*

$$1 \longrightarrow T^1 \cdot I \longrightarrow G \overset{\nu}{\longrightarrow} \mathbf{Sp}(n, \mathbf{R}) \longrightarrow 1.$$

We first note that if $g \in \ker \nu$, then $g\pi(h)g^{-1} = \pi(h)$ for $h \in H_n$. Since π is irreducible, this implies that $g = \lambda I$ and $|\lambda| = 1$. Hence to complete the proof we need only show that ν is surjective.

Set

$$M = \left\{ \left[\begin{array}{c|c} A & 0 \\ \hline 0 & {}^t A^{-1} \end{array} \right] \middle| A \in \mathbf{GL}(n, \mathbf{R}) \right\},$$

$$N = \left\{ \begin{bmatrix} I & X \\ 0 & I \end{bmatrix} \middle| X \in M_n(\mathbf{R}), \ {}^t X = X \right\}, \qquad J = \begin{bmatrix} 0 & -I \\ I & 0 \end{bmatrix}.$$

It is well known that $\mathbf{Sp}(n, \mathbf{R})$ is generated by $N \cup M \cup \{J\}$. The equality 1.4(2) says that

$$(1) \qquad \nu(\mathcal{F}) = J.$$

If $A \in \mathbf{GL}(n, \mathbf{R})$ and $f \in L^2(\mathbf{R}^n)$, define $\alpha(A)$ by

$$(\alpha(A)f)(z) = |\det A|^{1/2} f({}^t A z), \qquad z \in \mathbf{R}^n.$$

Then it is easily seen that

(2) $$\alpha(A) \in G \quad \text{and} \quad \nu(\alpha(A)) = \begin{bmatrix} A & 0 \\ 0 & {}^t A^{-1} \end{bmatrix}.$$

If $X \in M_n(\mathbf{R})$ and ${}^t X = X$, then for $f \in L^2(\mathbf{R}^n)$ we set

$$\mu(X)f(z) = \exp(i\langle Xz, z\rangle/2)f(z).$$

Then

(3) $$\mu(X) \in G \quad \text{and} \quad \nu(\mu(X)) = \begin{bmatrix} I & X \\ 0 & I \end{bmatrix}.$$

This completes the proof of the proposition.

1.7. Since an extension of a Lie group by a Lie group is a Lie group, 1.6 implies that G is a Lie group. Since $\nu \colon G \to \mathbf{Sp}(n, \mathbf{R})$ is continuous, ν is a Lie group homomorphism. Let ν_* be the differential of ν.

Let $\mathfrak{sp}(n, \mathbf{R})$ denote the Lie algebra of $\mathbf{Sp}(n, \mathbf{R})$. Then $\nu_* \colon [\mathfrak{g}, \mathfrak{g}] \to \mathfrak{sp}(n, \mathbf{R})$ is a Lie algebra isomorphism.

1.8. DEFINITION. The *metaplectic group* is the commutator group; $\mathbf{Mp}(n, \mathbf{R})$, of G.

1.9. LEMMA. *Set $j = \nu\big|_{\mathbf{Mp}(n,\mathbf{R})}$. Then*

$$j \colon \mathbf{Mp}(n, \mathbf{R}) \to \mathbf{Sp}(n, \mathbf{R})$$

is a finite covering.

Since j_* is bijective, $\ker j$ is discrete. But $\ker j \subset T^1 I$. Hence $\ker j$ is finite.

1.10. We look upon $(\mathbf{Mp}(n, \mathbf{R}), j)$ as an abstract covering group of $\mathbf{Sp}(n, \mathbf{R})$. The realization $\mathbf{Mp}(n, \mathbf{R}) \subset U(L^2(\mathbf{R}^n))$ will be denoted $(W, L^2(\mathbf{R}^n))$. It is called the *oscillator* (sometimes Weil, Shale, harmonic) *representation* of $\mathbf{Mp}(n, \mathbf{R})$.

1.11. LEMMA. *The space $L^2(\mathbf{R}^n)^\infty$ of C^∞ vectors for $(W, L^2(\mathbf{R}^n))$, with the C^∞ topology, is isomorphic to $\mathcal{S}(\mathbf{R}^n)$.*

Set $\mathfrak{p}_n(\mathbf{R}) = \{X \in M_n(\mathbf{R}) \mid {}^t X = X\}$. Set $\xi(X, A, Y)\phi = \mathcal{F}\mu(X)\mathcal{F}^{-1}\alpha(A)\mu(Y)\phi$ for $X, Y \in \mathfrak{p}_n(\mathbf{R})$, $A \in \mathbf{GL}(n, \mathbf{R})$ and $\phi \in \mathcal{S}(\mathbf{R}^n)$. An easy calculation shows that ξ is of class C^∞. By computing the differential of ξ, one also sees that the topology of $\mathcal{S}(\mathbf{R}^n)$, as a subspace of $L^2(\mathbf{R}^n)^\infty$, is the usual topology. $W(U(\mathfrak{g}))$ contains the operators $(\sum x_i^2)^k (\sum \partial^2/\partial x_i^2)^l$, $k, l \in \mathbf{N}$. Hence $L^2(\mathbf{R}^n)^\infty \subset \mathcal{S}(\mathbf{R}^n)$.

1.12. We identify the Lie algebra of $\mathbf{Mp}(n, \mathbf{R})$ with $\mathfrak{sp}(n, \mathbf{R})$. We denote by Exp the exponential mapping of $\mathfrak{sp}(n, \mathbf{R})$ into $\mathbf{Mp}(n, \mathbf{R})$ and by exp the exponential mapping of $\mathfrak{sp}(n, \mathbf{R})$.

Let E_{ij} denote the $n \times n$ matrix with a 1 in the i, j position and all other entries 0. Set

(1) $$h_j = i \left[\begin{array}{c|c} 0 & E_{jj} \\ \hline -E_{jj} & 0 \end{array} \right].$$

Then $ih_j \in \mathfrak{sp}(n, \mathbf{R})$. It is easy to see that

If $\phi \in \mathcal{S}(\mathbf{R}^n)$, then

(2) $$\frac{d}{dt} W(\mathrm{Exp}(ith_j))\phi\big|_{t=0} = iH_j\phi, \quad \text{where } 2H_j\phi = \left(\frac{\partial^2}{\partial x_j^2} - x_j^2 \right)\phi.$$

Set $T = \{\text{Exp}(\sum t_j(ih_j)) \mid t_j \in \mathbf{R}\}$ and $T_0 = \nu(T)$. Then T and T_0 are Cartan subgroups of $\mathbf{Mp}(n, \mathbf{R})$ and $\mathbf{Sp}(n, \mathbf{R})$ respectively.

Set $P^j(\mathbf{R}^n)$ equal to the space of all complex valued polynomial functions on $\mathbf{R}^n$ of degree less than or equal to j, and $P(\mathbf{R}^n) = \bigcup P^j(\mathbf{R}^n)$.

1.13. LEMMA. *Set $\psi_0(x) = (2\pi)^{-n/2} \exp(-\langle x, x \rangle/2)$ for $x \in \mathbf{R}^n$. Then*

$$W(T)\psi_0 P^j(\mathbf{R}^n) \subset \psi_0 P^j(\mathbf{R}^n)$$

for all $j = 0, 1, 2, \ldots$.

We note that if $\phi \in P^i(\mathbf{R}^n)$ for some i, then

$$(1) \qquad H_j(\psi_0\phi) = -\frac{1}{2}\psi_0 \left(\phi + 2x_j \frac{\partial \phi}{\partial x_j} - \frac{\partial^2 \phi}{\partial x_j^2} \right).$$

Hence $H_j\psi_0 P^k(\mathbf{R}^n) \subset \psi_0 P^k(\mathbf{R}^n)$ for all $k = 0, 1, \ldots$. Let $(\ ,\ )$ denote the L^2-inner product on $L^2(\mathbf{R}^n)$. If $\psi, \phi \in \mathcal{S}(\mathbf{R}^n)$, then it is easily seen that

$$(H_j\phi, \psi) = (\phi, H_j\psi).$$

Hence H_j diagonalizes on $\psi_0 P^k(\mathbf{R}^n)$ with real eigenvalues. If $h \in \psi_0 P^k(\mathbf{R}^n)$ and if $H_j h = \lambda h$, then by 1.12(2)

$$(2) \qquad \frac{d}{dt}(W(\text{Exp}(ith_j))h, \phi) = i\lambda(W(\text{Exp}(ith_j))h, \phi) \quad \text{for } \phi \in \mathcal{S}(\mathbf{R}^n).$$

This implies that $W(\text{Exp}(ith_j))h = e^{i\lambda t}h$. The lemma follows.

1.14. We also note that 1.13(1) implies

$$(1) \qquad H_j\psi_0 = -\tfrac{1}{2}\psi_0, \quad \text{for } j = 1, \ldots, n.$$

1.15. LEMMA. $(\mathbf{Mp}(n, \mathbf{R}), j)$ *is a twofold covering group of* $\mathbf{Sp}(n, \mathbf{R})$.

Indeed, $j^{-1}(T_0) = T$. Hence $\ker j \subset T$. If $t \in T$ and $j(t) = I$, then $W(t) = \xi(t)I$. If $t \in T$, then $t = \text{Exp}(\sum_j t_j(ih_j))$. If $\nu(t) = I$, then $t_j = 2\pi k_j$ with $k_j \in \mathbf{Z}$. Now 1.14(1) implies that

$$(1) \qquad W(t)\psi_0 = \exp\left(-\frac{i}{2} \sum_{j=1}^{n} 2\pi k_j \right) \psi_0.$$

This implies that $\xi(t)^2 = 1$ if $t \in \ker j$. Hence $\ker j \subset \{\pm I\}$. Obviously there is $t \in \ker j$ so that $\xi(t) = -1$. This proves the lemma.

1.16. Define

$$A_j^\pm = \frac{1}{2} \left(\frac{\partial}{\partial x_j} \pm x_j \right).$$

Then:

$$(1) \qquad [A_i^+, A_j^-] = -\tfrac{1}{2}\delta_{ij}I,$$

$$(2) \qquad A_j^+ A_j^- + A_j^- A_j^+ = H_j,$$

$$(3) \qquad [H_j, A_i^+] = \delta_{ij}A_i^+, \qquad [H_j, A_i^-] = -\delta_{ij}A_i^-,$$

$$(4) \qquad A_j^+ \psi_0 = 0,$$

(5) $[A_j^+, A_i^+] = [A_j^-, A_i^-] = 0.$

If $m = (m_1, \ldots, m_n)$, $m_i \in \mathbf{N}^n$, set $|m| = \sum m_i$ and $m! = m_1! \cdots m_n!$. Set $\psi_m = (m!)^{-1/2}(A_1^-)^{m_1} \cdots (A_n^-)^{m_n}\psi_0$. Then

(6) $$\sum_{|m| \leq k} \mathbf{C}\psi_m = \psi_0 P^k(\mathbf{R}^n).$$

The following result is an easy consequence of the formulae in this section.

1.17. LEMMA. (1) $\{\psi_k\}_{k \in (\mathbf{Z}^+)^n}$ is an orthonormal basis of $L^2(\mathbf{R}^n)$.
(2) $W(\mathrm{Exp}(i\sum t_j h_j))\psi_k = \exp(-\frac{i}{2}(\sum(2k_j + 1)t_j))\psi_k$.

1.18. PROPOSITION. The space of C^∞ vectors for $W|_T$ is $\mathcal{S}(\mathbf{R}^n)$ with the Schwartz topology. That is, $W|_T$ and W have the same C^∞ vectors.

It follows from 1.17(2) that

(1)

$$h = \sum a_m \psi_m \in L^2(\mathbf{R}^n) \text{ is a } C^\infty \text{ vector for } W|_T$$
$$\text{if and only if } |a_m| \leq C_r(1 + \langle m, m \rangle)^{-r},$$

for each $r > 0$.

(1) is the condition that $h \in \mathcal{S}(\mathbf{R}^n)$ (cf. [**94**]). Furthermore, if we set $\|h\|_s^2 = \sum(1 + \langle m, m \rangle)^s |a_m|^2$, then the norms $\|\cdots\|_s$ define both the Schwartz topology (cf. [**94**], V, 13) and the topology on the C^∞ vectors of $W|_T$ (see 0, 2.3).

2. The decomposition of the restriction of the oscillator representation to certain subgroups

2.1.   For $p + q = n$, $p \geq q \geq 0$, set

$$I_{p,q} = \left[\begin{array}{c|c} I_p & 0 \\ \hline 0 & -I_q \end{array}\right],$$

where I_p is the $p \times p$ identity matrix. If $g \in \mathbf{M}_n(\mathbf{C})$, set g^* equal to the conjugate transpose matrix of g.

We look upon $\mathbf{C}^n$ as $\mathbf{R}^{2n}$, and we write $z = x + iy$, $x, y \in \mathbf{R}^n$, as in 1.1(1). If $X \in \mathbf{M}_n(\mathbf{C})$, then we view X as being in $\mathbf{M}_{2n}(\mathbf{R})$ by neglecting the complex structure.

Let $\mathbf{U}(p, q)$ be the group of all $g \in \mathbf{M}_n(\mathbf{C})$ such that

$$gI_{p,q} \cdot g^* = I_{p,q}.$$

Set

$$Z_{p,q} = \begin{bmatrix} I_{p,q} & 0 \\ 0 & I_n \end{bmatrix}.$$

For $g \in \mathbf{U}(p, q)$, define

$$\psi(g) = Z_{p,q} \cdot g \cdot Z_{p,q}$$

as an element of $\mathbf{GL}(2n, \mathbf{R})$.

It is easily checked that

$$\psi \colon \mathbf{U}(p, q) \to \mathbf{Sp}(n, \mathbf{R}).$$

We also use the notation ψ for the corresponding Lie algebra homomorphism of $\mathfrak{u}(p, q)$ into $\mathfrak{sp}(n, \mathbf{R})$.

2.2. Let $h_j \in \mathfrak{sp}(n, \mathbf{C})$ be defined as in 1.12(1). Set $\mathfrak{h} = \sum \mathbf{C} h_j$. Then $\mathfrak{h} \cap \mathfrak{sp}(n, \mathbf{R})$ is a Cartan subalgebra. Set $\Phi(\mathfrak{sp}(n, \mathbf{C}), \mathfrak{h}) = \Phi$. Set

$$\varepsilon_i \left(\sum z_j h_j \right) = z_i.$$

Then

$$\Phi = \{\varepsilon_i - \varepsilon_j \mid i \neq j\} \cup \{\pm(\varepsilon_i + \varepsilon_j) \mid 1 \leq i, \ j \leq n\}.$$

Set

$$\Phi^+ = \{\varepsilon_i - \varepsilon_j \mid 1 \leq i < j \leq n\} \cup \{\varepsilon_i + \varepsilon_j \mid 1 \leq i \leq j \leq n\}.$$

Let $\mathfrak{t}$ denote the algebra of diagonal elements of $\mathfrak{u}(p, q)$. Set $z_j = E_{jj}$ (see 1.12). Then

$$\psi(z_j) = h_j \quad \text{if } j \leq p, \qquad \psi(z_j) = -h_j \quad \text{if } j > p.$$

Define $\eta_i \in \mathfrak{t}_c^*$ by

$$\eta_i(z_j) = \delta_{ij} \qquad (1 \leq i, \ j \leq n).$$

Then

$$\psi^*(\varepsilon_i) = \begin{cases} \eta_i, & 1 \leq i \leq p, \\ -\eta_i, & p < i \leq n. \end{cases}$$

Set $\mathfrak{u}_c = \mathfrak{u}(p, q) \otimes_{\mathbf{R}} \mathbf{C}$. Let $\Psi = \Phi(\mathfrak{u}_c, \mathfrak{t}_c)$, and $\Psi^+ = \psi^*(\Phi^+) \cap \Psi$. Then

$$\Psi^+ = \{\eta_i - \eta_j \mid 1 \leq i < j \leq p \text{ or } 1 \leq i \leq p < j \leq n\} \cup \{\eta_j - \eta_i \mid p < i < j \leq n\}.$$

Set $\mathfrak{u}^+ = \bigoplus_{\alpha \in \Psi^+} (\mathfrak{u}_c)_\alpha$.

2.3. Let $\mathfrak{s} = \mathfrak{sp}(n, \mathbf{R})$. Let $(W, L^2(\mathbf{R}^n))$ be as in §1. Let us also denote by W the representation of $\mathfrak{s}_c$ on $\mathcal{S}(\mathbf{R}^n)$. Using 1.16 and the direct computation of W on $\mathfrak{s}$, it is easy to show that

(1) $$W((\mathfrak{s}_c)_{\varepsilon_i + \varepsilon_j}) = \mathbf{C} A_i^+ A_j^+, \qquad 1 \leq i \leq j \leq n,$$

(2) $$W((\mathfrak{s}_c)_{\varepsilon_i - \varepsilon_j}) = \mathbf{C} A_i^+ A_j^-, \qquad 1 \leq i < j \leq n.$$

2.4. By going to $\widetilde{U}(p, q)$, a twofold covering of $U(p, q)$, we can lift

$$\psi \colon \mathbf{U}(p, q) \to \mathbf{Sp}(n, \mathbf{R})$$

to

$$\widetilde{\psi} \colon \widetilde{\mathbf{U}}(p, q) \to \mathbf{Mp}(n, \mathbf{R}).$$

Let $V(g) = W(\widetilde{\psi}(g))$ for $g \in \widetilde{U}(p, q)$. We note that, since $\psi(\mathfrak{t}_c) = \mathfrak{h}_c$, 1.18 implies that $(V, L^2(\mathbf{R}^n))$ has the same space of C^∞ vectors as W. We denote by V the corresponding representation of $\mathfrak{u}_c$ on $\mathcal{S}(\mathbf{R}^n)$.

2.5. LEMMA. (1) $(V, L^2(\mathbf{R}^n))$ *splits into a countable direct sum of inequivalent, irreducible, invariant subspaces.*

(2) *If $H \subset L^2(\mathbf{R}^n)$ is a closed invariant subspace under V, then $H \cap \psi_0 P(\mathbf{R}^n)$ is dense in H. If $H \neq (0)$, then*

$$H_\infty^{\mathfrak{u}^+} = \{f \in H^\infty \mid V(\mathfrak{u}^+)f = 0\} \neq (0).$$

(1) is already true for $W(T) \subset V(\widetilde{U}(p,q))$.

(2) Since $W(T) \subset V(\widetilde{U}(p,q))$, it is also clear that $H \cap \psi_0 P(\mathbf{R}^n)$ is dense in H. Using 2.2 and 2.3, it is easy to see that

$$(3) \qquad V(\mathfrak{u}^+) = \sum_{1 \le i < j \le p} \mathbf{C} A_i^+ A_j^- + \sum_{1 \le i \le p < j \le n} \mathbf{C} A_i^+ A_j^+ + \sum_{p < i < j \le n} \mathbf{C} A_i^+ A_j^-.$$

Order $\mathbf{N}^n$ as follows: $m > m'$ if $|m| > |m'|$ or if $|m| = |m'|$ and $m_i = m_i'$ for $i < j$, $m_j > m_j'$. Using (3), it is easy to see that

$$(4) \qquad V(\mathfrak{u}^+)\psi_m \subset \sum_{m > m'} \mathbf{C}\psi_{m'}.$$

Now $H \cap \psi_0 P(\mathbf{R}^n) = \bigoplus_{m \in S(H)} \mathbf{C}\psi_m$, with $S(H) \subset (\mathbf{Z}^+)^n$ a subset. Let $m \in S(H)$ be a minimal element of $S(H)$. Then (4) implies that $V(\mathfrak{u}^+)\psi_m = 0$.

2.6. LEMMA. *Set*

$$(\psi_0 P(\mathbf{R}^n))^{\mathfrak{u}^+} = \{ f \in \psi_0 P(\mathbf{R}^n) \mid V(\mathfrak{u}^+)f = 0 \}.$$

Then

$$(\psi_0 P(\mathbf{R}^n))^{\mathfrak{u}^+} = \sum_{k \ge 0} \mathbf{C}\psi_{0,\dots,0,k,0,\dots,0} + \sum_{k > 0} \mathbf{C}\psi_{0,\dots,0,k},$$

where the index k in the first sum is in the p-th position.

We leave it to the reader to check that $\psi_0 f \in (\psi_0 P(\mathbf{R}^n))^{\mathfrak{u}^+}$, $f \in P(\mathbf{R}^n)$, if and only if

$$(1) \qquad \frac{\partial^2 f}{\partial x_i \partial x_j} = 2x_j \frac{\partial f}{\partial x_i}, \qquad 1 \le i < j \le p,$$

$$(2) \qquad \frac{\partial^2 f}{\partial x_i \partial x_j} = 2x_j \frac{\partial f}{\partial x_i}, \qquad p < i < j \le n,$$

$$(3) \qquad \frac{\partial^2 f}{\partial x_i \partial x_j} = 0, \qquad 1 \le i \le p < j \le n.$$

Write $f = \sum_{l \le k} f_l(x_1, \dots, x_{p-1}, x_{p+1}, \dots, x_n)x_p^l$. Then f satisfying (1) for $1 \le i \le p$ implies

$$(4) \qquad \frac{\partial^2 f}{\partial x_i \partial x_p} = \sum_{l \le k} l \frac{\partial f_l}{\partial x_i} x_p^{l-1} = 2x_p \sum_{l \le k} \frac{\partial f_l}{\partial x_i} x_p^l.$$

Comparing coefficients of x_p in (4), we see that $\partial f_k/\partial x_i = 0$, $i \le p-1$. Hence f_k is independent of $x_1, \dots, x_{p-1}$. Arguing by downward induction, using (4), we see that f_l is independent of $x_1, \dots, x_{p-1}$ for $l = 0, \dots, k$.

Arguing the same way, expanding in terms of x_n and using (2), we see that f is a polynomial in x_p and x_n. (3) implies that $\partial^2 f/\partial x_p \partial x_n = 0$. Write $f(x_p, x_n) = \sum_{j=0}^r f_j(x_p)x_n^j$. Then

$$0 = \sum_j \frac{\partial f_j(x_p)}{\partial x_p} j x_n^{j-1}.$$

This implies that $\partial f_j(x_p)/\partial x_p = 0$ for $j > 0$. Hence $f(x_p, x_n) = h_1(x_p) + h_2(x_n)$. Clearly, if $h_1, h_2 \in \mathbf{C}[x]$, then $h_1(x_p) + h_2(x_n)$ satisfies (1), (2), (3). The lemma now follows.

2.7. Set

$$J_{p,q} = \psi(-iI) = -Z_{p,q}JZ_{p,q} = \sum_{j=1} ih_j - \sum_{j=p+1} ih_j.$$

Then

(1) $$W(\operatorname{Exp} tJ_{p,q})\psi_m = \exp\left(-i\left(\frac{p-q}{2} + \sum_{i=1}^{q} m_{i+p}\right)t\right)\psi_m.$$

Set

$$\xi_k(\operatorname{Exp}(tJ_{p,q})) = e^{-i((p-q)/2+k)t}, \qquad k \in \mathbf{Z},$$

and

$$L_k^2(\mathbf{R}^n) = \{f \in L^2(\mathbf{R}^n) \mid W(\operatorname{Exp}(tJ_{p,q}))f = \xi_k(\operatorname{Exp} tJ_{p,q})f\}.$$

Then, since $W(\operatorname{Exp}(tJ_{p,q})) \circ V(g) = V(g) \circ W(\operatorname{Exp}(tJ_{p,q}))$ for $g \in \widetilde{U}(p,q)$, we see that $V(g)L_k^2(\mathbf{R}^n) \subset L_k^2(\mathbf{R}^n)$ for $k \in \mathbf{Z}$, $g \in \widetilde{U}(p,q)$. Set $V_k(g) = V(g)|_{L_k^2(\mathbf{R}^n)}$.

2.8. LEMMA. *We have*

(1) $$L^2(\mathbf{R}^n) = \bigoplus_{k=-\infty}^{\infty} L_k^2(\mathbf{R}^n) \quad (\textit{orthogonal direct sum}),$$

(2) $(V_k, L_k^2(\mathbf{R}^n))$ *is an irreducible representation of* $\widetilde{U}(p,q)$.

PROOF. (1) is clear. To prove (2) we note that

$$W(\operatorname{Exp} tJ_{p,q})\psi_{0,\ldots,0,k,0,\ldots,0} = \xi_k(\operatorname{Exp} tJ_{p,q})\psi_{0,\ldots,0,k,0,\ldots,0}$$
$$\underset{p\text{th position}}{\uparrow}$$

and

$$W(\operatorname{Exp} tJ_{p,q})\psi_{0,\ldots,0,k} = \xi_{-k}(\operatorname{Exp} tJ_{p,q})\psi_{0,\ldots,0,k}.$$

This implies that $\dim(L_k^2(\mathbf{R}^n) \cap \psi_0 P(\mathbf{R}^n))^{\mathfrak{u}^+} = 1$ for $k \in \mathbf{Z}$. Lemma 2.5 now implies (2).

Before we go on we have one piece of unfinished business. We state it as a lemma.

2.9. LEMMA. $\psi\colon \mathbf{SU}(p,q) \to \mathbf{Sp}(n,\mathbf{R})$ *lifts to an injective homomorphism* $\widetilde{\psi}\colon \mathbf{SU}(p,q) \to \mathbf{Mp}(n,\mathbf{R})$.

In other words, the connected subgroup of $\widetilde{U}(p,q)$ with Lie algebra $\mathfrak{su}(p,q)$ is $\mathbf{SU}(p,q)$.

Let B denote the group of diagonal elements of $\mathbf{SU}(p,q)$. Then B has Lie algebra $^0\mathfrak{t} = \{z \in \mathfrak{t} \mid \operatorname{tr} z = 0\}$. Now

$$V\left(\exp\left(\sum i\theta_j z_j\right)\right)\psi_m = \exp\left(-i\left(\sum_{j=1}^{p} \frac{(2m_j+1)}{2}\theta_j - \sum_{j=1}^{q} \frac{(2m_{j+p}+1)}{2}\theta_{j+p}\right)\right)\psi_m.$$

If $\sum_1^n \theta_j = 0$, then

$$\sum_{j=1}^p \left(\frac{2m_j+1}{2}\right)\theta_j - \sum_{j=1}^q \left(\frac{2m_{j+p}+1}{2}\right)\theta_{j+p} = \sum_{j=1}^p m_j\theta_j - \sum_{j=1}^q m_{j+p}\theta_{j+p} + \sum_{j=1}^p \theta_j.$$

This implies that the weights of $W \circ \psi$ on ${}^0\mathfrak{t}$ are $\mathbf{SU}(p,q)$-integral. Since $\exp({}^0\mathfrak{t})$ contains the center of $\mathbf{SU}(p,q)$, the lemma follows.

2.10. The center of $\widetilde{U}(p,q)$ acts on $(V_k, L_k^2(\mathbf{R}^n))$ by scalars. Hence the restriction of V_k to $\mathbf{SU}(p,q)$ is still irreducible. We will look upon $(V_k, L_k^2(\mathbf{R}^n))$ as a representation of $\mathbf{SU}(p,q)$.

Set $K = \mathbf{U}(n) \cap \mathbf{SU}(p,q)$, $G = \mathbf{SU}(p,q)$. Set H_l equal to the space of K-finite vectors of $L_l^2(\mathbf{R}^n)$. Then clearly

(1) $$H_l = L_l^2(\mathbf{R}^n) \cap \psi_0 P(\mathbf{R}^n).$$

Let

$$\mathfrak{b} = \mathfrak{t} \cap \mathfrak{g}, \qquad \Psi_k = \Phi(\mathfrak{k}_c, \mathfrak{b}_c), \qquad \Psi_k^+ = \Psi_k \cap \Psi^+.$$

2.11. LEMMA. *Set $\Lambda_l = -l\eta_p + \eta_{p+1} + \cdots + \eta_{p+q}$ if $l \geq 0$, and $\Lambda_l = \eta_{p+1} + \cdots + \eta_{p+q-1} + (1-l)\eta_{p+q}$ if $l < 0$ and $q > 0$. Then the weights of $\mathfrak{b}$ on H_l are of multiplicity 1 and are of the form $\Lambda_l - Q$ with Q a sum of elements of Ψ^+.*

Set $\mathfrak{u}^- = \sum_{\alpha \in \Psi^+}(\mathfrak{u}_c)_{-\alpha}$. Then $H_l = U(\mathfrak{u}^-) \cdot H_l^{\mathfrak{u}^+}$. 2.6 implies that

$$H_l^{\mathfrak{u}^+} = \begin{cases} \mathbf{C}\psi_{0,\ldots,l,0,\ldots,0} & \text{if } l \geq 0, \\ \mathbf{C}\psi_{0,\ldots,0,-l} & \text{if } l < 0, \end{cases}$$

with l in the p-th position if $l \geq 0$. An easy computation shows that

$$V(h)\big|_{H_l^{\mathfrak{u}^+}} = \Lambda_l(h)I, \qquad h \in \mathfrak{b}_c.$$

This proves the lemma.

2.12. Set $\mathfrak{b_R} = \{h \in \mathfrak{b}_c \mid \alpha(h) \in \mathbf{R} \text{ for } \alpha \in \Psi\}$. Then the Weyl chamber in $\mathfrak{b_R}$ defined by Ψ^+ is given by the following inequalities:

$$\eta_1 > \cdots > \eta_p > \eta_{p+q} > \cdots > \eta_{p+1}.$$

Let $\mathfrak{p}$ be the $\mathfrak{k}$-invariant complement to $\mathfrak{k}$ in $\mathfrak{g}$. Set $\Psi_n = \Psi - \Psi_k$ and $\Psi_n^+ = \Psi_n \cap \Psi^+$. Then $\mathfrak{p}_c = \mathfrak{p}^+ \oplus \mathfrak{p}^-$ with $\mathfrak{p}^\pm = \sum_{\alpha \in \Psi_n^+}(\mathfrak{g}_c)_{\pm\alpha}$. It is easily checked that $\mathrm{Ad}(k)\mathfrak{p}_c^\pm \subset \mathfrak{p}_c^\pm$ for $k \in K$.

Set, as usual, $2\rho = \sum_{\alpha \in \Psi^+}\alpha$. Then

(1) $$V_l \text{ has infinitesimal character } \chi_{\Lambda_l+\rho} \text{ for } l \in \mathbf{Z}.$$

(1) follows from 2.11. We note that

(2) $$\rho = \sum_{i=1}^p (p+q-i)\eta_i + \sum_{i=1}^q (i-1)\eta_{i+p}.$$

This implies that if $l \geq 0$, then

(3) $$\Lambda_l + \rho = \sum_{i=1}^{p-1}(p+q-i)\eta_i + (q-l)\eta_p + \sum_{i=1}^q i\eta_{i+p}.$$

Also, if $l < 0$, then

$$(4) \qquad \Lambda_l + \rho = \sum_{i=1}^{p}(p+q-i)\eta_i + \sum_{i=1}^{q-1}i\eta_{i+p} + (q-l)\eta_{p+q}.$$

2.13. PROPOSITION. *Let (δ_l, F_l) be the irreducible representation of G with highest weight $-l\eta_{p+1}$ for $l \geq 0$. (That is, F_l is the l-th symmetric power of the standard representation of G on $\mathbf{C}^{p+q}$.) If $l \geq q > 0$, then*

$$H^q(\mathfrak{g}, K; H_l \otimes F_{l-q}^*) \neq (0).$$

The Weyl group $W(\mathfrak{g}_c, \mathfrak{b}_c)$ acts on $\mathfrak{b}_c^*$ by the permutations of the η_i. Let S_n be the permutation group on n letters. If $s \in S_n$, set $s\eta_i = \eta_{s^{-1}i}$. Let s be the permutation $(p, p+1, \ldots, p+q)$. Then

$$(1) \qquad s^{-1}(\Lambda_l + \rho) = (q-l)\eta_{p+1} + \rho.$$

This implies that H_l and F_{l-q} have the same infinitesimal character. Thus the results in Chapter II imply that if $l \geq q$

$$(2) \qquad \dim H^j(\mathfrak{g}, K; H_l \otimes F_{l-q}^*) = \dim \operatorname{Hom}_K(\Lambda^j\mathfrak{p} \otimes F_{l-q}, H_l).$$

Let W_λ be the irreducible representation of K with highest weight λ. We have

$$(3) \qquad \dim \operatorname{Hom}_K(W_{s\rho-\rho}, \Lambda^q\mathfrak{p}^-) = 1.$$

We note that $l(s) = q$. We leave it to the reader to check that $s\Psi^+ \supset \Psi_k^+$. (3) now follows from Lemma 3.5 in [**62**].

Moreover,

$$(4) \qquad \dim \operatorname{Hom}_K(W_{(q-l)s\eta_{p+1}}, F_{l-q}) = 1.$$

Indeed, $(q-l)s\eta_{p+1}$ is the highest weight of F_{l-q} relative to $s\Psi^+$. The equalities (3), (4) imply

$$\Lambda^q\mathfrak{p} \otimes F_{l-q} \supset W_{s\rho-\rho} \otimes W_{s(q-l)\eta_{p+1}}.$$

Thus

$$(5) \qquad \operatorname{Hom}_K(W_{s((q-l)\eta_{p+1}+\rho)-\rho}, \Lambda^q\mathfrak{p} \oplus F_{l-q}) \neq (0).$$

Since $s((q-l)\eta_{p+1} + \rho) - \rho = \Lambda_l$ and $W_{\Lambda_l} \subset H_l$ by 2.11, (5) and (2) imply the proposition.

2.14. COROLLARY (to the proof of 2.13).

$$(1) \qquad \dim \operatorname{Hom}_K(\Lambda^q\mathfrak{p}^- \otimes F_{l-q}, H_l) \geq 1$$

for $l \geq q > 0$. In particular if $q = 1$, then H_1 is equivalent with $J_{0,1}$ (in the notation of VI, 4.8).

REMARK. By [**62**, 3.7], (1) is in fact an equality.

3. The theta distributions

3.1. If $\phi \in \mathcal{S}(\mathbf{R}^n)$ define

$$(F\phi)(z) = \int_{\mathbf{R}^n} \phi(x) e^{-2\pi i \langle x, z \rangle} \, dx.$$

Then $F \colon \mathcal{S}(\mathbf{R}^n) \to \mathcal{S}(\mathbf{R}^n)$, and F extends to a bijective unitary operator on $L^2(\mathbf{R}^n)$.

Set $A = \alpha((2\pi)^{1/2} I)$ (see 1.6(2)). Then $AFA^{-1} = F$. For $g \in \mathbf{Mp}(n, \mathbf{R})$, define $\mathcal{W}(g) = AW(g)A^{-1}$. Set $\phi_m = A\psi_m$ for $m \in (\mathbf{Z}^+)^n$.

In particular, we note that if $X \in \mathfrak{p}_n(\mathcal{R})$ (see (1.11)), then

$$W\left(\mathrm{Exp}\begin{pmatrix} 0 & X \\ 0 & 0 \end{pmatrix}\right) \phi(z) = e^{\pi i \langle Xz, z \rangle} \phi(z).$$

3.2. If $L \subset \mathbf{R}^n$ is a lattice, define

$$L^* = \{r \in \mathbf{R}^n \mid \langle r, \tau \rangle \in \mathbf{Z} \text{ for all } \tau \in L\}.$$

If L is a lattice, then $T_L = \mathbf{R}^n / L$ is a torus. If $\gamma \in L^*$ and $x \in \mathbf{R}^n$, set $e_\gamma(x) = \exp(2\pi i \langle \gamma, x \rangle)$. Then $e_\gamma(x + \tau) = e_\gamma(x)$ for $\tau \in L$. Thus $e_\gamma \in \widehat{T_L}$. It is easy to see that $\widehat{T_L} = \{e_\gamma \mid \gamma \in L^*\}$. We give T_L the invariant measure that satisfies $\int_{\mathbf{R}^n} f(x) \, dx = \int_{T_L} f_L(t) \, dt$, where $f_L(x + L) = \sum_{\gamma \in L} f(x + \gamma)$ for, say, $f \in \mathcal{S}(\mathbf{R}^n)$. Let $m(L) = \mathrm{vol}(T_L)$ relative to dt.

3.3. THEOREM (Poisson summation). *If $f \in \mathcal{S}(\mathbf{R}^n)$, then*

$$\sum_{\gamma \in L^*} (Ff)(\gamma) = m(L) \sum_{\gamma \in L} f(\gamma).$$

If $f \in \mathcal{S}(\mathbf{R}^n)$ and $\phi \in L^2(T_L)$, define

$$(\lambda(f)\phi)(z) = \int_{\mathbf{R}^n} \phi(z - x) f(x) \, dx = \int_{T_L} \phi(z - t) f_L(t) \, dt = \int_{T_L} \phi(t) f_L(z - t) \, dt.$$

The standard theory of Fourier series (or the Peter-Weyl theorem) implies that $\lambda(f)$ is of trace class and

$$\mathrm{tr}\, \lambda_L(f) = m(L) f_L(0) = m(L) \sum_{\gamma \in L} f(\gamma).$$

On the other hand if $\gamma \in L^*$,

$$(\lambda(f)e_\gamma)(z) = e_\gamma(z) \int_{T_L} e_\gamma(t)^{-1} f_L(t) \, dt = e_\gamma(z)(Ff)(\gamma)$$

by the normalization of measures. Hence

$$\mathrm{tr}\, \lambda(f) = \sum_{\gamma \in L^*} (Ff)(\gamma).$$

3.4.   For a lattice $L \subset \mathbf{R}^n$ and $f \in \mathcal{S}(\mathbf{R}^n)$, define

$$\delta_L(f) = f_L(0) = \sum_{\gamma \in L} f(\gamma).$$

Clearly $\delta_L \in \mathcal{S}'(\mathbf{R}^n)$ (i.e., δ_L is a tempered distribution).

If $S \subset L$ is a sublattice and if $\chi \in (L/S)\hat{\ }$, define

$$\delta_{L,S,\chi}(f) = \sum_{\tau \in L/S} \chi(\tau) f_S(\tau).$$

The tempered distributions $\delta_{L,S,\chi}$ are the theta distributions alluded to in the title of this section.

3.5. LEMMA. *Let L and S be as in 3.4, and let $\chi \in (L/S)\hat{\ }$ and $f \in \mathcal{S}(\mathbf{R}^n)$. Then*

(1)
$$\delta_{L,S,\chi}(f) = \frac{1}{m(L)}(Ff)_{L^*}(-\mu),$$

where $\mu \in S^*$ is such that

$$\chi(\gamma + S) = e^{2\pi i \langle \gamma, \mu \rangle} \qquad (\gamma \in L).$$

It follows from the definition that

(2)
$$\delta_{L,S,\chi}(f) = \sum_{\tau \in L} \chi(\tau) f(\tau) = (\chi f)_L(0),$$

where χ is viewed as a character of L. But

(3)
$$F(e^{2\pi i \langle \nu, \cdot \rangle} f)(y) = (Ff)(y - \nu), \qquad \nu, y \in \mathbf{R}^n.$$

Hence the lemma follows from (2) and Poisson summation (3.3).

3.6.   If S is a sublattice of $\mathbf{Z}^n$ and $L = \mathbf{Z}^n$, then we denote $\delta_{\mathbf{Z}^n,S,\chi}$ by $\delta_{S,\chi}$, $\chi \in (\mathbf{Z}^n/S)\hat{\ }$. Set

$$\Gamma_{S,\chi} = \{\gamma \in \mathbf{Mp}(n,\mathbf{R}) \mid \delta_{S,\chi} \circ \mathcal{W}(\gamma) = \delta_{S,\chi} \text{ and } \nu(\gamma) \in \mathbf{Sp}(n,\mathbf{Z})\}.$$

3.7. THEOREM (Bass, Milnor, Serre [1]). *If $S \subset \mathbf{Z}^n$ is a sublattice and $\chi \in (\mathbf{Z}^n/S)\hat{\ }$, then $\nu(\Gamma_{S,\chi})$ contains a congruence subgroup of $\mathbf{Sp}(n,\mathbf{Z})$.*

Let $m \in \mathbf{Z}$, $m > 0$, be such that $mS^* \subset \mathbf{Z}^n$. Let $\mathfrak{p}_n(\mathbf{Z}) = \{X \in \mathfrak{p}_n(\mathbf{R}) \mid X$ has integral matrix entries$\}$. If $X \in \mathfrak{p}_n(\mathbf{Z})$ and $\gamma_1, \gamma_2 \in S^*$, then $\langle 2m^2 X \gamma_1, \gamma_2 \rangle \in 2\mathbf{Z}$. We compute

$$\delta_{S,\chi} \circ \left(\mathcal{W}\left(\text{Exp} \begin{bmatrix} 0 & 2m^2 X \\ 0 & 0 \end{bmatrix} \right) f \right) = \sum_{\tau \in \mathbf{Z}^n/S} \chi(\tau) \sum_{\gamma \in S} e^{\pi i \langle 2m^2 X(\gamma+\tau), \gamma+\tau \rangle} f(\gamma + \tau)$$

$$= \sum_{\tau \in \mathbf{Z}^n/S} \chi(\tau) \sum_{\gamma \in S} f(\gamma + \tau) = \delta_{S,\chi}(f).$$

Also taking into account 3.5, we get

$$\delta_{S,\chi} \circ \mathcal{W}\left(\mathrm{Exp}\begin{bmatrix} 0 & 0 \\ -2m^2X & 0 \end{bmatrix}\right) f = \delta_{S,\chi} \cdot F \cdot \mathcal{W}\begin{bmatrix} 0 & 2m^2X \\ 0 & 0 \end{bmatrix} F^{-1}f$$

$$= \sum_{\tau \in \mathbf{Z}^n} e^{\pi i \langle (2m^2X)(\tau+\mu),(\tau+\mu)\rangle}(F^{-1}f)(\tau+\mu)$$

$$= \sum_{\tau \in \mathbf{Z}^n} (Ff)(\tau-\mu) = \delta_{S,\chi}(f).$$

This implies that $\nu(\Gamma_{S,\chi})$ contains the group generated by the elements of the form $\begin{bmatrix} I & X \\ 0 & I \end{bmatrix}$ and $\begin{bmatrix} I & 0 \\ X & I \end{bmatrix}$ with $X \in 2m^2\mathfrak{p}_n(\mathbf{Z})$. It is shown in [1], p. 130, that these matrices generate a congruence subgroup of $\mathbf{Sp}(n, \mathbf{Z})$.

3.8. LEMMA. *Let* $f \in \mathcal{S}(\mathbf{R}^n)$. *If* $\delta_{S,\chi}(f) = 0$ *for all lattices* $S \subset \mathbf{Z}^n$, *and* $\chi \in (\mathbf{Z}^n/S)\hat{\ }$, *then* $f(\tau) = 0$ *for all* $\tau \in \mathbf{Z}^n$.

Let $\mu \in \mathbf{Q}^n$. Then there is $j \in \mathbf{Z}$ so that $\mu \in (j\mathbf{Z}^n)^*$. If $\chi_\mu(\tau) = e^{2\pi i \langle \mu,\tau\rangle}$, then $\chi_\mu \in (\mathbf{Z}^n/mj\mathbf{Z}^n)\hat{\ }$ for all $m = 1, 2, \cdots$. Set $S_m = mj\mathbf{Z}^n$.

(1) $\qquad$ If $f \in \mathcal{S}(\mathbf{R}^n)$, then $\delta_{S_m,\chi_\mu}(f) = \displaystyle\sum_{\tau \in \mathbf{Z}^n} e^{2\pi i \langle \mu,\tau\rangle} f(\tau)$.

This is clear, since $e^{2\pi i \langle \mu,\tau+\gamma\rangle} = e^{2\pi i \langle \mu,\tau\rangle}$ for $\gamma \in S_m$. Also,

(2) $\qquad$ If $\displaystyle\lim_{i\to\infty} \mu_i = \mu_0$, $\mu_0 \in \mathbf{R}^n$, then

$$\lim_{j\to\infty} \sum_{\tau \in \mathbf{Z}^n} e^{2\pi i \langle \mu_j,\tau\rangle} f(\tau) = \sum_{\tau \in \mathbf{Z}^n} e^{2\pi i \langle \mu_0,\tau\rangle} f(\tau).$$

(2) follows from the dominated convergence theorem. (1) and (2) imply

(3) $\qquad\qquad \displaystyle\sum_{\tau \in \mathbf{Z}^n} e^{2\pi i \langle \tau,\mu\rangle} f(\tau) = 0 \quad$ for all $\mu \in \mathbf{R}^n$.

Since $|f(\tau)| \leq C_k(1 + \|\tau\|^2)^{-k}$ for $k = 1, 2, \ldots$, the left-hand side of (3) is an absolutely convergent Fourier series representing 0; hence its coefficients, $f(\tau)$, are zero.

3.9. THEOREM. *If* $\Gamma \subset \mathbf{Mp}(n, \mathbf{R})$ *is a discrete subgroup, set* $\mathcal{S}'(\mathbf{R}^n)^\Gamma = \{\lambda \in \mathcal{S}'(\mathbf{R}^n) \mid \lambda \circ \mathcal{W}(\gamma) = \lambda, \gamma \in \Gamma\}$. *If* $\phi \in \mathcal{S}(\mathbf{R}^n)$ *is such that* $\lambda(\phi) = 0$ *for all* $\lambda \in \mathcal{S}'(\mathbf{R}^n)^\Gamma$ *and all* $\Gamma \subset \mathbf{Mp}(n, \mathbf{R})$ *such that* $\nu(\Gamma)$ *contains a congruence subgroup of* $\mathbf{Sp}(n, \mathbf{Z})$, *then* $\phi(\tau) = 0$ *for all* $\tau \in \mathbf{Z}^n$.

This follows from 3.7 and 3.8.

3.10.   The discussion in this section is strongly influenced by the many conversations the second named author has had with Roger Howe about the oscillator representation. In particular, the term theta distribution is due to Roger Howe.

4. The reciprocity formula

In this section G denotes a connected semi-simple Lie group with finite center, and K a maximal compact subgroup of G.

4.1.   If (π, H) is a unitary representation of G, then H^∞ denotes (as usual, see 0, 2.3) the space of C^∞ vectors for (π, H) with the C^∞ topology. $(H^\infty)^*$ denotes the space of continuous linear functionals on H^∞. If (π_i, H_i), $i = 1, 2$ are unitary representations of G, then $\mathrm{Hom}_G(H_1, H_2)$ denotes the space of all bounded linear operators $A \colon H_1 \to H_2$ such that

$$A \circ \pi_1(g) = \pi_2(g) \circ A$$

for $g \in G$.

4.2.   Let $\Gamma \subset G$ be a cocompact, discrete subgroup of G. Let π_Γ denote the right regular representation of G on $L^2(\Gamma \backslash G)$ (here we fix a bi-invariant measure dg on G, hence a right invariant measure $d(\Gamma g)$ on $\Gamma \backslash G$). We recall that the space of C^∞ vectors of $(\pi_\Gamma, L^2(\Gamma \backslash G))$ is precisely $C^\infty(\Gamma \backslash G)$ with the C^∞ topology (III, 7.9).

If (π, H) is a unitary representation of G, set $(H^\infty)^{*\Gamma} = \{\lambda \in (H^\infty)^* \mid \lambda \circ \pi(\gamma) = \lambda$ for $\gamma \in \Gamma\}$.

4.3.   THEOREM (Gelfand, Graev, Piatetski-Shapiro [42]). *Let (π, H) be an irreducible unitary representation of G. If $A \in \mathrm{Hom}_G(H, L^2(\Gamma \backslash G))$, set $\lambda_A(v) = A(v)(\Gamma \cdot 1)$ for $v \in H^\infty$ (this makes sense by 4.2). Then the map $A \mapsto \lambda_A$ is a bijection from $\mathrm{Hom}_G(H, L^2(\Gamma \backslash G))$ to $(H^\infty)^{*\Gamma}$.*

If $A \in \mathrm{Hom}_G(H, L^2(\Gamma \backslash G))$ and $\lambda_A = 0$, then $A(v)(\Gamma \cdot 1) = 0$ for $v \in H^\infty$. Hence, if $g \in G$,

$$0 = A(\pi(g)v)(\Gamma \cdot 1) = (\pi_\Gamma(g)A(v))(\Gamma \cdot 1) = A(v)(\Gamma \cdot g)$$

for all $v \in H^\infty$. Thus $A(H^\infty) = 0$. But then $A = 0$, since H^∞ is dense. This proves the injectivity of $A \mapsto \lambda_A$.

If $\lambda \in (H^\infty)^{*\Gamma}$, then set $A_\lambda(v)(\Gamma g) = \lambda(\pi(g)v)$ for $g \in G$, $v \in H^\infty$.

Then $A_\lambda \colon H^\infty \to C^\infty(\Gamma \backslash G)$, and $A_\lambda(\pi(g)v) = \pi_\Gamma(g)A_\lambda(v)$ for $g \in G$.

Let $\langle \, , \, \rangle$ denote the inner product on H and let $\langle \, , \, \rangle_\Gamma$ denote the inner product on $L^2(\Gamma \backslash G)$. Set

$$(v, w) = \langle A_\lambda(v), A_\lambda(w) \rangle_\Gamma.$$

Then $(\, , \,)$ defines a $\mathfrak{g}$-invariant inner product on H_0 (the K-finite vectors of H). Hence, since the K-isotypic components of H_0 are finite dimensional, we see that if $v, w \in H_0$, then

$$(v, w) = \langle Bv, w \rangle,$$

with $B \colon H_0 \to H_0$ a linear map such that $B(X \cdot v) = X \cdot B(v)$ for $X \in \mathfrak{g}$, $v \in H_0$. H_0 is an irreducible $(\mathfrak{g}, K)$-module. Hence $B = \mu I$ with $\mu \in \mathbf{R}$, $\mu \geq 0$.

This shows that if $v \in H_0$, then

$$(1) \qquad\qquad \langle A_\lambda(v), A_\lambda(v) \rangle_\Gamma = \mu \langle v, v \rangle.$$

(1) implies that $A_\lambda\big|_{H_0}$ extends to a bounded operator C from H to $L^2(\Gamma \backslash G)$. Since H_0 consists of analytic vectors for H, it follows that $C \in \mathrm{Hom}_G(H, L^2(\Gamma \backslash G))$. But $\lambda_C\big|_{H_0} = \lambda$; therefore $\lambda_C = \lambda$.

4.4. REMARK. Theorem 4.3 can be viewed as a consequence of III, 7.9, and two quite general facts. To see this, note first that the space $C^\infty(\Gamma\backslash G)$ of C^∞ complex valued functions on $\Gamma\backslash G$ may be viewed as the space of the induced representation Ind_Γ^G in the smooth category (III, 2.1), where $\mathbf{C}$ is viewed as a trivial Γ-module. The first part of the proof of 4.1 just establishes a special case of "Frobenius reciprocity" (see IX, 5.9), namely

$$(1) \qquad \mathrm{Hom}_G(H, C^\infty(\Gamma\backslash G)) = \mathrm{Hom}_\Gamma(H^\infty, \mathbf{C}) = (H^\infty)^{*\Gamma}.$$

(Here, H^∞ could be any admissible smooth G-module.)

The second part of the proof of 4.3 shows in fact more generally that if U, V are unitary G-modules, and U is irreducible (hence admissible), then

$$(2) \qquad \mathrm{Hom}_G(U, V) = \mathrm{Hom}_G(U^\infty, V^\infty).$$

The special case considered in 4.3 is

$$(3) \qquad \mathrm{Hom}_G(H, L^2(\Gamma\backslash G)) = \mathrm{Hom}_G(H^\infty, L^2(\Gamma\backslash G)^\infty).$$

Theorem 4.3 is then a consequence of (1), (3) and III, 7.9 (recalled in 4.2).

5. The imbedding of V_l into $L^2(\Gamma\backslash G)$

5.1. Let k be a totally real finite extension of $\mathbf{Q}$, and denote by $r + 1$ its degree. We assume $r \geq 1$, fix an imbedding of k into $\mathbf{R}$, and view k as a subfield of $\mathbf{R}$. Let $\Sigma = \{\sigma_1, \ldots, \sigma_{r+1}\}$ be the set of isomorphisms of k into $\mathbf{R}$, where $\sigma_{r+1} = \mathrm{id}$. Let $k' = k(i)$. We extend $\sigma \in \Sigma$ to the imbedding of k' into $\mathbf{C}$ which leaves i fixed.

Let n be a positive integer, h a non-degenerate Hermitian form on $U_{k'} = k'^n$ of signature (p, q) $(p \geq q > 0;\ p + q = n)$. We assume that for $\sigma \in \Sigma$, $\sigma \neq 1$, the form ${}^\sigma h$, given by $z, w \mapsto \sigma^{-1}(h(\sigma z, \sigma w))$, is definite.

5.2. Let $\mathcal{H}(k) = \{g \in \mathbf{SL}(n, k') \mid h(g{\cdot}z, g{\cdot}w) = h(z, w), (z, w \in U_{k'})\}$. It is the group of points over k of a k-form $\mathcal{H}$ of $\mathbf{SL}_n$. Now $h(z, w) = \mu(z, w) + \sqrt{-1}\beta(z, w)$, with μ a symmetric k-bilinear form with values in k and β a skew symmetric k-bilinear form with values in k. We regard $U_{k'}$ as a $2n$-dimensional vector space over k, and write U_k instead. Using a symplectic basis for β, we see that $\mathcal{H}(k) \subset \mathbf{Sp}(n, k)$, or more precisely that we have an imbedding, defined over k, of $\mathcal{H}$ in the symplectic group $\mathbf{Sp}_n$, viewed as a k-group.

5.3. Let $\mathrm{Res}_{k/\mathbf{Q}}$ denote *restriction of scalars* from k to $\mathbf{Q}$ (see Weil [**116**], Chap. 1). Then $\mathcal{G} = \mathrm{Res}_{k/\mathbf{Q}}(\mathcal{H})$ and $\mathrm{Res}_{k/\mathbf{Q}}(\mathbf{Sp}_n)$ are defined over $\mathbf{Q}$, and we have a canonical imbedding $\mathcal{G} = \mathrm{Res}_{k/\mathbf{Q}}\mathcal{H} \to \mathrm{Res}_{k/\mathbf{Q}}\mathbf{Sp}_n$. Moreover, the group $\mathcal{G}(\mathbf{Q})$ of rational points of $\mathcal{G}$ is equal to $\mathrm{Res}_{k/\mathbf{Q}}(\mathcal{H}(k))$. Let $U_\mathbf{Q}$ be U_k viewed as a $2n(r+1)$-dimensional vector space over $\mathbf{Q}$, and β_C the bilinear form on $U_\mathbf{Q}$ defined by β. It is antisymmetric non-degenerate, and we have $U_\mathbf{Q} = \mathrm{Res}_{k/\mathbf{Q}}U_k$, $\beta_\mathbf{Q} = \mathrm{Res}_{k/\mathbf{Q}}\beta$. Therefore $\mathcal{G}$ is naturally embedded in the group of automorphisms of $U_\mathbf{Q} \otimes_\mathbf{Q} \mathbf{C}$ leaving $\beta_\mathbf{Q}$ invariant, i.e. in $\mathbf{Sp}_N$, where $N = n(r+1)$.

Over $\mathbf{R}$, the group $\mathcal{G}$ is isomorphic to the product of the groups ${}^\sigma H$ $(\sigma \in \Sigma)$, where ${}^\sigma H$ is the group of automorphisms of $U_{k'} \otimes \mathbf{C}$ preserving ${}^\sigma h$. Therefore the group $\mathcal{G}(\mathbf{R})$ of real points of $\mathcal{G}$ is isomorphic to the product of $\mathbf{SU}(p, q)$ by r copies of $\mathbf{SU}(n)$. Of course ${}^\sigma\mathbf{Sp}_n$ is again $\mathbf{Sp}_n$; hence the group of real points of $\mathrm{Res}_{k/\mathbf{Q}}\mathbf{Sp}_n$ is the direct product of $r + 1$ copies of $\mathbf{Sp}(n, \mathbf{R})$. The imbedding $\mathcal{H} \hookrightarrow \mathbf{Sp}_n$ yields $\psi_{r+1}\colon \mathbf{SU}(p, q) \hookrightarrow \mathbf{Sp}(n, \mathbf{R})$ and, for $\sigma \in \Sigma$, $\sigma \neq \sigma_{r+1}$, the corresponding imbedding of ${}^\sigma\mathcal{H}$ into $\mathbf{Sp}_n$ yields $\psi_i\colon \mathbf{SU}(n) \hookrightarrow \mathbf{Sp}(n, \mathbf{R})$. The direct product of $(r+1)$ copies of $\mathbf{Sp}(n, \mathbf{R})$ is naturally contained in $\mathbf{Sp}(n, \mathbf{R})$; our given embedding $\mathcal{G}(\mathbf{R}) \hookrightarrow \mathbf{Sp}(N, \mathbf{R})$ is, up to conjugation over $\mathbf{R}$, the product ψ of the ψ_i, followed by that inclusion.

Let $e_1, \ldots, e_{2N}$ be a basis of $U_{\mathbf{Q}}$, so that $\beta_{\mathbf{Q}}$ is in standard form. We have $\mathcal{G}(\mathbf{Z}) = \{\gamma \in \mathcal{G}(\mathbf{Q}) \mid \psi(\gamma) \in \mathbf{Sp}(N, \mathbf{Z})\}$. Then $\mathcal{G}(\mathbf{Z})$ is an arithmetic subgroup of $\mathcal{G}(\mathbf{R})$ (see [9], 7.11, 7.12). Also $\psi: \mathcal{G}(\mathbf{Z}) \to \mathbf{Sp}(N, \mathbf{Z})$.

5.4. THEOREM (Borel and Harish-Chandra [14]). $\mathcal{G}(\mathbf{Z})$ *is a cocompact discrete subgroup of* $\mathcal{G}(\mathbf{R})$.

We have $\mathcal{G}(\mathbf{R}) = \bigtimes_{i=1}^{r+1} G_i$, where $G_i = \mathbf{SU}(n)$, $i \le r$, and $G_{r+1} = \mathbf{SU}(p, q)$. Let $p_i: \mathcal{G}(\mathbf{R}) \to G_i$ be the i-th projection. The definition of $\mathcal{G}(\mathbf{Q})$ implies that $p_i|_{\mathcal{G}(\mathbf{Q})}$ is injective for each i.

If $\gamma \in \mathcal{G}(\mathbf{Z})$ were not semisimple, then $p_i(\gamma)$ would be so for each $1 \le i \le r$. But $p_i(\gamma) \in G_i = \mathbf{SU}(n)$. Thus $\mathcal{G}(\mathbf{Z})$ consists of semisimple elements. The result now follows from [14].

5.5. Let $p_i: \mathcal{G}(\mathbf{R}) \to G_i$, $i \le r+1$, be as in the proof of 5.4. Set $p_{r+1}(\mathcal{G}(\mathbf{Z})) = \Gamma$. If $\omega \subset \mathbf{SU}(p, q)$ is a compact subset, then $p_{r+1}^{-1}(\omega) \subset \mathcal{G}(\mathbf{R})$ is compact. Thus Γ is a cocompact, discrete subgroup of $\mathbf{SU}(p, q)$.

Lemma 2.9 implies that $\psi: \mathcal{G}(\mathbf{R}) \to \mathbf{Sp}(N, \mathbf{R})$ lifts to $\widetilde{\psi}: \mathcal{G}(\mathbf{R}) \to \mathbf{Mp}(N, \mathbf{R})$. Indeed, we have $\widetilde{\alpha}: \bigtimes_{i=1}^{r+1} \mathbf{Mp}(n, \mathbf{R}) \to \mathbf{Mp}(N, \mathbf{R})$, and $\psi_i: G_i \to \mathbf{Sp}(n, \mathbf{R})$ lifts to $\widehat{\psi}_i: G_i \to \mathbf{Mp}(n, \mathbf{R})$, and we set $\widetilde{\psi} = \alpha \circ \bigtimes_{i=1}^{r+1} \widehat{\psi}_i$. Using this observation, we see that if W^j is the oscillator representation of $\mathbf{Mp}(n, \mathbf{R})$, $j = 1, \ldots, r+1$ and W is the oscillator representation of $\mathbf{Mp}(N, \mathbf{R})$, then $W \circ \widetilde{\psi}$ is equivalent with

$$\left(W^1 \circ \widehat{\psi}_1\right) \widehat{\otimes} \left(W^2 \circ \widehat{\psi}_2\right) \widehat{\otimes} \cdots \widehat{\otimes} \left(W^{r+1} \circ \widehat{\psi}_{r+1}\right) : (g_1, \ldots, g_{r+1}) \mapsto \bigotimes_{i=1}^{r+1} \left(W^i \circ \widehat{\psi}_i(g_i)\right).$$

Set $V^i = W^i \circ \widehat{\psi}_i$. Then V^i acts on the coordinates $x_{(i-1)n+1}, \ldots, x_{in}$, $y_{(i-1)n+1}, \ldots, y_{in}$.

It should be noted that the basis that splits ψ into a product is *not* the same as the basis for which our $\mathbf{Sp}(N, \mathbf{Z})$ is defined.

If $1 \le j \le r$, then $V^j = \widetilde{\bigoplus_{l \ge 0}} V_l^j$ with $\dim V_l^j < \infty$. (This corresponds to the case $q = 0$ in §2.) This implies that

$$V = \widetilde{\bigoplus_{l_1, \ldots, l_r \in \mathbf{N}}} V_{l_1}^1 \widehat{\otimes} \cdots \widehat{\otimes} V_{l_{r+1}}^{r+1}$$

as a representation of $\mathcal{G}(\mathbf{R})$. Set $V_{(l_1, \ldots, l_{r+1})}$ equal to $V_{l_1}^1 \widehat{\otimes} \cdots \widehat{\otimes} V_{l_{r+1}}^{r+1}$, and let $L^2(\mathbf{R}^N)_{(l_1, \ldots, l_{r+1})}$ be the representation space for $V_{(l_1, \ldots, l_{r+1})}$.

The results of §§1 and 2 easily imply

(1)   *The space of C^∞ vectors of $V_{(l_1, \ldots, l_{r+1})}$ is precisely $L^2(\mathbf{R}^N)_{(l_1, \ldots, l_{r+1})} \cap \mathcal{S}(\mathbf{R}^N)$ with the subspace topology.*

(2) $L^2(\mathbf{R}^N)_{(l_1, \ldots, l_{r+1})} \cap \psi_0 P(\mathbf{R}^N)$ *is dense in* $L^2(\mathbf{R}^N)_{(l_1, \ldots, l_{r+1})}$ *and in the C^∞ vectors for* $V_{(l_1, \ldots, l_{r+1})}$.

5.6. THEOREM. *If $l \in \mathbf{Z}$, then there exists a subgroup Γ' of finite index (indeed, a congruence subgroup) of Γ such that*

$$((V_l^{r+1})^\infty)^{*\Gamma'} \neq (0).$$

(Here, V_l^{r+1} is the representation of $\mathbf{SU}(p,q)$ denoted by V_l in §2.)

Fix $l_1, \ldots, l_r \in \mathbf{N}$. Let $H = L^2(\mathbf{R}^N)_{(l_1, \ldots, l_r, l)}$. Then $H^\infty = H \cap \mathcal{S}(\mathbf{R}^N)$, and $H \cap \psi_0 P(\mathbf{R}^N)$ is dense in H and H^∞. Let $\mathbf{Z}^N$ be the lattice associated with the basis for the $\mathbf{Sp}(N, \mathbf{Z})$ we are considering in this section. Let A be as in 3.1. If $\phi \in H \cap \psi_0 P(\mathbf{R}^N)$ and $A\phi(\tau) = 0$ for $\tau \in \mathbf{Z}^N$, then $\phi = 0$ (i.e. $\mathbf{Z}^N$ is Zariski dense in $\mathbf{R}^N$). Thus if $\phi \neq 0$, there exists $\Delta \subset \mathbf{Mp}(N, \mathbf{R})$ such that

(1) $\nu(\Delta)$ *contains a congruence subgroup of* $\mathbf{Sp}(N, \mathbf{Z})$.

(2)   *There is $\lambda \in \mathcal{S}'(\mathbf{R}^N)^\Delta$ so that $\lambda(A\phi) \neq 0$.*

See Theorem 3.9 for this result and the notation.

Set $\mu = \lambda \circ A$ restricted to H^∞. Let Ω be a congruence subgroup of $\mathcal{G}(\mathbf{Z})$ so that $\psi(\Omega) \subset \nu(\Delta)$. Then $\widetilde{\psi} \colon \Omega \to \Delta$. Hence $\mu \in (H^\infty)^{*\Omega}$. Let $\Gamma' = \pi(\Omega)$. Then $\Gamma' \subset \Gamma$ is a congruence subgroup. Fix $\phi = \phi_1 \otimes \cdots \otimes \phi_{r+1} \in \psi_0 P(\mathbf{R}^N) \cap H$ so that $\mu(\phi) \neq 0$. Define $\xi(f) = \mu(\phi_1 \otimes \cdots \otimes \phi_r \otimes f)$ for $f \in (V_l^{r+1})^\infty$. Then $\xi \in ((V_l^{r+1})^\infty)^{*\Gamma'}$, $\xi \neq 0$.

5.7. We now revert to our old notation: $G = \mathbf{SU}(p,q)$, $p \geq q > 0$. $\psi \colon G \to \mathbf{Sp}(n, \mathbf{R})$ $(n = p + q)$ and $\widetilde{\psi} \colon G \to \mathbf{Mp}(n, \mathbf{R})$ the lift of ψ. Let V_l and $V = W \circ \widetilde{\psi}$ be as in §2. However, we fix Γ as constructed above.

5.8. COROLLARY. *If $l \in \mathbf{Z}$, then there is a congruence subgroup Γ' of Γ (possibly depending on l) such that*

$$\mathrm{Hom}_G(V_l, L^2(\Gamma' \backslash G)) \neq 0.$$

This is just 4.3 combined with 5.6.

5.9. COROLLARY. *Let $l \in \mathbf{N}$, $l \geq q$, and let F_l be as in 2.13. Let Γ be as above. Then there is a congruence subgroup $\Gamma' \subset \Gamma$ such that $H^q(\Gamma'; F_{l-q}) \neq (0)$.*

This follows from 2.13, 5.8 and VII, 6.1.

5.10. We note that in this case the cohomology of Γ is bigraded in the same way as the cohomology of $\Gamma \backslash G \mid K$. We actually have $H^{0,q}(\Gamma'; F_{l-q}) \neq 0$ for $l \geq q$ (see VII, 6.2).

The results of this section are substantially due to D. Kazhdan [**70**]. He concentrated on the case $\mathbf{SU}(n, 1)$ and V_1. He also studied the significance of the V_{-j}, $j > 0$, for $\mathbf{SU}(2, 1)$ for Γ not necessarily cocompact. Kazhdan's proof of the pertinent results uses the global oscillator representation and strong approximation rather than Theorem 3.9.

CHAPTER IX

Continuous Cohomology and Differentiable Cohomology

Introduction

In most of the previous chapters we have been studying the relative Lie algebra cohomology spaces $H^*(\mathfrak{g}, K; V)$ with coefficients in a $(\mathfrak{g}, K)$-module. Our only case of interest is when V is the set of K-finite vectors in the space V^∞ of C^∞ vectors of a continuous G-module. In that case, by the van Est theorem, this space is also the space $H_d^*(G; V^\infty)$ of continuous (or differentiable) Eilenberg–Mac Lane cohomology of G with coefficients in V^∞. The relationship between cohomology of discrete subgroups and cohomology with coefficients in infinite dimensional representations described in VII can also be expressed in terms of continuous cohomology (and obtained directly by use of a suitable Shapiro lemma). Moreover, this relationship is also valid in the p-adic case, where there is no direct analog of the Lie algebra cohomology.

This chapter is devoted to the basic notions and results on continuous or differentiable cohomology. This is not a completely self-contained exposition, since, when convenient, we have referred to [35] or [60]. But a number of proofs have been included.

At this point, we are mainly interested in real Lie groups. However, in preparation for the p-adic or mixed case, we shall first develop continuous cohomology for locally compact groups (§§1 to 4), in the framework of [35] or [60, §2]. §§5, 6 are concerned with differentiable cohomology, which was initiated by W. T. van Est (see [104, 105] and earlier references given there). We shall largely follow the exposition of [60]. We have also borrowed from three lectures given by G. D. Mostow at the Institute for Advanced Study in Spring 1975, in particular for 5.4, 6.2, 6.3 and the proof of 5.2.

In the original version, we shifted from Lie algebra cohomology to differentiable cohomology because the latter theory had a Hochschild-Serre spectral sequence and a Shapiro lemma, both needed to compute cohomology with respect to an induced representation. As pointed out in the first version of this book, we noticed subsequently that analogues existed in the framework of Lie algebra cohomology. In the previous chapters, we have used those, so that we do not need the results of this chapter. We have kept it, however, since continuous and differentiable cohomology are of interest in various contexts, and also for the sake of the analogy with the p-adic case, whose treatment will be based on the use of such cohomology.

In this chapter, locally compact groups are assumed to be countable at infinity, topological vector spaces are Hausdorff, over **C**, *and locally convex.*

1. Continuous cohomology for locally compact groups

1.1. Let G be a locally compact group. By a *topological G-module*, or simply a G-module (π, V), we mean a topological vector space on which G acts via a continuous representation π. A G-morphism of two such G-modules is a continuous linear map which commutes with G. We let $\mathcal{C}_G$ or simply $\mathcal{C}$ if G is clear from the context, denote the category of topological G-modules and G-morphisms, and $\mathcal{C}_G^{qc}$ the full subcategory of quasi-complete G-modules.

For some of the main theorems, we shall assume that the G-modules under consideration are Fréchet spaces, i.e. are complete and metrizable. In fact, for our needs, it would be no essential loss in generality to assume this from the start, as far as real Lie groups are concerned.

1.2. If X, F are topological spaces, we let $C(X; F)$ denote the space of continuous maps of X into F, endowed with the compact open topology. Let F be a topological vector space, and let X be locally compact. Then $C(X; F)$ is quasi-complete if F is so [**22**, X7, Cor. 3], and is a Fréchet space if F is one and X countable at infinity, as follows from [**22**, X21, Cor. to Prop. 1]. If A and B are topological vector spaces, then $\mathrm{Hom}(A, B)$ denotes the space of continuous linear maps from A to B, endowed with the compact open topology. If $A, B \in \mathcal{C}_G$, then $\mathrm{Hom}(A, B)$ will be given the G-module structure defined by

$$(1) \qquad (xf)(a) = x(f(x^{-1} \cdot a)) \qquad (x \in G; \ a \in A; \ f \in \mathrm{Hom}(A, B)).$$

$\mathrm{Hom}_G(A, B)$ denotes the set of homomorphisms which commute with G. Both $\mathrm{Hom}(A, B)$ and $\mathrm{Hom}_G(A, B)$ are closed subspaces of $C(A; B)$. Similarly, if A is just a topological space on which G operates continuously, then $C(A; B)$ is endowed with the G-action defined by (1).

We recall that if $f \colon A \to B$ is a surjective continuous linear map of Fréchet spaces, then f induces a topological isomorphism of $A/\ker f$, endowed with the quotient topology, onto B [**23**, I, §3, n° 2, Thm. 1]. In particular, if

$$(2) \qquad 0 \longrightarrow A \overset{u}{\longrightarrow} B \overset{v}{\longrightarrow} C \longrightarrow 0$$

is an exact sequence of Fréchet spaces and continuous homomorphisms, then u is an isomorphism of A onto $u(A)$ and v induces an isomorphism of $B/u(A)$ onto C. Moreover, *if X is a topological space, then the associated sequence*

$$(3) \qquad 0 \longrightarrow C(X; A) \overset{u'}{\longrightarrow} C(X; B) \overset{v'}{\longrightarrow} C(X; C) \longrightarrow 0$$

is exact. This is obvious at $C(X; A)$ and $C(X; B)$. The surjectivity of v' follows from the fact that v admits a continuous (not necessarily linear) cross-section (Bourbaki, yet unpublished). In particular, if X is locally compact and countable at infinity (our only case of interest), (3) is again an exact sequence of Fréchet spaces. The surjectivity of v' in that case has already been pointed out, without proof, by A. Grothendieck in the footnote on p. 84 of [**45**].

1.3. Let $V \in \mathcal{C}_G$ and $q \in \mathbf{N}$. We let $C^q(G; V) = C(G^{q+1}; V)$, viewed as a G-module by means of the action

$$(1) \qquad (x \cdot f)(x_0, \ldots, x_q) = x(f(x^{-1} \cdot x_0, \ldots, x^{-1} \cdot x_q)) \qquad (x, x_0, \ldots, x_q \in G).$$

We let $F^q(G;V)$ be the same space, but with the action of G defined by right translations on G, i.e.

$$(2) \qquad (x \cdot f)(x_0, \ldots, x_q) = f(x_0 \cdot x, \ldots, x_q \cdot x) \qquad (x, x_0, \ldots, x_q \in G).$$

Since G is assumed to be countable at infinity, these spaces are Fréchet or quasi-complete spaces if V is so (1.2).

The map $\mu \colon F^0(G;V) \to C^0(G;V)$ defined by

$$(3) \qquad \mu(f)(x) = x \cdot f(x^{-1}) \qquad (x \in G; \ f \in F^0(G;V))$$

is readily seen to be a G-isomorphism. Since the canonical map

$$(4) \qquad C(G; C(G^q; V)) \to C(G^{q+1}; V)$$

is a topological isomorphism [**22**, X29, Thm. 3, Cor. 2], we get by iteration a G-isomorphism of $F^q(G;V)$ onto $C^q(G;V)$ $(q = 0, 1, \cdots)$ (see [**60**, §2]).

We let ε denote the maps $V \to F^0(G;V)$ and $V \to C^0(G;V)$ which assign respectively to v the function $x \mapsto x \cdot v$ and the constant function equal to v on G. These two injections are G-morphisms, which correspond to each other under μ.

1.4. The *standard homogeneous resolution* of $V \in \mathcal{C}$ is the (augmented) complex

$$(1) \qquad 0 \longrightarrow V \xrightarrow{\ \varepsilon\ } A^0(V) \xrightarrow{\ d_0\ } A^1(V)$$
$$\longrightarrow \cdots \longrightarrow A^q(v) \xrightarrow{\ d_q\ } A^{q+1}(V) \longrightarrow \cdots,$$

where $A^q(V) = C^q(G;V)$ and d_q is given by

$$(2) \qquad (d_q f)(x_0, \ldots, x_{q+1}) = \sum_i (-1)^i f(x_0, \ldots, \widehat{x}_i, \ldots, x_{q+1})$$
$$(x_i \in G; \ i = 0, \ldots, q+1).$$

The *q-th continuous cohomology group* $H^q_{\mathrm{ct}}(G;V)$ of G with coefficients in V is then, by definition, the q-th cohomology group of the complex

$$(3) \qquad A^0(V)^G \to \cdots \to A^q(V)^G \to \cdots.$$

The topological vector space $A^q(V)^G$ is isomorphic to $F^{q-1}(G;V)$ via the map $f \mapsto f'$, where

$$(4) \qquad f'(x_1, \ldots, x_q) = f(1, x_1, x_1 \cdot x_2, \ldots, x_1 \cdots x_q).$$

(By definition, $F^{-1}(G;V) = V$.) The complex (3) can then be written

$$(5)$$
$$V \xrightarrow{\ d'_0\ } F^0(G;V) \xrightarrow{\ d'_1\ } \cdots \longrightarrow F^q(G;V) \xrightarrow{\ d'_{q+1}\ } F^{q+1}(G;V) \longrightarrow \cdots,$$

where $F^q(G;V)$ is viewed as the space of elements of degree $q+1$, and the differential d'_q is given by

$$(6)$$
$$(d'_q f)(x_0, \ldots, x_q)$$
$$= x_0 \cdot f(x_1, \ldots, x_q) + \sum_{0 \leq i < q}{}' (-1)^{i+1} f(x_0, \ldots, x_i \cdot x_{i+1}, \ldots, x_q)$$
$$+ (-1)^{q+1} f(x_0, \ldots, x_{q-1}).$$

(5) is the complex of *non-homogeneous continuous cochains*. For all this, see [**60**, §2]. This is of course just the continuous analog of standard notions concerning the *Eilenberg-Mac Lane cohomology* of abstract groups [**78**, IV].

1.5.   Next we define these groups in the context of relative homological algebra [**59, 78**]. For this, as usual, we keep the objects of $\mathcal{C}$ but restrict the morphisms. We shall say that a G-morphism $f\colon A \to B$ is an s-*morphism* (*strong morphism*) if: (i) ker f and im f are closed topological direct summands; and (ii) f induces an isomorphism of $A/\ker f$ onto $f(A)$. The facts recalled in 1.2 imply that if A and B are Fréchet spaces, then (ii) follows from (i). In fact, for (ii) to hold, it suffices then that $f(A)$ be closed in B. A sequence of morphisms in G is *strong* (or an s-*sequence*) if all the morphisms are s-morphisms. An s-*exact sequence* is an exact sequence in which all morphisms are strong. If f is injective, then f is strong if (and only if) there exists a continuous linear map $h\colon B \to A$ such that $h \circ f = \mathrm{Id}$. In fact, it is easily checked that $f(A) = \ker(\mathrm{Id} - f \circ h)$ is closed and that $f \circ h$ and $\mathrm{Id} - f \circ h$ are projectors on $f(A)$ and $\ker h$ respectively.

An element $U \in \mathcal{C}_G$ is s-injective if, given a strong injection $A \to B$, every G-morphism $f\colon A \to U$ extends to a G-morphism $B \to U$ (neither is required to be strong). A continuously s-injective resolution (or, simply, an s-injective resolution of $V \in \mathcal{C}_G$) is an s-exact sequence:

$$(1) \qquad \begin{aligned} 0 \longrightarrow V \xrightarrow{\ \varepsilon\ } A^0 \xrightarrow{\ d_0\ } A^1 \\ \longrightarrow \cdots \longrightarrow A^q \xrightarrow{\ d_q\ } A^{q+1} \longrightarrow \cdots, \end{aligned}$$

in which the A^i's are s-injective. The fact that (1) is s-exact is equivalent with the existence of continuous linear maps

$$(2) \qquad \delta\colon A^0 \to V, \qquad e_q\colon A^q \to A^{q-1} \quad (q \geq 1)$$

such that

$$(3) \quad \delta \circ \varepsilon = \mathrm{Id}, \qquad \varepsilon \circ \delta + e_1 \circ d_0 = \mathrm{Id}, \qquad e_{q+1} \circ d_q + d_{q-1} \circ e_q = \mathrm{Id} \quad (q \geq 1).$$

Given such a resolution of V, and $U \in \mathcal{C}$, one defines (as usual) $\mathrm{Ext}^q_G(U, V)$ to be the q-th cohomology group of the complex $\{\mathrm{Hom}_G(U, A^i)\}$. In particular, $\mathrm{Ext}^q_G(\mathbf{C}, V)$, where $\mathbf{C}$ is viewed as the trivial G-module, is the q-th cohomology group of the complex $\{A^{iG}\}$ $(q + 0, 1, \cdots)$. Clearly,

$$(4) \qquad \mathrm{Ext}^0_G(U; V) = \mathrm{Hom}_G(U, V), \qquad \mathrm{Ext}^0_G(\mathbf{C}; V) = V^G.$$

It is standard that these groups do not depend on the s-injective resolution chosen, up to natural isomorphisms [**60**, §2]. That s-injective resolutions exist follows from the following lemma:

1.6. LEMMA.   *Let $V \in \mathcal{C}$. Then $F^0(G; V)$ is s-injective, and $\varepsilon\colon V \to F^0(G; V)$ (see 1.3) is a strong injection.   The homogeneous resolution of V (1.4(1)) is s-injective. It consists of Fréchet (resp. quasi-complete) spaces if V is one.*

The last assertion follows from 1.2. The others are proved in [**60**, §2]; see also [**35**]. In view of 1.4, this implies in particular

$$(1) \qquad \mathrm{Ext}^q_G(\mathbf{C}; V) = H^q_{\mathrm{ct}}(G; V) \qquad (q = 0, 1, 2, \cdots).$$

Since the topology of V is uniform, the natural bijections

(2)
$$Mp(U \times G^q, V) \overset{\sim}{\to} Mp(U, Mp(G^q, V)),$$
$$Mp(U \times G^q, V) \overset{\sim}{\to} Mp(G^q, Mp(U, V)),$$

where Mp refers to arbitrary maps, induce topological isomorphisms

(3) $C(U \times G^q; V) \overset{\sim}{\to} C(U; C(G^q; V)), \qquad C(U \times G^q; V) \overset{\sim}{\to} C(G^q; C(U; V)),$

[**22**, X §1, n° 4, Prop. 2]. From this it follows that we have a canonical isomorphism of topological vector spaces

(4)
$$\mathrm{Hom}(U, C^q(G; V)) = C^q(G; \mathrm{Hom}(U, V)),$$

which is easily checked to commute with G. Consequently, (1) generalizes to

(5)
$$\mathrm{Ext}_G^q(U, V) = H_{\mathrm{ct}}^q(G; \mathrm{Hom}(U, V)) \qquad (U, V \in \mathcal{C}_G; q \in \mathbf{N}).$$

REMARK. A quasi-complete G-module which is s-injective in $\mathcal{C}_G$ is of course s-injective in $\mathcal{C}_G^{\mathrm{qc}}$. Lemma 1.6 shows that $\mathcal{C}_G^{\mathrm{qc}}$ has enough injectives and that for $U, V \in \mathcal{C}_G^{\mathrm{qc}}$ the spaces $\mathrm{Ext}^*(U, V)$ and $H_{\mathrm{ct}}^*(G; V)$ may also be computed within $\mathcal{C}_G^{\mathrm{qc}}$ (without changing the topology of $H_{\mathrm{ct}}^*(G; V)$ defined below in 3.3). A similar remark is valid for Fréchet G-modules.

    1.7. LEMMA. *Let*

(1)
$$0 \to A \to B \to C \to 0$$

be an exact sequence in $\mathcal{C}$, *and for* $q \in \mathbf{N}$ *let*

(2)
$$0 \to F^q(G; A) \to F^q(G; B) \overset{u}{\longrightarrow} F^q(G; C) \to 0$$

be the canonically associated sequence of G-modules.

    (i) *If* (1) *is s-exact, then so is* (2).

    (ii) *If A, B, C are Fréchet spaces, then* (2) *is an exact sequence of Fréchet spaces.*

    (iii) *In both cases, u induces a topological isomorphism*

(3)
$$F^q(G; B)/F^q(G; A) \overset{\sim}{\to} F^q(G; C).$$

    Clearly, if $B = B' \oplus B''$ is the topological direct sum of two closed subspaces, then $F^q(G; B)$ is isomorphic to the topological direct sum of $F^q(G; B')$ and $F^q(G; B'')$, whence we get (i) and (iii) in this case. The other assertions follow from 1.2.

    We note that, in both cases, in (1) we can identify A with its image in B and B/A with C; hence (3) can also be written

(4)
$$F^q(G; B)/F^q(G; A) \overset{\sim}{\to} F^q(G; B/A).$$

    1.8.   Lemma 1.3 implies, under either set of assumptions, that the sequence

(1)
$$0 \to F^*(G; A) \to F^*(G; B) \to F^*(G; C) \to 0$$

of non-homogeneous complexes is exact. Therefore, in either case, there is associated to 1.7(1) a long exact sequence in continuous cohomology. Note also that, by 1.5,

(2)
$$H_{\mathrm{ct}}^q(G; V) = 0 \quad \text{for } q \geq 1,$$

if V is s-injective.

1.9. PROPOSITION. *Let $U, V \in \mathcal{C}_G$. If there exists an element z in the group algebra over $\mathbf{C}$ of the center $C(G)$ of G which acts as the identity on U and as the zero-morphism on V, then $\mathrm{Ext}_G^q(U, V) = 0$ for all $q \in \mathbf{Z}$.*

This is the analogue in $\mathcal{C}_G$ of I, 4.1 in $\mathcal{C}_{\mathfrak{g}, \mathfrak{k}}$. Both proofs given there extend to the present case. This is obvious for the second one. For the first one, interpret the groups $\mathrm{Ext}_G^q(U, V)$ as equivalence classes of long exact s-sequences from V to U, as in I, §3, following [**78**, III]. [Note that we did not have to introduce strong morphisms in I, because, the $(\mathfrak{g}, \mathfrak{k})$-modules being locally finite and semi-simple with respect to $\mathfrak{k}$ by definition, all morphisms of $(\mathfrak{g}, \mathfrak{k})$-modules are automatically strong with respect to the $\mathfrak{k}$-module structure.]

1.10. LEMMA. *Let G be compact. Then the functor $V \mapsto V^G$ from quasi-complete continuous G-modules to topological vector spaces is exact and strongly exact, and transforms strong morphisms to strong morphisms.*

Let $U, V \in \mathcal{C}_G$ be quasi-complete and $\phi \colon U \to V$ a continuous map. Define $\overline{\phi} \colon U \to V$ by

$$(1) \qquad \overline{\phi}(u) = \int_G g \cdot \phi(g^{-1} \cdot u) \, dg \qquad (u \in U),$$

where dg is the normalized Haar measure on G. Then $\overline{\phi}$ is continuous linear if ϕ is. It commutes with G and equals ϕ if ϕ commutes with G. If $W \in \mathcal{C}_G$ is quasi-complete and $\psi \colon V \to W$ is a linear continuous map, then

$$(2) \qquad \overline{(\psi \circ \phi)} = \overline{\psi} \circ \overline{\phi}, \quad \text{if either } \phi = \overline{\phi} \text{ or } \psi = \overline{\psi}.$$

Now let $f \colon U \to V$ be a surjective G-morphism. Then, by averaging over G, we see immediately that f induces a surjective map of U^G onto V^G. This implies that $V \mapsto V^G$ is exact. Let $e \colon V \to W$ be a G-morphism and assume that there exist continuous linear maps $a \colon V \to U$, $b \colon W \to V$ such that $f \circ a + b \circ e = \mathrm{Id}$. Then we also have $f \circ \overline{a} + \overline{b} \circ e = \mathrm{Id}$; therefore we can arrange that a and b commute with G. Then a (resp. b) maps V^G into U^G (resp. W^G into V^G), and the previous relation is still satisfied. This implies that $V \mapsto V^G$ is s-exact and transforms strong morphisms into strong morphisms. (This argument is borrowed from the proof of Lemma 7 in [**35**].)

1.11. PROPOSITION. *Let N be a closed normal subgroup of G.*
(i) If V is s-injective in $\mathcal{C}_G$, then V^N is s-injective in $\mathcal{C}_{G/N}$.
(ii) Let $U, V \in \mathcal{C}_G$ be quasi-complete. Assume that N is compact and acts trivially on U. Then

$$(1) \qquad \mathrm{Ext}_G^q(U, V) = \mathrm{Ext}_{G/N}^q(U, V^N) \qquad (q \in \mathbf{Z}).$$

In particular,

$$(2) \qquad H_{\mathrm{ct}}^q(G; V) = H_{\mathrm{ct}}^q(G/N; V^N) \qquad (q \in \mathbf{Z}).$$

The space V^N is stable under G, and the structure of G/N-module understood in (i) is of course the one inherited from the G-action. Since every G/N-module may be viewed as a G-module via the projection $G \to G/N$, the assertion (i) just follows from the definitions. Now let $V \in \mathcal{C}_G$ be quasi-complete. It has an s-injective resolution $0 \to V \to A^*$ by quasi-complete G-modules (1.6). By 1.10 and (i), the associated sequence $0 \to V^N \to A^{*^N}$ is an s-injective resolution of V^N in $\mathcal{C}_{G/N}$.

It follows therefore that $\operatorname{Ext}_G^q(U, V)$ (resp. $\operatorname{Ext}_{G/N}^q(U, V^N)$) is the q-th cohomology space of the complex

$$\operatorname{Hom}_G(U, A^*) \quad (\text{resp. } \operatorname{Hom}_{G/N}(U, {A^*}^N)).$$

However, since N acts trivially on U, the image of U in any G-module W under a G-morphism is contained in W^N, and so these two complexes are identical. This proves (1). Then (2) is a special case where U is the trivial one-dimensional G-module.

1.12. PROPOSITION. *Assume G to be compact. Let U be a Fréchet G-module and V a quasi-complete G-module. Then $\operatorname{Ext}_G^q(U, V) = 0$ for $q \geq 1$. In particular, $H_{\mathrm{ct}}^q(G; V) = 0$ for $q \geq 1$.*

Under our assumptions, $\operatorname{Hom}(U, V)$ is quasi-complete (cf. [**24**], III, §1, n° 1 and §3, n° 7, Cor. 2). By 1.6(5), it therefore suffices to prove the second assertion. The latter follows from 1.11, for $G = N$.

REMARK. The second assertion is proved in [**35**] (cf. Lemma 7) by the same argument.

2. Shapiro's lemma

2.1. Let H be a closed subgroup of G and $U \in \mathcal{C}_H$. We put

$$(1) \qquad I(U) = \operatorname{Ind}_H^G U = \{ f \in C(G; U) \mid f(hg) = h \cdot f(g) \ (g \in G; \ h \in H) \}.$$

It is a closed subspace of $C(G, U)$, hence a Fréchet or a quasi-complete space if U is one (1.2). If G acts trivially on U, and $H = \{1\}$, then $\operatorname{Ind}_H^G U = F^0(G; U)$. If U is a G-module, then the map α which associates to $f \in I(U)$ the function $\alpha(f)$ on G defined by $\alpha(f)(x) = x \cdot f(x^{-1})$ is easily seen to define a G-isomorphism of $I(U)$ onto $C(G/H; U)$. Its inverse is given by the same formula.

2.2. LEMMA. *Let H, U be as above and $V \in \mathcal{C}_G$. Then the map*

$$\operatorname{Hom}_G(V, I(U)) \to \operatorname{Hom}_H(V, U),$$

associated to the map $I(U) \to U$ given by $f \mapsto f(1)$, is a topological isomorphism.

For the proof, cf. [**35**, Lemma 2].

2.3. PROPOSITION. *Let H be a closed subgroup of G. Assume that the fibration of G by H admits a continuous local cross-section.*

(i) *Every s-injective G-module is s-injective as an H-module.*

(ii) *("Shapiro's lemma") Given $U \in \mathcal{C}_H$ and $V \in \mathcal{C}_G$, there are canonical isomorphisms*

$$(1) \qquad\qquad \operatorname{Ext}_G^q(V, I(U)) = \operatorname{Ext}_H^q(V; U) \qquad (q \in \mathbf{N}).$$

In particular,

$$(2) \qquad\qquad H_{\mathrm{ct}}^q(G; I(U)) = H_{\mathrm{ct}}^q(H; U) \qquad (q \in \mathbf{N}).$$

Since G is by assumption a countable union of compact subsets, the space G/H is paracompact; hence (i) is Lemma 3.4 of [**60**].

(ii) is proved in exactly the same way as Prop. 3 of [**35**]: one starts from the homogeneous resolution $C^*(G; I(U))$ of $I(U)$ (see 1.4) and shows that

$$(3) \qquad\qquad C^n(G; I(U)) = I(C^n(G; U)) \qquad (n \in \mathbf{N}).$$

By 2.2, we then have

(4) $\operatorname{Hom}_G(V; C^n(G; I(U))) = \operatorname{Hom}_H(V, C^n(G; U))$ $(n \in \mathbf{N})$.

Since these isomorphisms are natural, they yield an isomorphism of complexes

$$\{\operatorname{Hom}_G(V, C^n(G; I(U)))\} \xrightarrow{\sim} \{\operatorname{Hom}_H(V, C^n(G; U))\}.$$

By definition, the q-th cohomology group of the left-hand side is $\operatorname{Ext}^q_G(V; I(U))$. Since $C^n(G; U)$ is s-injective with respect to H (by (i)), the complex $\{C^n(G; U)\}$ provides an s-injective resolution of U in $\mathcal{C}_H$; hence the q-th cohomology group of the right-hand side is $\operatorname{Ext}^q_H(V, U)$. This proves (1).

2.4.   Assume N is a closed normal subgroup of G such that the fibration of G by N has continuous local cross-sections. Let $0 \to V \to A^*$ be an s-injective resolution of V in $\mathcal{C}_G$. By 2.3, it may be viewed as an s-injective resolution in $\mathcal{C}_N$; hence $H^*(N; V)$, identified to $H^*(A^{*^N})$, inherits a natural G/N-action. This G/N-module structure is continuous with respect to the quotient topology (as defined in 3.3). It does not depend on the s-injective resolution, in view of the existence of maps over the identity of V of any two such. Slightly more generally, let $0 \to V \to B^*$ be an s-resolution in $\mathcal{C}_G$ of V by modules which are s-injective in $\mathcal{C}_N$. Then the action of G/N induced from its action on B^{*^N} is the previous one. In fact, since $0 \to V \to B^*$ is an s-resolution in $\mathcal{C}_G$, there is a natural G-map $B^* \to A^*$ of complexes over the identity. It induces a G/N-map of complexes $B^{*^N} \to A^{*^N}$. Since both resolutions are s-injective in $\mathcal{C}_N$, this map induces an isomorphism of $H^*(B^N)$ onto $H^*(A^N)$, which clearly commutes with G/N.

2.5. PROPOSITION. *Let N be a closed normal subgroup of G. Assume that G/N is compact and that the fibration of G by N has continuous local cross-sections. Let $V \in \mathcal{C}_G$ be quasi-complete. Then*

(1) $H^q_{\mathrm{ct}}(G; V) = H^q_{\mathrm{ct}}(N; V)^{G/N}$ $(q \in \mathbf{Z})$.

Let $0 \to V \to A^*$ be an s-injective resolution of V by quasi-complete G-modules (1.6). By 2.3, it may be viewed as an s-injective resolution of V in $\mathcal{C}_N$; hence $H^*(N; V)$ is the cohomology of the complex A^{*N}. We have then, in view of 1.10,

(2)
$$H^q_{\mathrm{ct}}(N; V)^{G/N} = (H^q(A^{*^N}))^{G/N} = H^q((A^{*^N})^{G/N})$$
$$= H^q(A^{*^G}) = H^q_{\mathrm{ct}}(G; V)   (q \in \mathbf{Z}).$$

REMARK. If G/N is finite, then the fibration of G by N always has continuous cross-sections, and the second inequality in (2) is valid without assuming V to be quasi-complete. Then (1) is true for any $V \in \mathcal{C}_G$.

2.6. LEMMA. *Let K be a compact subgroup of G. Let $E \in \mathcal{C}^{\mathrm{qc}}_K$. Then $I^G_K(E)$ is s-injective in $\mathcal{C}^{\mathrm{qc}}_G$.*

Let $U, V \in \mathcal{C}^{\mathrm{qc}}_G$, and let $m: U \to V$ be a strong injection. Let $s: V \to U$ be a continuous linear map such that $s \circ m = \mathrm{Id}$. Let $\bar{s}$ be the average of s over K, as defined by 1.10(1). Then $\bar{s} \circ m = \mathrm{Id}$, and $\bar{s}$ commutes with K. Now let $\alpha: U \to I^G_K(E)$ be a continuous linear G-morphism. By Frobenius reciprocity (2.2) it corresponds canonically to a K-morphism $\alpha': U \to E$. Then $\beta' = \bar{s} \circ \alpha': V \to E$ is a K-morphism extending α', whence, by 2.2 again, it is a G-morphism $\beta: V \to I^G_K(E)$ extending α.

3. Hausdorff cohomology

3.1. Let C^* be a complex in $\mathcal{C}_G$ (we do not exclude trivial action, i.e., C^* may just be a complex of topological vector spaces, with continuous linear differentials). Then Z^q is closed in C^q, and $H^q(C) = Z^q/d(C^{q-1})$ may be given the quotient topology. It is Hausdorff if and only if $d(C^{q-1})$ is closed in Z^q or, equivalently, in C^q. If so, we shall view $H^q(C)$ as a topological vector space in this way, and shall say that $H^q(C)$ is Hausdorff, or that C has *Hausdorff cohomology* in dimension q. If this is true for all q's, then we say that $H^*(C)$ is Hausdorff or that C^* has Hausdorff cohomology. Since Z^q and $d(C^{q-1})$ are stable under G, $H^q(C)$ inherits an action of G, which is continuous with respect to the quotient topology. Thus, if $H^q(C)$ is Hausdorff, it is canonically in $\mathcal{C}_G$. Of course, $H^0(C)$ is always Hausdorff.

LEMMA. *Let A^*, B^* be two complexes of topological vector spaces, $f: A^* \to B^*$, a morphism, and $q \in \mathbf{N}$. Assume that $f^*: H^q(A^*) \to H^q(B^*)$ is bijective. If $H^q(B^*)$ is Hausdorff, so is $H^q(A^*)$.*

Let Z^q (resp. Z'^q) be the space of q-cocycles in B^q (resp. A^q). Let g be the canonical projection of Z^q onto $H^q(B^*)$. Then $g \circ f: Z'^q \to H^q(B^*)$ is a continuous linear map. In view of our assumptions, it is surjective and its kernel is $d(A^{q-1})$. The latter is then closed in Z'^q, whence the lemma.

3.2. LEMMA. *Let $V \in \mathcal{C}_G$ and $q \in \mathbf{N}$. Assume that there exists an s-injective resolution E^* of V such that $H^q(E^{*^G})$ is Hausdorff. Then any s-injective resolution F^* of V has the same property. The canonical isomorphism of $H^q(F^{*^G})$ onto $H^q(E^{*^G})$ associated to the identity map of V is topological.*

The identity map of V extends to a G-morphism $\widetilde{u}$ of F^* into E^* [**60**, §2], hence also to a morphism $u: F^{*^G} \to E^{*^G}$, which induces an isomorphism u^* of $H^*(F^{*^G}) \to H^*(E^{*^G})$ (loc. cit.). The previous lemma implies that $H^q(F^{*^G})$ is Hausdorff. The map u^* is a continuous bijective map of $H^q(F^{*^G})$ onto $H^q(E^{*^G})$. Similarly, a lifting of the identity of V to a map $E^* \to F^*$ yields to a bijective continuous map $v^*: H^q(E^{*^G}) \to H^q(F^{*^q})$. Since $u^* \circ v^*$ and $v^* \circ u^*$ are the identity, this proves the lemma.

3.3. In fact, the proof of the lemma shows that u^* is a topological isomorphism of $H^q(F^{*^G})$ onto $H^q(E^{*^G})$, both spaces being endowed with the quotient topology, regardless of whether they are Hausdorff or not whence the existence of a canonical topology on $H^q_{\mathrm{ct}}(G; V)$. If the condition of 3.2 is fulfilled, then we shall say that $H^q_{\mathrm{ct}}(G; V)$ is Hausdorff. $H^*_{\mathrm{ct}}(G; V)$ will be said to be Hausdorff if $H^q_{\mathrm{ct}}(G; V)$ is so for all q's.

3.4. LEMMA. *Assume that C^* is a complex of Fréchet spaces (and continuous linear maps) and that $H^q(C)$ is finite dimensional. Then $d_{q-1}(C^{q-1})$ is closed in C^q.*

The proof is the same as that of Prop. 6 in [**35**]. We repeat it for the sake of completeness. Let E be a subspace of Z^q which maps bijectively onto $H^q(C)$ under the natural projection $Z^q \to H^q(C)$. It is finite dimensional, hence closed in Z^q [**23**, I, §2, n° 3]. The obvious map $B^q \oplus E \to Z^q$, where $B^q = C^{q-1}/Z^{q-1}$ is endowed with the quotient topology, is continuous and bijective, hence an isomorphism (1.2), whence the lemma.

REMARK. The proof shows more precisely that the sequence

$$0 \longrightarrow C^{q-1}/Z^{p-1} \xrightarrow{d_{q-1}} Z^q \longrightarrow H^q(C) \longrightarrow 0,$$

is s-exact.

3.5. PROPOSITION. *Let $V \in \mathcal{C}_G$ and $q \in \mathbf{N}$. Assume that V is a Fréchet space and that $H^q(G;V)$ is finite dimensional. Then $H^q(G;V)$ is Hausdorff.*

In fact the standard homogeneous resolution consists of Fréchet spaces, and the condition of 3.1 is satisfied in view of 3.4.

4. Spectral sequences

We again assume familiarity with standard material on spectral sequences (cf., e.g., [**78**, XI] or [**43**, I, §4]). The spectral sequences considered here are all "first quadrant" spectral sequences associated to double complexes with positive degrees.

4.1. THEOREM. *Let*

$$(1) \qquad A^*: A^0 \xrightarrow{e_0} A^1 \longrightarrow \cdots \longrightarrow A^q \xrightarrow{e_q} A^{q+1} \longrightarrow \cdots$$

be a complex of acyclic G-modules and G-morphisms. If either (i) (1) *is an s-sequence, or* (ii) A^* *consists of Fréchet spaces and has Hausdorff cohomology* (3.1), *then there exists a spectral sequence (E_r) which abuts to the cohomology of the complex $A^{*G} = \{A^{q^G}\}$ and where*

$$(2) \qquad E_2^{p,q} = H_{\mathrm{ct}}^p(G; H^q(A)) \qquad (p,q \geq 0).$$

We note first that in both cases $H^q(A)$ is in $\mathcal{C}_G$ in a canonical way (3.1). It is this G-module structure which is meant in the right-hand side of (2).

Let $F^*(G; A^q)$ be the non-homogeneous complex of continuous A^q-valued cochains (see 1.4(5), (6)). Then the direct sum C^* of the $F^*(G; A^q)$ is a double complex in the usual way, with differentials induced by 1.4(6) and by the differentials of A^*. We have (see 1.4)

$$(1) \qquad C^{p,q} = F^{p-1}(G; A^q) \qquad (p,q \in \mathbf{N})$$

and $C^{p,q} = 0$ otherwise. We consider the two spectral sequences $('E_r)$, $(''E_r)$ associated to the filtrations defined by the partial degrees. If the degree in A is used (giving the "second filtration"), then

$$(2) \qquad ''E_0^{*,q} = F^*(G; A^q),$$

and the differential d_0'' of the spectral sequence is that of $F^*(G; A^q)$. Therefore, $''E_1^{p,q} = H_{\mathrm{ct}}^p(G; A^q)$. Since the A^q's are acyclic, we have $''E_1^{p,q} = 0$ if $p \neq 0$ and $''E_1^{0,q} = A^{q^G}$. Then d_1'' is induced by the differentials of A^*, whence

$$(3) \qquad \begin{aligned} &''E_2^{0,q} = H^q(A^{*G}) = {}''E_\infty^{0,q} = H^q(C^*), \\ &''E_r^{p,q} = 0 \qquad (r \geq 1;\ p \neq 0). \end{aligned}$$

We now consider the spectral sequence $('E_r)$ associated to the filtration by the degree in F^* (the "first filtration"). We have then

$$(4) \qquad 'E_0^{p,*} = F^p(G; A^*) \qquad (p \in \mathbf{N}).$$

We want to prove that

$$(5) \qquad 'E_1^{p,q} = F^p(G; H^q(A)) \qquad (p,q \in \mathbf{N}).$$

Let

(6) $$Z^q = \ker e_q, \qquad B^q = A^q/Z^q.$$

By 1.7 and our assumptions, the exact sequence

(7) $$0 \to Z^q \to A^q \to B^q \to 0 \qquad (q \in \mathbf{N})$$

yields a topological isomorphism

(8) $$F^p(G; A^q)/F^p(G; Z^q) \xrightarrow{\sim} F^p(G; B^q) \qquad (p, q \in \mathbf{N}).$$

If (1) is strong, then the injection $e_{q\ 1} \colon B^{q-1} \to Z^q$ is strong and

(9) $$0 \to B^{q-1} \to Z^q \to H^q(A) \to 0$$

is an exact s-sequence, where $H^q(A)$ is endowed with the quotient topology. Under assumption (ii) the subspace $e_{q-1}(B^{q-1})$ of Z^q is closed; hence e_{q-1} is an isomorphism of B^{q-1} onto its image, and (9) is again an exact sequence of Fréchet spaces, $H^q(A)$ being endowed with the quotient topology. Lemma 1.7 then yields

(10) $$F^p(G; Z^q) = (\ker d_0) \cap {}'E_0^{p,q},$$

(11) $$F^p(G; B^{q-1}) = F^p(G; e_{q-1}(B^{q-1})) = d_0({}'E_0^{p,q-1}),$$

(12) $$F^p(G; H^q(A)) = F^p(G; Z^q)/F^p(G; B^{q-1}).$$

This proves (5). The differential d_1' of $'E_1$ is then the differential of F^*, given by 1.4(6), whence

(13) $${}'E_2^{p,q} = H_{\mathrm{ct}}^p(G; H^q(A)) \qquad (p, q \in \mathbf{N}).$$

Since $('E_r)$ abuts to $H^*(C)$, and the latter is equal to $H^*(A^{*^G})$ by (3), the spectral sequence $('E_r)$ satisfies our conditions.

4.2. COROLLARY. *Let $V \in \mathcal{C}_G$ and let*

(1) $$0 \longrightarrow V \longrightarrow A^0 \xrightarrow{e_0} A^1$$
$$\longrightarrow \cdots \longrightarrow A^q \xrightarrow{e_q} A^{q+1} \longrightarrow \cdots$$

be a resolution of V by acyclic G-modules. Assume that (1) is strong or consists of Fréchet spaces. Then

(2) $$H_{\mathrm{ct}}^q(G; V) = H^q(A^{*^G}) \qquad (q \in \mathbf{N}).$$

We have $H^q(A^*) = 0$ for $q \geq 1$; hence the complex $A^* = \{A^i\}$ is Hausdorff. Moreover, e is an isomorphism of V onto $\ker e_0 = H^0(A^*)$: this is clear if e is strong, and follows from 1.2 if V and A^0 are Fréchet spaces. Therefore, we can apply 4.1. We then have

$$E_2^{p,q} = 0 \quad \text{for } q \neq 0, \quad E_2^{p,0} = H_{\mathrm{ct}}^p(G; V) \qquad (p, q \in \mathbf{N}),$$

and our assertion follows.

REMARK. This isomorphism is only one of vector spaces. To be more precise, the proof of 4.1 implies the existence of continuous bijective maps

$$H_{\mathrm{ct}}^*(G; V) \to H^*(C^*) \leftarrow (A^{*^G}),$$

given by the "edge homomorphisms" of the two spectral sequences considered there. If A^* is strong, then it maps into any s-injective resolution of V, whence also a continuous bijective map $H^*(A^{*^G}) \to H_{\mathrm{ct}}^*(G; V)$, and it follows, in particular, by the lemma in 3.1, that if $H_{\text{ct}}^*(G; V)$ is Hausdorff, then A^{*^G} has Hausdorff cohomology. According to P. Deligne, this last fact is also true if V is a Fréchet space and A^* just an acyclic resolution by Fréchet spaces. Then $H_{\text{ct}}^q(G; V)$ and $H^q(A^{*^G})$ are topologically isomorphic for all q's.

4.3. THEOREM. *Let N be a closed normal subgroup of G. Assume that the fibration of G by N admits a continuous local cross-section. Let $V \in \mathcal{C}_G$ be such that $H_{\text{ct}}^*(N; V)$ is Hausdorff (3.3). Assume either that V is a Fréchet space or that there exists an s-injective resolution A^* of V in $\mathcal{C}_G$ such that A^{*^N} is a strong complex. Then $H_{\text{ct}}^*(N; V)$ admits a natural structure of topological (G/N)-module (1.11), and there exists a spectral sequence (E_r), abutting to $H_{\text{ct}}^*(G; V)$, in which*

$$(1) \qquad E_2^{p,q} = H_{\text{ct}}^*(G/N; H_{\text{ct}}^*(N; V)) \qquad (p, q \in \mathbf{N}).$$

We let A^* be any s-injective resolution of V in $\mathcal{C}_G$ if V is a Fréchet space, and be as in the statement of the theorem otherwise. It is s-injective in $\mathcal{C}_N$ (2.3); therefore

$$(2) \qquad H^q(A^{*^N}) = H_{\text{ct}}^q(N; V) \qquad (q \in \mathbf{N}).$$

Moreover, A^{*^N} has Hausdorff cohomology, in view of 3.2 and our assumption. By 1.11(i) the module $(A^q)^N$ is s-injective in $\mathcal{C}_{G/N}$ $(q \in \mathbf{N})$. A fortiori it is (G/N)-acyclic (1.8(2)). Thus A^{*^N} is a complex of (G/N)-acyclic modules, which either is strong or consists of Fréchet spaces. In both cases, we may apply 4.1, with G/N and A^{*^N} playing the roles of G and C^*. Therefore there exists a spectral sequence (E_r) abutting to $H^*((A^{*^N})^{G/N})$, in which

$$(3) \qquad E_2^{p,q} = H_{\text{ct}}^p(G/N; H^q(A^{*^N})) \qquad (p, q \in \mathbf{N}).$$

In view of (2) and the obvious equality $(A^{*^N})^{G/N} = A^{*^G}$, this spectral sequence has the required properties.

REMARK. A somewhat stronger result is stated as Prop. 5 in [**35**], but the proof is incomplete.

5. Differentiable cohomology
and continuous cohomology for Lie groups

From now on, G is a Lie group. All manifolds are assumed to be smooth and countable at infinity.

5.1. If $V \in \mathcal{C}_G^\infty$ (cf. 0, 2.3), then, in agreement with 1.3, we let $C^\infty(G; V)$ be endowed with the G-module structure defined by $(x \cdot f)(g) = x \cdot f(x^{-1} \cdot g)$ $(g, x \in G)$, while $F^\infty(G; V)$ denotes the same space, but with G acting by right translations on the first argument, i.e., $xf(g) = f(g \cdot x)$ $(g, x \in G)$. They are differentiable G-modules, isomorphic under the map μ of 1.3.

An element $V \in \mathcal{C}_G^\infty$ is differentiably or smoothly (resp. continuously) s-injective if it is s-injective in $\mathcal{C}_G^\infty$ (resp. $\mathcal{C}_G$). Of course, the latter implies the former.

If $V \in \mathcal{C}_G^\infty$, then $F^\infty(G; V)$ (or, equivalently, $C^\infty(G; V)$) is smoothly s-injective [**60**, 5.1]. (Note that since V is Hausdorff by our standing assumption, the separability condition in 5.1 of [**60**] is automatically fulfilled.) As in 1.4, it follows that smoothly s-injective resolutions exist. In fact, the standard homogeneous resolution of 1.4, computed with smooth cochains, is one. We can then define the q-th differentiable cohomology group $H_d^q(G;V)$ as in 1.4(5), (6), using smooth cochains, and, as in 1.5, it can be computed by means of any smoothly s-injective resolution. Since smooth cochains are in particular continuous cochains, there is a natural map

$$\mu\colon H_d^*(G;V) \to H_{ct}^*(G;V) \qquad (V \in \mathcal{C}_G^\infty),$$

which is natural in V. If V is quasi-complete, then j^* is an isomorphism [**60**, Thm. 5.1]. This follows from the following lemma, which implies that the smooth standard resolution is also continuously s-injective.

5.2. LEMMA ([**60**, Lemma 5.2]). *Let $V \in \mathcal{C}_G^\infty$ be quasi-complete. Then $C^\infty(G;V)$ is continuously s-injective.*

Put $A = C^\infty(G;V)$. Since $C(G;A)$ is continuously s-injective (1.6), it suffices to show that there exists a continuous G-map $\mu\colon C(G;A) \to A$ such that $\mu \circ \varepsilon = \mathrm{id}_A$.

Fix a left invariant Haar measure dg on G. Let $\phi \in C_c^\infty(G)$ be a compactly supported smooth real valued function on G such that $\int_G \phi(g^{-1})\, dg = 1$. Given $f \in C(G;A)$, let $\alpha(f) = f * \phi$, i.e.

$$\alpha(f)(x) = \int_G \phi(y^{-1} \cdot x) f(y)\, dy \qquad (x \in G).$$

This defines a continuous G-map $\alpha\colon C(G;A) \to C(G;A)$ with image in $C^\infty(G;A)$, which is the identity on $\varepsilon(A)$. Let $\beta\colon C^\infty(G;A) \to A$ be defined by $(\beta f)(g) = f(g)(g)$. Then $\mu = \beta \circ \alpha$ satisfies our conditions.

5.3. Let X be a space on which G operates continuously. G is said to *operate properly* on X if the map $G \times X \to X \times X$ defined by $(g \cdot x) \mapsto (g \cdot x, x)$ is proper [**22**, III]. This implies in particular that the isotropy groups G_x $(x \in X)$ are compact and that the orbit space X/G is Hausdorff if X is (loc. cit.).

Now let M be a manifold on which G operates smoothly. A *differentiable slice* S at a given point $m \in M$ is a closed submanifold in a neighborhood of m with the following properties:

(i) $S \cap G \cdot m = \{m\}$, $G_m(S) = S$, and $G_m = \{g \in G \mid g \cdot S \cap S \neq \varnothing\}$.

(ii) The map $(g,s) \mapsto g \cdot s$ induces a diffeomorphism of $G \times_{G_m} S$ (G operating on the right on itself) onto $G \cdot S$, and $G \cdot S$ is an open neighborhood of $G \cdot m$ in M.

(iii) The map $(g,s) \mapsto g \cdot m$ induces a smooth G-equivariant retraction r_m of $G \cdot S$ onto $G \cdot m = G/G_m$.

Note that the definition of r_m makes good sense since, by (i), if $s \in S$, then $G_s \subset G_m$.

If $f \in C^\infty(G \cdot S)^G$, then its restriction $\overline{f}$ to S is in $C^\infty(S)^{G_m}$. We claim that the map $f \mapsto \overline{f}$ of $C^\infty(G \cdot S)^G$ into $C^\infty(S)^{G_m}$ is bijective. It is clearly injective. Let $\overline{f} \in C^\infty(S)^{G_m}$. By (ii), GS is the total space of a C^∞ fibration over G/G_m with structural group G_m and typical fiber S, which is locally trivial. In any local chart of the form $U \times S$ we extend $\overline{f}$ to a function f_U constant on the sets $U \times \{s\}$. Then these functions match to a G-invariant smooth function on $G \cdot S$ which restricts to $\overline{f}$ on S.

If G operates properly on M, then there is a differentiable slice at every point of M [**89**, 2.2.2]. In fact, M always has a smooth G-invariant Riemannian metric [**89**, 4.3.1], and we may take S such that GS is a tubular neighborhood of $G \cdot m$ [**89**, 2.2.3].

5.4. PROPOSITION (G. D. Mostow). *Let M be a smooth manifold on which G operates smoothly and properly. Let $V \in \mathcal{C}_G^\infty$ be a Fréchet (resp. quasi-complete)* space. Then the space $A^q(M;V)$ (cf. 0, 1.7) is a continuously s-injective Fréchet (resp. quasi-complete) G-module ($q \in \mathbf{N}$).

We already pointed out that $A^q(M;V)$ is a Fréchet (resp. quasi-complete) space (0, 1.7).

If M is the quotient of G by a compact subgroup, this is shown in [**60**, p. 385–6]. This case suffices in fact to prove van Est's theorem (5.6). We sketch Mostow's argument in the general case.

Assume first that there exist $m \in M$ and a differentiable slice S at m such that $G \cdot S = M$. Put $A = A^q(M;V)$. We know that $C^\infty(G;A)$ is continuously s-injective (5.2). It suffices therefore to show that there exists a continuous G-map $\mu \colon C^\infty(G;A) \to A$ such that $\mu \circ \varepsilon = \mathrm{id}_A$. Let dy be a Haar measure on G_m with total mass 1. Given $f \in C^\infty(G;A)$, define $\alpha(f)$ by

$$\alpha(f)(x) = \int_{G_m} f(x \cdot y)\, dy.$$

Then, α is a continuous G-map: $C^\infty(G;A) \to C^\infty(G/G_m;A)$. Given $x \in M$, $Y_x \in T_x(M)$, choose $g \in G$ such that $x \in g \cdot S$; put

$$\beta(f)(x, Y_x) = f(g)(x, Y_x) \qquad (f \in C^\infty(G;A)).$$

(This is well defined in view of 5.3(i).) It is then immediately checked that β maps $C^\infty(G/G_m;A)$ into A, and that $\mu = \beta \circ \alpha$ has the required properties.

We now consider the general case, and let $\pi \colon M \to G \backslash M$ be the canonical projection. Since M is paracompact, and the action is proper, so is $G \backslash M$. In view of 5.2, we can find a countable subset $Q \subset M$ and a differentiable slice S_m at $m \in Q$ such that the sets $\pi(S_m)$ ($m \in Q$) form a locally finite open cover of $G \backslash M$. Then the sets $M_m = G \cdot S_m$ form a locally finite open cover $\mathcal{U}$ of M by G-stable sets. By making use of a continuous partition of unity on $G \backslash M$ subordinated to the cover $\{\pi(S_m)\}_{m \in Q}$, we get first a continuous partition of unity on M by G-invariant functions, subordinated to the cover $\mathcal{U}$. But then, using the bijection $C^\infty(M_m)^G \to C^\infty(S_m)^{G_m}$ (cf. 5.2), we see that we can change it slightly to get a *smooth* partition of unity $(t_m)_{m \in Q}$ by G-invariant smooth functions subordinated to the cover $\mathcal{U}$. Then, the map $v \colon f \mapsto (t_m f)_{m \in Q}$ is a continuous G-map of $A^q(M;V)$ into the direct product E of the $A^q(M_m;V)$ ($m \in Q$). Since each factor is s-injective, E is also s-injective. It therefore suffices to exhibit a continuous G-map $w \colon E \to A^q(M;V)$ such that $w \circ v = \mathrm{Id}$. Since the M_m are G-invariant open submanifolds of M and form a locally finite cover, it is immediate that the map which assigns to $a = (a_m)$ ($a_m \in A^q(M_m;V)$, $m \in Q$) the sum of the a_m's is well defined and satisfies those conditions.

5.5. PROPOSITION. *Let M and V be as in 5.4. Assume that M is diffeomorphic to a Euclidean space. Then $0 \to V \to A^0(M;V) \to A^1(M;V) \to \cdots$ is a continuously s-injective resolution (1.5) of V by Fréchet (resp. quasi-complete) modules of $\mathcal{C}_G^\infty$.*

We already know that each $A^q(M,V)$ is continuously s-injective (5.4). Since V is at any rate quasi-complete, the usual proof of the Poincaré lemma in Euclidean space (see e.g. [**112**, 4.18]) works also for V-valued forms and provides a continuous contracting homotopy, i.e. continuous linear maps $e_q \colon A^q(M;V) \to A^{q-1}(M;V)$ ($q \geq 1$) and $\delta \colon A^0(M;V) \to V$ satisfying 1.5(3).

REMARK. In the case $M = G/K$, where G has finitely many connected components and K is a maximal compact subgroup, this is proved in [**60**, p. 385–6].

5.6. COROLLARY. *Assume that G has finitely many connected components, and let K be a maximal compact subgroup of G. Then:*
(i) $H_d^*(G; V)$ *is isomorphic to* $H^*(A^*(M; V)^G)$.
(ii) $H_d^*(G; V) = H^*(\mathfrak{g}, K; V)$. *If V is admissible, then $H_d^*(G; V)$ is Hausdorff and finite dimensional. The functor $V \mapsto H_d^*(G; V)$ commutes with inductive limits.*

(i) follows from 5.5 and the definitions (see 5.1). We can apply this to $M = G/K$. This yields (see I, 1.4)

$$
(1) \qquad\qquad A^*(G/K; V)^G = C^*(\mathfrak{g}, K; V),
$$

whence we get the first part of (ii). For $\delta \in \widehat{K}$, let V_δ be the isotypic subspace of V of type δ. Let S be a set of K-types occurring in $\Lambda^*(\mathfrak{g}/\mathfrak{k})$. It is finite. Then we have

$$
(2) \qquad C^*(\mathfrak{g}, K; V) \subset \mathrm{Hom}_K(\Lambda(\mathfrak{g}/\mathfrak{k}), V_S), \quad \text{where } V_S = \bigoplus_{\delta \in S} V_\delta.
$$

This shows that if V is admissible, then $C^*(\mathfrak{g}, K; V)$ is finite dimensional, and the second part of (ii) follows. The last assertion of (ii) follows from the first one and I, 1.2(4).

REMARK. (ii) is the well-known van Est theorem (see [**104**, Thm. 2], or [**60**, Thm. 6.1]), where it is in fact stated under somewhat more general assumptions on V).

5.7. COROLLARY. *Let E be an admissible quasi-complete smooth G-module, and assume that G acts trivially on V. Let $E \otimes V$ be the projective tensor product of E and V, and assume that $E \otimes V$ is quasi-complete. Then $H_d^*(G; E \otimes V)$ is Hausdorff and is isomorphic to $H_d^*(G; E) \otimes V$.*

(The notion of projective tensor product is briefly recalled in 6.1. If E is finite dimensional, this is the obvious topology which makes $E \otimes V$ into a topological direct sum of $\dim E$ copies of V; then $E \otimes V$ is obviously quasi-complete, and the conclusion of 5.7 holds.)

Let $M = G/K$. Then $A^*(M; E \otimes V)$ defines an s-injective resolution of $E \otimes V$ (5.5), and we have

$$
A^*(M; E \otimes V)^G = C^*(\mathfrak{g}, \mathfrak{k}; E \otimes V)^{K/K^0} = (\Lambda(\mathfrak{g}/\mathfrak{k})^* \otimes E \otimes V)^{K/K^0},
$$

where K^0 is the identity component of K (I, 1.4); hence, since K/K^0 acts trivially on V:

$$
A^*(M; E \otimes V)^G = (\Lambda(\mathfrak{g}/\mathfrak{k})^* \otimes E)^{K/K^0} \otimes V = C^*(\mathfrak{g}, K; E) \otimes V.
$$

Since $C^*(\mathfrak{g}, K; E)$ is finite dimensional (I, 2.2), it is then clear that $C^*(\mathfrak{g}, K; E) \otimes V$ has Hausdorff cohomology. Since we started from a continuously s-injective resolution, this means, by definition (3.3), that $H_d^*(G; E \otimes V)$ is Hausdorff, and implies also that it is equal to $H^*(\mathfrak{g}, K; E) \otimes V$, i.e., to $H_d^*(G; E) \otimes V$.

5.8. THEOREM. *Let N be a closed normal subgroup of G which has finitely many connected components. Let E be a finite dimensional G-module and $V \in \mathcal{C}_G^\infty$ a Fréchet differentiable G-module on which N acts trivially. Then $H^*(N; E \otimes V)$ is Hausdorff, isomorphic to $H^*(N; E) \otimes V$, admits a natural structure of differentiable*

Fréchet (G/N)-module, and there exists a spectral sequence (E_r) abutting $H^*(G; E \otimes V)$ and in which

$$E_2^{p,q} = H_d^p(G/N; H_d^q(N; E) \otimes V) \qquad (p, q \in \mathbf{N}).$$

Let $M = G/K$ be as in 5.6. Then $A^*(M; E \otimes V)$ provides a continuously s-injective resolution of $E \otimes V$ (5.5). The fibration of G by N has local cross-sections; therefore $A^q(M; V)^N$ is continuously s-injective in $\mathcal{C}_N$ (2.3).

By 5.7, $H^*(N; E \otimes V)$ is Hausdorff. By 2.3 it is the cohomology of C^{*^N}, where C^* is a continuously s-injective resolution of $E \otimes V$ in $\mathcal{C}_G$. The G/N-module structure on $H^*(N; E \otimes V)$ then stems from the natural action of G/N on C^{*^N}. Theorem 5.8 then follows from 5.7, applied to N and $E \otimes V$, and from 4.3.

REMARK. To determine the action of G/N on $H^*(N; E \otimes V)$ we may use any continuously s-injective resolution with respect to N of V in $\mathcal{C}_G$ (2.4). In particular, take as resolution $A^*(X; E \otimes V)$, where X is the space of maximal compact subgroups of N. Then

(1) $$A^*(X; E \otimes V)^N \simeq A^*(X; E)^N \otimes V,$$

as follows from the equality

$$A^*(X; E \otimes V)^N = C^*(\mathfrak{u}, L; E \otimes V) = C^*(u, L; E) \otimes V,$$

where $\mathfrak{u}$ is the Lie algebra and L a maximal compact subgroup of N. As a consequence, the action of G/N on $H^*(N; E) \otimes V$ is the tensor product of its actions on the two factors.

5.9. Induced modules. Shapiro's lemma. Let H be a closed subgroup of G and $U \in \mathcal{C}_H^\infty$. The induced module in the differentiable category is the space

$$I^\infty(U) = \operatorname{Ind}_H^G(U)^\infty = \{f \in C^\infty(G; U) \mid f(hg) = h \cdot f(g) \ (h \in H, \ g \in G)\}.$$

It is a differentiable G-module with respect to right translations and a Fréchet or a quasi-complete space if U is so.

If we consider Fréchet modules, there is no difficulty in seeing that §2 remains true if continuous functions are replaced by smooth ones and the compact open topology by the C^∞ topology. One has only to use 1.2 and to remark that if X, Y are manifolds and V a Fréchet space, then the canonical map $C^\infty(X, Y; V) \to C^\infty(X; C^\infty(Y; V))$ is a continuous linear bijection of Fréchet spaces, hence an isomorphism.

6. Further results on differentiable cohomology

To complete this discussion of differentiable cohomology, we prove the existence of a spectral sequence relating continuous cohomology and cohomology of invariant differential forms, which generalizes a result from [**105**], and give a further relation between continuous and differentiable cohomology. However, the results of this section will not be needed in the sequel.

6.1. We need some facts on topological tensor products, for which we refer to [**45, 46**]. If E, F are topological vector spaces (we recall that only locally convex Hausdorff spaces are considered here), then $E \otimes F$ will be endowed with the "*projective tensor product* topology" [**46**, I, §1, n° 3], and $E \overline{\otimes} F$ will denote the completion of $E \otimes F$ with respect to that topology. We recall that the latter is the finest locally convex topology such that the canonical bilinear map $E \times F \to E \otimes F$

is continuous, where $E \times F$ is endowed with the product topology. If $u \colon E \to E'$ and $v \colon F \to F'$ are continuous linear maps of topological vector spaces, then $u \otimes v$ is continuous. Its unique continuous extension to a map $E \mathbin{\overline{\otimes}} F \to E' \mathbin{\overline{\otimes}} F'$ is denoted $u \mathbin{\overline{\otimes}} v$.

The projective and completed projective tensor products are associative [**46**, I, p. 50]. They are also distributive with respect to finite sums. [If $E = E_1 \oplus E_2$, then the inclusions $E_i \to E$ and the projections $E \to E_i$ define bijective maps between $E \mathbin{\overline{\otimes}} F$ and $(E_1 \mathbin{\overline{\otimes}} F) \oplus (E_2 \mathbin{\overline{\otimes}} F)$ which are continuous and inverse to each other, hence isomorphisms.]

6.1.1.   If E and F are Fréchet spaces, then so is $E \mathbin{\overline{\otimes}} F$ [**46**, I, §1, n° 3, Prop. 5]. If E is finite dimensional, then $E \mathbin{\overline{\otimes}} F = E \otimes F$, and the topology is the one used in 5.7.

6.1.2.   Let $0 \to E' \to E \to E'' \to 0$ be an exact sequence of Fréchet spaces and F a Fréchet space. If either E or F is *nuclear* [**46**, II, §2, n° 1], then

$$0 \longrightarrow E' \mathbin{\overline{\otimes}} F \xrightarrow{u \mathbin{\overline{\otimes}} 1} E \mathbin{\overline{\otimes}} F \xrightarrow{v \mathbin{\overline{\otimes}} 1} E'' \mathbin{\overline{\otimes}} F \longrightarrow 0$$

is exact. Without the nuclearity assumption, it follows from [**46**, I, §1, n° 2, Prop. 3] that $u \mathbin{\overline{\otimes}} 1$ is injective, $v \mathbin{\overline{\otimes}} 1$ is surjective and $\mathrm{Im}(u \mathbin{\overline{\otimes}} 1) \subset \ker(v \mathbin{\overline{\otimes}} 1)$. The equality $\mathrm{Im}(u \mathbin{\overline{\otimes}} 1) = \ker(v \mathbin{\overline{\otimes}} 1)$, when either E or F is nuclear, follows from the corollary to Prop. 10 in [**46**, II, §3, n° 1]. Note that if E is nuclear, then so are E' and E'' [**46**, II, Thm. 3].

6.1.3.   Let F be a Fréchet space. If $C^* \colon C^0 \to C^1 \to \cdots$ is a complex of nuclear Fréchet spaces with Hausdorff cohomology (3.1), then so is $C^* \mathbin{\overline{\otimes}} F \colon C^0 \mathbin{\overline{\otimes}} F \to C^1 \mathbin{\overline{\otimes}} F \to \cdots$, and we have $H^*(C^* \mathbin{\overline{\otimes}} F) = H^*(C^*) \mathbin{\overline{\otimes}} F$.

This follows from 6.1.2 by splitting C^* into short exact sequences and using 1.2. It follows also that if C^* is acyclic, then so is $C^* \mathbin{\overline{\otimes}} F$.

6.1.4.   Let M be a smooth manifold, E a finite dimensional real or complex vector space, and $q \in \mathbf{N}$. Then $A^q(M; E)$ is a nuclear Fréchet space [**46**, II, §2, n° 3, Thm. 10]. If V is a complete space, then Example 1 in [**46**, II, §3, n° 3] implies that the natural map

$$(1) \qquad\qquad A^q(M; E) \mathbin{\overline{\otimes}} V \to A^q(M; E \otimes V)$$

is an isomorphism.

6.1.5.   Let M be a smooth manifold. Then the assignment $V \mapsto C^\infty(M; V)$ is an exact functor from Fréchet spaces to Fréchet spaces.

By 1.2, it suffices to prove that if

$$0 \to V' \to V \to V'' \to 0$$

is an exact sequence of Fréchet spaces, then

$$0 \to C^\infty(M; V') \to C^\infty(M; V) \to C^\infty(M; V'') \to 0$$

is exact. This follows from 6.1.2 (with $F = C^\infty(M)$) and 6.1.4.

6.1.6.   Let G_i be a Lie group and $(\pi_i, E_i) \in \mathcal{C}_{G_i}$ $(i = 1, 2)$. Then the tensor product representation of $G = G_1 \times G_2$ into $E_1 \otimes E_2$ is continuous and extends to a continuous representation $\pi_1 \mathbin{\overline{\otimes}} \pi_2$ in $E = E_1 \mathbin{\overline{\otimes}} E_2$ [**113**, 4.1.2.4]. If the E_i are smooth $(i = 1, 2)$, then E is smooth [**113**, 4.4.1.10, p. 259].

6.2. PROPOSITION. *Let M be a manifold with finite dimensional real cohomology. Let V be a Fréchet space and E a finite dimensional complex vector space.*

Then $A^*(M; V \otimes E)$ has Hausdorff cohomology, and we have

(1)
$$H^*(M; E) \otimes V = H^*(M; E \otimes V)$$
$$= H^*(A^*(M; E \otimes V)) = H^*(A^*(M; E)) \otimes V.$$

(The main point here is the second equality, which says that the de Rham theorem is valid for forms with values in a Fréchet space.)

The first equality follows from the universal coefficient theorem and the finite dimensionality of $H^*(M; \mathbf{R})$.

The space $H^*(M; E) = H^*(M; \mathbf{R}) \otimes E$ is finite dimensional; therefore $A^*(M; E) = A^*(M) \otimes E$ is a complex of Fréchet spaces with Hausdorff cohomology (3.5). By 6.1(3), so is the complex $A^*(M; E) \overline{\otimes} V$; this proves our first assertion and shows that we have $H^*(A(M; E) \overline{\otimes} V) = H^*(A(M; E)) \otimes V$. Of course, $H(A^*(M; E)) = H^*(M; E)$ by the usual de Rham theorem; hence the first and fourth terms of (1) are equal. Finally, it follows from 6.1.4 that we have

$$A^q(M; E \otimes V) = A^q(M) \overline{\otimes} (E \otimes V)$$
$$= (A^q(M) \otimes E) \overline{\otimes} V = A^q(M; E) \overline{\otimes} V \qquad (q \in \mathbf{N}),$$

whence the last equality in 1).

6.3. THEOREM (G. D. Mostow). *Let M be a manifold with finite dimensional real cohomology, on which G operates smoothly and properly. Let V be a Fréchet G-module. Then there exists a spectral sequence (E_r) which abuts $H^*(A(M; V)^G)$ and in which*

$$E_2^{p,q} = H_d^p(G; H^q(M; \mathbf{R}) \otimes V) \qquad (p, q \in \mathbf{N}).$$

We consider the sequence

(1)
$$0 \to V \to A^0(M; V) \to A^1(M; V) \to \cdots.$$

It is an augmented complex of Fréchet spaces, with Hausdorff cohomology (6.2). Each $A^q(M; V)$ is a continuously s-injective G-module (5.4); in particular, it is G-acyclic. Taking 6.2 into account, we see that the spectral sequence of 4.1 has the required properties.

REMARK. If M is a smooth principal G-bundle, this result is due to van Est [**105**, Thm. 4].

6.4. COROLLARY. *Assume that M is acyclic over $\mathbf{R}$. Then*

$$H_d^*(G; V) = H^*(A^*(M; V)^G).$$

In fact, the complex $A^*(M; V)$ is also acyclic (6.2); hence 6.3(1) yields a resolution of V by a complex of G-acyclic Fréchet spaces. We may apply 4.2 and 5.1, or remark that we have $E_2^{p,q} = 0$ for $q \neq 0$ in the spectral sequence of 6.3.

6.5. LEMMA. *Let $V \in \mathcal{C}_G$.*

(i) *If V is s-injective in $\mathcal{C}_G$, then V^∞ is s-injective in $\mathcal{C}_G^\infty$.*

(ii) *If V is quasi-complete, differentiable, and s-injective in $\mathcal{C}_G^\infty$, then V is continuously s-injective in $\mathcal{C}_G$.*

(iii) *The functor $V \mapsto V^\infty$ is exact in the category of Fréchet G-modules.*

PROOF. (i) Let $u: A \to B$ be a strong injection of differentiable Fréchet G-modules, and $f: A \to V^\infty$ a G-morphism. By assumption, f extends to a G-morphism $g: B \to V$. The latter induces a continuous G-morphism $g_\infty: B^\infty \to$

V^∞. Since B is differentiable, we have $B = B^\infty$ set-theoretically and topologically by definition (0, 2.3). Hence $\operatorname{Im} g \subset V^\infty$ and g, viewed as a map of B into V^∞, where V^∞ is endowed with its topology of differentiable G-modules (which is finer than the topology induced from V), is also continuous.

(ii) Since V is s-injective in $\mathcal{C}_G^\infty$, it is a topological direct G-summand in any differentiable G-module containing it, in particular in $C^\infty(G; V)$. Our assertion then follows from 5.2.

(iii) In view of 1.2, it suffices to show that $V \mapsto V^\infty$ preserves short exact sequences and that the only non-obvious part is the exactness on the right, i.e. if $f : U \to V$ is a surjective G-morphism of Fréchet G-modules, then $f_\infty : U^\infty \to V^\infty$ is also surjective. But this is proved in [**113**] (see 4.4.1.11, p. 260 in [**113**], taking into account that f induces an isomorphism of $U / \ker u$ onto V).

6.6. PROPOSITION (P. Blanc [**3**, 5.2]). *Let V be a Fréchet G-module. Then the natural map $H_{\mathrm{ct}}^q(G; V^\infty) \to H_{\mathrm{ct}}^q(G; V)$ is an isomorphism $(q \in \mathbf{N})$.*

Let $0 \to V \to F^0 \to F^1 \to \cdots$ be an s-injective resolution of V by Fréchet G-modules (see 1.5). Then by 6.5, $0 \to V^\infty \to F^{0\infty} \to F^{1\infty} \to \cdots$ is a resolution of V^∞ by Fréchet modules, which are s-injective in $\mathcal{C}_G^\infty$, and in particular acyclic. (It is not necessarily s-exact, though.) By 4.2 and 5.2, $H_d(G; V^\infty)$ is the cohomology of the complex $\{F^{\infty q^G}\}$. But $F^{q^G} \subset F^{\infty q}$, hence $F^{q^G} = F^{\infty q^G}$ $(q \in \mathbf{N})$, and therefore $H_{\mathrm{ct}}^*(G; V)$ is the cohomology of the same complex as $H_d^*(G; V^\infty)$. In view of 5.1, this proves our assertion.

6.7. REMARK. The above argument only shows that $H_{\mathrm{ct}}^q(G; V^\infty) \to H_{\mathrm{ct}}^q(G; V)$ is continuous and bijective. However, the theorem proved by P. Blanc (and communicated to us without proof in Spring 1977) asserts more precisely that this map is a topological isomorphism. This is established by showing that the non-homogeneous complexes $F^{*\infty}(G; V^{\infty G})$ and $F^*(G; V)^G$ are homotopy equivalent.

We note, however, that the proof given here implies easily that if $H_{\mathrm{ct}}^*(G; V)$ is Hausdorff, then $H_{\mathrm{ct}}^*(G; V^\infty)$ is also Hausdorff, and consequently topologically isomorphic to $H_{\mathrm{ct}}^*(G; V)$, since both spaces are then Fréchet spaces. In fact, if $H_{\mathrm{ct}}^*(G; V)$ is Hausdorff, then $A^{*\infty G} = A^{*G}$ has Hausdorff cohomology. Moreover, if $0 \to V^\infty \to D^*$ is an s-injective resolution in $\mathcal{C}_G$ by smooth modules, then $V^\infty \to V$ extends to a morphism $D^* \to A^*$ which factors through $D^* \to A^{*\infty}$ and induces a bijective continuous map of $H^*(D^{*G})$ onto $H^*(A^{*G})$. By 3.1, D^{*G} has then Hausdorff cohomology.

We next generalize 5.6.

6.8. PROPOSITION. *Let E be a smooth Fréchet G-module and F a complete space on which G acts trivially. Let $X = G/K$.*

(i) *We have $A^q(X; E) \overline{\otimes} F = A^q(X; E \overline{\otimes} F)$. The G-module $A^q(X; E) \overline{\otimes} F$ is smooth and s-injective in $\mathcal{C}_G$ $(q \in \mathbf{N})$.*

(ii) *Let E be admissible. Then*

$$(1) \qquad A^q(X; E \overline{\otimes} F)^G = A^q(X; E)^G \otimes F \qquad (q \in \mathbf{N}).$$

$H_{\mathrm{ct}}^*(G; E \overline{\otimes} F)$ *is Hausdorff, and we have*

$$(2) \qquad H_{\mathrm{ct}}^*(G; E \overline{\otimes} F) = H_{\mathrm{ct}}^*(G; E) \otimes F.$$

We note first that $E \overline{\otimes} F$ is a smooth G-module (6.1.6). The associativity of $\overline{\otimes}$ and 6.1.4 yield the topological isomorphisms

(3) $A^q(X; E) \overline{\otimes} F = (A^q(X) \overline{\otimes} E) \overline{\otimes} F = A^q(X) \overline{\otimes} (E \overline{\otimes} F) = A^q(X; E \overline{\otimes} F)$.

Together with 5.4, this proves (i).

Now assume E to be admissible. For $\delta \in \widehat{K}$ and W a quasi-complete G-module, let W_δ be the isotypic subspace of W of type δ. There is a projector π_δ of W on W_δ. In particular, E is the direct sum of E_δ, which is finite dimensional by assumption, and of $E'_\delta = \ker \pi_\delta$. Therefore $E \overline{\otimes} F$ is the topological direct sum of $E_\delta \overline{\otimes} F$ and $E'_\delta \overline{\otimes} F$. The projector π_δ annihilates $E'_\delta \otimes F$, hence also $E'_\delta \overline{\otimes} F$; therefore

(4) $(E \overline{\otimes} F)_\delta = E_\delta \otimes F$.

Then (1) follows from 5.6(1), (2). Since $A^*(X; E)^G$ is finite dimensional, it implies the remaining part of (ii).

6.9.   We conclude this section with a Künneth rule that generalizes the remark at the end of 6.1.3.

Let A^* and B^* be graded vector spaces whose graded components are Fréchet spaces and such that the A^i are nuclear. Let d_A (resp. d_B) denote continuous differentials on A^* (resp. B^*). We assume that for every $q \in \mathbf{Z}$

(1) $\{i \in \mathbf{Z} \mid A^i \text{ and } B^{q-i} \neq 0\}$ *is finite.*

This is obviously the case if the degrees of A^* or B^* are bounded, or if the degrees of A^* and B^* are both bounded above or below.

Let S^* denote the usual graded algebraic tensor product of complexes, $A^* \otimes B^*$, with the summand of total degree q equal to $\bigoplus_i A^i \otimes B^{q-i}$ (a finite direct sum by (1)) and with differential given by $\bigoplus_i (d_A \otimes \mathrm{Id} + (-1)^i \mathrm{Id} \otimes d_B)$. Similarly, $\overline{S}^*$ will denote the complex graded by the summands

(2) $$\overline{S}^q = \bigoplus_i A^i \overline{\otimes} B^{q-i}$$

endowed with the differential

(3) $$\bigoplus_i \left(d_A \overline{\otimes} \mathrm{Id} + (-1)^i \mathrm{Id} \overline{\otimes} d_B \right)$$

(see 6.1).

6.10.   THEOREM. *We keep the notation and assumptions of 6.9. If $H^i(A^*)$ is finite dimensional for each $i \in \mathbf{Z}$ (resp. A^* and B^* have Hausdorff cohomology), then*

(1) $H^*(\overline{S}^*) = H^*(A^*) \otimes H^*(B)$ (*resp.* $H^*(\overline{S}^*) = H^*(A^*) \overline{\otimes} H^*(B)$).

The argument is the standard one. We sketch it for the sake of completeness. Note that 3.5 implies that in the first case A^* has Hausdorff cohomology. The complex $\overline{S}^*$ is filtered by the subspaces

(2) $$F_p \overline{S}^* = \bigoplus_{q \geq p} A^* \overline{\otimes} B^q.$$

This sum is finite in each total degree by 6.9(1). We have

(3) $$F_p \overline{S}^* = A^* \overline{\otimes} B^q \oplus F_{p+1} \overline{S}^*.$$

There is thus a spectral sequence with

(4) $E_0 = \operatorname{Gr} \overline{S}^*, \qquad E_0^{p,q} = A^p \overline{\otimes} B^q \quad (p, q \in \mathbf{Z}).$

From 6.9(3) we see that $d_0 = d_A \overline{\otimes} \operatorname{Id}$. Thus, 6.1.2 and 6.1.3 imply that

(5) $F_p E_1 = \bigoplus_{q \geq p} H^*(A^*) \overline{\otimes} B^q \qquad (p \in \mathbf{Z}).$

By 6.9(3) and the definitions it follows that the differential on $H^p(A^*) \overline{\otimes} B^q$ is $(-1)^p \operatorname{Id} \overline{\otimes} d_B$. If $H^p(A^*)$ is finite dimensional, then the completed tensor product in (5) is equal to the algebraic tensor product, so we have

(6) $F_p E_1 = \bigoplus_{q \geq p} H^*(A^*) \otimes B^q$

and

(7) $E_2 = H^*(A^*) \otimes H^*(B^*).$

Set $Z^q = B^q \cap \ker d_B$. If B^* also has Hausdorff cohomology, then 6.1.2 implies that the exact sequence

(8) $0 \to d_B B^{q-1} \to Z^q \to H^q(B^*) \to 0$

yields the exact sequence

(9) $0 \to H^p(A^*) \overline{\otimes} d_B B^{q-1} \to H^p(A^*) \overline{\otimes} Z^q \to H^p(A^*) \overline{\otimes} H^q(B^*) \to 0.$

In the second case, we therefore have,

(10) $E_2 = H^*(A^*) \overline{\otimes} H^*(B^*).$

Let (E_r') be the spectral sequence constructed similarly for S^*. It satisfies (2)–(7), with the algebraic tensor product replacing the completed one in (2)–(5). Since $\overline{d}$ is an extension of d, the inclusion $\mu \colon S^* \to \overline{S}^*$ induces a morphism (μ_r) of spectral sequences. In the first case (6) shows that $\mu_1 \colon E_1' \to E_1$ is an isomorphism, hence so is $\mu_r \colon E_r' \to E_r$ for all $r \geq 2$. In the second case μ_2 is the natural map

$$H^*(A^*) \otimes H^*(B^*) \hookrightarrow H^*(A^*) \overline{\otimes} H^*(B^*)$$

and d_2 is a continuous extension of d_2'. Since (E_r') is the usual Künneth spectral sequence, $d_r' = 0$ for $r \geq 2$. We therefore see that in both cases $d_r = 0$ for $r \geq 2$. Hence $E_2 = E_\infty$, and the theorem follows.

CHAPTER X

Continuous and Differentiable Cohomology for Locally Compact Totally Disconnected Groups

In this chapter, G is a locally compact group which is countable at infinity and totally disconnected (every point has a neighborhood basis consisting of compact open neighborhoods). Unless otherwise stated, G is assumed to be metrizable. We keep the conventions of IX.

1. Continuous and smooth cohomology

1.1. For brevity a group satisfying the above conditions will be called a *t.d. group*. We note that any t.d. group H has a fundamental set of neighborhoods of the identity consisting of a decreasing sequence of compact open subgroups, which may be chosen to be normal if H is moreover compact. We also recall that, by a theorem of E. Michael [**84**], the fibration of H by a closed subgroup always has continuous cross-sections.

1.2. Let V be a real or complex vector space. Assume V to be the union of an increasing sequence of subspaces $(V_n)_{n \in \mathbf{N}}$, where each V_n is a topological vector space (hence Hausdorff, locally convex in view of our general conventions) and the inclusion maps $g_{mn} \colon V_m \to V_n$ are isomorphisms onto a closed subspace for all $m, n \in \mathbf{N}$ $(m \leq n)$. The *inductive limit topology* on V is the unique locally convex topology such that a linear map $f \colon V \to W$ into a topological vector space is continuous if and only if its restriction to the V_m's is continuous. It is also the finest locally convex topology such that the inclusion maps $V_m \to V$ are continuous $(m \in \mathbf{N})$ [**23**, II, §4, nos. 4, 6]. It is strict and induces the given topology on each V_m. The space V is complete (resp. quasi-complete) if the V_n are [**24**, III, §2, nos. 4, 5].

If the V_n's are finite dimensional, then the *inductive limit topology* is the finest locally convex topology. In fact, if $f \colon V \to W$ is a bijective linear map on a topological vector space W, then its restriction to each V_n is necessarily continuous; hence f is continuous.

1.3. Let $(\pi, V) \in \mathcal{C}_G$. An element $v \in V$ is *smooth* if it is fixed under an open subgroup of G. The space V^∞ of smooth vectors is stable under G. We let π_∞ be the restriction of π to V^∞. The space V^∞ is also the space of K-finite vectors, where K is any compact open subgroup of G. If V is quasi-complete (our only case of interest), then V^∞ is dense in V, and V^∞ is locally finite and semisimple as a K-module, for any compact open subgroup K of G. The representation (π_∞, V^∞) is continuous if V^∞ is endowed with the discrete topology, as is usually done. However, V^∞ is then not a topological vector space over $\mathbf{C}$ (unless the latter is endowed with the discrete topology!). We want to view it as a topological G-module and shall give it the inductive limit topology with respect to the subspaces V^K (K a compact open subgroup), these being endowed with the topology induced from that of V. Since the V^K have a countable cofinal set, we are in the case of 1.1. It follows in particular that the inclusion $V^\infty \to V$ is continuous, and that V^∞ is quasi-complete (resp. complete) if V is.

More generally $(\pi, V) \in \mathcal{C}_G$ is said to be smooth if $V = V^\infty$ as a topological G-module. We let $\mathcal{C}_G^\infty$ be the category of smooth G-modules and continuous G-morphisms.

If a vector space V is the union of an increasing sequence of subspaces V_n, then the finest locally convex topology of V is the strict inductive limit of that of the V_n's. Therefore, if G acts on a vector space V so that every $v \in V$ is fixed under an open subgroup of G, then V, endowed with the finest locally convex topology, is a smooth G-module.

A G-module $(\pi, V) \in \mathcal{C}_G$ is *admissible* if V^K is finite dimensional for every open subgroup K of G. Thus V is admissible if and only if V^∞ is; in that case the topology of V^∞ is the finest locally convex topology (1.2).

If $(\pi, V) \in \mathcal{C}_G^\infty$, then the *Hecke algebra* $\mathcal{H}(G)$ (under convolution) of compactly supported locally constant functions on G acts in the following way on V:

$$\pi(f) \cdot v = \int_G f(g) \pi(g) \cdot v \, dg$$

($f \in \mathcal{H}(G)$, $v \in V$, dg a Haar measure on G). The endomorphisms $\pi(f)$ ($f \in \mathcal{H}(G)$) are *continuous*. To prove this, it suffices to see that if f is the characteristic function of a compact open set C and K an open compact subgroup of G, then $\pi(f) \colon V^K \to V$ is continuous. There exist a subgroup L of finite index of K and a finite subset S of G such that C is the disjoint union of the $g \cdot L$ ($g \in S$). Then $\pi(f)$ is given on V^K by

$$(1) \qquad \qquad \pi(f) \cdot v = c \cdot \sum_{g \in S} \pi(g) \cdot v \qquad (v \in V^K)$$

(where c is the volume of L with respect to the Haar measure underlying the definition of the Hecke algebra), hence is continuous.

Let $V, W \in \mathcal{C}_G$, and let $f \colon V \to W$ be a G-morphism. It induces a G-morphism $f_\infty \colon V^\infty \to W^\infty$. The map is *continuous*: in fact, for every compact open subgroup, f_∞ maps V^K into W^K and coincides with f on V^K. Hence the restriction of f_∞ to V^K is continuous, whence our assertion (1.1). Therefore $V \mapsto V^\infty$ defines a functor from $\mathcal{C}_G$ to $\mathcal{C}_G^\infty$.

1.4. LEMMA. *Let $V \in \mathcal{C}_G$ be s-injective. Then V^∞ is s-injective in $\mathcal{C}_G^\infty$. For any $W \in \mathcal{C}_G^\infty$, the G-module $F^0(G; W)^\infty$ is s-injective, and $\varepsilon \colon W \to F^0(G; W)^\infty$ (cf. IX, 1.3) is a strong injection.*

The first assertion is proved exactly as IX, 6.5(i). Together with IX, 1.6, this implies that if $W \in \mathcal{C}_G^\infty$, then $F^0(G; W)^\infty$ is s-injective. The map $f \mapsto f(1)$ then provides a splitting for ε; hence ε is strong.

It follows that $\mathcal{C}_G^\infty$ contains enough injectives. For $V \in \mathcal{C}_G^\infty$, we let $H_d^q(G; V)$ be the q-th cohomology space of G with coefficients in V, computed in $\mathcal{C}_G^\infty$, and refer to $H_d^*(G; V)$ as to the *smooth* or *differentiable cohomology* of G with respect to V.

More generally, if $U, V \in \mathcal{C}_G^\infty$, we let

$$(1) \qquad \qquad \operatorname{Ext}_d^q(U, V) \quad \text{or} \quad \operatorname{Ext}_{G,d}^q(U, V)$$

be the q-th derived functor in $\mathcal{C}_G^\infty$ of $\operatorname{Hom}_G(U, V)$ ($q \in \mathbf{Z}$).

1.5. PROPOSITION. *The functor $V \mapsto V^\infty$ from quasi-complete continuous G-modules to smooth G-modules is exact and strongly exact, and transforms s-injective modules to s-injective modules and strong morphisms into strong morphisms.*

Let V, W be quasi-complete G-modules and $f: V \to W$ a surjective G-morphism. Using averages, one sees immediately that $f: V^K \to W^K$ is surjective for every compact open subgroup K of G; hence f_∞ is surjective. This implies that $V \mapsto V^\infty$ is exact for quasi-complete G-modules. Now let

$$(1) \qquad \cdots \longrightarrow V_{i-1} \xrightarrow{d_{i-1}} V_i \xrightarrow{d_i} V_{i+1} \longrightarrow \cdots$$

be an s-exact sequence of quasi-complete G-modules and G-morphisms. We want to prove that the associated sequence

$$(2) \qquad \cdots \longrightarrow V_{i-1}^\infty \xrightarrow{d_{i-1,\infty}} V_i^\infty \xrightarrow{d_{i,\infty}} V_{i+1}^\infty \longrightarrow \cdots,$$

which we already know to be exact, is s-exact. By assumption, there exist continuous linear maps $e_i: V_i \to V_{i-1}$ such that

$$(3) \qquad \qquad d_{i-1} \circ e_i + e_{i+1} \circ d_i = \operatorname{Id}_{V_i}, \quad \text{for all } i\text{'s.}$$

Fix a compact open subgroup K of G. The argument of IX, 1.10, shows that we can arrange that the e_i's commute with K. But then e_i transforms K-finite vectors into K-finite vectors, hence V_i^∞ into V_{i-1}^∞. Its restriction to V_i^L (L a compact open subgroup of K) is continuous; hence $e_{i,\infty}: V_i^\infty \to V_{i-1}^\infty$ is continuous. These maps then provide a splitting of (2), as an exact sequence of topological vector spaces. This argument also shows that f_∞ is strong if f is strong. Finally, if V is s-injective in $\mathcal{C}_G$, then V^∞ is s-injective in $\mathcal{C}_G^\infty$ by Lemma 1.4.

1.6. PROPOSITION. *Let $V \in \mathcal{C}_G$ be quasi-complete and $U \in \mathcal{C}_G^\infty$. Then $\operatorname{Ext}_{\mathrm{ct}}^q(U, V)$, $\operatorname{Ext}_{\mathrm{ct}}(U, V^\infty)$ and $\operatorname{Ext}_d(U, V^\infty)$ are canonically isomorphic ($q \in \mathbf{Z}$). The spaces $H_{\mathrm{ct}}^q(G; V)$, $H_{\mathrm{ct}}^q(G; V^\infty)$ and $H_d^q(G; V^\infty)$, endowed with their canonical topologies (IX, 3.3), are canonically isomorphic ($q \in \mathbf{Z}$).*

It suffices to prove the assertions concerning $\operatorname{Ext}_{\mathrm{ct}}(U, V)$, $\operatorname{Ext}_d(U, V^\infty)$, $H_{\mathrm{ct}}^q(G; V)$ and $H_d^q(G; V^\infty)$ ($d \in \mathbf{Z}$). Let $0 \to V \to A^*$ be an s-injective resolution of V in $\mathcal{C}_G$. By 1.4, the associated sequence $0 \to V^\infty \to A^{*\infty}$ of smooth G-modules is an s-injective resolution of V^∞ in $\mathcal{C}_G^\infty$. The space $\operatorname{Ext}_{\mathrm{ct}}^q(U, V)$ (resp. $\operatorname{Ext}_d^q(U, V^\infty)$, resp. $H_{\mathrm{ct}}^q(G; V)$, resp. $H_d^q(G; V^\infty)$) is then the q-th cohomology space of the complex

$$\operatorname{Hom}_G(U, A^*) \quad (\text{resp. } \operatorname{Hom}_G(U, A^{*\infty}), \text{ resp. } A^{*G}, \text{ resp. } (A^{*\infty})^G).$$

But since U is smooth, its image in a G-module under any continuous G-morphism is contained in the space of smooth vectors. Therefore the first two complexes are identical. Similarly, since the fixed points of G in a G-module are smooth, the two last complexes, viewed as complexes of topological vector spaces, are identical. The proposition follows.

1.7. PROPOSITION. *Let H be a closed subgroup of G. Let $U \in \mathcal{C}_H^\infty$, $V \in \mathcal{C}_G^\infty$. Then there are canonical isomorphisms*

(1)
$$\operatorname{Ext}_{G,d}^q(V, I(U)^\infty) = \operatorname{Ext}_{H,d}^q(V, U) \qquad (q \in \mathbf{Z}).$$

In particular,

(2)
$$H_d^q(G; I(U)^\infty) = H_d^q(H, U) \qquad (q \in \mathbf{Z}).$$

This follows from 1.1, 1.6 and IX, 2.3.

1.8. LEMMA. *Let H be a closed subgroup of G such that G/H is compact. Let $E \in \mathcal{C}_H$ be admissible. Then $I_H^G(E)$ is admissible, and the inclusion $i\colon E^\infty \to E$ induces an isomorphism of $I_H^G(E^\infty)^\infty$ onto $I_H^G(E)^\infty$.*

Let K be a compact open subgroup of G. Let $f\colon G \to E$ be in $I_H^G(E)^K$. Let $x \in G$. Then we have, for $k \in K \cap xKx^{-1} \cap H = L$,

$$f(x) = f(xk) = f(xkx^{-1}x) = \sigma(xkx^{-1}) \cdot f(x);$$

hence $f(x)$ is fixed under L, i.e., f takes its values in E^∞. Therefore i induces a bijection of $I_H^G(E^\infty)^K$ into $I_H^G(E)^K$.

Fix a set S of representatives of $H\backslash G/K$. It is finite since G/H is compact. An element of $I_H^G(E)^K$ is completely determined by its values on S. The above argument shows the existence of a compact open subgroup M of H such that if $f \in I_H^G(E)^K$, then $f(s) \in E^M$ for all $s \in S$. Therefore $I_H^G(E)^K$ may be identified to a subspace of the direct sum of finitely many copies of E^M, hence is finite dimensional, and the map $I_H^G(E^\infty)^K \to I_H^G(E)^K$ is an isomorphism. This being true for any K, the lemma follows.

1.9. LEMMA. *Let K be a compact subgroup of G and $E \in \mathcal{C}_K^\infty$. Then $I_K^G(E)$ (resp. $I_K^G(E)^\infty$) is s-injective in $\mathcal{C}_G$ (resp. $\mathcal{C}_G^\infty$).*

This follows from 1.5 and the same argument as in IX, 2.6, using 1.7(1) for $q = 0$ instead of IX, 2.2, once it is established that if $f\colon V \to U$ is a continuous linear map ($U, V \in \mathcal{C}_G^\infty$), then the definition of $\bar{f}$ given by 1.10(1) of IX makes sense and yields a continuous K-morphism, even if V and U are not quasi-complete. But this is easily seen: Since V and U are smooth, it is clear that the integral is in fact a finite sum, hence is well defined. It remains to check that if L is a compact open subgroup of G, then $\bar{f}$ is continuous on V^L. We may assume L to be a normal subgroup of finite index of K. If S is a set of representatives of K/L in K, then, for a suitable Haar measure dx on L, we have

$$\bar{f}(v) = \sum_{g \in S} \int_L g \cdot x \cdot f(g^{-1} \cdot v)\, dx,$$

for all $v \in V^L$. The continuity then follows from the fact that the Hecke algebra $\mathcal{H}_G$ of G operates on U by continuous endomorphisms (1.3).

We shall conclude this section by some remarks on a situation which will occur for reductive groups or products of reductive groups.

1.10. Let Y be a locally finite polysimplicial complex (cf. [**30**, 1.1]; in fact, only products of simplicial complexes will be used later). We let $\mathcal{Y}_q$ be the set of q-dimensional cells ($q \in \mathbf{N}$), $\mathcal{Y}$ the union of the $\mathcal{Y}_q$'s, $C_q(Y)$ the space of finite q-chains with complex coefficients and $C_*(Y)$ the direct sum of the $C_q(Y)$. We assume the cells oriented in the usual manner and let $\partial\colon C_q(Y) \to C_{q-1}(Y)$ be the boundary operator. It commutes with all automorphisms of Y. We assume

(∗) *The group G operates on Y by automorphisms. The set $\mathcal{Y}/G$ is finite. For every $c \in \mathcal{Y}$, the subgroup G_c of elements in G fixing c pointwise is a compact open subgroup of G.*

We let St_c be the subgroup of elements in G which leave c stable. It contains G_c as a normal subgroup of finite index and is also compact open in G.

For $V \in \mathcal{C}_G$, let $C^q(Y;V) = C(\mathcal{Y}_q;V)$ be the space of V-valued q-cochains on Y, acted upon by G as usual (IX, 1.2), and $C^*(Y;V)$ the direct sum of the $C^q(Y;V)$. We have $C^q(Y;V) = \operatorname{Hom}(C_q(Y),V)$, and we let $d\colon C^q(Y;V) \to C^{q+1}(Y;V)$ be adjoint to ∂. Here $\mathcal{Y}_q$ is viewed as a discrete set, and the compact open topology on $C^q(Y;V)$ is just the topology of simple convergence. Let $\varepsilon\colon V \to C^0(Y;V)$ be the usual augmentation, which assigns to $v \in V$ the constant function on $\mathcal{Y}_0$ with value v.

1.11. Lemma. *Let $V \in \mathcal{C}_G^{\mathrm{qc}}$ (resp. $V \in \mathcal{C}_G^\infty$). Then:*

(i) *The G-module $C^q(Y;V)$ (resp. $C^q(Y;V)^\infty$) is s-injective in $\mathcal{C}_G^{\mathrm{qc}}$ (resp. $\mathcal{C}_G^\infty$) ($q \in \mathbf{N}$).*

(ii) *Assume Y to be acyclic over $\mathbf{C}$. Then*

(1) $$0 \longrightarrow V \xrightarrow{\ \varepsilon\ } C^0(Y;V) \longrightarrow \cdots \longrightarrow C^q(Y;V) \longrightarrow \cdots$$

(2)
$$(\textit{resp. } 0 \longrightarrow V \xrightarrow{\ \varepsilon\ } C^0(Y,V)^\infty \longrightarrow \cdots \longrightarrow C^q(Y;V)^\infty \longrightarrow \cdots)$$

is an s-injective resolution of V in $\mathcal{C}_G^{\mathrm{qc}}$ (resp. $\mathcal{C}_G^\infty$).

Let A_q be a set of representatives of $\mathcal{Y}_q/G$ in $\mathcal{Y}_q$ ($q \in \mathbf{N}$). By assumption it is finite. We have

(3) $$C^q(Y;V) = \bigoplus_{c \in A_q} C(G \cdot c; V).$$

It therefore suffices to show that each term on the right-hand side is s-injective. Assume first that $\mathrm{St}_c = G_c$. Then

(4) $$C(G \cdot c; V) = C(G/G_c; V) = I_{G_c}^G(V)$$

(IX, 2.1), and (i) follows from IX, 2.6 (resp. 1.9). This case would in fact suffice later. For the sake of completeness, we discuss the more general one. Let $\chi_c\colon \mathrm{St}_c \to \mathbf{C}^*$ be the character equal to $+1$ (resp. -1) on $g \in \mathrm{St}_c$ if g does not change (resp. changes) the orientation of c. Let $\mathbf{C}_c$ be $\mathbf{C}$ acted upon by St_c via χ_c. We let St_c/G_c act on the right on G/G_c by right translations. This is a free action, which commutes with G. View $C(G/G_c;V)$ as a (St_c/G_c)-module via the action on G/G_c. Then it is easily seen that we have an isomorphism of G-modules

(5) $$C(G \cdot c; V) = \operatorname{Hom}_{\mathrm{St}_c/G_c}(\mathbf{C}_c, C(G/G_c;V)) = (\mathbf{C}_c \otimes C(G/G_c;V))^{\mathrm{St}_c/G_c}.$$

The left-hand side is then a direct G-summand of $C(G/G_c;V)$, which is s-injective by IX, 2.6 and 1.9, whence (i) in general.

Assume now Y to be acyclic over $\mathbf{C}$. Since $C_*(Y)$ is a free chain complex, it admits then a contracting homotopy; i.e. we have linear maps $h'_q\colon C_q(Y) \to C_{q+1}(Y)$ ($q \in \mathbf{N}$) and $\delta'\colon \mathbf{C} \to C_0(Y)$ such that

(6)
$$h'_{q-1} \circ \partial_q + \partial_{q+1} \circ h'_q = \mathrm{Id} \qquad (q \geq 1),$$
$$\partial_1 \circ h'_0 + \delta' \circ \varepsilon' = \mathrm{Id}, \qquad \varepsilon' \circ \delta' = \mathrm{Id},$$

where $\varepsilon' \colon C_0(Y) \to \mathbf{R}$ is the augmentation which assigns to a chain the sum of its coefficients. It is then clear that the maps

$$(7) \qquad h_q \colon C^q(Y;V) \to C^{q-1}(Y;V) \quad (q = 1, 2, \cdots), \qquad \delta \colon C^0(Y;V) \to V,$$

transposed to h'_{q-1} and $\varepsilon' \otimes 1 \colon C_0(Y) \otimes V \to V$ respectively, are continuous and satisfy the conditions IX, 1.5(3); together with 1.5, this proves (ii).

1.12. LEMMA. *Assume there exists Y satisfying the conditions of 1.10(∗) and 1.11(ii). If $V \in \mathcal{C}_G^{\mathrm{qc}}, \mathcal{C}_G^{\infty}$, then $H_{\mathrm{ct}}^*(G; V) = H^*(C^*(Y; V)^G)$. If V is admissible, or, more generally, if V^{G_c} is finite dimensional for all $c \in Y$, then $C^*(Y; V)^G$ is finite dimensional, while $H_{\mathrm{ct}}^*(G; V)$ is Hausdorff and finite dimensional. The functor $V \mapsto H_{\mathrm{ct}}^*(G; V)$ commutes with inductive limits.*

The first assertion follows from 1.11. Let $c \in Y$. Then

$$(1) \qquad\qquad\qquad C(G/G_c; V)^G = V^{G_c};$$

therefore, in the notation of 1.11(5), we have

$$(2) \qquad\qquad\qquad C(G \cdot c; V)^G = (\mathbf{C}_c \otimes V^{G_c})^{\mathrm{St}_c / G_c}.$$

Together with 1.11(3), this shows that $C^*(Y; V)^G$ is finite dimensional if all the V^{G_c} are finite dimensional. The second assertion is then clear.

By 1.11(3) and 1.12(2), $C^*(Y, V)^G$ may be viewed as a complex of vector spaces over $\mathcal{Y}/G$ which depends functorially on V. Since $\mathcal{Y}/G$ is finite, the last assertion follows.

2. Cohomology of reductive groups and buildings

In IX, we saw that, for a semi-simple group G, the differentiable cohomology can be computed by means of differential forms on the symmetric space G/K of maximal compact subgroups of G. Here, analogously, we show that in the p-adic case, the cohomology can be computed using cochains on the Bruhat-Tits building. For simply connected groups, this was first pointed out in [**35**].

2.1. From now on, k is a non-Archimedean local field with finite residue field. k-groups are denoted by script letters, and their groups of k-rational points by the corresponding roman capitals. The latter groups are viewed as topological groups, with the topology defined by that of k. They are t.d. groups.

$\mathcal{G}$ will denote a connected reductive k-group, $\mathcal{Z} = \mathcal{C}(\mathcal{G})^0$ the greatest central torus of $\mathcal{G}$, $\mathcal{G}' = \mathcal{D}\mathcal{G}$ the derived group of $\mathcal{G}$ and $\widetilde{\mathcal{G}}'$ the universal covering of $\widetilde{\mathcal{G}}'$ [**124**, 24.1]. Set $\widetilde{\mathcal{G}} = \mathcal{Z} \times \widetilde{\mathcal{G}}'$, and let $\sigma \colon \widetilde{\mathcal{G}} \to \mathcal{G}$ be the natural projection. σ is a central isogeny ([**19**]; [**124**, §22]). Hence $\sigma(\widetilde{\mathcal{G}})$ is a closed normal subgroup of G such that $G/\sigma(\widetilde{\mathcal{G}})$ is compact and commutative of finite exponent [**20**, 3.19]. We shall also say that a t.d. group H is a *p-adic reductive group* if $H = \mathcal{H}(k')$, where k' is as k and $\mathcal{H}$ is a reductive k'-group.

2.2. As usual, $X^*(\mathcal{G})_k$ denotes the group of rational characters defined over k of $\mathcal{G}$, i.e. of k-morphisms of $\mathcal{G}$ into $\mathbf{GL}_1$. It is a finitely generated free commutative group whose rank is equal to the k-rank of $\mathcal{Z}$, i.e. the dimension of the greatest k-split subtorus of $\mathcal{Z}$. The restriction of $\alpha \in X^*(\mathcal{G})_k$ to G is a continuous homomorphism into k^*. Composed with the normalized absolute value $|\cdot|_k$ on k, it gives a continuous homomorphism $|\alpha|_k$ of G into the multiplicative group $\mathbf{R}_+^*$ of strictly positive real numbers. We let

$$(1) \qquad\qquad {}^0G = \bigcap_{\alpha \in X^*(\mathcal{G})_k} \ker |\alpha|_k.$$

The subgroup ${}^0G$ is normal, open, contains $\mathcal{D}G$ and all compact subgroups of G, and the quotient $G/{}^0G$ is finitely generated and free abelian. If $\mathcal{D}G$ has k-rank zero, then G has a greatest compact subgroup, and ${}^0G$ is that group. Let $\mathcal{T} = \mathcal{G}/\mathcal{D}G$ and $\pi\colon G \to T$ the canonical projection. By [20, 3.19], there exists a compact set C in G such that $G = C \cdot Z \cdot \mathcal{D}G'$, and $\pi(G)$ is closed cocompact in T. Therefore $\pi(G) \cap {}^0T$ is compact and $\pi(G)/(\pi(G) \cap {}^0T)$ is a free commutative subgroup of finite index in $T/{}^0T$. It follows easily that ${}^0G = \pi^{-1}({}^0T)$, hence also that ${}^0G/\mathcal{D}G'$ is compact. Moreover, the rank of $G/{}^0G$ is equal to the k-rank of $\mathcal{T}$. But $\pi\colon \mathcal{Z} \to \mathcal{T}$ is a central k-isogeny, and hence preserves the k-rank [19]. Therefore the rank of $G/{}^0G$ is equal to that of $X^*(\mathcal{G})_k$.

Given $\alpha \in X^*(\mathcal{G})_k$, let $v(\alpha)\colon G \to \mathbf{Z}$ be given by

$$(2) \qquad\qquad v(\alpha)(g) = \operatorname{ord} |\alpha(g)|_k = \log_q |\alpha(g)|_k,$$

where q is the order of the residue field of k. We get in this way a homomorphism $X^*(\mathcal{G})_k \to \operatorname{Hom}(G/{}^0G, \mathbf{Z})$. It is injective because, if α is not trivial, then $\alpha(G)$ contains k^{*m} for some $m \neq 0$. Both groups having the same rank, we see that v induces an isomorphism

$$(3) \qquad\qquad v\colon X^*(\mathcal{G})_k \otimes_{\mathbf{Z}} \mathbf{C} \xrightarrow{\sim} \operatorname{Hom}(G/{}^0G, \mathbf{Z}) \otimes_{\mathbf{Z}} \mathbf{C}.$$

We let $X(G)$ be the group of *characters* of G, as a topological group, i.e. of continuous homomorphisms of G into $\mathbf{C}^*$. An element $\chi \in X(G)$ is *unramified* if it is trivial on ${}^0G$. We let $X_{\mathrm{nr}}(G)$ be the group of unramified characters of G. Then

$$(4) \qquad\qquad X_{\mathrm{nr}}(G) = \operatorname{Hom}(G/{}^0G, \mathbf{C}^*).$$

The characters $|\alpha|_k$ are unramified, whence an embedding $X(\mathcal{G})_k \to X_{\mathrm{nr}}(G)$. It follows from (3) that $X_{\mathrm{nr}}(G) = X^*(\mathcal{G})_k \otimes_{\mathbf{Z}} \mathbf{C}^*$. The group ${}^0G$ could also be defined as the intersection of the kernels of the homomorphisms $g \mapsto |\chi(g)|$ $(\chi \in X(G))$, where $|\cdot|$ is now the ordinary absolute value on $\mathbf{C}$.

2.3.   We let $Y(\mathcal{G})$ or simply Y be the *Bruhat-Tits building* of $\mathcal{G}$ over k [17, 30, 95, 102]. If $\mathcal{G}$ is semi-simple, almost k-simple, then Y is a simplicial complex of dimension equal to the k-rank of $\mathcal{G}$. If $\mathcal{G}$ is semi-simple, then Y is the product of the buildings of its almost k-simple factors, viewed as simplicial complexes, and is a polysimplicial complex. The latter structure is associated to the Tits system of *parahoric subgroups* of G. If $\mathcal{G}$ is a torus, then G modulo its greatest compact subgroup ${}^0G$ is a finitely generated free commutative group, and $Y = (G/{}^0G) \otimes_{\mathbf{Z}} \mathbf{R}$. The group G acts by translations; hence there exist simplicial structures invariant under G, although not a canonical one. We assume one to be chosen. In general, $Y(\mathcal{G})$ is the product of the buildings of $\mathcal{Z}$ and $\mathcal{D}\mathcal{G}$ and is always contractible. The buildings for $\widetilde{\mathcal{G}}$ and $\mathcal{G}$ are the same. There is on $Y(\mathcal{Z})$ a simplicial structure invariant under G. We choose one and endow $Y(\mathcal{G})$ with the product of that structure and the canonical polysimplicial structure on $Y(\mathcal{G}')$. For our purposes it would suffice to consider the case $G = \widetilde{G}$, or even $G = \widetilde{G}'$, where the action of G and the quotient $Y(G)/G$ are more easily described.

A *chamber* in Y is the product of a chamber in $Y(\mathcal{G}')$, i.e. a polysimplex of maximal dimension, by $Y(\mathcal{Z})$. An automorphism τ of Y is *special* if any polysimplex stable under τ is pointwise fixed under τ. The subgroup G_0 of special automorphisms of G is a closed normal subgroup of finite index of G, containing $\sigma(\widetilde{G}')$, and $Y(\mathcal{G})/G_0$ is a finite polysimplicial complex [in general, $Y(\mathcal{G})/G$ may be viewed as a finite simplicial complex if we pass to a suitable subdivision of the given polysimplicial structure]. If G is semi-simple, then G is the semi-direct product of G_0 by a finite subgroup leaving a given chamber stable.

2.4. THEOREM. *Let Y be the Bruhat-Tits building of $\mathcal{G}$. Let $V \in \mathcal{C}_G^{\mathrm{qc}}$ (resp. $V \in \mathcal{C}_G^{\infty}$). Then $0 \to V \xrightarrow{\varepsilon} C^*(Y;V)$ (resp. $0 \to V \xrightarrow{\varepsilon} C^*(Y;V)^{\infty}$) is an s-injective resolution of V in $\mathcal{C}_G^{\mathrm{qc}}$ (resp. $\mathcal{C}_G^{\infty}$). The space $H_{\mathrm{ct}}^q(G;V)$ is equal to $H^q(C^*(Y;V)^G)$ for all q's, and is equal to zero if $q > \mathrm{rk}_k \mathcal{G}$. Let B be an Iwahori subgroup of G. Then $C^*(Y;V)^G$ is finite dimensional if V^B is finite dimensional, in particular if V is admissible, and then $H_{\mathrm{ct}}^*(G;V)$ is Hausdorff and finite dimensional. If $V^B = \{0\}$, then $H_{\mathrm{ct}}^*(G;V) = 0$. The functor $V \mapsto H_{\mathrm{ct}}^*(G;V)$ commutes with inductive limits.*

The building Y is contractible, in particular acyclic. Thus all assumptions of 1.10 and 1.11 are fulfilled. Moreover, $C^q(Y;V) = 0$ if $q > \dim Y = \mathrm{rk}_k \mathcal{G}$. The first two assertions then follow from 1.12.

There exists a chamber C of Y such that $G_C = B$. Since every cell d of Y is a face of some chamber, and G is transitive on the chambers, it follows that for any $d \in \mathcal{Y}$, G_d contains a conjugate of B; hence $\dim V^{G_d} \leq \dim V^B$ for all $d \in Y$. Using 1.11(3) and 1.12(2), we see that $C^*(Y;V)^G$ is finite dimensional (resp. zero-dimensional) if V^B is. Together with 1.12, this proves the other assertions of the theorem.

2.5. Let $\mathcal{G}$ be semi-simple and simply connected. Then G acts on Y by special automorphisms; hence 1.12(1) is fulfilled. Let B be an Iwahori subgroup and C the chamber pointwise fixed under B. Then C may be identified with Y/G, and the complex $C^*(Y;V)^G$ takes a simple form. If s is a face of C, let V_s be the fixed point set of G_s in V. If $s \subset t$, then we have an inclusion $j_{t,s} \colon V_s \subset V_t$. We let $\mathcal{F}_V$ be the simplicial sheaf on C defined by the V_s and the above inclusions. Then

$$(1) \qquad\qquad C^q(Y,V)^G = \bigoplus_{\dim s = q} V_s \qquad (q \in \mathbf{N}),$$

and for $v \in V_S$ we have

$$(2) \qquad\qquad dv = \bigoplus_{\substack{t \supset s \\ \dim t = q+1}} [t:s] j_{t,s}(v),$$

where the $[t:s]$ are the incidence coefficients in C (see [**17**, p. 216] for a similar construction). We then have isomorphisms

$$(3) \quad C^*(Y;V)^G = C^*(C;\mathcal{F}_V), \quad H_d^*(G;V) = H^*(C;\mathcal{F}_V) \qquad (V \in \mathcal{C}_G^{\mathrm{qc}}, \mathcal{C}_G^{\infty}).$$

REMARK. This last result is due to W. Casselman and D. Wigner [**35**, Thm. 2]. In that paper, it is also proved that $C^*(Y;V)$ is an acyclic resolution of V.

2.6. PROPOSITION. *Let $E, F \in \mathcal{C}_G^{\infty}$. Assume that E is admissible and that G acts trivially on F. Endow $V = E \otimes F$ with its natural topology of smooth module.*

Then $H_d^*(G;V)$ is Hausdorff, and we have

(1) $$H_d^*(G;V) = H_d^*(G;E) \otimes F,$$

(2) $$H_d^q(G;F) = \Lambda^q \operatorname{Hom}(G/^0G, \mathbf{C}) \otimes F \qquad (q \in \mathbf{N}).$$

By 1.6 and 2.4, $H_d^*(G;V)$ is the cohomology of the complex $C^*(Y;V)^G$. Any G-invariant cochain is determined by its values on a given set of representatives of $\mathcal{Y}/G$; since $\mathcal{Y}/G$ is finite, any such cochain has its values in the product of E by a finite dimensional subspace of F, whence

$$C^*(Y;V)^G = C^*(Y;E)^G \otimes F.$$

This implies (1) and also, since $C^*(Y;E)^G$ is finite dimensional, that $H_d^*(G;V)$ is Hausdorff. To prove (2), it remains to consider the case where $V = E = \mathbf{C}$ is the trivial one-dimensional module.

First let G be semi-simple and simply connected. Then $\mathcal{F}_V$ is the constant sheaf with value E. By 1.6 and 2.5, $H_d^*(G;E)$ is the cohomology of the polysimplex C with complex coefficients, and our assertion is clear. If G is semi-simple, then 2.1; IX, 1.11; and IX, 2.5, reduce us to the simply connected case.

The group $^0G/\mathcal{D}G'$ is compact (2.2). Since

$$H_d^*(\mathcal{D}G';E) = H^0(\mathcal{D}G';E) = E,$$

by the case already treated, these equalities also hold for ^0G by IX, 2.5. We then apply IX, 4.3, to G and $N = {}^0G$, 1.6, and the following well-known fact.

2.7. LEMMA. *Let L be a finitely generated commutative free group and E a finite dimensional vector space on which L acts trivially. Then*

$$H^q(L;E) = \Lambda^q \operatorname{Hom}(L,\mathbf{Z}) \otimes_{\mathbf{Z}} E \qquad (q \in \mathbf{Z}).$$

This follows, e.g., from the fact that the left-hand side is the cohomology of a torus with fundamental group L and coefficients in E.

3. Representations of reductive groups

We collect here some facts about representations. General references for these are [**32, 34**].

3.1. We fix a maximal k-split torus $\mathcal{S}$ of $\mathcal{G}$ and a minimal parabolic k-subgroup $\mathcal{P}$ containing $\mathcal{S}$. Then $\mathcal{P}$ is a semi-direct product $\mathcal{P} = \mathcal{M} \cdot \mathcal{N}$, where $\mathcal{M} = \mathcal{Z}(\mathcal{S})$ is connected reductive and $\mathcal{N}$ is the unipotent radical of $\mathcal{P}$. The derived group of $\mathcal{M}$ has k-rank zero; hence M has a greatest compact subgroup ^0M. We let $W = N(S)/Z(S)$ be the Weyl group of $\mathcal{G}$ with respect to $\mathcal{S}$. We fix a good maximal compact subgroup K and an Iwahori subgroup B adapted to S. This means, in particular, that $^0M = M \cap K$, B fixes a chamber in the apartment of the building $X(\mathcal{G})$ stabilized by S, the group K contains representatives of W, and $G = K \cdot P$.

3.2. The unramified principal series. Let δ denote the modular function of P. Let χ be an unramified character of M (2.2). We view it as a character of P, trivial on N. Then $PS(\chi)$ is the normalized induced representation

$$PS(\chi) = \{f \in C^\infty(G,P) \mid f(pg) = \chi(p)\delta^{1/2}(p)f(g) \ (g \in G, \ p \in P)\}.$$

It is a smooth admissible representation of G. It may also be defined as the space $I_P^G(\mathbf{C}_\phi)^\infty$ of smooth vectors in the representation induced from $\mathbf{C}$ acted upon by P via $\phi = \chi\delta^{1/2}$. Restricting these functions to K, one sees easily that it may be viewed as the space of smooth vectors in a continuous representation of G in a Hilbert space. The space $PS(\chi)^K$ (resp. $PS(\chi)^B$) has dimension one (resp. equal to the order of W). The representations $PS(\chi)$ form the unramified principal series of G, with respect to P.

If V is an admissible representation of G, we let V_N denote (as usual) the *Jacquet module* of V. The functor $V \mapsto V_N$ is exact [**32, 34**].

(1) The semi-simplification of $PS(\chi)_N$ as an S-module is isomorphic to $\bigoplus_{s \in W} \mathbf{C}_{\delta^{1/2} \cdot s\chi}$ [**34, 32**, Thm. 3.5].

This, combined with Frobenius reciprocity, implies:

(2) $PS(\chi)$ has a finite Jordan-Hölder series.

(3) If $\chi' \notin W(\chi)$, then $PS(\chi)$ and $PS(\chi')$ have no constituent in common.

(4) If $s\chi \neq \chi$ for $s \neq 1$, $s \in W$, then $PS(\chi)$ has a *unique* non-zero irreducible subrepresentation, which we denote W_χ.

(4) was communicated to us by A. Silberger as a well-known observation. To prove (4) we note that if V is a G-module, then, by Frobenius reciprocity, $\operatorname{Hom}_G(V, PS(\chi)) = \operatorname{Hom}_M(V_N, \mathbf{C}_{\delta^{1/2} \cdot \chi})$; hence V_N contains an S-module isomorphic to $\mathbf{C}_{\delta^{1/2} \cdot \chi}$. If $PS(\chi)$ were to contain two distinct irreducible submodules V_1, V_2, it would contain $V_1 \oplus V_2$, whence $(V_1 \oplus V_2)_N \subset PS(\chi)_N$, which would contradict (1).

The assertion (4) immediately implies

(5) If $s\chi \neq \chi$ for $s \neq 1$, $s \in W$, then $PS(\chi)$ has a multiplicity free Jordan-Hölder series, and each constituent is equivalent to some $W_{s\chi}$ ($s \in W$).

(6) An irreducible, admissible module V is a constituent (resp. submodule) of some $PS(\chi)$ if and only if $V^B \neq (0)$ ([**34**], [**13**, pp. 248–249], [**32**, p. 138]).

4. Cohomology with respect to irreducible admissible representations

The results of this section are due to W. Casselman [**33**]. The proofs below are rather different from the original ones (which have not been published).

4.1. LEMMA. *Let $\mathcal{U}$ be a connected unipotent group over k and E a Fréchet space on which U acts trivially. Then $H_{\mathrm{ct}}^q(U; E) = 0$ for $q \geq 1$.*

The group U is the union of an increasing sequence of compact open subgroups, as is easily seen by embedding U into a group of unipotent upper triangular matrices. Let R be one of them. First assume U to be commutative. Then U/R is a discrete commutative torsion group. Moreover, by IX, 1.11,

(1) $$H_{\mathrm{ct}}^q(U; E) = H_{\mathrm{ct}}^q(U/R; E) \qquad (q \in \mathbf{N}).$$

It then suffices to show that the Eilenberg–Mac Lane cohomology of a commutative torsion group L in a vector space W over a field F of characteristic zero, over which L acts trivially, is zero in dimensions ≥ 1. For this, one uses the relation

(2) $$H^q(L; W) = \operatorname{Hom}(H_q(L; F), W),$$

and the fact that $H_q(L; F)$ is the inductive limit of the homology groups $H_q(J; F)$, where J runs through the finite subgroups of L, which are well known to be zero for $q \geq 1$.

This proves the lemma for U commutative. In the general case, one argues by induction on the length of the derived series of U: let V be the last non-trivial derived group of U. We have proved that $H^q(V; E) = 0$ for $q \geq 1$; hence $H^*(V; E)$ is trivially Hausdorff. Since the fibration of U by V has continuous cross-sections (1.1), we may use the spectral sequence of IX, 4.3, and get

$$E_2^{p,q} = H^p(U/V; H_{\mathrm{ct}}^q(V; E)) = \begin{cases} 0 & \text{if } q \geq 1, \\ H^p(U, V; E) & \text{if } q = 0; \end{cases}$$

the induction assumption then yields $E_2^{p,q} = 0$ for $(p, q) \neq (0, 0)$, whence the lemma.

4.2. PROPOSITION. *Let Q be a parabolic k-subgroup of $\mathcal{G}$ and $Q = M_Q \cdot N_Q$ a Levi decomposition of Q. Let (σ, E) be an admissible Fréchet M_Q-module. Let $V = I_Q^G(E^\infty)^\infty$, where E^∞ is viewed as a Q-module on which N_Q acts trivially. Then*

(1) $$H_d^*(G; V) = H_d^*(M_Q; E^\infty).$$

If M_Q has a central element z such that $\sigma(z) = c \cdot \mathrm{Id}$ with $c \neq 1$, then $H_d^*(G; V) = 0$.

The quotient G/Q is compact. Therefore (1.8), V may be identified with W^∞, where $W = I_Q^G(E)$. In view of 1.6, (1) is then equivalent to

(2) $$H_{\mathrm{ct}}^*(G; W) = H_{\mathrm{ct}}^*(M_Q; E).$$

By IX, 2.2, we have

(3) $$H_{\mathrm{ct}}^*(G; W) = H_{\mathrm{ct}}^*(Q; E).$$

By 4.1, $H_{\mathrm{ct}}^q(N; E) = 0$ for $q \geq 1$; hence $N_{\mathrm{ct}}^*(N_Q; E)$ is trivially Hausdorff. Since the fibration of Q by N_Q has continuous cross-sections, we may use the spectral sequence of IX, 4.3. This spectral sequence degenerates since E is acyclic for N_Q and yields the isomorphism

(4) $$H_{\mathrm{ct}}^*(Q; E) = H_{\mathrm{ct}}^*(M_Q; E).$$

(2) follows from (3) and (4). The second assertion is then a consequence of IX, 1.9.

4.3. THEOREM. *Let $\chi \in X_{\mathrm{nr}}(M)$, $\chi \notin W(\delta^{1/2})$, and let V be a subquotient of $PS(\chi)$. Then $H_d^q(G; V) = 0$ $(q \in \mathbf{Z})$.*

We prove the theorem by induction on q and on the length of a Jordan-Hölder series for V.

For $q \leq -1$, there is nothing to prove. Fix $q \geq 0$ and assume the theorem for $q - 1$, all V, and all unramified χ not in $W(\delta^{1/2})$. First let V be an irreducible submodule of $PS(\chi)$ and $V' = PS(\chi)/V$. The cohomology sequence associated to the exact sequence

(1) $$0 \to V \to PS(\chi) \to V' \to 0$$

yields the exact sequence

(2) $$H_d^{q-1}(G; V') \to H_d^q(G; V) \to H_d^q(G; PS(\chi)).$$

By the induction assumption, the left-hand term is zero. Since $\delta^{-1/2}$ is contained in $W(\delta^{1/2})$, the character $\chi \cdot \delta^{1/2}$ is not trivial. Since $PS(\chi) = I_P^G(\chi \cdot \delta^{1/2})^\infty$, 4.2 shows that the third term in (2) is also zero. Hence so is the middle term.

If now V is a constituent of $PS(\chi)$, it may be identified with a submodule of $PS(w(\chi))$ for some $w \in W$, by 3.2(3), (6). Of course, $w(\chi) \notin W(\delta^{1/2})$; hence the previous argument also gives $H_d^q(G; V) = 0$. If V is a subquotient of $PS(\chi)$ and V' is a G-submodule of V, then we have the exact sequence

$$(3) \qquad H_d^q(G; V') \to H_d^q(G; V) \to H_d^q(G; V/V'),$$

so that our assertion now follows by induction on the length of V.

4.4. It therefore remains to discuss the cohomology of the constituents of $PS(\chi)$, for $\chi \in W(\delta^{1/2})$. They do not depend on χ. It suffices therefore to consider one of these representations, for instance $PS(\delta^{-1/2})$, which is $I_P^G(\mathbf{C})^\infty$, i.e. the space $C^\infty(P\backslash G)$ of locally constant complex valued functions on $P\backslash G$, acted upon by right translations.

Without essential loss of generality, we assume G to be semi-simple.

4.5. LEMMA. *Let $\chi \in X_{\mathrm{nr}}(G)$ be unramified and regular. Let $V_0, \ldots, V_m$ be G-submodules of $V = PS(\chi)$. Then*

$$(1) \qquad V_0 \cap (V_1 + \cdots + V_m) = (V_0 \cap V_1) + \cdots + (V_0 \cap V_m).$$

If $W_0, \ldots, W_m$ are M-submodules of V_N, then the analogue for the W_i's of (1) is obviously satisfied. Since the right-hand side of (1) is contained in the left-hand side, (1) then follows from the observations in 3.2.

4.6. Let $\mathcal{Q}$ be a parabolic k-subgroup of $\mathcal{G}$. The representation $I_Q^G(1)$ is just the representation by right translations of G on the space $C^\infty(Q\backslash G)$. If $Q' \supset Q$, there is a natural injection $\pi_{Q'Q} \colon I_{Q'}^G(1) \to I_Q^G(1)$. We let U_Q be the submodule spanned by the $\pi_{Q'Q}(I_{Q'}^G)$, where Q' runs through the parabolic subgroups of G containing Q strictly, and $V_Q = I_Q^G(1)/U_Q$. If $Q = G$, then $V_Q = I_Q^G(1)$ is the space of constant functions on G. If $Q = P$, then V_Q is the *Steinberg* or *special* module [**13, 17, 33**].

LEMMA. *$I_Q^G(1)$ (resp. U_Q) has a composition series whose successive quotients are the $V_{Q'}$ ($Q' \supset Q$) (resp. $Q' \supsetneq Q$), each occurring with multiplicity one.*

It suffices to prove this when $Q \supset P$. We identify the G-modules under consideration with submodules of $C^\infty(P\backslash G)$. Clearly, if $Q, Q' \supset P$, then $I_Q^G(1) \cap I_{Q'}^G(1) = I_R^G(1)$, where R is the smallest parabolic subgroup containing Q and Q'. The lemma then follows from 4.5 by an easy induction on the parabolic rank prk Q of Q (recall that prk Q is the k-rank of the radical of Q).

4.7. PROPOSITION. *Let $\mathcal{Q}$ be a parabolic k-subgroup of $\mathcal{G}$. Then $H_d^q(G; V_Q) = \mathbf{C}$ if $q = \mathrm{prk}\, Q$, and is zero otherwise.*

It suffices to consider the case where $Q \supset P$. Let $_k\Phi$ be the set of k-roots of $\mathcal{G}$ with respect to the maximal k-split torus $\mathcal{S}$ of $\mathcal{P}$, and Δ the set of simple roots for the ordering associated to $\mathcal{P}$. The set of parabolic k-subgroups containing $\mathcal{P}$ is parametrized by Δ: for $I \subset \Delta$, we let $\mathcal{P}_J$ be the parabolic subgroup containing $\mathcal{P}$ and $\mathcal{Z}(\mathcal{S}_J)$, where

$$(1) \qquad \mathcal{S}_J = \left(\bigcap_{\alpha \in J} \ker \alpha \right)^0.$$

The smallest parabolic k-subgroup containing P_J and $P_{J'}$ is then $P_{J \cup J'}$. Let us write I_J for $I_{P_J}^G(1)$ and V_J, U_J for V_Q, U_Q if $Q = P_J$. Then $I_J \cap I_{J'} = I_{J \cup J'}$.

In view of 2.2(3) and 2.6(2), Prop. 4.2 implies

$$(2) \qquad H_d^q(G; I_J) \overset{\sim}{\to} \Lambda^q X^*(\mathcal{P}_J)_k \otimes \mathbf{C} \qquad (q \in \mathbf{Z}).$$

If $J' \supset J$, then $I_{J'} \hookrightarrow I_J$, and there is a restriction map $X^*(\mathcal{P}_{J'})_k \to X^*(\mathcal{P}_J)_k$. It is easily checked that (2) is compatible with the homomorphisms induced by these maps.

Fix a scalar product on $X^*(\mathcal{S})_k \otimes_{\mathbf{Z}} \mathbf{Q}$ invariant under the Weyl group, and let (z_α) be the basis dual to Δ. For $J \subset \Delta$, write z^J for the exterior product of the z_α $(\alpha \in J)$ taken in some order. We have $X^*(\mathcal{S})_k \otimes \mathbf{Q} = X^*(\mathcal{P})_k \otimes \mathbf{Q}$. It is standard that the restriction map $X^*(\mathcal{P}_J)_k \to X^*(\mathcal{P})_k$ identifies $X^*(\mathcal{P}_J)_k \otimes_{\mathbf{Z}} \mathbf{C}$ with the subspace spanned by the z_α $(\alpha \in \Delta - J)$. This identification being made, we can replace 4.7 by the more precise statement

(*) Let $J \subset \Delta$. Then the natural homomorphism $\nu \colon H_d^*(G; I_J) \to H_d^*(G; V_J)$ induces an isomorphism of the one-dimensional space $\mathbf{C} \cdot z^J$ onto $H_d^*(G; V_J)$.

For $\Delta = J$, i.e. $P = G$, this is obvious. We then use induction on the cardinality of $\Delta - J$, i.e. on the parabolic rank of $\mathcal{P}_J$. So fix J and assume our assertion true for $J' \supsetneqq J$. We consider the exact sequence

$$(3) \qquad \cdots \to H_d^q(G, U_J) \overset{\mu}{\to} H_d^q(G; I_J) \overset{\nu}{\to} H_d^q(G; V_J) \to H_d^{q+1}(G; U_J) \to \cdots.$$

Let $s = \mathrm{Card}(\Delta - J)$. Then, by (2), $H_d^*(G; I_J)$ is an exterior algebra on s generators. If $J' \supset J$, then the image of $H_d^*(G; I_{J'}) \to H_d^*(G; I_J)$ is the exterior algebra over $\Delta - J'$; hence μ is surjective in dimensions $\neq s$. But, by the induction assumption, 4.6, and repeated use of the exact cohomology sequence, we see that $\dim H_d^q(G; U_J) \leq \dim H_d^q(G; I_J)$ for all q's and is zero for $q \geq s$. It follows that μ is an isomorphism in dimensions $\neq s$ and is zero in dimension q. Consequently ν is the zero map for $q \neq s$, and is an isomorphism for $q = s$.

4.8. Our next goal is to prove that the V_J's are irreducible. Our proof uses Macdonald's explicit computation of the C-functions for the unramified principal series and Lemma 4.10, which was communicated to us by A. Silberger. We first note that if $J = \varnothing$ (resp. $J = \Delta$), then V_J is the Steinberg (resp. trivial) representation; hence V_J is irreducible in these cases (cf. [**13**, §6]).

4.9. LEMMA. *Suppose that* $\mathrm{rk}_k(\mathcal{G}) = 1$. *Then* $I_P^G(\delta^{1/2}\delta^t)$ *is equivalent with* $I_{\overline{P}}^G(\delta^{-1/2}\delta^t)$ *for* $|t| \neq \frac{1}{2}$.

We will need the following well-known fact. Let L be a locally compact group, and let $Q \subset L$ be a closed subgroup. Let $x \in L$. Let (σ, H_σ) be a continuous representation of Q. Put

$$\sigma^x(q) = \sigma(x^{-1}qx), \qquad q \in xQx^{-1}.$$

Then

$$(1) \qquad I_Q^L(\sigma) \cong I_{xQx^{-1}}^L(\sigma^x)$$

under the linear isomorphism $(Tf)(y) = f(x^{-1}y)$, $f \in I_Q^L(\sigma)$, $y \in L$.

We also note that there is $x \in G$ so that $xPx^{-1} = \overline{P}$ and $\delta^x = \delta^{-1}$. In light of (1), it is therefore enough to prove that

$$(2) \qquad PS(\delta^t) \cong PS(\delta^{-t}); \quad \text{for } |t| \neq \tfrac{1}{2}.$$

(2) is obvious for $t = 0$. We may therefore assume that $t \neq 0$. If χ is an unramified character of S, then set

$$(3) \qquad C(\chi) = \begin{cases} \dfrac{1 - q_\alpha^{-1}\chi(a_\alpha)}{1 - \chi(a_\alpha)}, & \text{if } \Phi(\mathcal{P}, \mathcal{S}) = \{\alpha\} \\ \dfrac{(1 + q_{\alpha/2}^{-\frac{1}{2}}\chi(a_\alpha))(1 - q_{\alpha/2}^{-\frac{1}{2}}q_\alpha^{-1}\chi(a_\alpha))}{1 - \chi(a_\alpha)^2}, & \text{if } \Phi(\mathcal{P}, \mathcal{S}) = \{\alpha, \alpha/2\}. \end{cases}$$

Here $a_\alpha \in S$ and q_α (resp. $q_{\alpha/2}$) are as in [**32**, p. 141]. We note that [**32**, (24e)] implies that $C(\delta^{1/2}) = 0$. From their definitions, q_α and $q_{\alpha/2}$ are strictly positive. Since $\delta^t(a) > 0$ for $t \in \mathbf{R}$, $\alpha \in S$, we see that as a function of $t \in \mathbf{R}$, $t \neq 0$, $C(\delta^t)$ has a unique zero. Hence

$$(4) \qquad\qquad If \ |t| \neq 0, \tfrac{1}{2}, \ then \ C(\delta^t)C(\delta^{-t}) \neq 0.$$

Now (2) for $|t| \neq 0, \tfrac{1}{2}$ follows from [**32**, Corollary 3.6].

Note. Under the hypothesis of the above lemma, $PS(\delta^t)$ is actually irreducible (see [**32**, Theorem 3.10]).

We now drop the above restriction on $\mathrm{rk}_k(\mathcal{G})$. If $J \subset \Delta(\mathcal{P}, \mathcal{S})$, then put

$$W^J = \{s \in W \mid J = s^{-1}\Phi(\mathcal{P}, \mathcal{S}) \cap \Delta(\mathcal{P}, \mathcal{S})\}.$$

The W^J define a partition of W into 2^l subsets ($l = \mathrm{rk}_k(\mathcal{G})$).

4.10. LEMMA. *Let $J \subset \Delta(P, S)$. Then $PS(s_1\delta^{1/2}) \equiv PS(s_2\delta^{1/2})$ for $s_1, s_2 \in W^J$.*

Let $s \in W$ and $Q = s^{-1} \cdot P \cdot s$. Fix an element x in the Weyl chamber corresponding to Q in $X^*(\mathcal{S}) \otimes_{\mathbf{Z}} \mathbf{R}$. Then $s \in W^J$ if and only if $(x, \alpha) > 0$ for $\alpha \in J$ and $(x, \alpha) < 0$ for $\alpha \in \Delta - J$. The closures of the Weyl chambers satisfying this condition form a convex set. It therefore suffices to prove the lemma when $Q_1 = s_1^{-1} \cdot P \cdot s_1$ and $Q_2 = s_2^{-1} \cdot P \cdot s_2$ are adjacent. There exists then $\alpha \in \Phi(\mathcal{P}, \mathcal{S})$ such that $\alpha \in \Delta(\mathcal{Q}, \mathcal{S})$, $-\alpha \in \Delta(\mathcal{Q}_2, \mathcal{S})$, and hence $\alpha \notin \Delta(\mathcal{P}, \mathcal{S})$. Set

$$(Q, S) = (Q_1, S)_{\{\alpha\}} = (Q_2, S)_{\{-\alpha\}},$$

and let $Q = M_Q \cdot N_Q$ be the standard Levi decomposition of Q. Then $\mathrm{rk}_k(^0M_Q) = 1$. Put $^*Q = {}^0M_Q \cap Q_1$. Then

$$^0M_Q \cap Q_2 = {}^*\overline{Q} \quad \text{and} \quad \mathcal{S} = ({}^*\mathcal{S}) \cdot \mathcal{S}_Q,$$

as usual. We have clearly

$$(1) \qquad\qquad PS(s_i\delta^{1/2}) = I_{Q_i}^G(\delta_{Q_i}^{1/2} \cdot \delta^{1/2}) \qquad (i = 1, 2).$$

Induction in stages implies

$$(2) \qquad \begin{aligned} I_{Q_1}^G\left(\delta_{Q_1}^{1/2}\delta^{1/2}\right) &= I_Q^G\left(I_{{}^*Q}^{{}^0M_Q}\left(\delta_{{}^*Q}^{1/2}\left(\delta^{1/2}|_{{}^*S}\right)\right) \otimes \left(\delta_Q^{1/2}\delta^{1/2}\right)\big|_{S_Q}\right) \\ I_{Q_2}^G\left(\delta_{Q_2}^{1/2}\delta^{1/2}\right) &= I_Q^G\left(I_{{}^*\overline{Q}}^{{}^0M_Q}\left(\delta_{{}^*\overline{Q}}^{1/2}\left(\delta^{1/2}|_{{}^*S}\right)\right) \otimes \left(\delta_Q^{1/2}\delta^{1/2}\right)\big|_{S_Q}\right). \end{aligned}$$

But $\delta^{1/2}|_{{}^*S} = \delta_{{}^*Q_1}^t$ with $t > \tfrac{1}{2}$, since α is not in $\Delta(\mathcal{P}, \mathcal{S})$. Hence 4.9 implies:

$$(3) \qquad\qquad I_{{}^*Q}^{{}^0M_Q}\left(\delta_{{}^*Q}^{1/2}\delta^{1/2}|_{{}^*S}\right) \equiv I_{{}^*\overline{Q}}^{{}^0M_Q}\left(\delta_{{}^*\overline{Q}}^{1/2}\delta^{1/2}|_{{}^*S}\right).$$

The lemma now follows from (2) and (3).

4.11. THEOREM (Casselman [**33**]). *For every $J \subset \Delta(P, S)$, the G-module V_J is irreducible.*

Set $l = \mathrm{rk}_k(\mathcal{G})$. Then 4.10 and 3.3(4) imply that there are at most 2^l pairwise inequivalent elements in the set $\{W_{s\delta^{1/2}} \mid s \in W\}$. Therefore, by 3.3(5), a Jordan-Hölder series for $PS(\delta^{1/2})$ is of length at most 2^l. On the other hand the V_J's are $\neq 0$ (say by 4.7), and 4.6 shows that $PS(\delta^{1/2})$ has a composition series whose successive quotients are the V_J's. This composition series is then a Jordan-Hölder series, whence the theorem.

4.12. THEOREM. *Let V be an irreducible admissible representation of G such that $H^*_{\mathrm{ct}}(G; V) \neq 0$. Then there exists a parabolic subgroup Q of G such that V is isomorphic to the G-module V_Q (see 4.6). The dimension of $H^q_{\mathrm{ct}}(G; V)$ is one if $q = \mathrm{prk}\, Q$, and zero otherwise.*

By 3.4 and 3.2(6), $V \subset PS(\chi)$ for some unramified χ. We then have $\chi \in W(\delta^{1/2})$ by 4.3. The theorem now follows from 4.7 and 4.11.

In XI, 2.15, we shall see that if V_Q has compact kernel, then V_Q is not unitarizable unless Q is minimal, i.e. V_Q is the Steinberg representation.

5. Forgetting the topology

5.1. We now go back to the setup of §1, and again let G be a t.d. group. Let (π, V) be a representation of G in a vector space (no topology). The *smooth vectors* and V^∞ are defined as in §1. The space V^∞ is stable under G, and V is said to be *smooth* if $V = V^\infty$, *admissible* if moreover V^L is finite dimensional for all compact open subgroups of G. We let $\mathcal{C}^f_G$ be the category of complex vector spaces on which G operates smoothly and of linear G-maps. Let $\alpha \colon \mathcal{C}^\infty_G \to \mathcal{C}^f_G$ be the forgetful functor which ignores the topology. Given $V \in \mathcal{C}^f_G$, let $\beta(V)$ be V endowed with its finest locally convex topology. Then $\beta(V)$ is in $\mathcal{C}^\infty_G$ (1.3), whence a functor $\beta \colon \mathcal{C}^f_G \to \mathcal{C}^\infty_G$. We have $\alpha \circ \beta = \mathrm{Id}$. A linear map between two spaces endowed with the finest locally convex topology is always continuous and strong. Therefore, if $V \in \mathcal{C}^\infty_G$ is s-injective, then $\alpha(V)$ is injective. This shows first that $\mathcal{C}^f_G$ has enough injectives. In fact, if $V \in \mathcal{C}^f_G$, then the union of the spaces of maps

$$(1) \qquad F(G, V)^\infty = \bigcup_L \mathbf{Mp}(G/L, V) \quad (L \text{ a compact open subgroup of } G)$$

is injective. If $V \in \mathcal{C}^\infty_G$ is admissible, its topology is the finest locally convex topology. If $W \in \mathcal{C}^\infty_G$ has the finest locally convex topology, then any linear map of W into a topological vector space is continuous; therefore

$$(2) \qquad \mathrm{Ext}^q_d(U, V) = \mathrm{Ext}^q_e(\alpha(U), \alpha(V)) \qquad (U, V \in \mathcal{C}^\infty_G, q \in \mathbf{N}),$$

if U has the finest locally convex topology (in particular, if U is admissible), and

$$(3) \qquad \mathrm{Ext}^q_e(U, V) = \mathrm{Ext}^q_d(\beta(U), \beta(V)) \qquad (U, V \in \mathcal{C}^f_G, \, q \in \mathbf{N}).$$

In particular,

$$(4) \qquad \begin{array}{c} H^q_d(G; V) = H^q_e(G; \alpha(V)), \qquad H^q_e(G; W) = H^q_d(G; \beta(W)) \\ (V \in \mathcal{C}^\infty_G, \; W \in \mathcal{C}^f_G, \; q \in \mathbf{N}), \end{array}$$

where Ext^*_e and H^*_e refer to the derived functors of Hom_G in $\mathcal{C}^f_G$.

The category $\mathcal{C}_G^f$ is obviously an abelian category [**47**], and we can therefore avail ourselves of some standard results valid in such categories. In particular, we have

5.2. PROPOSITION. *Let A^* be a complex in $\mathcal{C}_G^f$ whose elements are G-acyclic. Then there exists a spectral sequence abutting $H^*(A^{*G})$, in which $E_2 = H_e^*(G; H^*(A^*))$. In particular, if A^* defines a G-acyclic resolution of $V \in \mathcal{C}_G^f$, then $H_e^*(G; V) = H^*(A^{*^G})$.*

Cf. [**47**, 2.4, Rem. 3]. The proof is in fact the one of IX, 4.1, rid of the topology.

5.3. PROPOSITION. *Let N be a closed normal subgroup of G. Assume that the following is true*:

(∗)   *If $V \in \mathcal{C}_G^f$ is injective, then V is N-acyclic.*

Let $V \in \mathcal{C}_G^f$. Then there exists a spectral sequence (E_r) abutting $H_e^*(G; V)$, in which $E_2 = H_e^*(G/N; H_e^*(N; V))$.

This follows from 5.2 in the same way as IX, 4.3, follows from IX, 4.1, taking into account the obvious fact that if $W \in \mathcal{C}_G^f$ is injective, then W^N is injective in $\mathcal{C}_{G/N}^f$.

5.4. LEMMA. *Let G be the direct product of two closed subgroups N and M. Assume that N is a p-adic reductive group (2.1). Then 5.3(∗) holds.*

PROOF. If $V \in \mathcal{C}_G^f$ is injective, it is a direct summand of $F(G; V)^\infty$. It suffices therefore to show that if $W \in \mathcal{C}_G^f$, then $F(G; W)^\infty$ is acyclic in $\mathcal{C}_N^f$. Let $L \supset L'$ be compact open subgroups of M. Then we have an obvious inclusion

(1) $i_{L', L} \colon F(N, F(M/L, W))^\infty \to F(N, F(M/L', W))^\infty$.

It follows immediately from the definitions that

(2) $F(G, W)^\infty = \mathrm{dir} \lim_L F(N; F(M/L, W))^\infty$,

where L runs through the compact open subgroups of M and the direct limit is taken with respect to the above maps $i_{L', L}$. This is an isomorphism of N-modules. Each module $F(N, F(M/L, W))^\infty$ is injective in $\mathcal{C}_N^f$, hence in particular acyclic. Since N is a reductive p-adic group, $F(G, W)^\infty$ is then also N-acyclic by 2.4.

REMARK. This proof shows that 5.3(∗) is valid if the functor $V_1 \to H_e^*(N; V)$ commutes with inductive limits (e.g. if N is as G in 6.3). It was pointed out to us by W. Casselman that $F(G, W)^\infty$ is not equal to $F(N, F(M; W)^\infty)^\infty$, as had been erroneously stated in an earlier version. That would have proved $F(G, N)^\infty$ to be N-injective (which might still be true).

5.5.   Recall that if A is an algebra and V an A-module, then V is said to be *non-degenerate* if $V = A \cdot V$. This condition is of course of interest only if A has no unit element. We shall consider the case where $A = \mathcal{H}(G)$ is the Hecke algebra of G (1.3). It has no unit element (unless G is discrete), but it is *idempotented* [**39**]; i.e. it has a set of idempotents e such that $\mathcal{H}(G)$ is the union of the $e \cdot \mathcal{H}(G) \cdot e$, namely the normalized characterized functions e_L of the compact open subgroups of G. Therefore a $\mathcal{H}(G)$-module V is non-degenerate if and only if it is the union of the fixed point sets of the e_L's. Also, we have $\mathcal{H}(G) \cdot \mathcal{H}(G) = \mathcal{H}(G)$; hence, if

V is any module over $\mathcal{H}(G)$, it has a greatest nondegenerate submodule, namely $\mathcal{H}(G) \cdot V$. Any smooth G-module is in a natural way an $\mathcal{H}(G)$-module which is non-degenerate. Conversely, any such $\mathcal{H}(G)$-module is associated in this way to a G-module: one shows easily that if $v \in V$, $g \in G$ and L is a sufficiently small compact open subgroup of G, then $\chi_0(gL) \cdot v$ (where $\chi_0(gL)$ is the normalized characteristic function of $g \cdot L$) is independent of L, and then one defines $g \cdot v$ to be that element. It follows that $\mathcal{C}_G^f$ can also be defined as the category of non-degenerate $\mathcal{H}(G)$-modules, and Ext_e^* as the derived functors of $\mathrm{Hom}_{\mathcal{H}(G)}$. One can then define injectives in terms of $\mathcal{H}(G)$. In particular, if $V \in \mathcal{C}_G^f$, then $\mathrm{Hom}_{\mathbf{C}}(\mathcal{H}(G), V)^\infty$ is injective and V imbeds canonically into it (see XII, §0).

5.6. Now let G be a reductive p-adic group (2.1), and Y be the Bruhat-Tits building of G. If $V \in \mathcal{C}_G^f$, then 5.1 and 2.4 show that

$$(1) \qquad\qquad H_e^*(G; V) = H^*(C^*(Y; V)^G).$$

Assume that $V = E \otimes F$, where $E, F \in \mathcal{C}_G^f$ and G acts trivially on F. Then 2.6(1) and its proof are also valid in the present case (the latter did not use the admissibility of E). Hence we also have

$$(2) \qquad\qquad H_e^*(G; V) = H_e^*(G; E) \otimes F.$$

6. Cohomology of products

6.1. THEOREM. *Let G_1 be a p-adic reductive group (2.1), G_2 a t.d. group, $V_i \in \mathcal{C}_{G_i}^f$ ($i = 1, 2$), and $V = V_1 \otimes V_2$. Assume either that V_1 is admissible or that G_2 is a p-adic reductive group. Then*

$$(1) \qquad\qquad H_e^*(G; V) = H_e^*(G_1; V_1) \otimes H_e^*(G_2; V_2).$$

Let $0 \to V_1 \overset{\varepsilon_1}{\to} A^*$ be the resolution of V_1 by the complex of V_1-valued cochains on the building of G_1 (2.4), $0 \to V_2 \overset{\varepsilon_2}{\to} B^*$ an injective resolution of V_2 in $\mathcal{C}_{G_2}^f$, and $C^* = A^* \otimes B^*$. The complex C^* is acyclic, as is seen from the Künneth rule. Therefore

$$(2) \qquad\qquad 0 \longrightarrow V \xrightarrow{\varepsilon_1 \otimes \varepsilon_2} C^*$$

is a resolution of V. We want to prove

$$(3) \qquad\qquad C^{r,s} = A^r \otimes B^s \text{ is } G\text{-acyclic} \qquad (r, s \in \mathbf{N}).$$

By 5.3 and 5.4 there exists a spectral sequence (E_r) abutting $H_e^*(G; C^{r,s})$ and in which

$$(4) \qquad\qquad E_2^{p,q} = H_e^p(G_2; H_e^q(G_1; C^{r,s})) \qquad (p, q \in \mathbf{N}).$$

It suffices therefore to show that

$$(5) \qquad\qquad E_2^{p,q} = 0, \quad \text{if } (p, q) \neq (0, 0).$$

By 5.6 and 5.1(4), we have

$$(6) \qquad H_e^q(G_1; C^{r,s}) = H_d^q(G_1; C^{r,s}) = H_d^q(G_1; A^r) \otimes B^s \qquad (q \in \mathbf{N}).$$

Since A^r is injective, for $r, s \in \mathbf{N}$ this yields

$$(7) \qquad \begin{aligned} H_e^q(G_1; C^{r,s}) &= 0 \qquad (q \geq 1), \\ H_e^0(G_1; C^{r,s}) &= (A^r)^{G_1} \otimes B^s, \end{aligned}$$

Assume now that V_1 is admissible. Then $(A^r)^{G_1}$ is finite dimensional (2.4); hence, as a G_2-module, $H_e^q(G_1; C^{r,s})$ is the direct sum of finitely many copies of B^s, therefore is injective in $\mathcal{C}_{G_2}^f$, and (5) follows.

Assume now that G_2 is also a p-adic reductive group. Then, we may again apply 5.1 and 5.6 and get

$$(8) \qquad H_e^q(G_2; (A^r)^{G_1} \otimes B^s) = H_e^p(G_2; B^s) \otimes (A^r)^{G_1} \qquad (p \in \mathbf{N}),$$

whence again (5). Thus C^* defines a resolution of V by acyclic G-modules; therefore (5.2)

$$(9) \qquad\qquad H^*(G; V) = H^*(C^{*^G}).$$

But, clearly

$$(10) \qquad\qquad C^{*^G} = A^{*^{G_1}} \otimes B^{*^{G_2}};$$

hence, by the usual Künneth rule for tensor products of complexes over fields,

$$(11) \qquad H^*(C^{*^G}) = H^*(A^{*^{G_1}}) \otimes H^*(B^{*^{G_2}}) = H_e^*(G_1; V_1) \otimes H_e^*(G_2; V_2).$$

6.2. COROLLARY. *Let $E_i \in \mathcal{C}_{G_i}$ $(i = 1, 2)$. Assume that E_i is a Fréchet (resp. unitary) module $(i = 1, 2)$, and let $E = E_1 \overline{\otimes} E_2$ (resp. $E = E_1 \widehat{\otimes} E_2$) be the completed projective (resp. Hilbert) tensor product of E_1 and E_2. If E_1 is admissible, then*

$$(1) \qquad\qquad H_{\mathrm{ct}}^*(G; E) = H_{\mathrm{ct}}^*(G_1; E_1) \otimes H_{\mathrm{ct}}^*(G_2; E_2).$$

By 1.6 and 5.1, in (1) we may replace E, E_i by E^∞, E_i^∞ $(i = 1, 2)$ and H_{ct}^* by H_e^*. In view of 6.1, it suffices then to prove that

$$(2) \qquad\qquad E^\infty = E_1^\infty \otimes E_2^\infty.$$

Let L be a compact open subgroup of G_1. Then E_1 is the topological direct sum of $E_1^L = e_L \cdot E$ and of the kernel N_L of the projector e_L onto E_1^L (cf. 5.5 for e_L). Therefore E is the topological direct sum of $N \overline{\otimes} E_2$ (resp. $N \widehat{\otimes} E_2$) and $E_1^L \overline{\otimes} E_2$ (resp. $E_1^L \widehat{\otimes} E_2$). Since E_1^L is finite dimensional, the last tensor product is in fact an ordinary tensor product. The space $N \otimes E_2$ is annihilated by e_L. Therefore e_L is also zero on the completion of $N \otimes E_2$, whence

$$(3) \qquad\qquad E^L = E_1^L \otimes E_2.$$

As a consequence, if M is a compact open subgroup of G_2, we have

$$(4) \qquad\qquad E^{L \times M} = E_1^L \otimes E_2^M,$$

whence (2).

6.3. PROPOSITION. *Let $m \in \mathbf{N}$. Let k_i be a locally compact non-Archimedean field, $\mathcal{G}_i$ a connected reductive k_i-group, $r_i = \mathrm{rk}_{k_i} \mathcal{G}_i$ and $G_i = \mathcal{G}_i(k_i)$ $(i = 1, \dots, m)$. Let $G = G_1 \times \cdots \times G_m$ and r the sum of the r_i's. Let X_i be the Bruhat-Tits building of G_i (2.2) and X the product of the X_i's. Then X satisfies the conditions 1.10($*$) and 1.11(ii). In particular,*

$$(1) \quad H_{\mathrm{ct}}^q(G; V) = 0 \quad (resp. \ H_e^q(G; V) = 0) \quad for \ q > r, \ V \in \mathcal{C}_G^{\mathrm{qc}} \ (resp. \ V \in \mathcal{C}_G^f),$$

and $H_{\mathrm{ct}}^*(G; V)$ (resp. $H_e^*(G; V)$) is finite dimensional if V is moreover admissible.

The first assertion is obvious, since X/G is the product of the quotients X_i/G_i and a product of acyclic (or contractible) spaces is acyclic (or contractible). By 1.11 the complex $C^*(X; V)$ of V-valued cochains on X provides an s-injective resolution of V. Since X is r-dimensional, it vanishes above dimension r, whence (1). The last assertion follows from 1.12.

6.4. Complement. Under the assumptions of 6.3, we also have:

(i) *The space $H_d^*(G; V)$ is finite dimensional and Hausdorff if V is admissible, or, more generally, if the fixed point set V^B of an Iwahori subgroup B of G is finite dimensional.*

(ii) *The functor*

$$V \mapsto H_d^*(G; V), \quad V \in \mathcal{C}_G^\infty \qquad (\text{resp. } V \mapsto H_e^*(G; V), \quad V \in \mathcal{C}_G^f),$$

commutes with inductive limits.

In fact, (i) and the assertion (ii) for smooth cohomology follow from 1.12 as in 2.4; then (ii) for H_e^* is a consequence of 1.5(4).

6.5. Remarks on the cohomology theories used in this chapter. If G is a t.d. group, it is usual to give the discrete topology to a complex vector space on which G acts smoothly (in the sense of 5.1). It is then a continuous G-module (and, conversely, continuity of a G-action with respect to the discrete topology of V implies smoothness). Contrary to this custom, here we have viewed smooth modules as topological vector spaces. There are two main reasons for this. First, it allowed us to use the general results of Chapter IX on continuous cohomology; second, it will be useful in Chapter XII to define a notion of smooth module for products of real Lie groups and t.d. groups. However, in §5 we "forgot the topology" and went over to an algebraic setting, chiefly to be able to prove a Künneth theorem. This seems rather roundabout, and it may be asked whether the recourse to topology was really necessary in the first place and whether it would not have been possible instead to work directly in the algebraic framework and give an independent treatment before relating H_e^* with smooth cohomology. One chief obstacle to doing this at present is that we do not know whether 5.3($*$) holds in $\mathcal{C}_G^f$ in general. This prevents us from showing the existence of a Hochschild-Serre spectral sequence for H_e^*, and we do not know how to prove Proposition 4.2 directly in that case (in 4.2, the assumption that E is a Fréchet space, which may seem irrelevant in the context of t.d. groups, was made so that we could use the spectral sequence of IX, 4.3). Note that the latter was also used in the proof of Lemma 4.1, although it might be easier there than for Proposition 4.2 to give a direct proof not using topology. At any rate, one is known for $q = 1$.

CHAPTER XI

Cohomology with Coefficients
in $\Pi_\infty(G)$: The p-adic Case

The main goals of this chapter are to prove the p-adic analogues of the results of Chapter IV and the non-unitarizability of the V_Q's $(Q \neq G, P_0)$ (cf. X, 4.6). After having recalled some results of Harish-Chandra in §1, we show in §2 how the ideas of Chapter IV can be used to carry out a classification of irreducible admissible representations of p-adic reductive groups similar to the one of Langlands in the real case (this has been done independently by A. Silberger [99]). Proposition 2.15 describes the G-modules V_Q (defined and proved irreducible in X, §4) in terms of this classification (2.15).

§3 introduces, in analogy with V, a class $\Pi_\infty(G)$ of irreducible admissible representations of the p-adic reductive group G and shows that it contains the irreducible admissible representations with compact kernel which are unitary (or, more generally, uniformly bounded). If G is almost simple, it follows from the definition of $\Pi_\infty(G)$ and 2.15 that, except in the extreme cases of the trivial and the Steinberg representations, V_Q is not in $\Pi_\infty(G)$—in particular, is not unitary. This then completes the proof of Casselman's results on the continuous cohomology of p-adic reductive groups with coefficients in unitary representations (3.9). We also note that if G has compact center and $\pi \in \Pi_\infty(G)$, then the matrix entries of π are in some space L^p $(0 < p < \infty)$ and vanish at infinity if moreover π has compact kernel. This last fact generalizes a theorem of R. Howe [63] on unitary representations.

§4 gives a more direct proof of the non-unitarizability of the V_Q's $(Q \neq P_0)$ with compact kernel, based on Howe's original theorem and on some facts proved in [34]. It can be read independently of the first three sections.

In this chapter, k is assumed to be a non-Archimedean local field with residue field of order $q < \infty$. We use the conventions of X, 2.1. G will denote the group of k-rational points of a reductive algebraic group $\mathcal{G}$ defined over k. We fix a minimal parabolic subgroup $\mathcal{P}_0 = \mathcal{M}_0\mathcal{N}_0$ defined over k.

1. Some results of Harish-Chandra

The following results of Harish-Chandra were for the most part not published by him. Proofs have been given by him in various seminars at the Institute for Advanced Study. A survey can be found in [53], and an exposition in [147].

1.1. Let $A_0 \subset M_0$ be a maximal k-split torus. As in the real case, we will look at standard p-pairs (P, A). That is, $\mathcal{P} \supset \mathcal{P}_0$ and $\mathcal{A} \subset \mathcal{A}_0$, where $\mathcal{P}$ is a parabolic subgroup of $\mathcal{G}$ defined over k and $\mathcal{A}$ is a maximal k-split torus in the center of $\mathcal{M}$ $(P = M \cdot N)$.

If $\mathcal{A}$ is a split torus over k, then we set $\mathfrak{a}^* = X(\mathcal{A})_k \otimes_{\mathbf{Z}} \mathbf{R}$ and $\mathfrak{a}_c^* = \mathfrak{a}^* \otimes_{\mathbf{R}} \mathbf{C} = X(\mathcal{A})_k \otimes_{\mathbf{Z}} \mathbf{C}$. If (P, A) is a standard p-pair, we look upon $\mathfrak{a}^*$ as a subspace of $\mathfrak{a}_0^*$ in the usual way.

We put on $\mathfrak{a}_0^*$ an inner product, $(\ ,\ )$, that is invariant under the action of the Weyl group of (G, A_0).

If $\Phi(\mathcal{P}, \mathcal{A})$ is the root system of $(\mathcal{P}, \mathcal{A})$ and $\alpha \in \Phi(\mathcal{P}, \mathcal{A})$, then we also use the notation α for the element μ of $\mathfrak{a}^*$ such that

$$q^{\mu(a)} = |\alpha(a)|, \quad \text{for } a \in A.$$

Here $|\ |$ is the usual absolute value on k.

The Weyl group acting on $\mathfrak{a}_0^*$ is then just the group generated by the orthogonal reflections s_α, $\alpha \in \Phi(P_0, A_0)$.

Let Z be the split component of G. Then we have $\mathfrak{z}^* = \{x \in \mathfrak{a}_0^* \mid (x, \alpha) = 0, \alpha \in \Phi(P_0, A_0)\}$.

1.2. If $K_0 \subset G$ is a compact open subgroup of G, we denote by $C_c^\infty(K_0 \backslash G / K_0)$ the space of all compactly supported, locally constant, K_0-bi-invariant functions on G.

Let $A(G)$ denote the space of all locally constant functions f on G such that for each compact open subgroup $K_0 \subset G$

$$(1) \qquad\qquad\qquad \dim C_c^\infty(K_0 \backslash G / K_0) * f < \infty,$$

$$(2) \qquad\qquad\qquad \dim f * C_c^\infty(K_0 \backslash G / K_0) < \infty.$$

It can be shown that condition (1) for all K_0 implies condition (2) for all K_0, and vice-versa.

We fix a compact open subgroup $K \subset G$ such that $P_0 K = G$.

If (τ, V) is a finite dimensional double unitary representation of K, we let

$$A(G, \tau) = \{f \in A(G) \otimes V \mid f(k_1 g k_2) = \tau(k_1) f(g) \tau(k_2) \ (g \in G,\ k_1, k_2 \in K)\}.$$

1.3. THEOREM (Harish-Chandra). *If $f \in A(G, \tau)$ and if r denotes the right regular representation of G on $C^\infty(G) \otimes V$, then $r(Z)f$ spans a finite dimensional vector space.*

1.4. As in the real case, if (P, A) is a standard p-pair, then $K_M = M \cap K$. If $t \geq 1$ we define

$$A^+(t) = \{a \in A \mid |\alpha(a)| \geq t\}, \quad \alpha \in \Phi(P, A), \qquad A^+ = \bigcup_{t > 1} A^+(t).$$

δ_P denotes the modular function of P.

1.5. THEOREM (Harish-Chandra). *If $f \in A(G, \tau)$ and (P, A) is a standard p-pair, then there exists a unique $f_P \in A(M, \tau|_{K_M})$ such that if $\Omega \subset M$ is a compact subset, then there is a $t \geq 1$ such that*

$$\delta_P(ma)^{1/2} f(ma) = f_P(ma)$$

for $a \in A^+(t)$ and $m \in \Omega$.

1.6.   1.3 implies that $r(A)f_P$ spans a finite dimensional subspace of $C^\infty(M) \otimes V$. Since A is abelian, we see that

$$(1) \qquad\qquad f_P = \sum_\chi f_{P,\chi},$$

where the sum is over the characters $\chi \colon A \to \mathbf{C}^*$ of A and

$$(2) \qquad\qquad (r(a) - \chi(a))^d f_{P,\chi} = 0 \qquad (a \in A),$$

for suitable d.

We set $d_{f,\chi}$ equal to the minimum d necessary in (2).

Recall that a representation (π, H) of G is said to be admissible if it is smooth and if $H^{K_0} = \{v \in H \mid \pi(k)v = v, k \in K_0\}$ is finite dimensional for every compact open subgroup $K_0 \subset G$. It is a basic theorem of Bernshtein [**2**] and Harish-Chandra that the underlying smooth representation of an irreducible unitary representation of G is admissible.

We will assume that admissible representations of G are on pre-Hilbert spaces, so that K acts unitarily and G acts continuously on the completion.

1.7.   Let (π, H) be an admissible finitely generated representation of G. Let $W \subset H$ be a finite dimensional K-stable subspace of H. Let $E \colon H \to W$ be the orthogonal projection. Let τ be the usual double representation of K on $\mathrm{End}(W)$. That is,

$$\tau(k_1) T \tau(k_2) = \pi(k_1) T \pi(k_2) \qquad (k_1, k_2 \in K,\ T \in \mathrm{End}(W)).$$

Set $\Psi_{W,\pi}(g) = \Psi(g) = E\pi(g)E$ for $g \in G$. Then $\Psi \in A(G, \tau)$, since π is admissible. For (P, A) a standard p-pair, set $E_W(P, \pi) = \{\chi \mid \Psi_{P,\chi} \neq 0\}$.

1.8.   THEOREM (Harish-Chandra). *We keep the notation of 1.7. There exist a finite subset $E(P, \pi)$ of the set of characters of A and $d_\pi \in \mathbf{N}$ such that $E_W(P, \pi) \subset E(P, \pi)$ and $d_\pi \geq d_{\Psi_{W,\pi,\chi}}$ for each finite dimensional K-stable subspace W of H.*

1.9.   We now assume that (π, H) is an irreducible admissible representation of G. Let $(\widetilde{\pi}, \widetilde{H})$ denote the *smooth dual* of (π, H) (i.e. $\widetilde{H}$ is the space of smooth vectors in the contragredient representation). If $\widetilde{v} \in \widetilde{H}$ and $v \in H$, then there is a K-stable subspace $W \subset H$ so that

$$\langle \pi(g)v, \widetilde{v} \rangle = \langle \Psi_{W,\pi}(g)v, \widetilde{v} \rangle.$$

Hence, if (P, A) is a fixed standard p-pair, then for $m \in M$ and $\chi \in E(P, \pi)$ we can define

$$P_\chi(m \colon v, \widetilde{v}) = \langle \Psi_{P,\chi}(m)v, \widetilde{v} \rangle.$$

(1) $m \mapsto P_\chi(m \colon v, \widetilde{v})$ *is in $A(M)$ for each fixed $v \in H$ and $\widetilde{v} \in \widetilde{H}$, and $v, \widetilde{v} \mapsto P_\chi(m \colon v, \widetilde{v})$ is bilinear.*

(2) *If $k_1, k_2 \in K_P$, then $P_\chi(k_1 m k_2 \colon v, \widetilde{v}) = P_\chi(m \colon \pi(k_2)v, \widetilde{\pi}(k_2)^{-1}\widetilde{v})$ for $m \in M$, $v \in H$, $\widetilde{v} \in \widetilde{H}$.*

(3) $(r(a) - \chi(a))^{d_{\pi,\chi}} P_\chi(\cdot \colon v, \widetilde{v}) = 0$ *for $a \in A$.*

(4) *If $\Omega \subset M$ is compact, there is $t \geq 1$ depending only on $v, \widetilde{v}$ and Ω such that*

$$\delta_P(ma)^{1/2} \langle \pi(ma)v, \widetilde{v} \rangle = \sum_\chi P_\chi(ma \colon v, \widetilde{v})$$

for $a \in A^+(t)$ and $m \in \Omega$.

(1)–(4) are just restatements of 1.6, 1.8 and 1.5.

We denote by $(\overline{P}, A)$ the opposite p-pair to (P, A).

1.10. LEMMA. *We keep the notation of 1.9. If $n \in N$ and $\overline{n} \in \overline{N}$, $m \in M$, $v \in H$, $\widetilde{v} \in \widetilde{H}$, then*

$$P_\chi(m \colon \pi(\overline{n})v, \widetilde{\pi}(n)\widetilde{v}) = P_\chi(m \colon v, \widetilde{v}).$$

Let $\Omega \subset M$ be compact, and let t be so large that if $g = ma$, $m \in \Omega$, $a \in A^+(t)$, then

$$\widetilde{\pi}(g\overline{n}g^{-1})\widetilde{v} = \widetilde{v}.$$

(This is possible since $\widetilde{v}$ is smooth.)

Then if $t_1 \geq t$ and t_1 is as in 1.9(4) and $g = ma$, $a \in A^+(t_1)$, $m \in \Omega$, then

$$\delta_P(g)^{1/2}\langle \pi(g)\pi(\overline{n})v, \widetilde{v}\rangle = \delta_P(g)^{1/2}\langle \pi(g)v, \widetilde{\pi}(g\overline{n}g^{-1})^{-1}\widetilde{v}\rangle$$
$$= \delta_P(ma)^{1/2}\langle \pi(ma)v, \widetilde{v}\rangle.$$

Hence

$$\sum_\chi P_\chi(ma \colon \pi(\overline{n})v, \widetilde{v}) = \sum_\chi P_\chi(ma \colon v, \widetilde{v})$$

for $a \in A^+(t_1)$, $m \in \Omega$. Now use uniqueness and A-finiteness.

The proof for $n \in N$ is similar and is left to the reader.

The following lemma is also an easy consequence of the definitions.

1.11. LEMMA (Notation as in 1.9). *If m_1, $m \in M$, $v \in H$, $\widetilde{v} \in \widetilde{H}$, then*

$$P_\chi(m_1 \colon \pi(m)v, \widetilde{v}) = \delta_P(m)^{-1/2} P_\chi(m_1 m \colon v, \widetilde{v}).$$

1.12. THEOREM (Harish-Chandra, Jacquet). *Let (P, A) be a standard p-pair that is minimal subject to the condition that $E(P, \pi) \neq \varnothing$. If $(P, A) = (G, Z)$, then every matrix coefficient of π restricted to ${}^0G$ is compactly supported.*

1.13. Let $\mathfrak{a}_0$ be the dual space of $\mathfrak{a}_0^*$. We define $H \colon A_0 \to \mathfrak{a}_0$ by

$$|\chi(a)| = q^{\nu(H(a))},$$

where χ is a real valued character of A_0 and ν is the corresponding element of $\mathfrak{a}_0^*$. Then $H(ab) = H(a) + H(b)$.

If (P, A) is a standard p-pair, then the real span of $H(A) \subset \mathfrak{a}_0$ is denoted by $\mathfrak{a}$, and $\mathfrak{a}^*$ is identified with the real dual space of $\mathfrak{a}$. Set

$${}^0E(P, \pi) = \{\nu \in \mathfrak{a}^* | \text{ there is } \chi \in E(P, \pi) \text{ so that } |\chi(a)| = q^{\nu(H(a))}, \ a \in A\}.$$

Let $\mathcal{F} = \sum \mathbf{R}\alpha$, the sum over $\alpha \in \Delta(P_0, A_0)$. Let $\Delta(P_0, A_0) = \{\alpha_1, \ldots, \alpha_l\}$, and let $\beta_1, \ldots, \beta_l \in \mathcal{F}$ be defined by $(\beta_i, \alpha_j) = \delta_{ij}$.

If (P, A) is a standard p-pair, then there is a subset $F \subset \{1, \ldots, l\}$ so that $\mathfrak{a}^* = \mathfrak{z}^* \oplus \sum_{i \notin F} \mathbf{R}\beta_i$. Set $(P, A) = (P_F, A_F)$ and $\mathcal{F}_P = \sum_{i \notin F} \mathbf{R}\beta_i$.

1.14. THEOREM (Harish-Chandra). *Let (π, H) be an irreducible admissible representation of G. Suppose that for some standard p-pair (P_F, A_F) minimal subject to $E(P_F, \pi) \neq \varnothing$ we have $(\nu, \beta_i) < 0$ for all $\nu \in {}^0E(P_F, \pi)$ and $i \notin F$. Then every matrix coefficient of (π, H), restricted to ${}^0G$, is square integrable.*

1.15. If (π, H) is a finitely generated, admissible representation of G, then the set $E(G, \pi)$ is called the set of *central exponents* of π. We note that if π is irreducible, then $E(G, \pi)$ consists of exactly one element.

1.16. For a p-pair (P, A) in G, $P = MN$, an admissible finitely generated representation σ of ${}^0M$ and a character χ of A, we let $(\pi_{P,\sigma,\chi}, I_{P,\sigma,\chi})$ denote the representation $\mathrm{Ind}_P^G(\delta_P^{1/2}\sigma_\chi)$; here

$$\sigma_\chi(ma) = \sigma(m)\chi(a) \qquad (m \in {}^0M, \ a \in A).$$

1.17. THEOREM. *Let (π, II) be an irreducible admissible representation of G. Let (P, A) be a standard p-pair, and let $\chi \in E(P, \pi)$. Then there exists an irreducible admissible representation (σ, H_σ) of ${}^0M$ such that (π, H) is equivalent with a subrepresentation of $I_{\overline{P},\sigma,\chi}$.*

Let $\widetilde{v} \in \widetilde{V}$ be such that $P_\chi(\cdot : \cdot, \widetilde{v}) \not\equiv 0$. Set $\lambda(v)(g)(m) = P_\chi(m, \pi(g)v, \widetilde{v})$. Then $\lambda(v) \in C^\infty(M \times G)$ and $\lambda \neq 0$. Furthermore,

(1) $$\lambda(v)(g) \in A(M) \quad \text{for } g \in G.$$

(2) $$\lambda(v)(mg) = \delta_P(m)^{-1/2}r(m)(\lambda(v)(g)), \qquad m \in M, \ g \in G.$$

(This is Lemma 1.11.)

(3) $$\lambda(v)(gg_1) = \lambda(\pi(g_1)v)(g), \qquad g_1, g_2 \in G.$$

(4) $$\lambda(v)(\overline{n}g) = \lambda(v)(g), \qquad \overline{n} \in \overline{N}.$$

(This is Lemma 1.10.)

Let U be the space of all functions of the form $\lambda(v)$, $v \in H$. Then $(r(m)|_U, U)$ is an admissible representation of M. We note that $E(M, r(m)|_U) = \{\chi\}$. Let $\widetilde{\sigma}$ be an irreducible quotient of U. Let $q\colon U \to H_{\widetilde{\sigma}}$ be the corresponding projection. Then it is clear that

(5) $$q(\lambda(v)(\overline{n}mg)) = \delta_{\overline{P}}(m)^{1/2}\widetilde{\sigma}(m)q(\lambda(v))(g).$$

Set $\sigma = \widetilde{\sigma}|_{{}^0M}$. Then $\widetilde{\sigma} = \sigma_\chi$.

Hence if $T(v) = q(\lambda(v))$, then $T\colon H \to I_{\overline{P},\sigma,\chi}$ is a G-intertwining operator. Since T is non-zero by construction, T is injective.

2. The Langlands classification (*p*-adic case)

2.1. We retain the notation of §1. In particular we have $\mathfrak{a}_0^*, \alpha_1, \ldots, \alpha_l \in \mathfrak{a}_0^*$ and $\beta_1, \ldots, \beta_l$. If $\nu \in \mathfrak{a}_0^*$, we use the notation ${}^0\nu$ for the orthogonal projection of ν onto $\mathcal{F} = \sum \mathbf{R}\alpha_i = \sum \mathbf{R}\beta_i$.

If $\nu, \mu \in \mathcal{F}$, we say that $\nu \geq \mu$ if $(\nu - \mu, \beta_i) \geq 0$ for all i. Noting that $(\alpha_i, \alpha_j) \leq 0$ for $i \neq j$, we can apply the results of IV.6 to $\mathcal{F}$, $(\ ,\ )$ and $\alpha_1, \ldots, \alpha_l$. Since we are off by a minus sign from the situation in IV.3, we recapitulate the results needed here.

If $F \subset \{1, \ldots, l\}$, we set $S_F = \{\lambda \in \mathcal{F} \mid \lambda = \sum_{i \notin F} x_i\beta_i - \sum_{i \in F} y_i\alpha_i, \ x_i > 0, y_i \geq 0\}$. Then

(1) $\mathcal{F}$ is the disjoint union of the S_F where $F \subset \{1, \ldots, l\}$.

(2) If $\nu \in S_F$, set $\nu_0 = \sum_{i \notin F} x_i\beta_i$ if $\nu = \sum_{i \notin F} x_i\beta_i - \sum_{i \in F} y_i\alpha_i$. If $\nu, \mu \in \mathcal{F}$ and $\nu \geq \mu$, then $\nu_0 \geq \mu_0$.

(3) If $\nu \in \mathcal{F}$ and $\nu \in S_F$, then, by (1), F is unique. We denote it by $F(\nu)$.

If (P, A) is a standard p-pair, then $(P, A) = (P_F, A_F)$, $F \subset \{1, \ldots, l\}$. Then $\mathfrak{a}^* = \mathfrak{z}^* \oplus \mathcal{F}_P$ is an orthogonal direct sum, where $\mathcal{F}_P = \sum_{i \notin F} \mathbf{R}\beta_i$. Thus if $\nu \in \mathfrak{a}^*$, then $^0\nu \in \mathcal{F}_P$.

(4) *If $P = P_F$ and $\nu \in \mathcal{F}_P$, then $F(\nu) \supset F$.*

We note that $(\alpha_i^F, \alpha_j^F) \leq 0$ for $i \neq j$, $i, j \notin F$. Set $F = \{r + 1, \ldots, l\}$. Then using the results of IV.6 we see that $\nu = -\sum_{i \in J} s_i \alpha_i^F + \sum_{i \notin J \cup F} t_i \beta_i$, $t_i > 0$, $s_i \geq 0$, with $J \subset \{1, \ldots, r\}$. But $\alpha_i^F = \alpha_i + \sum_{j \in F} c_{ji}\alpha_j$, $c_{ji} \geq 0$, for $i \notin F$. Hence $\nu = -\sum_{i \in J \cup F} s_i' \alpha_i + \sum_{i \notin J \cup F} t_i \beta_i$, $t_i > 0$, $s_i' \geq 0$. Thus $F(\nu) = J \cup F \supset F$. This proves (4).

2.2. Let (π, H) be an irreducible admissible representation of G. Let (P, A) be a standard p-pair, minimal subject to the condition $E(P, \pi) \neq \varnothing$.

2.3. THEOREM. *We keep the notation of 2.2. If for each $\nu \in {}^0E(P, \pi)$ we have $(\nu, \beta_i) \leq 0$, $i = 1, \ldots, l$, then there exist a standard p-pair (Q, B) with $Q \supset P$, $B \subset A$, $Q = M_Q N_Q$, a square integrable representation σ of ^0M_Q, and a character χ of A_Q such that $\chi|_{B \cap {}^0G}$ is unitary and such that π is equivalent with a subrepresentation of $I_{\overline{P}, \sigma, \chi}$.*

Let $\nu \in {}^0E(P, \pi)$ be a maximal element. Suppose $P = P_F$. Then $^0\nu = -\sum_{i \notin F} x_i \alpha_i^F$, $x_i \geq 0$. Let $J = \{i \notin F \mid x_i \geq 0\}$. Set $H = J \cup F$. Let $Q = P_H$, $B = A_H$. Then $^0\nu|_{\mathfrak{b}} = {}^0\mu$, with $\mu \in {}^0E(Q, \pi)$. Let $\chi \in E(Q, \pi)$ be such that $\nu_\chi = \mu$. Let σ be as in 1.17. Then π is equivalent with a subrepresentation of $\pi_{\overline{Q}, \sigma, \chi}$, and $\chi|_{{}^0G \cap B}$ is unitary. Also $E(M_Q \cap P, \sigma_\chi) \subset E(G, \pi)$. Hence 1.14 implies σ is square integrable as a representation of ^0M_Q. This proves the theorem.

2.4. We now assume that K is a "good K" for (P_0, A_0). We define

$$\Xi(g) = \int_K \delta_{P_0}(kg)^{1/2}\, dk,$$

for $g \in {}^0G$; here $\delta_{P_0}(pk) = \delta_{P_0}(p)$ for $p \in P_0$, $k \in K$.

If (P, A) is a standard p-pair, we set $^*P = {}^0M \cap P_0$, $K_P = {}^0M \cap K$. Then K_P is a good K for ^0M. Moreover, $^*P = {}^0M_0 {}^*A^*N$ and $^*N \cdot N = N_0$.

We extend $\Xi_{0_M}(m) = \int_{K_P} \delta_{{}^*P}(kg)^{1/2}\, dk$ to ^0G by $\Xi_{0_M}(namk) = \delta_P(a)^{1/2}\Xi_{0_M}(m)$, $n \in N$, $a \in A$, $m \in {}^0M$, $k \in K$. Then, just as in IV, 3.7, we have

$$(1) \qquad \int_K \Xi_{0_M}(kg)\, dk = \Xi(g), \quad g \in {}^0G.$$

(2) *Set $\Xi_{0_M, \nu}(namk) = q^{\nu(H(a))}\Xi_{0_M}(nam)$, $n \in N$, $a \in A$, $m \in {}^0M$, $k \in K$. Then, if $(\nu, \alpha) > 0$ for $\alpha \in \Phi(P, A)$, the integral*

$$\int_{\overline{N}} \Xi_{0_M, \nu}(\overline{n}g)\, d\overline{n}$$

converges.

We say that an irreducible, admissible representation (π, H) of G is *tempered* if π satisfies the hypothesis of 2.3.

(3) *An irreducible, admissible representation (π, H) is tempered if and only if for any coefficient $c_{u,v}$, u, v in H_K,*

$$|c_{u,v}(g)| \leq C\Xi(g), \quad \text{for } g \in {}^0G.$$

In light of the results of §1 and 2.3 this is proved in the same way as in the real case (see IV, 3.6).

2.5. Let (P, A) be a standard p-pair. We say $a \underset{P}{\to} \infty$ if $\|H(a)\| \to \infty$ and there is $\varepsilon > 0$ so that $\alpha(H(a)) \geq \varepsilon\|H(a)\|$ for all $\alpha \in \Phi(P, A)$.

The proof of the following result is identical with the proofs of IV, 4.3(1), (2); and IV, 4.5; IV, 4.6.

2.6. PROPOSITION. *Let (P, A) be a standard p-pair. Let σ be a tempered representation of ^0M. Let χ be a character of A such that $|\chi(a)| = q^{\nu(H(a))}$, and let $(\nu, \alpha_i) > 0$ for $i \notin F$ ($P = P_F$). Then*

(1) *If $f \in I_{P,\sigma,\chi}$, then $(j(\chi)f)(g) = \int_{\overline{N}} f(\overline{n}g)\,d\overline{n}$ converges absolutely and uniformly in g on compacta.*

(2) *$j(\chi)\colon I_{P,\sigma,\chi} \to I_{\overline{P},\sigma,\chi}$ intertwines $\pi_{P,\sigma,\chi}$ and $\pi_{\overline{P},\sigma,\chi}$, and $j(\chi) \neq 0$.*

(3) $\displaystyle\lim_{a\underset{P}{\to}\infty} \delta_P(a)^{1/2}\chi(a)^{-1}\langle\pi(am)f_1, f_2\rangle = \langle\sigma_\chi(m)(j(\chi)f_1)(1), f_2(1)\rangle$ *for $f_1, f_2 \in I_{P,\sigma,\chi}$, $m \in {}^0M$.*

(4) *$j(\chi)I_{P,\sigma,\chi}$ is irreducible, and if $f \notin \mathrm{Ker}\, j(\chi)$, $f \in I_{P,\sigma,\chi}$, then f is cyclic for $\pi_{P,\sigma,\chi}$.*

2.7. COROLLARY. *Let P, σ, χ be as in 2.6. If $W \subset I_{\overline{P},\sigma,\chi}$ is an irreducible, non-zero G-invariant subspace, then*

$$W = j(\chi)I_{P,\sigma,\chi} = J_{P,\sigma,\chi}.$$

The proof is identical with that of IV, 4.8.

2.8. LEMMA. *Let P, σ, χ be as in 2.6. Let (Q, B) be a p-pair minimal subject to the condition that $E(Q, \pi_{P,\sigma,\chi}) \neq \varnothing$. Let $|\chi(a)| = q^{\nu(H(a))}$ for $a \in A$. If $\lambda \in {}^0E(Q, \pi_{P,\sigma,\chi})$, then $\lambda \leq \nu$.*

The proof of this lemma is identical with that of IV, 4.9.

2.9. If (P, A) is a standard p-pair, σ a tempered representation of ^0M, χ a character of A such that $|\chi(a)| = q^{\nu(H(a))}$, and $(\nu, \alpha) > 0$ for $\alpha \in \Phi(P, A)$, then P, σ, χ will be called a set of *Langlands data*. The representation $J_{P,\sigma,\chi}$ (see 2.7) will be called the *Langlands representation* or *quotient* associated with the Langlands data P, σ, χ.

2.10. THEOREM. *Let P, σ, χ and Q, μ, η be Langlands data. If $J_{P,\sigma,\chi}$ is equivalent with $J_{Q,\mu,\eta}$, then $P = Q$, $\sigma = \mu$ and $\chi = \eta$.*

The proof is essentially the same as the proof of IV, 4.10. We leave it to the reader to make the appropriate changes.

2.11. THEOREM. *Let (π, H) be an irreducible, admissible representation of G. Then there exist Langlands data P, σ, χ such that π is equivalent with $J_{P,\sigma,\chi}$.*

The proof is essentially the same as the proof of IV, 3.9, in light of 1.14. We note that the reader must keep in mind the fact that our exponents are off by a minus sign from the corresponding exponents in the real case.

This result completes our sketch of the Langlands classification of the p-adic case.

2.12. Let (π, H) be an irreducible admissible representation of G. Then there exist Langlands data P, σ, χ, uniquely determined by π, so that π is equivalent with $J_{P,\sigma,\chi}$ (2.10, 2.11). Let $|\chi(a)| = q^{\nu(H(a))}$, $a \in A$. Then we set $\nu = \lambda_\pi \in \mathfrak{a}_0^*$, and call λ_π the *Langlands parameter* associated with π. We note that π is tempered if and only if ${}^0\lambda_\pi = 0$ (in which case $P = G$, and χ is the central character).

2.13. LEMMA. *Let P, σ, χ be Langlands data. If (μ, H) is a constituent of $I_{P,\sigma,\chi}$ and if $\pi = J_{P,\sigma,\chi}$, then $\lambda_\mu \leq \lambda_\pi$, and equality occurs if and only if μ is $J_{P,\sigma,\chi}$.*

This lemma follows from 2.8, 2.6(3),(4) and the definition of $J_{P,\sigma,\chi}$.

2.14. We now use the notation and definitions of X, 4.6, 4.7, 4.8. If $\mathcal{M}$ is a reductive algebraic group defined over k so that $M = \mathcal{M}(k)$ has compact center, then we set $\mathrm{st}(M) = V_\varnothing$ as in X, 4.7. The purpose of this section is to identify the V_J, $J \subset \Delta$, in the Langlands classification.

If $J \subset \Delta$, we set $P_J = M_J N_J$ as usual and ${}^*\overline{P}_J = {}^0M_J \cap \overline{P}_\varnothing$. Then by definition

(1)
$$\mathrm{st}({}^0M_J) = I_{{}^*\overline{P}_J}^{{}^0M_J}(1) / \sum_{Q \supsetneq P_J} \pi_{{}^*\overline{P}_J, {}^*\overline{Q}} \left(I_{{}^*\overline{Q}}^{{}^0M_J}(1) \right).$$

Using induction in stages, we find that

(2)
$$I_{\overline{P}_J, \mathrm{st}({}^0M_J), \delta_{P_J}^{1/2}} = I_{\overline{P}_\varnothing}^G(1) / \sum_{Q \supsetneq P_J} \pi_{\overline{Q}, \overline{P}_J}(I_{\overline{Q}}^G(1)).$$

$\overline{I} = I_{\overline{P}_J, \mathrm{st}({}^0M_J), \delta_{P_J}^{1/2}}$ has a *unique* non-zero irreducible subrepresentation $J_{\overline{P}_J, \mathrm{st}({}^0M_J), \delta_{P_J}^{1/2}}$. (2) implies that $\overline{I}$ contains

$$V_{\overline{J}} = I_{\overline{P}_J}^G(1) / \sum_{Q \supsetneq P_J} \pi_{\overline{Q}, \overline{P}_J}(I_{\overline{Q}}^G(1)),$$

which is irreducible (see X, 4.11). This proves the following result.

2.15. PROPOSITION. *If $J \subset \Delta$, let $\overline{J} \subset \Delta$ be the subset so that $\overline{P}_J$ is conjugate to $P_{\overline{J}}$ in G. Then*

$$J_{\overline{P}_J, \mathrm{st}({}^0M_J), \delta_{P_J}^{1/2}} = V_{\overline{J}}.$$

3. Uniformly bounded representations and $\Pi_\infty(G)$

3.1. Let (π, H) be an irreducible admissible representation of G. We say that (π, H) is uniformly bounded if there is a constant C such that

(1)
$$\|\pi(g)\| \leq C$$

for all $g \in G$. (Recall that H is a Hilbert space by assumption (cf. 1.6).)

3.2. By $\Pi_\infty(G)$ we mean the set of equivalence classes of irreducible admissible representations (π, H) which are either tempered or of the form $\pi = J_{P,\sigma,\chi}$, where P, σ, χ are Langlands data, $\mathrm{Ker}\,\pi$ is compact, and the corresponding Langlands parameter satisfies

(1)
$$(\lambda_\pi, \beta_i) < (\rho_P, \beta_i)$$

for $i \notin F$ ($P = P_F$). Here ρ_P is defined by

$$\delta_P(a)^{1/a} = q^{\rho_P(H(a))} \qquad (a \in A).$$

3.3. THEOREM. *Let (π, H) be an admissible, irreducible, uniformly bounded representation of G with compact kernel. Then the class of (π, H^∞) is in $\Pi_\infty(G)$.*

H has an inner product $\langle\ ,\ \rangle$. We define $\pi^*(g)$ by the formula $\langle \pi(g)v, w \rangle = \langle v, \pi^*(g^{-1}w) \rangle$. Then (π^*, H) is an admissible representation of G. We use (π^*, H^∞) rather than the admissible dual $(H^\infty)\widetilde{}$.

The results of §2 imply that there exist Langlands data P, σ, χ so that (π, H^∞) is equivalent with $J_{P,\sigma,\chi}$, $P = {}^0MAN$. Furthermore, the proof of the existence and uniqueness of P, σ, χ implies that

(1) $$\chi \in E(P, \pi).$$

If $Q = M_Q N_Q$ is a parabolic subgroup of G, set $H^\infty(N_Q)$ equal to the linear span of the vectors $\pi(n)v - v$, $v \in H^\infty$, $n \in N_Q$. Set $H^\infty_{N_Q} = H^\infty/H^\infty(N_Q)$. Then $H^\infty_{N_Q}$ is the Jacquet module of (π, H^∞) corresponding to N_Q. We set $\pi(m)v + H^\infty(N_Q) = \pi_{N_Q}(m)(v + H^\infty(N_Q))$. Then $(\pi_{N_Q}, H^\infty_{N_Q})$ is a finitely generated admissible representation of M_Q.

If (P_1, A_1) is a standard p-pair and if (P_2, A_2) is a standard p-pair dominating (P_1, A_1), set ${}^*\overline{N}_1 = \overline{N}_1 \cap M_2$ $(P_i = M_i N_i,\ i = 1, 2)$. Then $H^\infty_{\overline{N}_1} = (H^\infty_{\overline{N}_2})_{*\overline{N}_1}$.

Suppose that (Q, A_Q) is a standard p-pair. Let $\eta \in E(Q, \pi)$. Then 1.9 and 1.11 imply that

$$(H^\infty_{\overline{N}_Q})_\eta = \{v \in H^\infty_{\overline{N}_Q} \mid (\pi_{\overline{N}_Q}(a) - \eta(a)\delta_Q^{-1/2}(a))^d v = 0 \text{ for some } d\} \neq (0).$$

We also will need

(2) *If (P_i, A_i), $i = 1, 2$, are as above and if $\eta \in E(P_1, A_1)$, then $\eta|_{A_2} \in E(P_2, A_2)$.*

This is clear from the results in 1.9.

Let (P_1, A_1) be a standard p-pair, $P_1 \supset P$. Set $\overline{\pi} = \pi_{\overline{N}_1}$ on $H^\infty_{\overline{N}_1}$. If $a \in A_1$, then $\overline{\pi}(a) \in \text{Hom}_{M_1}(H^\infty_{\overline{N}_1}, H^\infty_{\overline{N}_1})$. Since $\dim \text{Hom}_{M_1}(H^\infty_{\overline{N}_1}, H^\infty_{\overline{N}_1}) < \infty$, we see that there exist $a_1, \ldots, a_r \in A_1$ so that $\overline{\pi}(a_1), \ldots, \overline{\pi}(a_r)$ is a basis of the linear span of $\overline{\pi}(A_1)$.

Let $p(m\colon v, w) = \sum P_\eta(m\colon v, w)$, $m \in M$, $v, w \in H^\infty$, be as in 1.9. Then for $a \in A^+(t)$, t sufficiently large,

(3) $$\delta_{P_1}(ma)^{1/2}\langle \pi(ma)v, w \rangle = p(ma\colon v, w).$$

Set $q(m\colon v, w) = \delta_{P_1}(m)^{-1/2}p(m\colon v, w)$. Then

(4) $$q(m\colon v, w) = q(1\colon \pi(m)v, w), \qquad m \in M_1,$$

(5) $$q(m\colon \pi(\overline{n})v, \pi^*(n)w) = q(m\colon v, w), \qquad \overline{n} \in \overline{N}_1,\ n \in N,\ m \in M_1.$$

Set $Q(\overline{v}, \overline{w}) = q(1\colon v, w)$, $\overline{v} = v + H^\infty(\overline{N}_1)$, $\overline{w} = w + H^\infty(N_1)$.

Then $q(a\colon v, w) = Q(\overline{\pi}(a)\overline{v}, \overline{w})$. Let $\eta \in E(P_1, \pi)$. Then there exist $x_1, \ldots, x_r \in \mathbf{C}$ such that the function

$$a \mapsto \mu(a\colon v, w) = \sum x_i q(aa_i\colon v, w)$$

on A_1 satisfies

(6) $$\mu(a\colon v, w) = \delta_{P_1}(a)^{-1/2}\eta(a)\mu(1\colon v, w) \quad \text{and} \quad \mu(a\colon v, w) \not\equiv 0.$$

Indeed, there is $B = \sum x_i \pi_{\overline{N}_1}(a_i)$ such that $B \neq 0$ and

$$\overline{\pi}(a)B = \delta_{P_1}(a)^{-1/2}\eta(a)B.$$

Now if $a \in A^+(t)$, t sufficiently large, then $\langle \pi(aa_i)v, w \rangle = q(aa_i\colon v, w)$. We have

$$|q(aa_i\colon v, w)| \leq C\|v\|\,\|w\|.$$

This implies there is $C_\mu > 0$ satisfying the following condition:

(7) Fix $v, w \in H^\infty$. Then there is $t > 1$ so that if $a \in A_1^+(t)$, then

$$|\mu(a\colon v, w)| \leq C_\mu\|v\|\,\|w\|,$$

(C_μ depends on μ and, because of uniform boundedness, *not* on v, w).

Also $\mu(a\colon v, w) = \delta_{P_1}(a)^{-1/2}\eta(a)\mu(1\colon v, w)$. Hence

(8) If $a \in A_1^+(t)$ (as in (7)), then

$$\delta_{P_1}(a)^{-1/2}|\eta(a)|\,|\mu(1\colon v, w)| \leq C_\mu\|v\|\,\|w\|.$$

Fixing v, w so that $\mu(1\colon v, w) \neq 0$, we see that

$$\delta_{P_1}(a)^{-1/2}|\eta(a)|$$

is bounded on $A_1^+(t)$ for t large. This implies that if $\nu \in \mathfrak{a}_1^*$ is such that $|\eta(a)| = q^{\nu(H(a))}$, then $(\nu - \rho_{P_1})(H(a)) \leq 0$ for $a \in A^+$.

Suppose that there exists $a \in A^+(t)$ for some $t > 1$ such that $(\nu - \rho_{P_1})(H(a)) = 0$. Then, if $v, w \in H^\infty$, there is k depending on v, w such that

$$\delta_{P_1}^{-1/2}(a^k)|\eta(a^k)|\,|\mu(1\colon v, w)| \leq C_\mu\|v\|\,\|w\|.$$

But $\delta_{P_1}^{-1/2}(a^k)|\eta(a^k)| = (\delta_{P_1}^{-1/2}(a)|\eta(a)|)^k = 1$. Hence

$$|\mu(1\colon v, w)| \leq C_\mu\|v\|\,\|w\|$$

for all $v, w \in H^\infty$.

Moreover, $\mu(1\colon v, w) = \langle Bv, w \rangle$, $B \in \text{End}(H^\infty)$ and $\|Bv\| \leq C_\mu\|v\|$ by the above. Hence B extends to a bounded operator on H.

Since $\mu(1\colon \pi(\overline{n})v, \pi^*(n)w) = \mu(1\colon v, w)$, $\overline{n} \in \overline{N}_1$, $n \in N_1$, we see that

$$B \circ \pi(n) = \pi(n) \circ B = B$$

for $\overline{n} \in \overline{N}_1$, $n \in N_1$.

Also $\pi(a) \circ B = \eta(a)\delta_{P_1}(a)^{-1/2}B$, $a \in A_1$. Arguing as in the proof of IV, 5.3, we see that $\pi(\overline{n}) \circ B = B$, $\overline{n} \in N_1$. Let $R \subset G$ be the subgroup of G generated by $\pi(\overline{N}_1)$ and $\pi(N_1)$. If $v \in BH$, then $\pi(x)v = v$, $x \in R$. Set $H^R = \{v \in H \mid \pi(x)v = v, x \in R\}$. Then, since R is normal in G [**18**, 6.25], H^R is G-invariant. Also, (π, H) is irreducible. Hence $H^R = H$ or $H^R = (0)$.

If $H^R = (0)$, then $B = 0$; hence $\mu = 0$, and we have contradicted our assumption about ν. Otherwise $H^R = H$. But then $R \subset \text{Ker}\,\pi$, which is contrary to our assumption. This proves

(9) Let (P_1, A_1) be a standard p-pair with $P_1 \supset P$. If $\eta \in E(P, \pi)$ and $a \in A^+(t)$ for some $t > 1$, then

$$|\delta_{P_1}(a)^{-1/2}\eta(a)| < 1.$$

(9) applies to $\chi|_{A_1}$ by (2). Let λ be such that $|\chi(a)| = q^{\lambda(H(a))}$ ($a \in A$). If $a \in \text{Cl}(A^+)$, $a \neq 1$, then we have $(\rho_P - \lambda)(H(a)) > 0$. This implies that $J_{P,\sigma,\chi} \in \Pi_\infty(G)$.

3.4. LEMMA. *Suppose that G has a compact center. Let (π, H) be in $\Pi_\infty(G)$. There exists $t > 0$ such that if $v, w \in H$, then*

$$|\langle \pi(g)v, w \rangle| \leq C\Xi(g)^t$$

for all $g \in G$.

The proof is identical to the proof of IV, 5.3. The following results also are proved by the same methods as in the real case.

3.5. PROPOSITION. *Suppose that G has a compact center. If (π, H) is in $\Pi_\infty(G)$, then the matrix entries of π vanish at infinity.*

This result for (π, H) unitary is Howe's theorem ([**63**]) in the p-adic case.

3.6. PROPOSITION. *Suppose that G has a compact center. If (π, H) is in $\Pi_\infty(G)$, then there is $p \in (0, \infty)$ such that the matrix entries of π are in L^p.*

3.7. We now apply these results to the modules V_J of X, 4.7, 4.8. We note that if $J \neq \Delta$, then $\dim V_J > 1$. Theorem 2.15 says that

(1) $$V_J \equiv J_{P_J, \mathrm{st}(^0 M_J), \delta_{P_J}^{1/2}}.$$

Hence, if $J \neq \varnothing, \Delta$, then $V_j \notin \Pi_\infty(G)$.

The following result now follows from X, 4.7, and X, 4.3.

3.8. THEOREM. *Assume that G has compact center. If $V \in \Pi_\infty(G)$ and $H_{\mathrm{ct}}^*(G, V) \neq (0)$, then $V = \mathrm{st}(G)$ and*

$$H_{\mathrm{ct}}^q(G, V) = \begin{cases} (0), & q \neq l, \\ \mathbf{C}, & q = l. \end{cases}$$

3.9. THEOREM (Casselman [**33**]). *Let $\mathcal{G}$ be semi-simple, and let (π, V) be an irreducible, admissible, unitary representation of $\mathcal{G}$.*

(a) *If $\mathcal{G}$ is simple, then $H_{\mathrm{ct}}^q(\mathcal{G}; V) = (0)$ unless $q = 0$ and V is the trivial representation, or $q = l$ and $V = \mathrm{st}(\mathcal{G})$, in which cases $H_{\mathrm{ct}}^q(\mathcal{G}; V) = \mathbf{C}$.*

(b) *If π has compact kernel, then $H_{\mathrm{ct}}^q(\mathcal{G}; V) = (0)$ unless $q = l$ and $V = \mathrm{st}(\mathcal{G})$, in which case $H^q(\mathcal{G}; V) = \mathbf{C}$.*

3.10. 3.9 had been proved earlier by the first named author, under the condition of large residue field, using Garland's methods.

4. Another proof of the non-unitarizability of the V_J's

As was pointed out in X, 4.12, the only item missing there to complete the determination of the continuous cohomology with coefficients in an irreducible unitary representation was the non-unitarizability of the V_J's which are not a product of trivial and Steinberg representations. A more precise result has been deduced here from 3.3, whose proof made use of the Langlands classification. 3.3 itself is a sharpening of a theorem of Howe (3.5). This theorem was proved originally ([**63**], see also [**64**]) directly, without any recourse to classification. For the benefit of the reader who would like to bypass the latter but is willing to assume Howe's theorem, we indicate here how to prove the non-unitarizability of the V_J's from Howe's theorem and some general facts on representations to be found in [**34**].

The notation is that of X, §§3, 4. It suffices to consider the case where G is almost simple. We have then to prove that V_J is not unitarizable if $J \neq \varnothing, \Delta$. We write ρ for $\delta^{1/2}$.

4.1.   Given two disjoint subsets I, J of Δ, we set

(1) $W(I, J) = \{w \in W \mid w(\alpha) > 0 \text{ for } \alpha \in I, \ w(\alpha) < 0 \text{ for } \alpha \in J\}.$

4.2. LEMMA. *Let $J \subset \Delta$. Then the Jacquet module $(V_J)_N$, viewed as an S-module, has the direct sum decomposition*

$$(V_J)_N = \bigoplus_{w \in W(J, \Delta - J)} \mathbf{C}_{w(\rho^{-1}) \cdot \rho}.$$

According to [**34**, 8.1.1], we have

(1) $$(I_L)_N = \bigoplus_{w \in W(L, \varnothing)} \mathbf{C}_{w(\rho^{-1}) \cdot \rho}.$$

The lemma then follows from the exactness of the Jacquet module functor and from X, 4.5, 4.6.

4.3.   The smooth dual of an admissible representation (π, V) is denoted $(\widetilde{\pi}, \widetilde{V})$. It is admissible [**34**, 2.1.10]. The smooth dual of the Jacquet module $\widetilde{V}_N$ for M may be canonically identified with the Jacquet module V_{N^-}, where $\mathcal{N}^-$ is the unipotent radical of the minimal parabolic group $\mathcal{P}^-$ containing $\mathcal{M}$ and opposite to $\mathcal{P}$ [**34**, 4.2.2]. For $\varepsilon > 0$, let

(1) $$A^-(\varepsilon) = \{a \in S \mid |a^\alpha| \leq \varepsilon \text{ for all } \alpha \in \Delta\}.$$

We have then the following lemma, due to W. Casselman [**34**, 4.2.3]:

LEMMA. *Let $v \in V$, $\widetilde{v} \in \widetilde{V}$. Let u (resp. $\widetilde{u}$) be the canonical image of v in V_N (resp. $\widetilde{v}$ in $\widetilde{V}_{N^-}$). There exists $\varepsilon > 0$ such that*

$$\langle \pi(a)v, \widetilde{v} \rangle = \langle \pi_N(a)u, \widetilde{u} \rangle \qquad (a \in A^-(\varepsilon)).$$

On the right-hand side, π_N refers to the representation of M in V_N, and the pairing is that of [**34**, 4.2.2].

4.4. LEMMA. *Let $J \subset \Delta$, $J \neq \varnothing, \Delta$, and $J' = \Delta - J$. Let $w_{J'}$ be the longest element in $W_{J'}$. Then there exist strictly positive integers m_α $(\alpha \in J')$ such that*

(1) $$w_{J'}(\rho^{-1}) \cdot \rho = \prod_{\alpha \in J'} |\alpha^{m_\alpha}|.$$

The element $w_{J'}$ transforms the positive roots of $M_{J'}$ with respect to S into the negative roots, and, since it is in the Weyl group of $M_{J'}$, it permutes the weights of S in the unipotent radical $N_{J'}$ of $P_{J'}$. As a consequence

(2) $$w_{J'}(J') = -J', \qquad w_{J'}(\alpha) > 0 \quad \text{if } \alpha \in J.$$

Let δ_1 be the product of the characters $|\alpha|$, where α runs through the positive roots of $M_{J'}$ with respect to S, and let δ_2 be the product of the weights of S in $N_{J'}$, each character being counted with its multiplicity in δ. Then

(3) $$\delta = \delta_1 \cdot \delta_2, \quad w_{J'}(\delta_1) = \delta_1^{-1}, \quad w_{J'}(\delta_2) = \delta_2,$$

whence

(4) $$w_{J'}(\rho^{-1}) \cdot \rho = \delta_1.$$

4.5. THEOREM. *Let $J \subset \Delta$, $J \neq \varnothing, \Delta$. Then V_J is not unitarizable.*

We already know that V_J is irreducible (X, 4.11). Moreover, since $J \neq \Delta$, the G-module V_J is not trivial. In view of Howe's theorem (3.5), it suffices therefore to show:

(∗) *There exist $v \in V_J$, $\widetilde{v} \in \widetilde{V}_J$ and an unbounded sequence of elements $g_n \in G$ ($n = 1, 2, \cdots$) such that $\langle \pi_J(g_n)v, \widetilde{v} \rangle$ does not tend to zero when $n \to \infty$.*

We revert to the notation of 4.4. Since $J' \neq \Delta$, the set of elements

$$C = \{c \in S \mid |c^\alpha| \leq 1 \; (\alpha \in J), \; |c^\alpha| = 1 \; (\alpha \in J')\}$$

is unbounded. Let $\sigma = w_{J'}(\rho^{-1}) \cdot \rho$. By 4.4(2), $w_{J'} \in W(J, J')$; hence (4.2), σ is a character of S in $(V_J)_N$. Let u be a non-zero element of $\mathbf{C}_\sigma$, and let $\widetilde{u} \in (\widetilde{V}_J)_{N^-}$ be such that $\langle u, \widetilde{u} \rangle \neq 0$. Let v (resp. $\widetilde{v}$) be an element of V_J (resp. $\widetilde{V}_J$) which maps unto u (resp. $\widetilde{u}$) under the canonical projection. Let ε be as in 4.3, and fix an element $a_0 \in A^-(\varepsilon)$. Then $a_0 C \subset A^-(\varepsilon)$, and we have, by 4.4 and 4.2,

$$\langle \pi_J(a_0 c)v, v \rangle = \langle \pi_{J,N}(a_0 c)u, \widetilde{u} \rangle = |(a_0 \cdot c)^\sigma| \langle u, \widetilde{u} \rangle$$

for all $c \in C$. It follows from 4.4 and the definition of C that $c^\sigma = 1$; hence

$$\langle \pi_J(a_0 c)v, \widetilde{v} \rangle = |a_0^\sigma| \langle u, \widetilde{u} \rangle$$

is independent of c and non-zero. Since C is unbounded, this proves (∗).

CHAPTER XII

Differentiable Cohomology for Products of Real Lie Groups and T.D. Groups

In this chapter we consider direct products (finite or restricted) of Lie groups and t.d. groups. The chief examples, and the motivation for doing this, are finite products of the type $\prod_{v \in S} \mathcal{G}(k_v)$, where $\mathcal{G}$ is a reductive group over a global field, S a finite number of places of k, and k_v the completion of k at v. We shall also incidentally consider the adèle groups of reductive groups. We shall however first discuss differentiable G-modules and cohomology under more general assumptions. Here too, it is convenient to go over to K-finite vectors in order to get into a basically algebraic situation. As in X, this leads naturally to the consideration of a category of non-degenerate modules over an idempotented algebra. Section 0 is devoted to some simple remarks about homological algebra in such categories.

0. Homological algebra over idempotented algebras

0.1. An algebra R over a field F is idempotented if it has a countable set of idempotents e such that R is the union of the sets eRe. We have then $R = R \cdot R$. A module M over R is non-degenerate if $M = R \cdot M$. This is equivalent to requiring that $M = \bigcup_e e \cdot M$. If M is an R-module, then $R \cdot M = M_f$ is the greatest non-degenerate submodule of M. If M is a nondegenerate R-module and N is an R-submodule, then N is non-degenerate, as follows from the existence of the idempotents. Therefore the category $\mathcal{C}_R^f$ of non-degenerate R-modules is an abelian category. We note also that $M \mapsto M_f$ is an exact functor, because, if e is idempotent, then $M \mapsto e \cdot M$ is obviously exact.

0.2. The "adjoint associativity" of $\otimes$ and Hom is proved in the standard texts under the blanket assumptions that rings or algebras have a unit (see e.g. [**31**, II, §5]). However the units are not used in the proofs, and we take it for granted that it holds without that assumption.

0.3. LEMMA. (i) *Let $A \in \mathcal{C}_R^f$. Then the map $\alpha \colon (r, a) \mapsto r \cdot a$ induces an R-isomorphism of $R \otimes_R A$ onto A.*

(ii) *The map $\mu \colon A \to \operatorname{Hom}_F(R, A)$ defined by assigning to $a \in A$ the function $\tilde{a} \colon r \mapsto r \cdot a$ is an injective R-morphism and induces an R-isomorphism of A onto $\operatorname{Hom}_R(R, A)_f$.*

(i) Since $R \cdot A = A$, the map α is surjective. We construct an inverse β to α. Let $\alpha \in A$, and let e be an idempotent which fixes a; set $\beta(a) = e \otimes a$. If e' is an idempotent such that $e \in e' \cdot R \cdot e'$, then

$$(1) \qquad e \cdot e' = e' \cdot e = e, \qquad e' \cdot a = e \cdot a = a,$$

from which it follows that $e' \otimes a = e \otimes a$. Thus $\beta(a)$ is independent of the choice of an idempotent fixing a. Routine computations show that α and β are R-morphisms which are inverse to each other.

(ii) We define a map $\nu \colon \operatorname{Hom}_R(R, A)_f \to A$ by $\nu(g) = g(e)$, where e is an idempotent fixing g. Again, one checks this is independent of e, and that μ and ν are R-morphisms inverse to each other.

0.4. LEMMA. *Let $V \in \mathcal{C}_R^f$. Then $\operatorname{Hom}_F(R, V)_f$ is an injective module in $\mathcal{C}_R^f$.*

This follows from 0.2 and 0.3 in the usual way: If $A \in \mathcal{C}_R^f$, then

(1)
$$\operatorname{Hom}_R(A, \operatorname{Hom}_F(R, V)_f) = \operatorname{Hom}_R(A, \operatorname{Hom}_F(R, V))$$
$$= \operatorname{Hom}_F(A \otimes_R R, V) = \operatorname{Hom}_F(A, V).$$

If $A \to B$ is an injective R-morphism, then $\operatorname{Hom}_F(B, V) \to \operatorname{Hom}_F(A, V)$ is surjective; hence, by the naturality of (1), so is

(2)
$$\operatorname{Hom}_R(B, \operatorname{Hom}_F(R, V)_f) \to \operatorname{Hom}_R(A, \operatorname{Hom}_F(R, V)_f),$$

which proves the lemma.

0.5. The canonical map $\mu \colon V \to \operatorname{Hom}_F(R, V)_f$ now defines an injection of V into an injective module. Hence injective resolutions can be constructed in the usual way.

1. Differentiable cohomology

1.1. *In this section, $G = G_1 \times G_2$ is the direct product of a real Lie group G_1 (with finitely many connected components, as usual) and a t.d. group G_2, in the sense of Chapter X.*
In particular, G is locally compact, countable at infinity, and metrizable.

1.2. Let $V \in \mathcal{C}_G$. An element $v \in V$ is said to be *smooth* or *differentiable* if every vector in $G \cdot v$ is smooth for G_1 and for G_2. Let V^∞ (resp. V^{∞_1}, resp. V^{∞_2}) be the space of vectors in V which are smooth with respect to G (resp. G_1, resp. G_2). This space is stable under G, and we let π_∞ (resp. π_{∞_1}, resp. π_{∞_2}) be the restriction of π to it. The space V^∞ is then the union of the subspaces $(V^{\infty_1})^L$, where L runs through the compact open subgroups of G_2. The space V^{∞_1} is endowed with the C^∞ topology with respect to G_1 (0, 2.3), $(V^{\infty_1})^L$ with its topology of closed subspace of V^{∞_1}, and V^∞ with the strict inductive limit topology of the $(V^{\infty_1})^L$. As in X, it is a strict inductive limit topology of an increasing sequence of closed subspaces. We have $(V^\infty)^L = (V^{\infty_1})^L$; hence V^∞ is the strict inductive limit of the closed subspaces $(V^\infty)^L$ (L a compact open subgroup of G_2). If $G_1 = \{1\}$ or $G_2 = \{1\}$, we get back the definitions of (0, 2.3) and X, 1.3, respectively. By definition, $V^\infty = (V^{\infty_1})^{\infty_2}$ topologically. The canonical inclusions $V^\infty \to V^{\infty_1} \to V$ are continuous G-maps. If V is quasi-complete (resp. complete), then so are V^{∞_1}, V^∞, V, and these inclusions have dense image. If V is a Fréchet space, then so is V^{∞_1}, while V^{∞_2} and V^∞ are strict inductive limits of sequences of Fréchet spaces.

1.3. A G-module V is *smooth* or *differentiable* if $V = V^\infty$, also topologically, i.e. if every $v \in V$ is smooth and V is the strict inductive limit of the V^L (L a compact open subgroup of G_2). We let $\mathcal{C}_G^\infty$ (resp. $\mathcal{C}_G^{\infty_1}$, resp. $\mathcal{C}_G^{\infty_2}$) be the category of continuous G-modules which are smooth with respect to G (resp. G_1, resp. G_2) and continuous G-morphisms. The map $V \mapsto V^\infty$ (resp. $V \mapsto V^{\infty_1}$, resp. $V \mapsto V^{\infty_2}$) is a functor from $\mathcal{C}_G$ to $\mathcal{C}_G^\infty$ (resp. $\mathcal{C}_G^{\infty_1}$, resp. $\mathcal{C}_G^{\infty_2}$).

1.4. Fix a maximal compact subgroup K_1 of G_1. A continuous quasi-complete G-module V is admissible if for every $\delta \in \widehat{K_1}$ and every compact open subgroup L of G_2, the space of L-fixed vectors V_δ^L in the isotypic component V_δ is finite dimensional. This is equivalent to each of the following conditions: (i) for every $\delta \in \widehat{K_1}$, the space V_δ is an admissible G_2-module; (ii) for every compact open subgroup L of G_2, the space V^L is an admissible G_1-module; (iii) for every compact open subgroup L of G_2 and every $\delta \in (K_1 \times L)^{\widehat{}}$, the isotypic subspace V_δ is finite dimensional.

A vector $v \in V$ is K-finite for one group K of the form $K_1 \times L$, with L compact open in G_2, if and only if it is so for all such subgroups. The space $V_{(K)}$ of K-finite vectors is dense in V. The space $V^\infty \cap V_{(K)}$ is a $(\mathfrak{g}_1, K_1) \times G_2$ module. If V is admissible, then $V_{(K)} \subset V^\infty$ and V^∞ is also admissible, with the same isotypic subspaces as V.

1.5. PROPOSITION. *Let* $\alpha = \infty, \infty_1, \infty_2$. *Let* $V \in \mathcal{C}_G$.

(i) *If V is s-injective, then V^α is s-injective in $\mathcal{C}_G^\alpha$.*

(ii) *The category $\mathcal{C}_G^\alpha$ has enough s-injective modules. Every quasi-complete G-module in $\mathcal{C}_G^\alpha$ admits an s-injective resolution in $\mathcal{C}_G^\alpha$ by quasi-complete modules.*

(iii) *The functor $V \mapsto V^\alpha$ is exact in the category of Fréchet G-modules. The functor $V \mapsto V^{\infty_2}$ from quasi-complete modules in $\mathcal{C}_G$ (resp. $\mathcal{C}_G^{\infty_1}$) to G-modules in $\mathcal{C}_G^{\infty_2}$ (resp. $\mathcal{C}_G^\infty$) is s-exact.*

(i) is proved in exactly the same way as IX, 6.5(i). It implies that $F(G, V)^\alpha$ is s-injective in $\mathcal{C}_G^\alpha$. Since $V \to F(G, V)$ is a strong injection, it follows that $V^\alpha \to F(G, V)^\alpha$ is also strong if $V = V^\alpha$, whence (ii), taking into account the fact that $F(G, V)$ is quasi-complete if V is.

(ii) The second assertion is proved in the same way as X, 1.5. Combined with IX, 6.5(iii), it implies the first assertion.

1.6. PROPOSITION. (i) *Let* $V \in \mathcal{C}_G^{\infty_1}$ *be quasi-complete and s-injective. Then V is s-injective in $\mathcal{C}_G$. Every quasi-complete (resp. Fréchet) module $W \in \mathcal{C}_G^{\infty_1}$ admits an s-resolution in $\mathcal{C}_G^{\infty_1}$ by quasi-complete (resp. Fréchet) G-modules which are s-injective in $\mathcal{C}_G$.*

(ii) *Let* $V \in \mathcal{C}_G^\infty$ *be s-injective. Then V is s-injective in $\mathcal{C}_G^{\infty_1}$.*

PROOF. (i) The second part follows from the first. As in IX, 6.5(ii), it suffices to prove the latter for $C^{\infty_1}(G; W)$ ($W \in \mathcal{C}_G^{\infty_1}$). In this case, the argument is basically the same as that of IX, 5.2, except that we take $\phi \in C_c^{\infty_1}(G_1)$ and let it operate on f by convolution on the right with respect to the G_1-coordinate. More explicitly, we set

$$\alpha(f)(x_1, x_2) = \int_{G_1} \phi(y^{-1} \cdot x_1) \cdot f(y, x_2)\, dy, \qquad (x_1 \in G_i, \ i = 1, 2).$$

(ii) The module V is a G-direct summand of $C^\infty(G; V)$, a fortiori a direct G_1-summand of $C^\infty(G; V)$. It suffices therefore to show that if $E \in \mathcal{C}_G^\infty$ and

$A = C^\infty(G; E)$, then there exists a continuous G_1-map $\delta: C^{\infty_1}(G_1; A) \to A$ such that $\delta \circ \varepsilon = \mathrm{Id}$, where ε is the standard inclusion. It is readily seen that the map δ defined by

$$\delta(f)((x_1, x_2)) = f(x_1)((x_1, x_2)) \qquad (x_i \in G_i, \ i = 1, 2),$$

satisfies those conditions.

1.7. We let $\mathrm{Ext}^*_{d_i}$ $(i = 1, 2)$ and Ext^*_d be the derived functors of Hom_G in $\mathcal{C}^{\infty_i}_G$ $(i = 1, 2)$ and $\mathcal{C}^\infty_G$ respectively, and similarly $H^*_{d_i}$ and H^*_d the corresponding cohomology spaces.

1.8. PROPOSITION. (i) *Let* $U, V \in \mathcal{C}^\infty_G$ *be quasi-complete. Then* $\mathrm{Ext}^q_{d_1}(U, V)$, $\mathrm{Ext}^q_{d_2}(U, V)$, $\mathrm{Ext}^q_d(U, V)$ *and* $\mathrm{Ext}^q_{\mathrm{ct}}(U, V)$ *are canonically isomorphic. The spaces* $H^q_d(G; V)$, $H^q_{d_1}(G; V)$, $H^q_{d_2}(G; V)$ *and* $H^q_{\mathrm{ct}}(G; V)$, *endowed with their natural topologies* (IX, 3.3), *are canonically isomorphic* $(q \in \mathbf{Z})$.
(ii) *If* $U, V \in \mathcal{C}^{\infty_i}_G$ *are quasi-complete, then* $\mathrm{Ext}^q_{d_i}(U, V)$ *is canonically isomorphic to* $\mathrm{Ext}^q_{\mathrm{ct}}(U, V)$, *and* $H^q_{d_i}(G; V)$ *is canonically isomorphic, as a topological vector space, to* $H^q_{\mathrm{ct}}(G; V)$ $(i = 1, 2, \ q \in \mathbf{Z})$.

(i) Let $0 \to V \to A^*$ be an s-injective resolution of V in $\mathcal{C}^{\infty_1}_G$. By 1.5 and 1.6, it is then an s-injective resolution in $\mathcal{C}_G$, and $0 \to V \to A^{*^{\infty_2}}$ is an s-injective resolution in $\mathcal{C}^{\infty_2}_G$ and in $\mathcal{C}^\infty_G$. It follows, as in X, 1.6, that the Ext spaces or the cohomology spaces in (i) are all computed from the same complex. The proof of (ii) is similar.

1.9. PROPOSITION. *Let* V *be a Fréchet* G-*module. Then the inclusion* $V^\infty \to V$ *induces an isomorphism* $H^*_d(G; V^\infty) \to H^*_{\mathrm{ct}}(G; V)$.

By 1.5, there exists an s-injective resolution $0 \to V^{\infty_1} \to A^*$ of V^{∞_1} by Fréchet modules in $\mathcal{C}^{\infty_1}_G$. Then $0 \to V^\infty \to A^{*^\infty}$ is an s-injective resolution of V^∞ in $\mathcal{C}^\infty_G$ by 1.6. Since $A^{*^{\infty G}} = A^{*G}$, we see again that $V^\infty \to V^{\infty_1}$ induces a topological isomorphism $H^*_d(G; V^\infty) \to H^*_{\mathrm{ct}}(G; V^{\infty_1})$.
Now let $0 \to V \to B^*$ be an s-injective resolution of V in $\mathcal{C}_G$ by Fréchet modules. Then $0 \to V^{\infty_1} \to B^{*^{\infty_1}}$ is a resolution by Fréchet spaces (IX, 6.5), which are s-injective in $\mathcal{C}_G$ (1.6). By IX, 4.2, we have $H^*_{\mathrm{ct}}(G; V^{\infty_1}) = H^*(B^{*^{\infty_1 G}}) = H^*(B^{*G})$; hence $H^*_{\mathrm{ct}}(G, V^{\infty_1}) = H^*_{\mathrm{ct}}(G; V)$.

1.10. REMARK. The isomorphism of 1.9 is the composition of two maps

$$H^*_d(G; V^\infty) \xrightarrow{\alpha} H^*_{\mathrm{ct}}(G; V^{\infty_1}) \xrightarrow{\beta} H^*_{\mathrm{ct}}(G; V),$$

the first of which is a topological isomorphism. An argument quite analogous to that of IX, 6.7, shows that β is topological if $H^*_{\mathrm{ct}}(G; V)$ is Hausdorff. It may be that the proof in [**3**] could be adapted to prove this in general.

2. Modules of K-finite vectors

In this section, G_1 *and* G_2 *are as in 1.1. We fix a maximal compact subgroup* K_1 *of* G_1, *and set* $R = U(\mathfrak{g}_1)$, $S = U(\mathfrak{k}_1)$.

2.1.   We first rephrase the van Est theorem (IX, 5.6) in different terms, since this will be convenient for combining the real and the t.d. cases.

Let $\mathcal{H}(\mathfrak{g}_1, K_1)$ be the *Hecke algebra* of left and right K_1-finite distributions on G_1 with support in K_1 [**39, 67**]. It is generated by R and the algebra A_{K_1} of K_1-finite measures on K_1. Any smooth representation (π, V) of G_1 in a quasi-complete space V extends to a smooth representation of $\mathcal{H}(\mathfrak{g}_1, K_1)$. For $\delta \in \widehat{K}_1$, let e_δ be the idempotent such that $\pi(e_\delta)$ is the projection on the isotypic subspace V_δ of type δ for any (π, V). The algebra $\mathcal{H}(\mathfrak{g}_1, K_1)$ is an idempotented algebra, with the finite linear combinations of the e_δ as set of idempotents. Any $(\mathfrak{g}_1, K_1)$-module may be viewed as a non-degenerate $\mathcal{H}(\mathfrak{g}_1, K_1)$-module (cf. 0.1), and conversely. Thus the category $\mathcal{C}_{\mathfrak{g}_1, K_1}$ (I, §5) may also be viewed as the category of non-degenerate $\mathcal{H}(\mathfrak{g}_1, K_1)$-modules, and the derived functors of $\mathrm{Hom}_{\mathfrak{g}_1, K_1}$ as those of $\mathrm{Hom}_{\mathcal{H}(\mathfrak{g}_1, K_1)}$. If V is as above, then the space $V_{(K_1)}$ of K_1-finite vectors is equal to $\mathcal{H}(\mathfrak{g}_1, K_1) \cdot V$ and is the greatest non-degenerate submodule of V. We shall also write V_{f_1} for $V_{(K_1)}$. The assignment $V \to V_{f_1}$ is a functor from $\mathcal{C}_{G_1}^{\infty, \mathrm{qc}}$ to $\mathcal{C}_{\mathfrak{g}_1, K_1}$. It is exact: to see this, the main point is to check that if $V \to W$ is surjective $(V, W \in \mathcal{C}_{G_1}^{\infty, \mathrm{qc}})$, then $V_\delta \to W_\delta$ is surjective for all $\delta \in \widehat{K}_1$; but this is clear since $V_\delta = e_\delta \cdot V$, $W_\delta = e_\delta \cdot W$ and $e_\delta \cdot e_\delta = e_\delta$. The van Est theorem implies that this map preserves cohomology.

The final remark of (I, 2.5) applies also to $(\mathfrak{g}_1, K_1)$-modules: any module $V \in \mathcal{C}_{\mathfrak{g}_1, K_1}$ has an injective resolution A^* such that $A^{*\mathfrak{g}_1, K_1} = C^*(\mathfrak{g}_1, K_1; V)$; in particular, $A^{*\mathfrak{g}_1, K_1}$ is finite dimensional if V is admissible.

2.2.   A $((\mathfrak{g}_1, K_1) \times G_2)$-module V is a vector space which is a $(\mathfrak{g}_1, K_1)$-module, a smooth G_2-module (X, 5.1) and such that these actions commute. Such a module is admissible if for every $\delta \in \widehat{K}_1$ and every compact open subgroup L of G_2, the space V_δ^L of L-fixed vectors in the isotypic component V_δ of type δ is finite dimensional. We let $\mathcal{C}_{G, K_1}^f$ or simply $\mathcal{C}_G^f$ be the category of $(\mathfrak{g}_1, K_1) \times G_2$-modules, and linear maps commuting with $\mathfrak{g}_1$, K_1, G_2. It is an abelian category. Let

$$(1) \qquad \mathcal{H}(G) = \mathcal{H}(\mathfrak{g}_1, K_1, G_2) = \mathcal{H}(\mathfrak{g}_1, K_1) \otimes \mathcal{H}(G_2),$$

and call $\mathcal{H}(G)$ the *Hecke algebra* of G. It is idempotented in the obvious way, by the tensor products of idempotents of the two factors.

In view of 2.1 and X, 5.3, we see that $\mathcal{C}_G^f$ can also be defined as the category of non-degenerate $\mathcal{H}(G)$-modules.

2.3.   Given a module V over $\mathcal{H} = \mathcal{H}(G)$ (resp. $\mathcal{H}_i = \mathcal{H}(G_i)$), we let $V_f = \mathcal{H} \cdot V$ (resp. $V_{f_i} = \mathcal{H}_i \cdot V$) be the greatest non-degenerate $\mathcal{H}$- (resp. $\mathcal{H}_i$-) submodule $(i = 1, 2)$. In particular, V_{f_1} is the space of K_1-finite vectors. By 0.4, the category $\mathcal{C}_G^f$ has enough injectives. We let $\mathrm{Ext}_{e, G}^*$ denote the derived functors of $\mathrm{Hom}_{\mathfrak{g}_1, K_1, G_2}$ in $\mathcal{C}_G^f$, and H_e^q the corresponding q-th cohomology space $(q \in \mathbf{Z})$.

LEMMA.   *Let $V \in \mathcal{C}_G^f$ be injective. Then it is acyclic in $\mathcal{C}_{G_1}^f$.*

The proof is quite similar to that of X, 5.4. As in that proof, it suffices to show that if $W \in \mathcal{C}_G^f$, then $\mathrm{Hom}_{\mathbf{C}}(\mathcal{H}, W)_f$ is acyclic in $\mathcal{C}_{G_1}^f$. If $L \supset L'$ are compact open subgroups of G_2, we have a canonical injection

$$(1) \qquad i_{L', L} \colon \mathrm{Hom}_{\mathbf{C}}(\mathcal{H}_1, \mathrm{Hom}_{\mathbf{C}}(\mathcal{H}_2, W)^L)_{f_1} \to \mathrm{Hom}_{\mathbf{C}}(\mathcal{H}_1, \mathrm{Hom}_{\mathbf{C}}(\mathcal{H}_2, W)^{L'})_{f_1},$$

and it follows from the definitions that, as an $\mathcal{H}_1$-module,

(2) $$\mathrm{Hom}_{\mathbf{C}}(\mathcal{H}, W)_f = \mathrm{dir}\lim \mathrm{Hom}_{\mathbf{C}}(\mathcal{H}_1, \mathrm{Hom}_{\mathbf{C}}(\mathcal{H}_2, W)^L)_{f_1},$$

where the direct limit is taken with respect to the maps $i_{L',L}$. The left-hand side is then a direct limit of injective, hence acyclic, G_1-modules and is therefore acyclic by IX, 5.6.

2.4. Let $V \in \mathcal{C}_G^\infty$ be quasi-complete. It may also be viewed as an $\mathcal{H}$-module, hence as an $\mathcal{H}_1$-module. It is already non-degenerate with respect to $\mathcal{H}_2$. Therefore $V_f = V_{f_1}$ may be defined as the space of K_1-finite vectors in V. It is then also the space of K-finite vectors for any compact subgroup K that is the product of K_1 by a compact open subgroup of G_2. The assignment $V \mapsto V_f$ is a functor from $\mathcal{C}_G^{\infty,\mathrm{qc}}$ to $\mathcal{C}_G^f$. As in 2.1, we see that it is exact.

2.5. LEMMA. *If $V \in \mathcal{C}_G^\infty$ is quasi-complete and s-injective, then V_f is acyclic in $\mathcal{C}_G^f$.*

By 2.3 and X, 5.3, $H_e^*(G, V_f)$ is the abutment of a spectral sequence in which

(1) $$E_2^{p,q} = H_e^p(G_2; H(\mathfrak{g}_1, K_1; V_f)) \qquad (p, q \in \mathbf{N}).$$

But

(2) $$H^*(\mathfrak{g}_1, K_1; V_f) = H_d^*(G_1; V)$$

(IX, 5.6), and V is s-injective in $\mathcal{C}_{G_1}^\infty$ by 1.6. Therefore

$$H^q(\mathfrak{g}_1, K_1; V_f) = 0 \quad (q \neq 0), \qquad H^0(\mathfrak{g}_1, K_1; V_f) = V^{G_1}.$$

Since V^{G_1} is s-injective in $\mathcal{C}_{G_2}^\infty$, it follows that $E_2^{p,q} = 0$ for $(p, q) \neq (0, 0)$, whence the lemma.

2.6. PROPOSITION. *Let $V \in \mathcal{C}_G^\infty$ be quasi-complete. Then*

(1) $$H_d^q(G; V) = H_e^q(G; V_f) \qquad (q \in \mathbf{N}).$$

Let $0 \to V \to A^*$ be a resolution of V by s-injective modules in $\mathcal{C}_G^\infty$. Then, by 2.3 and 2.4, $0 \to V_f \to A_f^*$ is an acyclic resolution of V_f. Then we have

(2) $$H_d^*(G; V) = H^*(A^{*G}), \qquad H_e^*(G; V_f) = H^*(A_f^{*\mathfrak{g}_1, K_1, G_2}),$$

the first equality by definition, the second by 2.3 as in X, 5.2. But, clearly,

(3) $$A^{*G} = A_f^{*\mathfrak{g}_1, K_1, G_2},$$

whence the proposition.

3. Cohomology of products

3.1. THEOREM. *Let G_1 be a Lie group, G_2 a t.d. group, $G = G_1 \times G_2$. Let $V_i \in \mathcal{C}_{G_i}^f$ $(i = 1, 2)$ and $V = V_1 \otimes V_2$. Assume V_1 to be admissible. Then*

(1) $$H_e^*(G; V) = H_e^*(G_1; V_1) \otimes H_e^*(G_2; V_2).$$

There exists an injective resolution $0 \to V_1 \to A^*$ of V_1 in $\mathcal{C}_{G_1}^f$ such that $A^{*\mathfrak{g}_1, K_1}$ is finite dimensional (2.1). In view of this, the proof of X, 6.1, is valid without change in the present case.

3.2. COROLLARY. *Let $E_1 \in \mathcal{C}_{G_1}$ be an admissible Fréchet (resp. unitary) G_1-module. Let $E_2 \in \mathcal{C}_{G_2}$ be quasi-complete (resp. unitary), and let $E = E_1 \overline{\otimes} E_2$ (resp. $E = E_1 \widehat{\otimes} E_2$) be the completed projective (resp. Hilbert) tensor product of E_1 and E_2. Then*

(1)
$$H^*_{\mathrm{ct}}(G; E) = H^*_{\mathrm{ct}}(G_1; E_1) \otimes H^*_{\mathrm{ct}}(G_2; E_2).$$

The argument is the same as that of X, 6.2, except that e_L is replaced by the projector $e_\delta \colon E_1 \to E_{1,\delta}$ ($\delta \in \widehat{K}_1$, K_1 a maximal compact subgroup of G_1).

3.3. THEOREM. *Let $m \in \mathbf{N}$. Let k_i be a locally compact non-Archimedean field, $\mathcal{G}_i$ a connected reductive k_i-group and r_i the k_i-rank of $\mathcal{G}_i$ ($i = 1, \ldots, m$). Let $G_i = \mathcal{G}_i(k_i)$, $G = G_1 \times \cdots \times G_m$ and $r(G) = \sum r_i$. Let V be a unitary irreducible representation of G with compact kernel. Then*

(1)
$$H^q_{\mathrm{ct}}(G; V) = 0 \quad \text{for } q \neq r(G).$$

The groups G_i are of type I [**2**]. Therefore V can be written as a Hilbert tensor product

(2)
$$V = \widehat{\bigotimes} V_i \qquad (V_i \in \mathcal{C}_{G_i}, \text{ irreducible, unitary, admissible}).$$

The kernel of $G_i \to \mathbf{GL}(V_i)$ is compact ($i = 1, \ldots, m$); hence

(3)
$$H^q_{\mathrm{ct}}(G_i; V_i) = 0 \quad \text{for } q \neq r_i,$$

by Casselman's theorem (XI, 3.9). The result then follows from the Künneth rule (X, 6.3).

REMARK. This result was also known to W. Casselman.

3.4. We assume some familiarity with the adèle language and of adèle groups (see e.g. [**116**]).

k *is a global field, Σ (resp. Σ_∞, resp. Σ_f) the set of places (resp. infinite places, resp. finite places) of k. For $s \in \Sigma$, the completion of k at s is denoted k_s ($s \in \Sigma$). A or A_k (resp. A_f) is the ring of adèles (resp. finite adèles) of k. Let $\mathcal{G}$ be a connected k-group. Then $\mathcal{G}(k_s)$ will be denoted G_s.*

3.5. PROPOSITION. *Let (π, V) be an irreducible unitary representation of $G(A)$ whose kernel in $G(A_f)$ is compact. Then $H^*_{\mathrm{ct}}(G(A); V) = 0$.*

The groups G_s are all of type I; therefore [**42**, Chap. 3, §3, no. 3], V can be written as a restricted infinite Hilbert tensor product $V = \widehat{\bigotimes}_s V_s$, where V_s is an irreducible unitary representation of G_s ($s \in \Sigma$). In particular, we can also write

(1)
$$V = V_\infty \widehat{\otimes} V_f, \quad \text{where } V_\infty = \widehat{\bigotimes}_{s \in \Sigma_\infty} V_s, \; V_f = \widehat{\bigotimes}_{s \in \Sigma_f} V_s$$

are irreducible unitary representations of $G_\infty = \prod_{s \in \Sigma_\infty} G_s$ and of $G(A_f)$ respectively. By 3.2,

(2)
$$H^*_{\mathrm{ct}}(G(A); V) = H^*_{\mathrm{ct}}(G_\infty; V_\infty) \otimes H^*_{\mathrm{ct}}(G(A_f); V_f).$$

It suffices therefore to consider the case where G_∞ is in the kernel of V, i.e. when $V = V_f$ is in fact an irreducible unitary representation of $G(A_f)$ with compact kernel.

Let S be a finite subset of Σ_f and $S' = \Sigma_f - S$. Let $G_S = \prod_{s \in S} G_s$, and let $G_{S'}$ be the restricted product of the G_s ($s \in S'$). Then $G(A_f) = G_S \times G_{S'}$, and we can also write

$$(3) \qquad\qquad V = \left(\widehat{\bigotimes}_{s \in S} V_S \right) \widehat{\otimes} V_{S'},$$

where $V_{S'}$ is the restricted Hilbert tensor product of the V_s ($s \in S'$).

By X, 6.2, we have

$$(4) \qquad\qquad H^*_{\mathrm{ct}}(G(A_f); V) = H^*_{\mathrm{ct}}(G_S; V_S) \otimes H^*_{\mathrm{ct}}(G_{S'}; V_{S'}).$$

By 3.3, the first factor of the right-hand side is zero in dimensions different from the sum $r(S) = \sum_{s \in S} r_S$, where r_S is the k_S-rank of $\mathcal{G}$, viewed as a k_s-group. Therefore

$$(5) \qquad\qquad H^q(G(A_f); V) = 0 \quad \text{for } q < r(S).$$

But the group $\mathcal{G}$ is quasi-split over k_s for almost all s's; hence the k_s-rank of $\mathcal{G}_s$ is ≥ 1 for almost all s's. Therefore, given a positive integer N, there exists S such that $r(S) > N$. Our assertion follows.

CHAPTER XIII

Cohomology of Discrete
Cocompact Subgroups

The main goal of this chapter is to prove some results on the cohomology of discrete cocompact subgroups of p-adic reductive groups and of products of such groups by real reductive groups. We shall first consider more generally the case where the ambient group is of the type considered in XII, and gradually specialize to our main case of interest.

1. Subgroups of products of Lie groups and t.d. groups

In this section $G = G_1 \times G_2$, where G_1 is a real Lie group and G_2 a t.d. group (X, §1). Γ is a discrete subgroup of G and (ρ, E) a finite dimensional representation of Γ.

1.1. We view Γ as operating on the left on G. If Γ operates on a space V, a map $f: G \to V$ is therefore Γ-equivariant if it satisfies the condition

(1) $$f(\gamma \cdot g) = \gamma \cdot f(g) \quad \text{for all } \gamma \in \Gamma, \; g \in G.$$

As usual, we let $I(E) = I_\Gamma^G(E)$ be the set of continuous Γ-equivariant maps from G to E, and view it as a G-module via right translations. It is a Fréchet G-module. By (IX, 2.3), we have

(2) $$H^*(\Gamma; E) = H_{\mathrm{ct}}^*(G; I(E)).$$

By XII, 1.9, 2.6,

(3) $$H_{\mathrm{ct}}^*(G; I(E)) = H_d^*(G; I(E)^\infty) = H_e^*(G; I(E)_f^\infty).$$

1.2. Now assume Γ to be *cocompact* and (ρ, E) to be *unitary*. The group G is then necessarily unimodular. As in VII, we let dx be a Haar measure on G and the associated measure on $\Gamma\backslash G$, and $(\, , \,)_E$ the scalar product on E. If $u, v \in I(E)$, then $g \mapsto (u(g), v(g))_E$ is left-invariant under Γ; hence

(1) $$(u, v) = \int_{\Gamma\backslash G} (u(x), v(x))_E \, dx \qquad (u, v \in I(E))$$

defines a scalar product on $I(E)$, invariant under G. We let $I_2(E)$ or $I_{\Gamma,2}^G(E)$ denote the completion of $I(E)$ with respect to the norm defined by (1). It may be identified with the space of measurable cross-sections of the vector bundle $G \times_\Gamma E \to \Gamma\backslash G$, over $\Gamma\backslash G$, with typical fiber E, and structural group Γ acting by means of ρ. By XII, 1.9, 2.6, again,

(2) $$H_{\mathrm{ct}}^*(G; I_2(E)) = H_d^*(G; I_2(E)^\infty) = H_e^*(G; I_2(E)_f^\infty).$$

We now claim that

(3) $$I_2(E)^\infty = I(E)^\infty.$$

If $G = G_1$ is a Lie group, this was proved in III, §7. Let $G = G_2$ be of t.d. type, and L a compact open subgroup of G. Then it is clear from the definition that

(4) $I_2(E)^L = I(E)^L = \{f \colon G/L \to E \mid f(\gamma \cdot x) = \gamma \cdot f(x) \ (\gamma \in \Gamma, \ x \in G/L)\},$

whence we get (3) in this case.

We now consider the general case. For a compact open subgroup L of G_2, let

(5) $$\Gamma_L = \Gamma \cap (G_1 \times L).$$

The orbits of $G_1 \times L$ in $\Gamma\backslash G$ are open, disjoint, hence compact, and finite in number. The orbit map $g \mapsto x \cdot g$ is open and induces a homeomorphism

(6) $$^{x^{-1}}(\Gamma_{x_L})\backslash(G_1 \times L) \xrightarrow{\sim} x \cdot (G_1 \times L).$$

In particular, Γ_L is cocompact in $G_1 \times L$. Since L is compact, the projection $\mathrm{pr}_1 \colon G \to G_1$ is proper on $G_1 \times L$; hence $\Gamma'_L = \mathrm{pr}_1(\Gamma_L)$ is a discrete cocompact subgroup of G_1. There exists a finite set $C \subset G_2$ such that G is the disjoint union of the double cosets $\Gamma \cdot c \cdot (G_1 \times L)$ $(c \in C)$. We have then a natural $(G_1 \times L)$-equivariant homeomorphism

(7) $$\Gamma\backslash G = \coprod_{c \in C} (^{c^{-1}}\Gamma \cap (G_1 \times L))\backslash(G_1 \times L).$$

Let

$$\Gamma'_c = \mathrm{pr}_1(^{c^{-1}}\Gamma \cap (G_1 \times L)).$$

Since $c \in G_2$, we also have

(8) $$\Gamma'_c = \mathrm{pr}_1(\Gamma_{c_L}).$$

Given a function f on G, right-invariant under L, let f_c be the function on G_1 defined by

(9) $$f_c(x) = f(c \cdot x) \qquad (x \in G_1).$$

It is immediate to check that f is left-invariant under Γ if and only if f_c is left-invariant under Γ'_c for every $c \in C$. It follows then that (7) yields the following isomorphisms of G_1-modules:

(10) $$I_\Gamma^G(E)^L = \bigoplus_{c \in C} I_{\Gamma'_c}^{G_1}(E),$$

(11) $$I_\Gamma^G(E)^{L,\infty_1} = \bigoplus_{c \in C} I_{\Gamma'_c}^{G_1}(E)^{\infty_1},$$

(12) $$I_{\Gamma,2}^G(E)^L = \bigoplus_{c \in C} I_{\Gamma'_c,2}^{G_1}(E),$$

where, as in XII, ∞_1 indicates C^∞ with respect to G_1. But

(13) $$I_{\Gamma'_c,2}^{G_1}(E)^{\infty_1} = I_{\Gamma'_c}^{G_1}(E)^{\infty_1},$$

by (III, 7.9). Then we have

(14) $$I_2(E)^{L,\infty_1} = I(E)^{L,\infty_1}$$

for every L, whence (3).

1.3. PROPOSITION. *We have*

$$(1) \qquad\qquad H^*(\Gamma; E) = H^*_{\mathrm{ct}}(G; I^G_{\Gamma,2}(E)).$$

The dimension of $H^0(\Gamma; E)$ is equal to the multiplicity of the trivial representation in $I_2(E)$.

The relation (1) follows from 1.1(2), (3) and 1.2(2), (3). The map which associates to $e \in E^\Gamma$ the constant function $\widetilde{e}$ on G with value e induces an isomorphism of E^Γ onto the space C of constant functions in $I_2(E)$. We can write $I_2(E)$ as the direct sum of C and of its orthogonal complement D. We have

$$(1) \qquad\qquad H^*_{\mathrm{ct}}(G; I_2(E)) = H^*_{\mathrm{ct}}(G; C) \oplus H^*_{\mathrm{ct}}(G; D),$$

$$(2) \qquad\qquad H^0_{\mathrm{ct}}(G; I_2(E)) = I_2(E)^G = C,$$

whence the second assertion.

1.4. The theorem of [**42**, 3, §3] quoted in XII, §3, also applies to the present case, and shows that $I_2(E)$ can be written as a Hilbert direct sum

$$(1) \qquad\qquad I^G_{\Gamma,2}(E) = \widehat{\bigoplus_{\pi \in \widehat{G}}} m(\pi, \Gamma, E) H_\pi$$

of irreducible unitary G-modules with finite multiplicities.

1.5. PROPOSITION. *Assume that $H^*(\Gamma; E)$ is finite dimensional. Then*

$$(1) \qquad\qquad H^*(\Gamma; E) = \bigoplus_{\pi \in G} m(\pi, \Gamma, E) H^*_{\mathrm{ct}}(G; H_\pi).$$

PROOF. By 1.3,

$$(2) \qquad\qquad H^*(\Gamma; E) = H^*_{\mathrm{ct}}\left(G; \widehat{\bigoplus_\pi} m(\pi, \Gamma, E) H_\pi\right).$$

As in VII, 3.2, we have to replace the Hilbert direct sum $\widehat{\bigoplus}$ by an ordinary algebraic direct sum. The proof is the analogue in the present framework of that of VII, 3.2. Let Q be the set of $\pi \in \widehat{G}$ for which $m(\pi, \Gamma, E) \neq 0$. It is countable. For any finite subset S of Q, there is the direct sum decomposition

$$(3) \qquad I_2(E) = \bigoplus_{\pi \in S} m(\pi, \Gamma, E) H_\pi \oplus \left(\widehat{\bigoplus_{\pi \in S'}} m(\pi, \Gamma, E) H_\pi\right) \qquad (S' = Q - S).$$

Hence

$$(4) \qquad H^*(\Gamma; E) = \bigoplus_{\pi \in S'} m(\pi, \Gamma, E) H^*_{\mathrm{ct}}(G; H_\pi) \oplus H^*_{\mathrm{ct}}\left(G; \widehat{\bigoplus_{\pi \in S'}} m(\pi, \Gamma, E) \cdot H_\pi\right).$$

Consequently, there are only finitely many $\pi \in Q$ for which $H^*(G; H_\pi) \neq 0$. Since the last term of (4) is also finite dimensional, it suffices to prove the following lemma.

1.6. LEMMA. *Let T be a countable set of irreducible unitary representations (π, H_π) of G, and V the Hilbert direct sum of the H_π's. Assume that $H^*_{\mathrm{ct}}(G; H_\pi) = 0$ for all $\pi \in T$, and that $H^*_{\mathrm{ct}}(G; V)$ is finite dimensional. Then $H^*_{\mathrm{ct}}(G; V) = 0$.*

For a unitary G-module W, let $F^*(W)$ be the non-homogeneous complex with coefficients in W (IX, 1.4(5)). It consists of Fréchet spaces. Since $H^*_{\mathrm{ct}}(G; V)$ is finite dimensional, it follows, as in VII, 3.3, that $dF^{q-1}(V)$ is closed in $Z^q = F^q(V) \cap \ker d$. For S finite in T, let pr_S be the orthogonal projection of V onto the direct sum H_S of the H_π ($\pi \in S$), and also the corresponding projection

$$(1) \qquad \mathrm{pr}_S \colon F^*(V) \to F^*(H_S) = \bigoplus_{\pi \in S} F^*(H_\pi).$$

The topology on $F^q(V)$ is that of uniform convergence on compact sets. If f is a continuous V-valued function on a compact space C and $S' \supset S$, then

$$(2) \qquad \|f(c) - \mathrm{pr}_{S'} f(c)\|^2 \leq \|f(c) - \mathrm{pr}_S f(c)\|^2, \quad \text{for all } c \in C.$$

This implies that any element $x \in F^*(V)$ is the limit of its projections $\mathrm{pr}_S x$, as S tends to T. The argument is then the same as in VII, 3.3.

2. Products of reductive groups

2.1. From now on we assume that the G_i's are reductive and introduce slightly different conventions, more adapted to the S-arithmetic case. We let S be a finite set, and for $s \in S$ we assume given a local field k_s and a connected reductive k_s-group $\mathcal{G}_s$. We let $G = \prod_s G_s$, where, as usual, $G_s = \mathcal{G}_s(k_s)$. Let S_∞ (resp. S_f) be the set of $s \in S$ for which k_s is Archimedean (resp. non-Archimedean), r_s the k_s-rank of $\mathcal{G}_s$ and

$$(1) \qquad r_\infty = \sum_{s \in S_\infty} r_s, \qquad r_f = \sum_{s \in S} r_s, \qquad r = r_f + r_\infty.$$

We also assume that if $S_\infty \neq \varnothing$, then all k_s are of characteristic zero. The groups G_s are of type I; hence any irreducible unitary representation (π, H) of G decomposes uniquely as a Hilbert tensor product

$$(2) \qquad (\pi, H_\pi) = \widehat{\bigotimes_s}(\pi_s, H_{\pi_s}),$$

where (π_s, H_{π_s}) is an irreducible unitary (hence admissible) representation of G_s [**42**, Chap. 3, §3, Lemma 1].

We also set

$$(3) \qquad G_\infty = \prod_{s \in S_\infty} G_s, \qquad G_f = \prod_{s \in S_f} G_s,$$

and write (2) as

$$(4) \qquad (\pi, H_\pi) = (\pi_\infty, H_{\pi,\infty}) \, \widehat{\otimes} \, (\pi_f, H_{\pi,f}),$$

where

$$(5) \qquad (\pi_\infty, H_{\pi_\infty}) = \widehat{\bigotimes_{s \in S_\infty}}(\pi_s, H_{\pi_s}), \qquad (\pi_f, H_{\pi_f}) = \widehat{\bigotimes_{s \in S_f}}(\pi_s, H_{\pi_s}).$$

If $T \subset S$ we put $G_T = \prod_{s \in T} G_t$ and denote by μ_T the projection of G on G_T.

[The use of the subscript ∞ here conflicts with the notation for representations in C^∞ vectors. We trust this will not cause any confusion.]

As before, Γ is a discrete cocompact subgroup of G and (ρ, E) a finite dimensional unitary representation of Γ.

2.2. PROPOSITION. *Let H_π and H_{π_s} be as in 2.1(2). Then*

(1)
$$H_{\mathrm{ct}}^*(G; H_\pi) = \bigotimes_s H_{\mathrm{ct}}^*(G_s; H_{\pi_s}).$$

If π has compact kernel and $r > 0$, then $H_{\mathrm{ct}}^q(G; H_\pi) = 0$ for $q < r$.

PROOF. By XII, 3.2 and repeated application of X, 6.2, we have

(2)
$$H_{\mathrm{ct}}^*(G; H_\pi) = H_{\mathrm{ct}}^*(G_\infty; H_{\pi_\infty}) \otimes \left(\bigotimes_{s \in S_f} H_{\mathrm{ct}}^*(G_s; H_{\pi_s}) \right).$$

By IX, 5.6, 6.6, we can replace the first factor on the right-hand side by relative Lie algebra cohomology with coefficients in $(H_{\pi_\infty})^\infty$. We then use the Künneth rule I, 1.3, switch back to continuous cohomology, and get

(3)
$$H_{\mathrm{ct}}^*(G_\infty; H_{\pi_\infty}) = \bigotimes_{s \in S_\infty} H_{\mathrm{ct}}^*(G_s; H_{\pi_s}).$$

This proves the first assertion. The second now follows from the vanishing theorems (V, 3.3, and XI, 3.9).

2.3. LEMMA. *Assume that k_s is non-Archimedean for all $s \in S$. Then $I_{\Gamma,2}^G(E)$ is an admissible G-module.*

Let L be a compact open subgroup of G. Then, as was already pointed out in 1.2(4), $I_{\Gamma,2}^G(E)^L$ may be identified with the space of Γ-equivariant maps of G/L into E. Such a map is completely determined by its values on a set of representatives of $\Gamma \backslash G/L$. Since $\Gamma \backslash G$ is compact, this set is finite; hence $I_2(E)^L$ is finite dimensional.

2.4. THEOREM. *We keep the assumptions and conventions of 2.1. Then $H^*(\Gamma; E)$ is finite dimensional, and we have*

(1)
$$H^*(\Gamma; E) = \bigoplus_{\pi \in \widehat{G}} m(\pi, \Gamma, E) \left(\bigotimes_s H_{\mathrm{ct}}^*(G_s; H_{\pi_s}) \right).$$

We prove first that $H^*(\Gamma; E)$ is finite dimensional. If $S_f = \varnothing$, this was shown in VII, 3.2. If $S_\infty = \varnothing$, this follows from 1.3, 2.3 and X, 6.3. So let S_∞ and S_f both be non-empty. The fields k_s are then of characteristic zero. The group Γ has a finite presentation [**17**, 6.2]. By embedding the k_i's into **C**, we see that Γ has a faithful linear representation over **C**. It then has a torsion-free normal subgroup Γ' of finite index [**9**, 17.7]. The space $H^*(\Gamma'; E)$ is finite dimensional by [**17**, 6.2(ii)]; hence

(2)
$$H^*(\Gamma; E) = (H^*(\Gamma'; E))^{\Gamma/\Gamma'}$$

is also finite dimensional. We then have, by 1.5,

(3)
$$H^*(\Gamma; E) = \bigoplus_{\pi \in \widehat{G}} m(\pi, \Gamma, E) \cdot H_{\mathrm{ct}}^*(G; H_\pi).$$

Since G_s is reductive, any irreducible unitary representation of G_s is admissible $(s \in S)$. Therefore (1) follows from (3) and 2.2.

2.5. REMARKS. (1) Recall that if L is a compact group and (ρ, V) an irreducible quasi-complete L-module, then $H^i_{\mathrm{ct}}(L; V)$ is equal to 0 for $i \geq 1$ and is equal to V^L in dimension zero (IX, 1.12). Therefore, if G_s is compact, the only terms which can contribute to the right-hand side of 2.4(1) are those in which H_{π_s} is the trivial one-dimensional G_s-module.

(2) Let T be the set of $s \in S$ for which G_s is not compact. Then $\Gamma' = \ker \mathrm{pr}_T$ is a finite normal subgroup of Γ and $\Gamma'' = \mathrm{pr}_T(\Gamma)$ is a discrete cocompact subgroup of G_T. We have, by the Hochschild-Serre spectral sequence,

$$(1) \qquad\qquad H^*(\Gamma; E) = H^*(\Gamma''; E^{\Gamma'}).$$

There is therefore no essential loss of generality in assuming that $G = G_T$.

We now specialize this to the case where $S = S_f$ consists of one element.

2.6. THEOREM. *Assume that $G = \mathcal{G}(k)$, where k is a non-Archimedean local field and $\mathcal{G}$ a connected semi-simple, almost k-simple, group over k. Let $r = \mathrm{rk}_k(\mathcal{G})$. Then*

$$(1) \qquad\qquad H^i(\Gamma; E) = 0, \quad \text{for } i \neq 0, r,$$

and $\dim H^0(\Gamma; E)$ *(resp.* $\dim H^r(\Gamma; E)$*) is equal to the multiplicity of the trivial (resp. Steinberg) representation of G in $I_2(E)$.*

This is obvious (and follows from IX, 1.12) if G is compact. So assume G noncompact, i.e. $r \geq 1$. As in X, 2.1, let $\widetilde{\mathcal{G}}$ be the universal covering of $\mathcal{G}$, and $\sigma \colon \widetilde{G} \to G$ the canonical central isogeny. Let $Q = \sigma(\widetilde{G})$. It is cocompact and cocommutative (X, 2.1).

We already know that $\dim H^0(\Gamma; E)$ is the multiplicity of the trivial representation π_0 of G (1.3). Also, 1.3 and X, 2.4, imply that $H^i(\Gamma; E) = 0$ for $i > r$. By X, 2.6, the trivial representation does not contribute to higher cohomology; therefore

$$(2) \qquad H^i(\Gamma; E) = \bigoplus_{\pi \in \widehat{G}, \pi \neq \pi_0} m(\pi, \Gamma, E) H^i_{\mathrm{ct}}(G; H_\pi) \qquad (i \geq 1).$$

Let $\pi \in \widehat{G}$ and assume that $\ker \pi$ is non-compact. We want to prove that $\ker \pi$ contains Q. The group Q is simple modulo its center [**100**]; therefore, if $Q \not\subset \ker \pi$, then $\ker \pi \cap Q$ is finite and central in G. Since G/Q is commutative, $\ker \pi$ would then be nilpotent and its Zariski closure in $\mathcal{G}$ would be a normal infinite nilpotent algebraic subgroup, which is absurd. Thus $Q \subset \ker \pi$. By IX, 1.11 and X, 2.6, applied to $\widetilde{G}$, we have then

$$(3) \qquad\qquad H^i_{\mathrm{ct}}(Q; H_\pi) = 0 \qquad (i \geq 1).$$

But, by IX, 2.5,

$$(4) \qquad\qquad H^*_{\mathrm{ct}}(G; H_\pi) = H^*_{\mathrm{ct}}(Q; H_\pi)^{G/Q},$$

and hence $H^i_{\mathrm{ct}}(G; H_\pi) = 0$ $(i \geq 1)$. If now π has a compact kernel, then by XI, 3.9, $H^i_{\mathrm{ct}}(G; H_\pi) = 0$ unless $i = r$, and π is the Steinberg representation, in which case it is one-dimensional. The theorem follows.

2.7. REMARK. For $G = \widetilde{G}$, this theorem was proved by H. Garland under the assumption that the residue field of k is sufficiently big [**40**], and announced by W. Casselman [**33**] in general. This work had been motivated by a conjecture of J.-P. Serre [**98**], stating that if (σ, F) is a *rational representation* of $\mathcal{G}$ defined over k, then $H^i(\Gamma; F) = 0$ for $0 < i < r$. It was pointed out to one of us by G. Prasad that if $r \geq 2$, and k is of characteristic zero, a deep result of G. A. Margulis allows one to derive this conjecture from 2.6. We shall outline this argument in a more general case later, and see that, in that case, the theorem is in fact true for any finite dimensional representation of Γ over a field of characteristic zero (3.7).

3. Irreducible subgroups of semi-simple groups

3.1.   We keep the conventions of 2.1, and moreover assume $\mathcal{G}_s$ to be semi-simple. Let $\widetilde{\mathcal{G}}_s$ be the universal covering of $\widetilde{\mathcal{G}}_s$, $\sigma_s \colon \widetilde{\mathcal{G}}_s \to \mathcal{G}_s$ the canonical isogeny, $\widetilde{G}_s = \widetilde{\mathcal{G}}_s(k_s)$ $(s \in S)$, $\widetilde{G}$ the product of the $\widetilde{G}_s$ and σ the product of the $\sigma_s \colon \widetilde{G}_s \to G_s$. We recall that $\widetilde{G}_s$ is connected if k_s is Archimedean.

A *standard normal subgroup* N of G is a closed normal subgroup of the form $N = \prod_s N_s$, where N_s is the group of rational points of a connected normal k_s-subgroup of $\mathcal{G}_s$. We say that Γ is *irreducible* if its intersection with every proper standard normal subgroup is finite. In the Archimedean case, this notion implies the similar notion of VII, 4.1 (and differs from it only in minor ways).

If k is any field, and $\mathcal{G}$ an almost k-simple k-group, then there exist a finite separable extension k' of k and an absolutely almost k'-simple k'-group $\mathcal{G}'$ such that $\mathcal{G} = R_{k'/k}\mathcal{G}'$ [**18**, 6.21]. We then have $\mathcal{G}(k) = \mathcal{G}'(k')$ (also for the underlying k- and k'-topology if both k and k' are local fields [**116**, Chap. I]). Since $\widetilde{G}_s$ is a direct product of almost k_s-simple k_s-groups, we see that, if $G = \widetilde{G}$, we can always assume the $\mathcal{G}_s$ to be absolutely almost k_s-simple without loss of generality.

3.2. LEMMA. *The group* $\widetilde{\Gamma} = \sigma^{-1}(\Gamma)$ *is discrete cocompact in* $\widetilde{G}$. *It is irreducible if* Γ *is. Let* V *be a vector space of characteristic zero on which* Γ *acts. Then*

$$(1) \qquad H^*(\Gamma; V) = H^*(\sigma(\widetilde{\Gamma}); V)^{\Gamma/\sigma(\widetilde{\Gamma})}, \qquad H^*(\widetilde{\Gamma}; V) = H^*(\sigma(\widetilde{\Gamma}); V).$$

The first assertion is an obvious generalization of [**11**, 3.4]. We repeat the argument for the sake of completeness: Let $Q = \sigma(\widetilde{G})$, $Q_s = \sigma_s(\widetilde{G}_s)$ $(s \in S)$. The group Q is the product of the Q_s; therefore [**20**, 3.19] implies that Q is closed, normal, cocompact, cocommutative in G, and that G/Q has finite exponent. The group Γ is finitely generated [**17**, 6.2(i)]; therefore its image in G/Q is finite, and hence $\Gamma \cap Q$ has finite index in Γ and is cocompact in G or Q. Since σ has finite kernel, the first assertion follows. If $\widetilde{N}$ is a standard normal subgroup of $\widetilde{G}$, then $\sigma(\widetilde{N})$ is cocompact in a standard normal subgroup N of G, as follows from the definition and [**20**, 3.19]. This implies the second assertion.

The kernel N of $\sigma \colon \widetilde{\Gamma} \to \Gamma$ is finite and acts trivially on V; therefore the second equality of (1) follows from a trivial application of the Hochschild-Serre spectral sequence (contained in IX, 1.11). The group $\sigma(\widetilde{\Gamma})$ is equal to $\Gamma \cap Q$, hence normal of finite index in Γ, whence the first equality of (1) (IX, 2.5).

3.3. LEMMA. *Let* Γ *be irreducible. Assume that* $G = \widetilde{G}$ *and has no non-trivial compact standard normal subgroup. Fix* $T \subset S$, $T \neq S$. *Then* $\Gamma_T = \mathrm{pr}_T(\Gamma)$ *is dense in* G_T.

By the remark at the end of 3.1, we may assume that $\mathcal{G}_s$ is absolutely almost k_s-simple for all $s \in S$. Our assumption on G then implies that G_s is non-compact (i.e. $r_s \geq 1$) for every s. Assume first that k_t is Archimedean for all $t \in T$. The group Γ_T is not discrete in G_T (by a standard argument, cf. e.g. [**156**], pp. 597–598), and $\overline{\Gamma}_T \cdot G_{S-T}$ is the closure of $\Gamma \cdot G_{S-T}$ in G. Therefore $G_T/\overline{\Gamma}_T$ is compact. It follows from [**5**] that Γ_T is Zariski dense in G_T; hence $\overline{\Gamma}_T$ is a Lie group whose Lie algebra $\mathfrak{h}$ is an ideal of the Lie algebra $\mathfrak{g}_T$ of G_T. Since the projection of Γ_T on any factor of G_T is non-discrete, we get $\mathfrak{h} = \mathfrak{g}_T$, whence $\overline{\Gamma}_T = G_T$. Assume now that the set T' of $t \in T$ for which k_t is Archimedean is non-empty and $\neq T$. Let $T'' = T - T'$. We prove first that $G_{T'} \subset \overline{\Gamma}_T$. Let L be a compact open subgroup of $G_{T''}$, and $\Gamma_L = \Gamma \cap (L \times (G_{S-T''}))$. The group Γ_L is discrete cocompact in $G_{S-T''} \times L$; hence its projection $\Gamma_{L,S-T''}$ in $G_{S-T''}$ is discrete cocompact. The previous argument shows that $\mathrm{pr}_{T'}(\Gamma_{L,S-T''})$ is dense in $G_{T'}$; hence we have, in G_T,

$$(1) \qquad L \cdot \overline{\Gamma}_T \supset L \cdot \overline{\Gamma}_{L,T} = L \cdot \overline{\Gamma}_{L,T'} \supset L \cdot G_{T'}.$$

Since L can be chosen arbitrarily small, this implies that $G_{T'} \subset \overline{\Gamma}_T$. Now we have $\overline{\Gamma}_T = G_{T'} \cdot \overline{\Gamma}_{T''}$, which reduces us to the case where k_t is non-Archimedean for all $t \in T$. Assume first that T consists of one element t. The quotient $G_t/\overline{\Gamma}_t$ is equal to the quotient of G by the closure of $\Gamma \cdot G_{S-T}$, hence is compact and carries an invariant measure. The main theorem of [**91**] then implies our assertion. If $\mathrm{Card}\, T \geq 2$, fix $t \in T$, and let $T' = T - \{t\}$. Let L be a compact open subgroup of $G_{T'}$. Then one argues as before that $\mathrm{pr}_t(\Gamma_L)$ is dense in G_t. We have

$$L \cdot \overline{\Gamma}_T \supset L \cdot \overline{\Gamma_{L,T}} = L \cdot \overline{\mathrm{pr}_t(\Gamma_L)} = L \cdot G_t.$$

Since L may be taken arbitrarily small, it follows that $G_t \subset \overline{\Gamma}_T$.

3.4. LEMMA. *Assume Γ to be irreducible, $G = \widetilde{G}$, and G_s to be almost k_s-simple for all $s \in S$. Let (π, H_π) be an irreducible unitary representation of G which occurs in $I_2(E)$ and is such that $H^*_\bullet(G; H) \neq 0$. Let $(\pi, H) = \widehat{\bigotimes}_s (\pi_s, H_s)$ be its canonical decomposition, where (π_s, H_s) is an irreducible unitary representation of $G_s(s \in S)$. Assume that it has a non-compact kernel. Then (π, H) is the trivial representation.*

As in VII, 4.2, we see first that E may be assumed to be irreducible. By 2.2

$$(1) \qquad H^*_{\mathrm{ct}}(G; H) = \bigotimes_s H^*_{\mathrm{ct}}(G_s; H_s).$$

We already know that if G_t is compact, then (π_t, H_t) is trivial (2.5(1)). This reduces us to the case where G_t is non-compact for all $t \in S$. There exists $s \in S$ such that $N_s = \ker \pi_s$ is not compact. But G_s is simple modulo its center, as follows from [**100**] and the fact that the Kneser-Tits conjecture is true over local fields. Hence $G_s = N_s$, and π_s is trivial.

The argument is now quite analogous to that of VII, 4.2: H_π is a space of Γ-equivariant functions $G \to E$ which are right-invariant under G_s, hence also left-invariant under G_s. Let $S' = S - \{s\}$. One shows, as in VII, 4.2, that ρ defines a representation of $\Gamma_{S'}$ which is continuous in the topology induced from that of $G_{S'}$; hence it extends to a unitary representation of $\overline{\Gamma}_{S'}$. Since $\overline{\Gamma}_{S'}$ is equal to $G_{S'}$ by 3.3, it is then the trivial representation. As in loc. cit., it follows that the elements of H_π are left-invariant under $G_s \cdot \Gamma_{S'}$, hence under the closure of that subgroup, which is equal to G by 3.3, whence the lemma.

3.5. THEOREM. *Let $G = \widetilde{G}$ and Γ be irreducible. Then* (cf. 2.1 for the notation)

(1) $\quad H^q(\Gamma; E) = H^q_{ct}(G_\infty; E^\Gamma) \oplus \bigoplus_\pi{}' m(\pi, \Gamma, E) \cdot H^{q-r_f}_{ct}(G_\infty; H_{\pi_\infty}) \qquad (q \in \mathbf{Z}),$

where E^Γ is viewed as a trivial G_∞-module, and $\bigoplus'$ is extended over the $\pi \in \widehat{G}$ for which π_∞ has trivial infinitesimal character, and π_f is the tensor product of the special representations of the G_s $(s \in S_f)$.

We start from 2.4. By 3.4, the only representations which can contribute to the right hand side are the trivial representation and those in which all factors H_{π_s} are infinite dimensional. By 1.3, the trivial representation occurs with multiplicity equal to $\dim E^\Gamma$. Moreover, by XII, 3.2, and X, 2.6, we have

(2) $\qquad\qquad\qquad H^*_{ct}(G; V) = H^*_{ct}(G_\infty; V),$

if V is the trivial representation. This accounts for the first term on the right-hand side of (1). Now let π be infinite dimensional. We can write

(3) $\qquad\qquad H^*_{ct}(G; H_\pi) = H^*_{ct}(G_\infty; H_{\pi_\infty}) \otimes \left(\bigotimes_{s \in S_f} H^*_{ct}(G_s; H_{\pi_s}) \right).$

The first factor on the right-hand side can be replaced by

(4) $\qquad\qquad H^*_d(G_\infty; (H_{\pi_\infty})^\infty) = H^*(\mathfrak{g}_\infty, K_\infty; (H_{\pi_\infty})^\infty),$

where K_∞ is a maximal compact subgroup of G_∞ (IX, 5.6, 6.6), hence can be $\neq 0$ only if the infinitesimal character of π_∞ is trivial (I, 5.3). Let $s \in S_f$. Then by Casselman's theorem (XI, 3.9), $H^q_{ct}(G_s, H_{\pi_s})$ is zero unless $q = r_s$ and H_{π_s} is the special representation, in which case it is one-dimensional. The theorem follows.

REMARK. This result was stated in [**12**], and was also known to W. Casselman. The proof alluded to in [**12**] is different.

3.6. PROPOSITION. *Assume Γ to be irreducible.*
(i) *If $S_\infty = \varnothing$, then $H^q(\Gamma; E) = 0$ for $q \neq 0, r$.*
(ii) *If $G = \widetilde{G}$ and $r \geq 1$, then $H^q(\Gamma; E)$ is canonically isomorphic to $H^q_{ct}(G_\infty; E^\Gamma)$ for $q < r$.*

(i) 3.2 allows one to reduce the proof to the case $G = \widetilde{G}$, where it follows from 3.5.

(ii) By 3.4, the $\bigoplus'$ in the right-hand side of 3.5(1) is over representations with compact kernel. By the vanishing theorem (V, 3.3), each term in that sum is zero if $q - r_f < r_\infty$, i.e. if $q < r$.

REMARK. If $G = \widetilde{G}$, the proof of 3.6(i) also shows that $\dim H^r(\Gamma; E)$ is the multiplicity in $I_2(E)$ of the representation $\widehat{\bigotimes}_s H_s$, where H_s is trivial if G_s is compact, and special otherwise.

3.7. PROPOSITION. *Assume that k_s is non-Archimedean of characteristic zero for all $s \in S$, that $r \geq 2$, and that Γ is irreducible. Let (τ, F) be a finite dimensional representation of Γ over a field k of characteristic zero. Then*

(1) $\qquad\qquad\qquad H^i(\Gamma; F) = 0, \quad \text{for } q \neq 0, r.$

Identify $GL(F)$ with $\mathbf{GL}_n(k)$ by choosing a basis (e_i) of F over k. Since Γ is finitely generated, there exists a subfield k_0 of k which is finitely generated over $\mathbf{Q}$ and contains the coefficients of the matrices $\tau(\gamma)$ $(\gamma \in \Gamma)$. Let F_0 be the vector space over k_0 spanned by the e_i's. Since k_0 is finitely generated over $\mathbf{Q}$, it can be embedded in $\mathbf{C}$. We have then

(2) $H^*(\Gamma; F) = H^*(\Gamma; F_0) \otimes_{k_0} k, \qquad H^*(\Gamma; F_0 \otimes_{k_0} \mathbf{C}) = H^*(\Gamma; F_0) \otimes_{k_0} \mathbf{C}.$

This reduces us to the case where $k = \mathbf{C}$. Moreover, using 3.2, we may assume that $G = \widetilde{G}$. For $s \in S$ let p_s be the characteristic of the residue field of k_s. Then k_s is a finite separable extension of $\mathbf{Q}_{p_s}$,

(3) $$G_s = \mathcal{G}'_s(\mathbf{Q}_{p_s}), \quad \text{where } \mathcal{G}'_s = R_{k_s/\mathbf{Q}_{p_s}} \mathcal{G}_s,$$

and $\mathcal{G}_s$ is almost k_s-simple if and only if $\mathcal{G}'_s$ is almost $\mathbf{Q}_{p_s}$-simple [18, 6.21]. Therefore we may assume that $k_s = \mathbf{Q}_{p_s}$ and that $\mathcal{G}_s$ is almost k_s-simple $(s \in S)$. Let T be the set of $s \in S$ for which G_s is not compact, and $N = \Gamma \cap \ker \mathrm{pr}_T$. Then N is finite, and hence (IX, 1.11)

$$H^*(\Gamma; F) = H^*(\Gamma_T; F^N).$$

Therefore it suffices to consider the case where $T = S$, and we may assume the G_s to be noncompact.

We first consider the case where F is irreducible under Γ. Let $\mathcal{H}$ be the Zariski closure of $\tau(\Gamma)$ and $\mathcal{H}^0$ the identity component of $\mathcal{H}$. We claim that $\mathcal{H}^0$ is semi-simple. If not, it is a reductive group, which admits a non-trivial rational homomorphism α onto $\mathbf{C}^*$. Let $\Gamma' = \Gamma \cap \tau^{-1}(\mathcal{H}^0 \cap \tau(\Gamma))$. The group $\tau(\Gamma')$ is Zariski dense in H^0; hence $\alpha(\tau(\Gamma'))$ is Zariski dense in $\mathbf{C}^*$. On the other hand, by 3.6, $H^1(\Gamma'; \mathbf{C}) = 0$. Hence the commutator subgroup of Γ' has finite index in Γ', and $\alpha(\tau(\Gamma'))$ is finite, a contradiction.

Let $\mathcal{M}$ be a simple factor of the adjoint group $\mathrm{Ad}\,\mathcal{H}^0$ of $\mathcal{H}^0$. Let $\mu: \Gamma' \to \mathrm{Aut}(\mathcal{M})$ be the composition of τ, of the isogeny $\mathcal{H}^0 \to \mathrm{Ad}\,\mathcal{H}^0$, and of the projection of $\mathrm{Ad}\,\mathcal{H}^0$ onto $\mathcal{M}$. The group $\mu(\Gamma')$ is Zariski dense in $\mathcal{M}$. Since there is no continuous homomorphism of k_s into $\mathbf{C}$ for $s \in S$, a fundamental theorem of Margulis ([79]; see also [101, Thm. 2]) implies that $\mu(\Gamma')$ is relatively compact. This being true for every simple factor of $\mathrm{Ad}\,\mathcal{H}^0$, we see that $\tau(\Gamma')$ is relatively compact, hence so is $\tau(\Gamma)$. But then there exists a positive non-degenerate invariant Hermitian form on F, and we are reduced to 3.6.

This proves 3.7 when F is irreducible. The general case follows by induction on the length of a Jordan-Hölder series for F and use of the long exact sequence in cohomology.

3.8. S-arithmetic subgroups of anisotropic groups. Let k be a global field. We adopt the notation of XII, 3.4. Let $\mathcal{G}$ be a connected k-group and S a finite set of places of k, which contains the Archimedean ones if k is a number field. Let $\mathfrak{o}$ be the ring of integers of k and $\mathfrak{o}_S$ the ring of elements of k integral outside S. Identify $\mathcal{G}$ to a matrix group. A subgroup Γ of $\mathcal{G}(k)$ is S-arithmetic if it is commensurable with the group $\mathcal{G}(\mathfrak{o}_S)$ of elements in $\mathcal{G}(k)$ whose coefficients are in $\mathfrak{o}_S$ and whose determinant is invertible in $\mathfrak{o}_S$ [98, 2.4]. The group Γ, embedded diagonally in $G_S = \prod_{s \in S} G_s$, is a discrete subgroup. If $\mathcal{G}$ is anisotropic over k, i.e. if $\mathrm{rk}_k(\mathcal{G}) = 0$, then Γ is cocompact in G_S. (See [7] for number fields, [48] for function fields.) If $\mathcal{G}$ is semi-simple and simply connected, the groups Γ and $G = G_S$ satisfy the condition of 3.5, and so 3.5 and 3.6 hold. If k is a number field and G_∞ is compact, then, for any S, the projection of an S-arithmetic group Γ in G_{S_f} satisfies the conditions imposed on Γ in 3.7 (with $G = G_{S_f}$); hence the conclusion of 3.7 holds for Γ. In fact, the results of Margulis [**79**] show that this is the most general situation covered by 3.7.

3.9. PROPOSITION. *Let k be a global field, and G a connected semi-simple k-group of k-rank zero.*

(i) *If k is of characteristic zero, then $H^*(\mathcal{G}(k); \mathbf{C})$ is canonically isomorphic to $H_{\mathrm{ct}}^*(\widetilde{G}_\infty; \mathbf{C})^{G_\infty}$.*

(ii) *If k has non-zero characteristic, then $H^q(\mathcal{G}(k); \mathbf{C}) - 0$ for $q \neq 0$.*

In this statement and below, $\mathbf{C}$ is viewed as a trivial module.

For $S \subset \Sigma_f$, let $r(S)$ be as in XII, 3.5. As remarked there, $r(S)$ tends to infinity if Card S does. We let S run through an increasing sequence of finite subsets of Σ, whose union is Σ, and all containing Σ_∞ if k has characteristic zero. Identify $\mathcal{G}$ to a matrix group over k. Then

(1)
$$\mathcal{G}(k) = \lim_{\to S} \mathcal{G}(\mathfrak{o}_S);$$

hence

(2)
$$H_*(\mathcal{G}(k); \mathbf{C}) = \lim_S H_*(\mathcal{G}(\mathfrak{o}_S); \mathbf{C}).$$

Note that, for any discrete group M and $q \in \mathbf{Z}$, the dual space to $H_q(M; \mathbf{C})$ is $H^q(M; \mathbf{C})$. The assertions (ii) and (i) for $\mathcal{G}$ simply connected then follow from these remarks and 3.6. If $\mathcal{G}$ is not simply connected, let

$$\widetilde{\Gamma}_S = \sigma^{-1}(\mathcal{G}(\mathfrak{o}_S)), \qquad L_S = \mathcal{G}(\mathfrak{o}_S)/\sigma(\widetilde{\Gamma}_S) = \mathcal{G}(\mathfrak{o}_S)/(\mathcal{G}(\mathfrak{o}_S) \cap \sigma(\widetilde{G}_S)).$$

Fix $q \in \mathbf{Z}$. Then 3.2 and 3.5 show that, for S big enough,

(3)
$$H^q(\mathcal{G}(\mathfrak{o}_S); \mathbf{C}) = H_{\mathrm{ct}}^q(G_\infty; \mathbf{C})^{L_S},$$

where L_S is viewed as the quotient of the projection of $\mathcal{G}(\mathfrak{o}_S)$ in G_∞ by the projection of $\sigma(\widetilde{\Gamma}_S)$ in G_∞. But $\mathcal{G}$ satisfies the weak approximation at infinity, i.e., $\mathcal{G}(k)$ meets every connected component (ordinary topology) of G_∞ and $\widetilde{G}_\infty$ is connected. For S big enough, L_S is then equal to G_∞/G_∞^0, and our assertion follows.

REMARK. 3.9(i) for $\mathcal{G}$ simply connected was proved jointly by H. Garland and one of us, and stated in [**41**]; it was also known to W. Casselman. A proof was already given in [**12**].

4. The Γ-module E is the restriction of a rational G-module

In the previous section we considered an S-arithmetic extension of the case considered in VII, §4. Now we want to discuss the parallel generalization of VII, §6, in the context of 3.8, where E is a complex finite dimensional G-module and, hence, k is a number field. This is the more important case for applications. The results are similar, of course, but cannot be deduced formally from those of §3, except when $E = \mathbf{C}$ is the trivial representation.

We shall need the following lemma.

4.1. LEMMA. *Let F be a local field, $\mathcal{L}$ a reductive group defined over F, $L = \mathcal{L}(F)$ and L' an open subgroup of finite index of L. If (π, H) is an irreducible unitary representation of L, then $\pi' = \pi|_{L'}$ is the direct sum of finitely many irreducible representations of L'.*

Let K be a maximal compact subgroup of L. Then $K' = K \cap L'$ is of finite index in K. Frobenius reciprocity implies that if $\tau \in \widehat{K}'$, then there are only a finite number of elements $\gamma \in \widehat{K}$ such that τ is a subrepresentation of γ. Since (π, H) is admissible as a K-representation, it follows that it is admissible as a K'-representation. Hence π' splits into a direct sum of irreducible representations of L' with finite multiplicities. Since L/L' is finite, this sum is clearly finite.

4.2. We keep the general assumptions and notation of §§2 and 3, except that now E is a rational representation of G_∞. If it is irreducible, then $E = \bigotimes_{s \in S_\infty} E_s$, where E_s is an irreducible rational representation of $\mathcal{G}$. We have

$$(1) \qquad\qquad I(E) = I(\mathbf{C}) \otimes \mathbf{E}$$

and, as before, we can write $I(\mathbf{C})$ as a Hilbert direct sum of unitary irreducible G-modules (π, H_π), each with some multiplicity $m(\pi, \Gamma)$.

To fix the notation we recall that

$$(2) \qquad\qquad (\pi, H) = \bigotimes_{s \in S} (\pi_s, H_{\pi_s}),$$

where (π_s, H_{π_s}) is an irreducible representation of G_s.

In the sequel, we assume Γ to be irreducible and $\mathcal{G}$ to be absolutely almost simple over k_s ($s \in S$).

4.3. LEMMA. *We keep the assumptions of 4.2. Let (π, H_π) be an irreducible representation of G occuring in $I(\mathbf{C})$ which has a non-compact kernel. If either $\mathcal{G} = \widetilde{\mathcal{G}}$ or $H^*_{\mathrm{ct}}(G; H^\infty_\pi \otimes E) \neq 0$, then π_s is the trivial representation, except possibly when s is Archimedean and G_s is compact.*

We may assume E to be irreducible. We first see from 2.1(2), 4.1(1), (2), and the Künneth rules in I, 1.3, XII, 3.2, and X, 6.2, that

$$(1) \qquad H^*_{\mathrm{ct}}(G, H^\infty \otimes E) = \bigotimes_{s \in S_\infty} H^*_{\mathrm{ct}}(G_s; H^\infty_s \otimes E_s) \otimes \bigotimes_{s \in S_f} H^*_{\mathrm{ct}}(G_s; H^\infty_s).$$

By 2.5(1),

$$(2) \qquad\qquad H^*_{\mathrm{ct}}(G_t; H^\infty_t) = 0$$

if $t \in S_f$ and G_t is compact; and

$$(3) \qquad \begin{aligned} H^i(G_t; H^\infty_t \otimes E_t) &= 0 \qquad (i \geq 1), \\ H^0_{\mathrm{ct}}(G_t; H^\infty_t \otimes E_t) &= (H^\infty_t \otimes E_t)^{G_t} \end{aligned}$$

if $t \in S_\infty$ and G_t is compact.

Leaving aside the compact factors, we are reduced to the case where all G_s are non-compact.

Assume first that $\mathcal{G} = \widetilde{\mathcal{G}}$. By [100], each G_s is simple modulo its center; hence there exists t such that $G_t \subset \ker \pi$. By assumption, H^∞ is realized as a space of functions on G right-invariant under G_t and left-invariant under Γ. Since G_t is normal, they are also left-invariant under G_t. Since $\mathcal{G}$ is assumed to be simply connected, strong approximation is valid and implies that $G_t \cdot \Gamma$ is dense in G. Therefore the elements of H^∞ are left-invariant under G, and hence are constant functions.

Let us now drop the assumption $\mathcal{G} = \widetilde{\mathcal{G}}$. We therefore assume now that the cohomology spaces in (1) are non-zero. Let σ be the product of the isogenies σ_s

(see 3.1) and let $G' = \sigma(G)$. By 3.19 in [20], G' is an open normal subgroup and G/G' is finite and commutative. Clearly

(4) $$G' \cap \ker \pi \neq \{1\}.$$

By 4.1, the restriction π' of π to G' is fully reducible, and a finite sum of irreducible admissible G'-modules. If (π_1, H_1) is one of them, then, by irreducibility, H is spanned by finitely many transforms of it. As a consequence H is a direct sum of finitely many irreducible G'-modules with isomorphic kernels. In view of our assumption on $\ker \pi$, these kernels are all non-compact. As before, we see from [100] that $G'_s \subset \ker \pi_s$ for all s; hence $G' \subset \ker \pi$. Thus, H_{π_s} may be viewed as an irreducible representation of the finite commutative group G_s/G'_s; hence it is one-dimensional. If s is non-Archimedean, this forces H_{π_s} to be trivial (XI, 3.9).

Let s be Archimedean. We are dealing with relative Lie algebra cohomology with coefficients in a finite dimensional representation. Since it is not zero by assumption, $H_{\pi_s} \otimes E_s$ must contain the trivial representation, i.e. E_s is contragredient to H_s, and therefore contains G'_s in its kernel. But G'_s is Zariski dense in $\mathcal{G}$ and E_s is a rational representation. Consequently E_s is the trivial representation, and so is H_{π_s}.

4.4. THEOREM. *Under the assumptions of* 4.2,

$$H^*(\Gamma; E) = H^*_{\mathrm{ct}}(G_\infty; E^\Gamma) \oplus \bigoplus_\pi{}' m(\pi, \Gamma) \cdot H^*_{\mathrm{ct}}(G_\infty; H_{\pi_\infty} \oplus E)[-r_f].$$

This follows from 4.3 and 4.3(3) in exactly the same way as 3.5 was deduced from 3.4.

4.5. From the above, we see, as in §3, that 3.6 holds in the present situation.

4.6. Here and in VII, we have mostly limited ourselves to two cases for the coefficient Γ-module (ρ, E): it is either unitary or the restriction of a rational representation $G \to \mathrm{GL}(E)$. However, if $\mathrm{rk}_s(G) \geq 2$, which we assume here, it is not that far from the general case of an arbitrary finite dimensional complex representation of Γ. In fact, given one, the Zariski closure H of $\rho(\Gamma)$ is always semi-simple ([79], VII, 3.10, p. 278). Then, if either $\mathcal{G}$ is simply connected, or H is of adjoint type, and $\rho(\Gamma)$ is not relatively compact, then ρ extends to a rational homomorphism $G \to H$, hence to a representation $G \to \mathrm{GL}(E)$ by VIII, 5.13(c), p. 233 in [79]. If $\rho(\Gamma)$ is relatively compact, then (ρ, E) is a unitary Γ-module, so we are back to the two cases already considered. [Moreover, as pointed out earlier (VII, 2.9), the latter one could also be subsumed to that of a rational G-module, by adding a compact factor to G.] On the other hand, this is not quite the general situation, and some condition such as $\mathcal{G}$ simply connected has to be added. To see this, consider the case where G has a non-trivial (finite) central subgroup N, and let $q: G \to G' = G/N$ be the natural projection. Assume Γ to be torsion free. Then the restriction q_Γ of q to Γ is an isomorphism of Γ onto a subgroup Γ' of G', which is of the same type in G' as Γ is in G. Then q_Γ^{-1}, composed with a finite dimensional rational representation of G, defines a Γ-module which cannot be extended to G', since, Γ being Zariski-dense in G, such an extension would yield a rational morphism of G' onto G, which is absurd.

CHAPTER XIV

Non-cocompact S-arithmetic Subgroups

In VII and XIII we have limited ourselves mostly to discrete cocompact subgroups. However, the most important case for applications is that of non-cocompact S-arithmetic subgroups, in particular arithmetic groups. In this chapter, we shall indicate how some of the results established for the cocompact case extend to this context. In particular, the exposition describes the chain of ideas that leads to the removal of the rank condition in XIII: 4.4.

1. General properties

1.1. We let k, S, $\mathcal{G}$, $\widetilde{\mathcal{G}}$, G_S be as in XIII, 3.1, but assume moreover that k is a *number field* and (for convenience) that $\mathcal{G}$ is almost absolutely simple over k. We let $\Gamma \subset G_S$ be an S-arithmetic subgroup (XIII, 3.8). There we assumed that $\mathrm{rk}_k(\mathcal{G}) = 0$, which was equivalent to Γ being cocompact in G_S. In this appendix, we assume that $\mathrm{rk}_k(\mathcal{G}) \geq 1$ (see 0, 3.0). Then Γ is not cocompact, but of finite covolume. The assumption $\mathrm{rk}_k \mathcal{G} \geq 1$ is equivalent to $\mathcal{G}$ containing proper parabolic subgroups defined over k. It also implies that G_s is not compact for all $s \in S$.

In view of the results of G. A. Margulis [**142**], if $r(S) \geq 2$, then, up to commensurability, any discrete irreducible subgroup of finite covolume of G_S is S-arithmetic. This reference implies that in the present situation, an irreducible subgroup of finite covolume can be non-arithmetic only if $k = \mathbf{Q}$, $S = S_\infty$, and $\mathrm{rk}_\mathbf{R}(G_\infty) = 1$.

1.2. For the rest of this chapter, (ρ, E) denotes a finite dimensional rational representation of G_∞, and we shall discuss $H^*(\Gamma; E)$. If $S = S_\infty$, then it follows from VII, 2.5 and IX, 5.6 that

$$(1) \qquad H^*(\Gamma; E) = H_d^*(G_S; C^\infty(\Gamma \backslash G) \otimes E).$$

This formula remains valid in the general case under consideration if we define $C^\infty(\Gamma \backslash G_S)$ as in XII, §1, i.e. as the space of functions on $\Gamma \backslash G_S$ which are continuous, right-invariant under some compact open subgroup of G_{S_f} (depending on the function), and smooth with respect to G_∞. In the notation of XII, 1.2 this is V^∞, where $V = C(\Gamma \backslash G_S)$ is the space of complex valued continuous functions on $\Gamma \backslash G_S$.

The following result can be found in [**129**].

THEOREM. (1) Γ *is finitely presented, and its finite subgroups form a finite number of conjugacy classes.*

(2) $H^*(\Gamma; E)$ *is finite dimensional.*

2. Stable cohomology

In the non-cocompact case, $C^\infty(\Gamma \backslash G_S)$ is too unwieldy. The strategy has been to replace it by some G-invariant subspace C^* that might be more manageable, and to study the maps μ^* in cohomology induced by the inclusion. In some cases, it can be proved that μ^* is an isomorphism. In other cases the two cohomology spaces are obviously different, but the image of μ^* may provide some useful information (for example, equality in certain degrees).

2.1.   We first assume that $S = S_\infty$ (i.e. the arithmetic case). We also write G for $G_\infty = G_S$. Let K be a maximal compact subgroup of G and $X = G/K$. We also assume, for the sake of convenience, that Γ is torsion-free. The passage to the general case then follows by arguments as in VII, 2.2. We have three interpretations of $H^*(\Gamma; E)$, namely

$$H^*(A^*(X; E)^\Gamma) = H^*(\mathfrak{g}, K; C^\infty(\Gamma\backslash G) \otimes E) = H^*_d(G; C^\infty(\Gamma\backslash G) \otimes E)$$

(see VII, 2.2, 2.7). We shall pass freely from one to another.

2.2.   The purpose of this subsection is to describe an extension of VII, 4.3, 4.4, 6.3 using the methods of [**122**]. We will also give an alternate discussion with an exposition of L^2-cohomology in §3.

The inclusion $E \to C^\infty(\Gamma\backslash G) \otimes E$, which assigns to $x \in E$ the element $1 \otimes x$, defines (as in VII, 2.8) a map

$$j^* \colon H^*_d(G; E) \to H^*_d(G; C^\infty(\Gamma\backslash G) \otimes E) = H^*(\Gamma; E).$$

If Γ is cocompact, it is an isomorphism up to some degree $m(G; E)$, which is at least equal to $\mathrm{rk}_{\mathbf{R}}(G) - 1$ (VII, 4.3, 6.4). This remains true, but with possibly a smaller range. In the cocompact case, j^* was easily seen to be injective in all dimensions, either because a harmonic form on a compact manifold is not cohomologous to zero or because $\mathbf{C} \otimes E$ is a direct summand in $C^\infty(\Gamma\backslash G) \otimes E$, viewed as a G-module. The main point of the argument was then to show surjectivity up to some degree $m(G, E)$. The proof of this second point extends, with the same bound, by a rather simple trick pointed out by R. Langlands to H. Garland (see [**121**], 3.6). On the other hand, $\mathbf{C} \otimes E$ is not a direct G-summand anymore and, indeed, injectivity is not true in all dimensions. As the simplest example, take $k = \mathbf{Q}$, $G = \mathbf{SL}_2(\mathbf{R})$, and Γ a subgroup of finite index of $\mathbf{SL}_2(\mathbf{Z})$. Then $\Gamma\backslash X$ is a non-compact connected 2-manifold; hence $H^2(\Gamma\backslash X; \mathbf{Q}) = 0$. On the other hand, $H^2_d(G; \mathbf{R})$ is one-dimensional, represented by an invariant differential form defining an invariant volume on $\Gamma\backslash X$.

However (this is where really new ideas are necessary), it can be shown that injectivity holds at least up to some degree $c(G; E)$, which can be estimated in terms of roots and weights [**121, 122**]. If $E = \mathbf{C}$ is the trivial module, then $c(G; E)$ is at least equal to $\mathrm{rk}_k(\mathcal{G})/2$ and in many cases is $\geq \mathrm{rk}_k(\mathcal{G}) - 1$. In particular, it tends to infinity with $\mathrm{rk}_k(\mathcal{G})$, and so does $m(G)$. This result has been applied to many sequences of classical arithmetic groups. As an example, define $\mathbf{SL}(\mathbf{Z})$ to be the inductive limit of the groups $\mathbf{SL}_n(\mathbf{Z})$, where $\mathbf{SL}_n(\mathbf{Z})$ is embedded as the first $n \times n$ diagonal block in $\mathbf{SL}_m(\mathbf{Z})$ $(m \geq n)$. Then $H^*(\mathbf{SL}(\mathbf{Z}), \mathbf{Q})$ is an exterior algebra over generators x_i $(i = 1, 2, \ldots; d^0 x_i = 4i + 1)$. If E is an irreducible non-trivial G-module, then $H^*(\mathbf{SL}(\mathbf{Z}); E) = 0$. See [**121**] for further examples and applications to algebraic K-theory, and [**122**] for a sharpening of the bound on $c(G, E)$.

2.3.   The proof of injectivity involves establishing the existence of a G-invariant subcomplex $C^*_{\log}$ of $A^*(\Gamma\backslash X)$ such that (i) the inclusion $C^*_{\log} \subset A^\infty(\Gamma\backslash X)$ induces an isomorphism in cohomology, (ii) it contains the G-invariant forms, and (iii) it consists of square integrable forms for $i \leq c(G)$. This space is defined in terms of growth conditions at the corners of the compactification introduced in [**16**]. Then injectivity follows from the fact that on a complete Riemannian manifold, a L^2-harmonic form is not the coboundary of a square integrable form. The proof is exactly the same as the proof in the case of compact manifolds, once one has a version of the Stokes theorem ([**121**], 2.5) that is valid in $C^*_{\log}$.

2.4. REMARK. As indicated above, the complex $C^*_{\log}$ was originally defined in order to show that in the stable range the cohomology is given by G-invariant forms. It has also been used to prove that certain square integrable harmonic forms whose degrees are at the rank or higher also have non-zero image in the ordinary cohomology [**150**]. In particular, in the notation of VIII, 5.1 we may drop the assumption that the form $^\sigma h$ is definite for $\sigma \neq 1$, and derive a theorem completely analogous to VIII, 5.10. The full result would take us too far afield. A representative special case of [**150**], Theorem 8.3 is the following.

THEOREM. *Let $1 \leq q \leq p$. Then there exist congruence subgroups Γ_j ($j = 1, 2, \ldots$) of $\mathbf{SU}(p,q)[\mathbf{Z}[i]]$ such that*

$$\lim_{j \to \infty} \dim H^q(\Gamma_j, \mathbf{C}) = \infty.$$

3. The use of L^2 cohomology

The purpose of this section is to give an alternate discussion of the results described in 2.3, making consistent use of L^2 cohomology. We include this material here since the techniques have applications to other contexts.

3.1. Let M be a smooth Riemannian manifold. The Riemannian metric defines an invariant volume dv and a metric $(\ ,\ )$ on the exterior power $\Lambda^* T_x^*(M)$ of the cotangent space at each point. The square norm (ω, ω) of a smooth i-form ω on M is then $(\omega, \omega) = \int_M (\omega_x, \omega_x)\, dv_x$.

Let $A_2^i(M)$ be the space of i-forms with finite norm and $A_{(2)}^i(M)$ the subspace of i-forms ω such that $\omega \in A_2^i(M)$ and $d\omega \in A_2^{i+1}(M)$. The direct sum $A_{(2)}^*(M)$ of the $A_{(2)}^i(M)$ is a subcomplex of $A^*(M)$ stable under exterior differential, and, by (one) definition, the L^2-cohomology $H_{(2)}^*(M; \mathbf{C})$ of M is $H^*(A_{(2)}^*(M), d)$.

There is also an L^2-definition of this cohomology. Let $\overline{A}_{(2)}^i(M)$ be the Hilbert space completion of $A_{(2)}^i(M)$ with respect to the square norm $(\omega, \omega) + (d\omega, d\omega)$, and let $\overline{A}_{(2)}^*(M)$ be the direct sum of the $\overline{A}_{(2)}^i(M)$. The differential d extends to a bounded operator $\overline{d}$ on $A_{(2)}^*(M)$, increasing the degree by one, of square zero. It may be shown that the inclusion $A_{(2)}^*(M) \to \overline{A}_{(2)}^*(M)$ induces an isomorphism in cohomology (cf. [**132**] for a sketch; a more detailed version may be found in [**153**]). It can also be shown that if $H_{(2)}^i(M)$ is finite dimensional for some value of i, then it is spanned by L^2-harmonic forms.

These definitions can be extended to the case of forms with values in an Hermitian bundle with a flat connection, but we will give a direct definition in the case of interest to us.

3.2. We now come back to our situation. Let $L^2(\Gamma \backslash G)$ be the space of square integrable functions on $\Gamma \backslash G$. It is a unitary G-module under right translations. The space $L^2(\Gamma \backslash G)^\infty$ of smooth vectors is then the space of functions on $\Gamma \backslash G$ which, together with all derivatives by right-invariant differential operators, are square integrable. It is a $(\mathfrak{g}, K)$-module, and it has been shown that

$$H_{(2)}^*(\Gamma\backslash X; E) = H^*(\mathfrak{g}, K; L^2(\Gamma\backslash G)^\infty \otimes E) = H_d^*(G; L^2(\Gamma\backslash G)^\infty \otimes E).$$

(see [**123**]). We also write it as $H_{(2)}(\Gamma; E)$.

Note that on the left-hand side one starts with differential forms which, together with their exterior differential, have L^2 coefficients. On the right-hand side, we deal with an a priori much smaller complex, consisting of differential forms with coefficients which are L^2 as well as all their derivatives.

The L^2 cohomology is finite dimensional if $\mathrm{rk}_{\mathbf{R}}(G) = \mathrm{rk}(K)$ (in fact, under a somewhat more general condition), but may be infinite dimensional otherwise [**126**].

3.3. Let $L^2(\Gamma\backslash G)_d$ be the *discrete spectrum*, i.e., by definition, the closed subspace of $L^2(\Gamma\backslash G)$ spanned by the closed G-invariant irreducible subspaces of $L^2(\Gamma\backslash G)$, and let $L^2(\Gamma\backslash G)_{\mathrm{ct}}$ be its orthogonal complement, the so-called *continuous spectrum*. Then, clearly,

$$(1) \qquad H_{(2)}^*(\Gamma; E) = H_d^*(G; L^2(\Gamma\backslash G)_d^\infty \otimes E) \oplus H_d^*(G; L^2(\Gamma\backslash G)_{\mathrm{ct}}^\infty \otimes E).$$

It is for the first summand on the right hand side that there is an analogue of VII, 2.6.

From the theory of automorphic forms, it follows that $L^2(\Gamma\backslash G)_d$ is a Hilbert sum of irreducible G-modules with *finite* multiplicities. Therefore, $L^2(\Gamma\backslash G)_d$ contains only finitely many irreducible constituents H_i $(i \in I)$ such that $H_d^*(G; H_i^\infty \otimes E) \neq 0$, namely, those constituents with infinitesimal character equal to that of E^*. Then ([**127**], 5.6)

$$(2) \qquad H_d^*(G; L^2(\Gamma\backslash G)_d^\infty \otimes E) = \bigoplus_{i\in I} H_d^*(G; H_i^\infty \otimes E).$$

Assume now that G is connected, which is in particular the case if $\mathcal{G} = \widetilde{\mathcal{G}}$. Then VII, 4.2 holds. As a consequence, Theorem V.3.3 implies that $H_d^i(G; H_i^\infty \otimes E) = 0$ for $j < \mathrm{rk}_{\mathbf{R}}(G)$ if H_i is not the trivial representation, whence

$$(3) \qquad H_d^j(G; L^2(\Gamma\backslash G)_d \otimes E) = H_d^j(G; E) \quad \text{if } j < \mathrm{rk}_{\mathbf{R}}(G).$$

The range of this equality could be improved in some cases by using the tables in II, 10.3.

If we now interpret the left-hand side of (3) as a summand of $H_{(2)}^*(\Gamma\backslash X; \widetilde{E})$, where $\widetilde{E}$ is the local system defined by E, then it can be identified with the space of $\widetilde{E}$-valued L^2-harmonic forms ([**127**], 5.6).

3.4. With these preliminaries in hand, we now return to the context of §2. In [**154**], S. Zucker showed that $H_{(2)}^i(\Gamma; E) \to H^i(\Gamma; E)$ is an isomorphism for $i \le z(G; E)$, where $z(G; E)$ is a constant which can be estimated by means of roots and weights. Then, for $i \le z(G; E)$, the space $H_{(2)}^i(\Gamma; E)$ is finite dimensional, hence consists of L^2-harmonic forms. If $i < \mathrm{rk}_{\mathbf{R}}(G)$, it reduces to $H_d^i(G; E)$, and we get back the result of §2.

By definition, the image of

$$\nu^* \colon H_{(2)}^*(\Gamma; E) \to H^*(\Gamma; E)$$

consists of those classes which (in the identification with $H^*(A^*(\Gamma\backslash X, E))$) are represented by a square integrable cocycle. By a theorem of Kodaira (recalled in [121, 2.4]) each such cocycle is cohomologous to a harmonic one. Therefore the image of ν^* is the same as that of the summand $H_d^*(G; L^2(\Gamma\backslash G)_d \otimes E)$. So, as far as the cohomology of Γ is concerned, only the discrete spectrum in $L^2(\Gamma\backslash G)$ matters. The continuous spectrum has been determined by R. Langlands (see [126] for references). The cohomology with respect to it is either zero or infinite dimensional (loc. cit.).

4. *S*-arithmetic subgroups

4.1. We now pass to the (genuinely) *S*-arithmetic case, i.e., we assume that $S_f \neq \varnothing$. To avoid certain minor complications, we also assume that $\mathcal{G} = \widetilde{\mathcal{G}}$ is simply connected. We continue to use the notation $L^2(\Gamma\backslash G_S)_d$ for the discrete spectrum, i.e. the closure of the subspace of $L^2(\Gamma\backslash G_S)$ spanned by the irreducible G_S-submodules, and $L^2(\Gamma\backslash G_S)_{\mathrm{ct}}$ for its orthogonal complement. Then $H_d^*(G_S; L^2(\Gamma\backslash G_S)_d^\infty \otimes E)$ is given by a formula essentially identical to the right-hand side of XIII, 3.5(1).

To state it, we fix some notation. Let (π, H_π) be an irreducible representation of G_S. It can be written as $(\pi_\infty, H_\infty) \otimes (\pi_f, H_f)$. Assume it occurs in $L^2(\Gamma\backslash G_S)_d$. Then it has finite multiplicity (again as a consequence of the theory of automorphic forms).

Assume that π is not trivial. Then π_∞ and π_f are both infinite dimensional (XIII, 4.2). We have

$$H_d^*(G_S; H_\pi) = H_d^*(G_\infty; H_{\pi_\infty}^\infty \otimes E) \otimes H_d^*(G_{S_f}; H_{\pi_f}^\infty).$$

The second factor is not zero only if π_f is the tensor product of the Steinberg representations of the groups G_s ($s \in S_f$), and then its cohomology is of dimension one, concentrated in dimension r_f. On the other hand, $H_d^*(G_\infty; H_{\pi_\infty}^\infty \otimes E)$ can be non-zero only if π_∞ has the infinitesimal character of E^*. Then let $\widehat{G}(E, St)$ be the set of equivalence classes of irreducible unitary representations π of G_S such that π_∞ has the infinitesimal character of E^* and π_f is as above. Let $\pi \in \widehat{G}(E, St)$. Its multiplicity m_π is finite. The corresponding isotypic subspace can be written as $I_{\pi_\infty} \otimes H_{\pi_f}$, where I_{π_∞} is the direct sum of m_π copies of H_{π_∞}. Then we have

$$
\begin{aligned}
&H_d^*(G_S; L^2(\Gamma\backslash G_d)_d^\infty \otimes E) \\
(1) \qquad &= H_d^*(G_\infty; E) \oplus \bigoplus_{\pi \in \widehat{G}(E, St)} H_d^*(G_\infty; I_{\pi_\infty}^\infty \otimes E)[-r_f]
\end{aligned}
$$

(see [128], 6.5(11)).

Since $H_d^*(G_S; L^2(\Gamma\backslash G_S)_{\mathrm{ct}} \otimes E) = 0$ ([128], §7), the right-hand side of (1) represents the full L^2 cohomology of Γ.

4.2. By a theorem of [120] (which uses [136] in its proof), *the L^2 cohomology is the full cohomology*. In complete analogy with XIII, 3.5, we then have

$$
(2) \qquad H^*(\Gamma; E) = H_d^*(G_\infty; E) \oplus \bigoplus_{I \in \widehat{G}_{\mathrm{dis}}(E, St)} H_2^*(G_\infty; I_{\pi_2} \otimes E)[-r_f].
$$

Then, as was noted in [130], we may apply the argument of XIII, 3.9 to prove

4.3. THEOREM.

$$(3) \qquad\qquad H^*(\mathcal{G}(k); E) = H_d^*(G_\infty; E)$$

i.e. XIII, 3.9, but *without any restriction on* $\mathrm{rk}_k(\mathcal{G})$.

Bibliography

1. H. Bass, J. Milnor, and J.-P. Serre, *Solution of the congruence subgroup problem for* $\mathbf{SL}_n$, $\mathbf{Sp}_n$, Inst. Hautes Études Sci. Publ. Math. **33** (1967), 421–499.

2. I. N. Bernshtein, *All reductive p-adic groups are tame*, Funct. Anal. Appl. **8** (1975), 91–93.

3. P. Blanc, "Sur la cohomologie continue des groupes localement compacts", Thèse, Université Paris VII (1978); Ann. Sci. École Norm. Sup. (4) **12** (1979), 137–168.

4. H. Boerner, *Representations of groups*, North Holland, Amsterdam, 1967.

5. A. Borel, *Density properties of certain subgroups of semi-simple groups*, Ann. of Math. (2) **72** (1960), 179–188.

6. ———, *Compact Clifford-Klein forms of symmetric spaces*, Topology **2** (1963), 111–122.

7. ———, *Some finiteness properties of adele groups over number fields*, Inst. Hautes Études Sci. Publ. Math. **16** (1963), 1–30.

8. ———, *On the automorphisms of certain subgroups of semi-simple Lie groups*, Proc. Bombay Colloquium on Algebraic Geometry, 1968, Oxford Univ. Press, London, 1969, pp. 43–73.

9. ———, "Introduction aux groupes arithmétiques", Actualités Sci. Indust., no. 1341, Hermann, Paris (1969).

10. ———, "Représentations de groupes localement compacts", Lecture Notes in Mathematics, 276, Springer, 1972.

11. ———, *Cohomologie de certains groupes discrets et laplacien p-adique*, Sém. Bourbaki, 26e année (1973–74), Exp. 437, Lecture Notes in Mathematics, 431, Springer.

12. ———, *Cohomologie de sous-groupes discrets et représentations de groupes semi-simples*, Astérisque **32-33** (1976), 73–112.

13. ———, *Admissible representations of a semi-simple group over a local field with vectors fixed under an Iwahori subgroup*, Invent. Math. **35** (1976), 233–259.

14. A. Borel and Harish-Chandra, *Arithmetic subgroups of algebraic groups*, Ann. of Math. (2) **75** (1962), 485–535.

15. A. Borel and J.-P. Serre, *Théorèmes de finitude en cohomologie galoisienne*, Comment. Math. Helv. **39** (1964), 111–164.

16. ———, *Corners and arithmetic groups*, Comment. Math. Helv. **48** (1973), 436–491.

17. ———, *Cohomologie d'immeubles et de groupes S-arithmétiques*, Topology **15** (1976), 211–232.

18. A. Borel and J. Tits, *Groupes réductifs*, Inst. Hautes Études Sci. Publ. Math. **27** (1965), 55–150.

19. ———, *Compléments à l'article "Groupes réductifs"*, Inst. Hautes Études Sci. Publ. Math. **41** (1972), 253–276.

20. ———, *Homomorphismes "abstraits" de groupes algébriques simples*, Ann. of Math. (2) **97** (1973), 499–571.

21. N. Bourbaki, "Algèbre, 8, Modules et anneaux semi-simples", Actualités Sci. Indust., no. 1261, Hermann, Paris, 1958.

22. ———, "Topologie générale", Chap. 1 à 10, 2 vol., Hermann, Paris.

23. ———, "Espaces vectoriels topologiques", Chap. 1, 2, Actualités Sci. Indust., no. 1189, Hermann, Paris, 1969, 2è éd.

24. ———, "Espaces vectoriels topologiques", Chap. 3, 4, Actualités Sci. Indust., no. 1229, Hermann, Paris, 1964.

25. ———, "Groupes et Algèbres de Lie", Chap. I, Actualités Sci. Indust., no. 1258, Hermann, Paris, 1971.

26. ———, "Groupes et Algèbres de Lie", Chap. 2, 3, Actualités Sci. Indust., no. 1349, Hermann, Paris, 1972.

27. _____, "Groupes et Algèbres de Lie", Chap. 4, 5, 6, Actualités Sci. Indust., no. 1337, Hermann, Paris, 1968.

28. _____, "Groupes et Algèbres de Lie", Chap. 7, 8, Actualités Sci. Indust., no. 1364, Hermann, Paris, 1975.

29. F. Bruhat, "Lectures on Lie groups and representations of locally compact groups", Tata Institute of Fundamental Research, 1958 (Reissued 1968).

30. F. Bruhat and J. Tits, *Groupes réductifs sur un corps local*, I, Inst. Hautes Études Sci. Publ. Math. **41** (1972), 1–251.

31. H. Cartan and S. Eilenberg, "Homological algebra", Princeton University Press, 1956.

32. P. Cartier, *Representations of p-adic groups: A survey*, "Automorphic forms, representations and L-functions", Proc. Sympos. Pure Math. **33**, 1979, Amer. Math. Soc., part 1, 111–155.

33. W. Casselman, *On a p-adic vanishing theorem of Garland*, Bull. Amer. Math. Soc. **80** (1974), 1001–1004.

34. _____, "Introduction to the theory of admissible representations of p-adic reductive groups", unpublished manuscript.

35. W. Casselman and D. Wigner, *Continuous cohomology and a conjecture of Serre's*, Invent. Math. **25** (1974), 199–211.

36. P. Delorme, *1-cohomologie des représentations unitaires des groupes de Lie semi-simples and solvables. Produits tensoriels continus de représentations*, Bull. Soc. Math. France **105** (1977), 281–336.

37. J. Dixmier, "Les C^*-algèbres et leurs représentations", Gauthier-Villars, Paris, 1969.

38. T. Enright and N. R. Wallach, *The fundamental series of a real semi-simple Lie algebra*, Acta. Math. **140** (1978), 1–32.

39. D. Flath, *Decomposition of representations into tensor products*, "Automorphic forms, representations and L-functions", Proc. Sympos. Pure Math. **33**, 1979, Amer. Math. Soc., part 1, 179–183.

40. H. Garland, *p-adic curvature and the cohomology of discrete subgroups of p-adic groups*, Ann. of Math. (2) **97** (1973), 375–423.

41. _____, *On the cohomology of discrete subgroups of p-adic groups*, Proc. Internat. Congr. Math. Vancouver 1974, Vol. 1 (1975), 449–453.

42. I. M. Gelfand, M. I. Graev, and I. Piatetski-Shapiro, "Representation theory and automorphic functions", Saunders Math. Books, 1969.

43. R. Godement, "Théorie des faisceaux", Actualités Sci. Indust., no. 1252, Hermann, Paris, 1958.

44. W. Greub, S. Halperin, and R. Vanstone, "Connections, curvature and cohomology", Vol. III, Academic Press, New York, 1976.

45. A. Grothendieck, *Résumé des résultats essentiels dans la théorie des produits tensoriels topologiques et des espaces nucléaires*, Ann. Inst. Fourier Grenoble **4** (1952), 49–112.

46. _____, *Produits tensoriels topologiques et espaces nucléaires*, Mem. Amer. Math. Soc. **16** (1955), 331 pp.

47. _____, *Sur quelques points d'algèbre homologique*, Tohoku Math. J. **9** (1957), 119–221.

48. G. Harder, *Minkowskische Reduktionstheorie ueber Funktionenkörper*, Invent. Math. **7** (1969), 33–54.

49. Harish-Chandra, *Spherical functions on a semi-simple Lie group*, I, Amer. J. Math. **80** (1958), 241–310.

50. _____, *Some results on differential equations*, preprint, 1960; first published in his Collected papers, vol. III, Springer, 1984, pp. 7–48.

51. _____, *Discrete series for semi-simple Lie groups*, I, Acta. Math. **113** (1965), 241–318.

52. _____, *Discrete series for semi-simple Lie groups*, II, Acta. Math. **116** (1966), 1–111.

53. _____, *Harmonic analysis on reductive p-adic groups*, Proc. Sympos. Pure Math. **26**, 1973, Amer. Math. Soc., 167–192.

54. _____, *Harmonic analysis on real reductive groups*, I, *The theory of the constant term*, J. Funct. Anal. **19** (1975), 104–204.

55. _____, *Harmonic analysis on real reductive groups*, II, Invent. Math. **36** (1976), 1–55.

56. _____, *Harmonic analysis on real reductive groups*, III, Ann. of Math. (2) **104** (1976), 117–201.

57. H. Hecht and W. Schmid, *A proof of Blattner's conjecture*, Invent. Math. **31** (1976), 129–154.

58. S. Helgason, "Differential geometry and symmetric spaces", Academic Press, New York, 1962.

59. G. Hochschild, *Relative homological algebra*, Trans. Amer. Math. Soc. **82** (1956), 246–269.

60. G. Hochschild and G. D. Mostow, *Cohomology of Lie groups*, Illinois J. Math. **6** (1962), 367–401.

61. R. Hotta and R. Parthasarathy, *Multiplicity formulae for discrete series*, Invent. Math. **26** (1974), 133–178.

62. R. Hotta and N. R. Wallach, *On Matsushima's formula for the Betti numbers of a locally symmetric space*, Osaka J. Math. **12** (1975), 419–431.

63. R. Howe, *On the asymptotic behavior of matrix coefficients*, preprint.

64. R. Howe and C. Moore, *Asymptotic properties of unitary representations*, J. Funct. Anal. **32** (1979), 72–96.

65. J. I. Igusa, "Theta functions", Grundlehren der Mathematischen Wissenschaften 194, Springer-Verlag, 1973.

66. H. Jacquet, *Représentations des groupes linéaires p-adiques*, Theory of group representations and Fourier analysis (Proceedings of a conference at Montecatini, 1970), C.I.M.E., Edizioni Cremonese, Rome, 1971, 119–220.

67. H. Jacquet and R. P. Langlands, "Automorphic forms on $\mathbf{GL}(2)$", Lecture Notes in Mathematics, 114, Springer, 1970.

68. S. Kaneyuki and T. Nagano, *On certain quadratic forms related to symmetric Riemannian spaces*, Osaka Math. J. **14** (1962), 1–20.

69. D. A. Kazhdan, *Connection of the dual space of a group with the structure of its closed subgroups*, J. Funct. Anal. Appl. **1** (1967), 63–65.

70. _____, *Some applications of the Weil representation*, J. Analyse Math. **32** (1977), 235–248.

71. A. Knapp and E. Stein, *Intertwining operators for semisimple groups*, Ann. of Math. (2) **93** (1971), 489–578.

72. B. Kostant, *Lie algebra cohomology and the generalized Borel-Weil theorem*, Ann. of Math. (2) **74** (1961), 329–387.

73. _____, *On the existence and irreducibility of certain series of representations*, "Lie groups and their representations". Edited by I. M. Gelfand, John Wiley, New York, 1975, 231–330.

74. J.-L. Koszul, *Homologie et cohomologie des algèbres de Lie*, Bull. Soc. Math. France **78** (1950), 65–127.

75. H. Kraljević, *On representations of the group* $\mathbf{SU}(n,1)$, Trans. Amer. Math. Soc. **221** (1976), 433–448.

76. R. Langlands, *On the classification of irreducible representations of real algebraic groups*, famous preprint, Institute for Advanced Study, 1973; finally published in Representation theory and harmonic analysis on semisimple Lie groups, Edited by P. Sally and D. Vogan, Amer. Math. Soc., 1989, pp. 101–170.

77. J. Lepowsky, *Algebraic results on representations of semisimple Lie groups*, Trans. Amer. Math. Soc. **176** (1973), 1–44.

78. S. Mac Lane, "Homology", Grund. Math. Wiss., 114, Springer, 1963.

79. G. A. Margulis, *Discrete groups of motions of non-positively curved manifolds*, Proc. Internat. Congr. Math. Vancouver, 1974, Vol. 2 (1975), 21–34.

80. Y. Matsushima, *On Betti numbers of compact, locally symmetric Riemannian manifolds*, Osaka Math. J. **14** (1962), 1–20.

81. _____, *A formula for the Betti numbers of compact locally symmetric Riemannian manifolds*, J. Diff. Geom. **1** (1967), 99–109.

82. Y. Matsushima and S. Murakami, *On vector bundle valued harmonic forms and automorphic forms on symmetric spaces*, Ann. of Math. (2) **78** (1963), 365–416.

83. _____, *On certain cohomology groups attached to Hermitian symmetric spaces*, Osaka J. Math. **2** (1965), 1–35.

84. E. Michael, *Selected selection theorems*, Amer. Math. Monthly **63** (1956), 233–238.

85. D. Miličić, *Asymptotic behavior of matrix coefficients of the discrete series*, Duke Math. J. **44** (1977), 59–88.

86. J. J. Millson and M. S. Raghunathan, *Geometric construction of homology for arithmetic groups*, I, *Unit groups of an anisotropic quadratic form*, Geometry and analysis (papers dedicated to the memory of V. K. Patodi), Indian Acad. Sci., Bangalore, and Tata Inst.

Fund. Res., Bombay, 1980, pp. 103–123; also published in Proc. Indian Acad. Sci. Math. Sci. **90** (1981), 103–123.

87. G. D. Mostow, *Self-adjoint group*, Ann. of Math. (2) **62** (1955), 44–55.

88. M. S. Narisimhan and C. S. Seshadri, *Stable and unitary vector bundles on a compact Riemann surface*, Ann. of Math. (2) **82** (1965), 540–567.

89. R. Palais, *On the existence of slices for actions of non-compact Lie groups*, Ann. of Math. (2) **73** (1961), 295–323.

90. R. Parthasarathy, *Dirac operator and the discrete series*, Ann. of Math. (2) **96** (1972), 1–30.

91. G. Prasad, *Strong approximation for semi-simple groups over function fields*, Ann. of Math. (2) **105** (1977), 553–572.

92. M. S. Raghunathan, *On the first cohomology of discrete subgroups of semi-simple Lie groups*, Amer. J. Math. **87** (1965), 103–139.

93. _____, *Vanishing theorems for cohomology groups associated to discrete subgroups of semi-simple Lie groups*, Osaka J. Math. **3** (1966), 243–256.

94. M. Reed and B. Simon, "Methods of modern mathematical physics I, Functional analysis", Academic Press, New York, 1972.

95. G. Rousseau, "Immeubles des groupes réductifs sur les corps locaux", Thèse, Université de Paris XI, 1977=Publ. Math. Orsay, no. 221-77.68, Univ. Paris-XI, Orsay, 1977.

96. W. Schmid, *On the characters of the discrete series (the Hermitian symmetric case)*, Invent. Math. **30** (1975), 47–144.

97. _____, *Some properties of square-integrable representations of semi-simple Lie groups*, Ann. of Math. **102** (1975), 535–564.

98. J.-P. Serre, *Cohomologie des groupes discrets*, Prospects in Mathematics, Ann. of Math. Studies **70** (1971), 77–169.

99. A. Silberger, *The Langlands quotient theorem for p-adic groups*, Math. Ann. **236** (1978), 95–104.

100. J. Tits, *Algebraic and abstract simple groups*, Ann. of Math. (2) **80** (1964), 313–329.

101. _____, *Travaux de Margulis sur les sous-groupes discrets de groupes de Lie*, Sém. Bourbaki 28e année, 1975/76, Exp. 482, Lecture Notes in Mathematics 576, Springer, 1977.

102. _____, *Reductive groups over local fields*, "Automorphic forms, representations and L-functions", Proc. Sympos. Pure Math. 33, 1979, Amer. Math. Soc., part 1, 29–69.

103. P. Trombi, *On the growth of matrix entries of uniformly bounded representations*, preprint.

104. W. T. van Est, *On the algebraic cohomology concepts in Lie groups*, I, II, Nederl. Akad. Wetensch. Proc. Series A, **58** (1955), 225–233, 286–294.

105. _____, *A generalization of the Cartan-Leray spectral sequence*, I, II, *ibid.* **61** (1958), 399–413.

106. N. Wallach, *On regular singularities in several variables*, preprint.

107. _____, "Harmonic analysis on homogeneous spaces", Marcel Dekker, New York, 1973.

108. _____, "Symplectic geometry and Fourier analysis", Math. Sci. Press, Boston, 1977.

109. _____, *On the Enright-Varadarajan modules, a construction of the discrete series*, Ann. Sci. École Norm. Sup. (4) **9** (1976), 81–102.

110. _____, *Representations of semi-simple Lie groups*, Proc. Canad. Math. Congr., 1977, 154–245.

111. S. P. Wang, *The dual space of semi-simple Lie groups*, Amer. J. Math. **89** (1967), 124–132.

112. F. W. Warner, "Foundations of differentiable manifolds and Lie groups", Scott, Foreman and Co., Glenview, Illinois, 1971.

113. G. Warner, "Harmonic analysis on semi-simple Lie groups", I, Grund. Math. Wiss. 188, Springer, 1972.

114. _____, "Harmonic analysis on semi-simple Lie groups", II, Grund. Math. Wiss. 189, Springer, 1972.

115. A. Weil, "Variétés kähleriennes", Actualités Sci. Indust., no. 1267, Hermann, Paris, 1958.

116. _____, "Adeles and algebraic groups", (Notes by M. Demazure and T. Ono) Progr. Math., vol. 23, Birkhäuser, Boston, 1982.

117. _____, *Remarks on the cohomology of groups*, Ann. of Math. (2) **80** (1964), 149–177.

118. G. Zuckerman, *Tensor products of finite and infinite dimensional representations of semisimple Lie groups*, Ann. of Math. (2) **106** (1977), 295–308.

119. _____, *Continuous cohomology and unitary representations of real reductive groups*, Ann. of Math. (2) **107** (1978), 495–516.

Additional References for the Second Edition

120. D. Blasius, J. Franke and F. Grunewald, *Cohomology of S-arithmetic groups in the number field case*, Invent. Math. **116** (1994), 75–93.

121. A. Borel, *Stable real cohomology of arithmetic groups*, Ann. Sci. École Norm. Sup. (IV) **7** (1974), 235–272.

122. _____, "Stable cohomology of arithmetic groups, II. Manifolds and Lie groups", J. Hano et al (eds.), Progress in Math. **14**, 21–55, Birkhäuser, Boston, 1981.

123. _____, *Regularization theorems in Lie algebra cohomology*, Applications, Duke Math. J. **50** (1983), 605–623. *Correction and complement*, ibid. **60** (1990), 299–301.

124. _____, *Linear Algebraic Groups*, Second Edition, Graduate Texts in Mathematics, **126**, Springer-Verlag, New York, 1991.

125. _____, *Semisimple Lie groups and symmetric spaces*, to appear.

126. A. Borel and W. Casselman, *L^2-cohomology of locally symmetric spaces of finite volume*, Duke Math. J. **50** (1983), 625–647.

127. A. Borel and H. Garland, *Laplacian and discrete spectrum of an arithmetic group*, Amer. J. Math. **105** (1983), 309–335.

128. A. Borel, J.-P. Labesse and J. Schwermer, *On the cuspidal cohomology of S-arithmetic subgroups of reductive groups over number fields*, Comp. Math. **102** (1996), 1–40.

129. A. Borel and J.-P. Serre, *Cohomologie d'immeubles et de groupes S-arithmétiques*, Topology **15** (1976), 211–232.

130. A. Borel and J. Yang, *The rank conjecture for number fields*, Math. Res. Letters **1** (1994), 689–699.

131. J. Carmona, "Sur la classification des modules admissibles irréductibles", *Noncommutative harmonic analysis and Lie groups*, Lecture Notes in Mathematics, **1020**, Springer-Verlag, Berlin-New York, 1983, 11–34.

132. J. Cheeger, "On the Hodge theory of Riemannian pseudomanifolds", Proc. Symposia Pure Math. **36**, 91–146, Amer. Math. Soc., Providence, RI, 1980.

133. C. Chevalley, *The algebraic theory of spinors*, Columbia Univ. Press, New York, 1954.

134. T. Enright, *Relative Lie algebra cohomology and unitary representations of complex Lie groups*, Duke Math. J. **46** (1979), 513–525.

135. T. Enright and N. Wallach, *Notes on homological algebra and representations of Lie algebras*, Duke Math. J. **47** (1980), 1–15.

136. J. Franke, *Harmonic analysis in weighted L^2-spaces*, Ann. Sci. École Norm. Sup. **31** (1998), no. 2, 181–279.

137. H. Garland and J. Lepowsky, *Lie algebra homology and the Macdonald-Kac formula*, Invent. Math. **34** (1976), 37–76.

138. Harish-Chandra, "Some results on differential equations", Harish-Chandra Collected Papers III, Springer-Verlag, 1984, 7–48.

139. K. Johnson and N. Wallach, *Composition series and intertwining operators for the principal series.* I", Trans. Amer. Math. Soc. **229** (1977), 137–173.

140. A. Knapp and D. Vogan, *Cohomological induction and unitary representations*, Princeton University Press, Princeton, 1995.

141. S. Kumaresan, *On the canonical $\mathfrak{k}$-types in the unitary $\mathfrak{g}$-modules with non-zero relative cohomology*, Invent. Math. **59** (1980), 1–11.

142. G. A. Margulis, *Discrete subgroups of semisimple Lie groups*, Erg. Math. d. Math. Und Ihrer Grenzgeb. (3) 17, Springer, 1989.

143. N. Mok, T. T. Siu, and S. K. Yeung, *Geometric superrigidity*, Invent. Math. **113** (1993), 57–83.

144. K. Parthasarathy, R. Rao, and V. Varadarajan, *Representations of complex semisimple Lie groups and Lie algebras*, Ann. of Math. **85** (1967), 383–429.

145. R. Parthasarathy, *Criteria for unitarizability of certain highest weight representations*, Proc. Indian Acad. Sci. Sect. A, **81** (1980), 1–24.

146. W. Schmid, "Boundary value problems for group invariant differential equations", *The mathematical heritage of Élie Cartan* (Lyon, 1984), Astérisque 1985, Numëro Hors Série, 311–321.

147. A. Silberger, *Introduction to harmonic analysis on reductive p-adic groups*, Based on Lectures of Harish-Chandra at the Institute for Advanced Study, 1971–1973, Mathematical Notes, **23**, Princeton University Press, Princeton, 1979.

148. D. Vogan, *Unitarizability of certain series of representations*, Ann. of Math. **120** (1984), 141–187.

149. D. Vogan and G. Zuckerman, *Unitary representations with continuous cohomology*, Comp. Math. **53** (1984), 51–90.

150. N. R. Wallach, *Square integrable automorphic forms and cohomology of arithmetic quotients of* SU(p,q), Math. Ann. **266** (1984), 261–278.

151. _____, *Real reductive groups*. I, Academic Press, Boston, 1988.

152. _____, "Transfer of unitary representations between real forms", Contemp. Math. **177** (1994), 184–216.

153. B. Youssin, L^p *cohomology of cones and horns*, J. Diff. Geom. **39** (1994), 559–603.

154. S. Zucker, L^2-*cohomology of warped products and arithmetic groups*, Invent. Math. **70** (1982), 169–218.

155. G. Zuckerman, *Tensor products of finite and infinite dimensional representations semisimple Lie groups*, Ann. of Math. **106** (1977), 295–308.

156. A. Weil, *On discrete subgroups of Lie groups*. II, Ann. of Math. (2) **75** (1962), 578–602.

157. N. Wallach, *On the Selberg trace formula in the case of compact quotient*, Bull. Amer. Math. Soc. **82** (1976), 171–195.

158. D. DeGeorge and N. Wallach, *Limit formulas for multiplicities in* $L^2(\Gamma \backslash G)$, Ann. of Math. (2) **107** (1978), 133–150.

Index

Selected Titles in This Series